PE
Mechanical Engineering Practice Problems

Fourteenth Edition, Revised

Michael R. Lindeburg, PE

Report Errors for This Book

PPI is grateful to every reader who notifies us of a possible error. Your feedback allows us to improve the quality and accuracy of our products. Report errata at **ppi2pass.com**.

Digital Book Notice

All digital content, regardless of delivery method, is protected by U.S. copyright laws. Access to digital content is limited to the original user/assignee and is non-transferable. PPI may, at its option, revoke access or pursue damages if a user violates copyright law or PPI's end-user license agreement.

MECHANICAL ENGINEERING PRACTICE PROBLEMS
Fourteenth Edition

Current release of this edition: 9

Release History

date	edition number	revision number	update
Apr 2022	14	7	Minor corrections.
Oct 2022	14	8	Minor corrections.
Feb 2023	14	9	Minor corrections.

© 2023 Kaplan, Inc. All rights reserved.

All content is copyrighted by Kaplan, Inc. No part, either text or image, may be used for any purpose other than personal use. Reproduction, modification, storage in a retrieval system or retransmission, in any form or by any means, electronic, mechanical, or otherwise, for reasons other than personal use, without prior written permission from the publisher is strictly prohibited. For written permission, contact permissions@ppi2pass.com.

Printed in the United States of America.

PPI
ppi2pass.com

ISBN: 978-1-59126-665-5

Topics

- **Topic I:** Background and Support
- **Topic II:** Fluids
- **Topic III:** Thermodynamics
- **Topic IV:** Power Cycles
- **Topic V:** Heat Transfer
- **Topic VI:** HVAC
- **Topic VII:** Statics
- **Topic VIII:** Materials
- **Topic IX:** Machine Design
- **Topic X:** Dynamics and Vibrations
- **Topic XI:** Control Systems
- **Topic XII:** Plant Engineering
- **Topic XIII:** Economics
- **Topic XIV:** Law and Ethics

Where do I find help solving these Practice Problems?

Mechanical Engineering Practice Problems presents complete, step-by-step solutions for more than 575 problems to help you prepare for the Mechanical Engineering PE Exam. You can find all the background information, including charts and tables of data, that you need to solve these problems in the *Mechanical Engineering Reference Manual for the PE Exam*.

The *Mechanical Engineering Reference Manual for the PE Exam* may be purchased from PPI at **ppi2pass.com** or from your favorite retailer.

Table of Contents

Preface and Acknowledgments vii

Codes Used to Prepare This Book ix

How to Use This Book xi

Topic I: Background and Support
Systems of Units 1-1
Engineering Drawing Practice 2-1
Algebra 3-1
Linear Algebra 4-1
Vectors 5-1
Trigonometry 6-1
Analytic Geometry 7-1
Differential Calculus 8-1
Integral Calculus 9-1
Differential Equations 10-1
Probability and Statistical Analysis of
    Data 11-1
Numbering Systems 12-1
Numerical Analysis 13-1

Topic II: Fluids
Fluid Properties 14-1
Fluid Statics 15-1
Fluid Flow Parameters 16-1
Fluid Dynamics 17-1
Hydraulic Machines and Fluid Distribution 18-1
Hydraulic and Pneumatic Systems 19-1

Topic III: Thermodynamics
Inorganic Chemistry 20-1
Fuels and Combustion 21-1
Energy, Work, and Power 22-1
Thermodynamic Properties of Substances 23-1
Changes in Thermodynamic Properties 24-1
Compressible Fluid Dynamics 25-1

Topic IV: Power Cycles
Vapor Power Equipment 26-1
Vapor Power Cycles 27-1
Reciprocating Combustion Engine Cycles 28-1
Combustion Turbine Cycles 29-1
Advanced and Alternative Power-Generating
    Systems 30-1
Gas Compression Cycles 31-1
Refrigeration Cycles 32-1

Topic V: Heat Transfer
Fundamental Heat Transfer 33-1
Natural Convection, Evaporation, and
    Condensation 34-1
Forced Convection and Heat Exchangers 35-1
Radiation and Combined Heat Transfer 36-1

Topic VI: HVAC
Psychrometrics 37-1
Cooling Towers and Fluid Coolers 38-1
Ventilation 39-1
Fans, Ductwork, and Terminal Devices 40-1
Heating Load 41-1
Cooling Load 42-1
Air Conditioning Systems and Controls 43-1

Topic VII: Statics
Determinate Statics 44-1
Indeterminate Statics 45-1

Topic VIII: Materials
Engineering Materials 46-1
Material Properties and Testing 47-1
Thermal Treatment of Metals 48-1
Properties of Areas 49-1
Strength of Materials 50-1
Failure Theories 51-1

Topic IX: Machine Design
Basic Machine Design 52-1
Advanced Machine Design 53-1
Pressure Vessels 54-1

Topic X: Dynamics and Vibrations
Properties of Solid Bodies 55-1
Kinematics 56-1
Kinetics 57-1
Mechanisms and Power Transmission
    Systems 58-1
Vibrating Systems 59-1

Topic XI: Control Systems
Modeling of Engineering Systems 60-1
Analysis of Engineering Systems 61-1

Topic XII: Plant Engineering

Management Science 62-1
Instrumentation and Measurements 63-1
Manufacturing Processes 64-1
Materials Handling and Processing 65-1
Fire Protection Sprinkler Systems 66-1
Pollutants in the Environment 67-1
Storage and Disposition of Hazardous
    Materials 68-1
Testing and Sampling 69-1
Environmental Remediation 70-1
Electricity and Electrical Equipment 71-1
Illumination and Sound 72-1

Topic XIII: Economics

Engineering Economic Analysis 73-1

Topic XIV: Law and Ethics

Professional Services, Contracts, and Engineering
    Law 74-1
Engineering Ethics 75-1

Preface and Acknowledgments

Mechanical Engineering Practice Problems has humble beginnings. The first few editions were simple collections of weekly homework assignments with handwritten solutions for the students in my classes. The assignments contained targeted practice problems that I wrote to illustrate the most important concepts in my lectures, which coincided with the most important exam concepts. In those days, the number of assigned problems was a function of my perception of what an average employed engineer could accomplish in a week. I always assigned a few more problems than most people could finish, because I wanted everyone to be working to capacity. Still, the amount of work was limited, and many important engineering concepts remained unrepresented in the weekly homework.

In 2020, the National Council of Examiners for Engineering and Surveying (NCEES) changed the Professional Engineering (PE) mechanical engineering licensing exams from a pencil-and-paper exam to a computer-based one. This in itself is not a big change. However, in the pencil-and-paper exam you were permitted to bring your own reference material, including *Mechanical Engineering Reference Manual* (*Reference Manual*), into the exam room. In the computer-based exam, you may not. Instead, you will have on-screen access to a searchable electronic copy of the *NCEES PE Mechanical Reference Handbook* (*NCEES Handbook*). This is the only reference you may consult during the exam.

This drastically changes how you must study for the exam. It is no longer enough to learn how to solve exam problems using the *Reference Manual* and other familiar reference books that you may annotate, highlight, and glue tabs to as you study. Now you must learn how to quickly find the equations and data you need in one specific source, an unmarked electronic copy of the *NCEES Handbook*. That is the reason for this new edition. *Mechanical Engineering Practice Problems* has been revised to give you practice not just in solving exam problems, but solving them using the particular equations and data that are found in the *NCEES Handbook*. Equations and data are labeled with blue pointers to their locations in the *NCEES Handbook*, so that as you do practice problems, you are also learning to use the *NCEES Handbook* efficiently.

We have used this new edition as an opportunity to make updates to terminology and descriptions; clarify and reword explanations of calculation steps; update and remove obsolete information; update chapter nomenclature; improve consistency between chapters; and, the inevitable correction of author's errata.

Now, I'd like to introduce you to the "I couldn't have done it without you" crew. The talented team at PPI that produced this edition is the same team that produced the new edition of the *Mechanical Engineering Reference Manual*, and the team members deserve just as many accolades here as I gave them in the Acknowledgements of that book. As it turns out, producing two new books is not twice as hard as producing one new book. Considering all of the connections between the two books, producing this edition is greatly complicated by the need for consistency with the *Reference Manual*. Producing a set of interrelated books is not a task for the faint of heart.

New content contributors: Jared Anna, PE; Bryson M. Brewer, PhD, PE; Bradley Heath, MS, PE; Jonathan Kweder, PhD; Alex Lukatskiy, MS; Alan Mushynski, MS, PE; N. S. Nandagopal, MS, PE; Nathan R. Palmer, PE; Cumali Semetay, PhD, PE; Harry Tuazon, PE

Lead technical reviewer: David W. Burris, PE

Technical reviewers: Keith E. Elder, PE; N. S. Nandagopal, MS, PE; Nebojsa Sebastijanovic, PhD, PE

Calculation checker: Anil Acharya, PhD, PE

Content team: Bonnie Conner, Meghan Finley, Anna Howland, Amanda Werts

Editorial team: Bilal Baqai, Tyler Hayes, Indira Kumar, Scott Marley, Scott Rutherford, Michael Wordelman

Editorial operations director: Grace Wong

Project manager: Beth Christmas

Product management team: Ellen Nordman, Megan Synnestvedt

Production team: Tom Bergstrom, Bradley Burch, Kim Burton-Weisman, Nikki Capra-McCaffrey, Robert Genevro, Richard Iriye, Teresa Trego, Kim Wimpsett, and Stan Info Solutions

Publishing systems team: Jeri Jump, Sam Webster

Technical illustrations and cover design: Tom Bergstrom

Marketing team: John Golden, Jared Schulze

This edition has so much new material (and, accordingly, so many opportunities for showing my fallibility), that I'm already shaking in my boots. So, if you think you've found something questionable, or if you think there is a better way to solve a problem, I hope you'll help me out by visiting the PPI website at **ppi2pass.com/errata**. I like to learn new things, too. So, now it's your turn. Teach me.

Thanks, everyone!

Michael R. Lindeburg, PE

About the Author

Michael R. Lindeburg, PE, is one of the best-known authors of engineering textbooks and references. His books and courses have influenced millions of engineers around the world. Since 1975, he has authored over 40 engineering reference and exam preparation books. He has spent thousands of hours teaching engineering to students and practicing engineers. He holds bachelor of science and master of science degrees in industrial engineering from Stanford University.

Codes Used to Prepare This Book

The documents, codes, and standards that I used to prepare this new edition were the most current available at the time. In the absence of any other specific need, that was the best strategy for this book.

Engineering practice is often constrained by law or contract to using codes and standards that have already been adopted or approved. However, newer codes and standards might be available. For example, the adoption of building codes by states and municipalities often lags publication of those codes by several years. By the time the 2018 codes are adopted, the 2020 codes have been released. Federal regulations are always published with future implementation dates. Contracts are signed with designs and specifications that were "best practice" at some time in the past. Nevertheless, the standards are referenced by edition, revision, or date. All of the work is governed by unambiguous standards.

All standards produced by ASME, ASHRAE, ANSI, ASTM, and similar organizations are identified by an edition, revision, or date. In mechanical engineering, NCEES does not specify "codes and standards" in its lists of PE exam topics. My conclusion is that the NCEES mechanical engineering PE exam is not sensitive to changes in codes, standards, regulations, or announcements in the Federal Register. That is the reason that I referred to the most current documents available as I prepared this new edition.

How to Use This Book

The main purpose of *Mechanical Engineering Practice Problems* (*Practice Problems*) is to get you ready for the NCEES PE mechanical exams. Use it along with the other PPI PE mechanical study tools to assess, review, and practice until you pass your exam.

ASSESS

To pinpoint the subject areas where you need more study, use the diagnostic exams on the PPI Learning Hub (**ppi2pass.com**). How you perform on these diagnostic exams will tell you which topics you need to spend more time on and which you can review more lightly.

REVIEW

PPI offers a complete solution to help you prepare for exam day. Our mechanical engineering prep courses and *Mechanical Engineering Reference Manual* (*Reference Manual*) offer a thorough review for the PE mechanical exams. *Practice Problems* and the PPI Learning Hub quiz generator offer extensive practice in solving exam-like problems. *Mechanical Engineering HVAC and Refrigeration Practice Exam*, *Mechanical Engineering Machine Design and Materials Practice Exam*, and *Mechanical Engineering Thermal and Fluid Systems Practice Exam* provide practice exams that simulate the exam-day experience and let you hone your test-taking skills.

PRACTICE

Learn to Use the *NCEES PE Mechanical Reference Handbook*

Download a PDF of the *NCEES PE Mechanical Reference Handbook* (*NCEES Handbook*) from the NCEES website. As you study, take the time to find out where important equations and tables are located in the *NCEES Handbook*. Although you could print out the *NCEES Handbook* and use it that way, it will be better for your preparations if you use it in PDF form on your computer. This is how you will be referring to it and searching in it during the actual exam.

A searchable electronic copy of the *NCEES Handbook* is the only reference you will be able to use during the exam, so it is critical that you get to know what it includes and how to find what you need efficiently. Even if you know how to find the equations and data you need more quickly in other references, take the time to search for them in *NCEES Handbook*. Get to know the terms and section titles used in the *NCEES Handbook* and use these as your search terms.

In this book, each equation from the *NCEES Handbook* is given in blue and annotated with the title of the section the equation is found in, also in blue. Whenever data are taken from a figure or table in the *NCEES Handbook*, the title of the figure or table is given in blue. Get to know these titles as you study; they will give you search terms you can use to quickly find the equations and data you need, saving valuable time during the exam.

Using steam tables, h_1 389.0 Btu/lbm, $s_1 = 1.567$ Btu/lbm-°R, and $p_2 = 4$ psia. h_2 represents the enthalpy for a turbine that is 100% efficient. Since the turbine is isentropic, $s_1 = s_2$. Using steam tables, find the appropriate enthalpy and entropy values at state 2′ where 2′ = 4 psia. [Properties of Saturated Water and Steam (Temperature) - I-P Units]

$$h_f = 120.87 \text{ Btu/lbm}$$
$$s_f = 0.2198 \text{ Btu/lbm-°R}$$
$$h_{fg} = 1006.4 \text{ Btu/lbm}$$
$$s_{fg} = 1.6424 \text{ Btu/lbm-°R}$$

The steam quality at the turbine exhaust (state 2) for a 100% efficient turbine is found from the entropy relationship.

Properties for Two-Phase (Vapor-Liquid) Systems

$$s = s_f + x s_{fg}$$

$$x = \frac{s - s_f}{s_{fg}}$$

$$= \frac{1.567 \, \frac{\text{Btu}}{\text{lbm-°R}} - 0.2198 \, \frac{\text{Btu}}{\text{lbm-°R}}}{1.6424 \, \frac{\text{Btu}}{\text{lbm-°R}}}$$

$$= 0.82$$

Some equations given in blue in this book may have a variable or two that is different from the equation as it appears in the *NCEES Handbook*. There are a small number of variables that are treated inconsistently in the *NCEES Handbook*; to minimize possible confusion while studying, in this book these variables have been made consistent.

For example, pressure is represented by both p and P in different sections of the *NCEES Handbook*; in this book pressure is always represented by p. Similarly, in this book heat is always represented by Q; heat rate is $\dot{Q}$ in reference to thermodynamic cycles and q otherwise; velocity is always v; and elevation is always z. All the variables and subscripts used in a chapter are defined in the nomenclature list at the end of each chapter.

Equations in blue may also differ from their presentation in the *NCEES Handbook* because of the presence of the gravitational constant, g_c. The *NCEES Handbook* generally does not indicate whether an equation requires g_c when used with U.S. customary units. On the PE exam, then, you will need to know when and how to include g_c in a calculation without any help from the *NCEES Handbook*.

To show the correct use of g_c, equations in this book are given in two versions where appropriate, one for use with SI units and one for use with U.S. customary units, with g_c correctly included in the U.S. version. When you solve practice problems, however, you should use the *NCEES Handbook* as your only reference, identifying when and how to use g_c on your own. This is more trouble than looking up the equations in this book, but it will better prepare you for the actual exam.

Access the PPI Learning Hub

Although the *Reference Manual*, *Practice Problems*, and the three mechanical engineering *Practice Exams* can be used on their own, they are designed to work with the PPI Learning Hub. At the PPI Learning Hub, you can access

- a personal study plan, keyed to your exam date, to help keep you on track
- diagnostic exams to help you identify the subject areas where you are strong and where you need more review
- a quiz generator containing hundreds of additional exam-like problems that cover all knowledge areas on the PE mechanical exams
- two full-length NCEES-like, computer-based practice exams for each of the PE mechanical engineering disciplines, to familiarize you with the exam day experience and let you hone your time management and test-taking skills
- electronic versions of *Mechanical Engineering HVAC and Refrigeration Practice Exam*, *Mechanical Engineering Machine Design and Materials Practice Exam*, *Mechanical Engineering Thermal and Fluid Systems Practice Exam*, *Mechanical Engineering Reference Manual*, and *Mechanical Engineering Practice Problems*

For more about the PPI Learning Hub, visit PPI's website at **ppi2pass.com**.

Be Thorough

This book is primarily a companion to the *Mechanical Engineering Reference Manual*. As a tool for preparing for an engineering licensing exam, there are a few, but not very many, ways to use it. And, at least one of those ways isn't very good.

The big issue is whether you really work the practice problems or just skim over them. Some people think they can read a problem statement, think about it for about ten seconds, read the solution, and then say "Yes, that's what I was thinking of, and that's what I would have done." Sadly, these people find out too late that the human brain doesn't learn very efficiently by observation alone. Under pressure, these people remember very little. For real learning, you have to spend some time with the stubby pencil.

There are so many ways that a problem's solution can mess with your mind. Maybe the stumble is using your calculator, like pushing log instead of ln, or forgetting to set the angle to radians instead of degrees. Maybe it's rusty math. What are $\text{erf}(x)$, $\cosh(t)$, and $\ln e(x)$, anyway? How do you complete the square or factor a polynomial? Maybe it's in finding the data needed (e.g., the specific heat of ice cream) or a unit conversion (e.g., watts to horsepower). Maybe it's trying to determine if an equation expects L to be in feet or inches, or if the volumetric flow rate is in gallons per minute or cubic feet per second. Maybe it's the definition of a strange term. Is the retardance coefficient the same as Manning's roughness constant? Getting past these stumbles takes time.

And unfortunately, most people learn by doing and have to make a mistake at least once in order not to make it again. Since making a mistake while taking the exam isn't an optimal strategy, working with the stubby pencil in your company's break area is looking more and more attractive.

Even if you do decide to get your hands dirty and actually work the problems (as opposed to skimming through them), you'll have to decide how much reliance you place on the published solutions. You'll naturally probably want to maximize the number of problems you solve by spending as little time as you can on each problem. After all, optimization is the engineering way. Are you stuck on a problem? It's tempting to turn to a solution when you get slowed down by details or stumped by the subject material. However, I want you to struggle a little bit more than that—not because I want to see you suffer, but because the "objective function" to be optimized is your exam performance. There are no prizes for minimizing your study time. When you get stuck, do your own original research as if you didn't have a detailed solution a few pages away. Start with the *NCEES Handbook*, as that's the reference you'll be expected to use on the actual exam. If the *NCEES Handbook* proves difficult for you, use the *Mechanical Engineering Reference Manual* to help you find where in the *Handbook* you should look and/or to understand the concepts underlying the equations used. If neither of those works, the practice problem solutions will tell you exactly which headings to look under in the *NCEES Handbook*, helping you learn where to look the next time.

Learning something new is analogous to using a machete to cut a path through a dense jungle. By doing the work, you develop pathways that weren't there before. It's a lot different than just looking at the route on a map. You actually get nowhere by looking at a map. But cut that path once, and you're in business until the jungle overgrowth closes in again.

I chose each problem for a reason. If you skip problems, your review will be piecemeal. So, do the problems. All of them—even if you think you're not going to work in some subjects. Look up the references, and follow the links. Don't look at the answers until you've sweated a little. And, let's not have any whining. Please.

Systems of Units

Content in blue refers to the *NCEES Handbook*.

PRACTICE PROBLEMS

1. Most nearly, what is 250°F converted to degrees Celsius?

(A) 115°C

(B) 121°C

(C) 124°C

(D) 420°C

2. Most nearly, what is the Stefan-Boltzmann constant (0.1713×10^{-8} Btu/ft^2-hr-°R^4) converted from English to SI units?

(A) 5.14×10^{-10} W/m^2·K^4

(B) 0.95×10^{-8} W/m^2·K^4

(C) 5.67×10^{-8} W/m^2·K^4

(D) 7.33×10^{-6} W/m^2·K^4

3. A company is trying to determine the cheapest method for filling their container trucks. The first method uses a system that fills the container truck at 0.5 ft^3/sec and has a cost of $25 per hour. The second method uses a system that fills at 300 L per minute and has a cost of $10 per hour. The container truck holds 3000 gal. Which method would be the cheapest and at what cost?

(A) Method B ($3.35)

(B) Method B ($5.58)

(C) Method A ($5.58)

(D) Method A ($6.31)

4. A car with a mass of 1600 kg is traveling at 40 mph. The kinetic energy of the car is most nearly

(A) 110 Btu

(B) 240 Btu

(C) 5200 Btu

(D) 7800 Btu

5. How tall would a container with a diameter of 20 cm have to be to hold 2 gal of water?

(A) 15 cm

(B) 25 cm

(C) 35 cm

(D) 45 cm

SOLUTIONS

1. The conversion to degrees Celsius is

Temperature Conversions
$$°C = (°F - 32°F)/1.8$$
$$= \frac{250°F - 32°F}{1.8}$$
$$= 121.1°C \quad (121°C)$$

The answer is (B).

2. In SI units, the Stefan-Boltzmann constant is 5.67×10^{-8} W/m²·K⁴. [Fundamental Constants]

The answer is (C).

3. Method A

Convert the rate of filling to units of gal/min. 1 gal is equal to 0.134 ft³. [Measurement Relationships]

$$v = \frac{\left(0.5 \frac{ft^3}{sec}\right)\left(60 \frac{sec}{min}\right)}{0.134 \frac{ft^3}{gal}}$$
$$= 223.88 \text{ gpm}$$

The cost to fill the truck using method A is

$$\text{cost per truck} = \frac{(3000 \text{ gal})\left(25 \frac{\$}{hr}\right)}{\left(223.88 \frac{gal}{min}\right)\left(60 \frac{min}{hr}\right)}$$
$$= \$5.58$$

Method B

Convert the rate of filling to gpm. 1 liter is equal to 0.264 gal.

$$v = \left(300 \frac{L}{min}\right)\left(0.264 \frac{gal}{L}\right)$$
$$= 79.2 \text{ gpm}$$

The cost to fill the truck for method B is

$$\text{cost per truck} = \frac{(3000 \text{ gal})\left(10 \frac{\$}{hr}\right)}{\left(79.2 \frac{gal}{min}\right)\left(60 \frac{min}{hr}\right)}$$
$$= \$6.31$$

Method A is cheaper than method B at a cost of $5.58.

The answer is (C).

4. The equation for kinetic energy is

Kinetic Energy
$$KE = \frac{1}{2}mv^2$$

Convert the velocity to units of ft/sec. There are 5280 ft in 1 mi. [Measurement Relationships]

$$v = \left(\frac{40 \frac{mi}{hr}}{3600 \frac{sec}{hr}}\right)\left(5280 \frac{ft}{mi}\right)$$
$$= 58.67 \text{ ft/sec}$$

Convert the mass to units of lbm. There are 2.205 lbm in 1 kg. [Measurement Relationships]

$$m = (1600 \text{ kg})\left(2.205 \frac{lbm}{kg}\right)$$
$$= 3528 \text{ lbm}$$

Find the kinetic energy. The mass must be divided by the gravitational constant to convert from lbm to lbf. [Units] 1 Btu is equal to 778 ft-lbf. [Measurement Relationships]

Kinetic Energy
$$KE = \frac{1}{2}mv^2 = \left(\frac{1}{2}\right)\left(\frac{(3528 \text{ lbm})\left(58.67 \frac{ft}{sec}\right)^2}{\left(32.174 \frac{lbm\text{-}ft}{lbf\text{-}sec^2}\right)\left(778 \frac{ft\text{-}lbf}{Btu}\right)}\right)$$
$$= 242.58 \text{ Btu} \quad (240 \text{ Btu})$$

The answer is (B).

5. Convert the volume of water to cubic centimeters. 1 gal is equal to 0.134 ft³, and 1 ft is equal to 30.48 cm. [Measurement Relationships]

$$V = (2 \text{ gal})\left(0.134 \frac{ft^3}{gal}\right)\left(30.48 \frac{cm}{ft}\right)^3$$
$$= 7588.91 \text{ cm}^3$$

The height required is equal to the volume divided by the area of the cylinder base.

Right Circular Cylinder

$$V = \pi r^2 h$$
$$h = \frac{V}{\pi r^2}$$
$$= \frac{V}{\frac{\pi D^2}{4}}$$
$$= \frac{7588.91 \text{ cm}^3}{\frac{\pi (20 \text{ cm})^2}{4}}$$
$$= 24.156 \text{ cm} \quad (25 \text{ cm})$$

The answer is (B).

2 Engineering Drawing Practice

Content in blue refers to the *NCEES Handbook*.

PRACTICE PROBLEMS

1. How many orthographic views are necessary to determine whether two lines intersect if the lines are in a three-dimensional space and the closest approach or intercept are within the views?

(A) one
(B) two
(C) three
(D) four

2. A $2 \times 2 \times 2$ cube is drawn two different ways, as shown.

Which types of projections have been used?

(A) oblique and isometric
(B) isometric and principal
(C) oblique and orthographic
(D) cavalier and cabinet

3. The portion of a staircase shown is presented in which type of view?

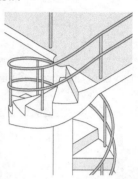

(A) isometric
(B) dimetric
(C) trimetric
(D) orthographic

4. Two views of an object are shown.

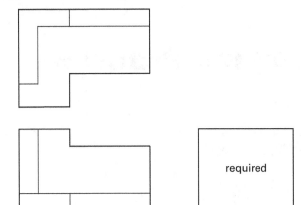

Which option represents the missing third view?

(A)
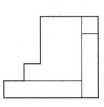

(B)
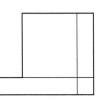

(C)
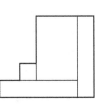

(D)
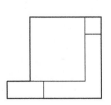

5. Two views of an object are shown.

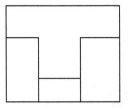 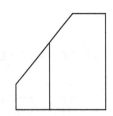

Which option represents the missing third view?

(A)
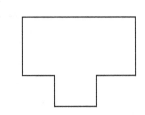

(B)
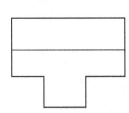

(C)
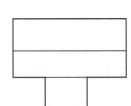

(D)
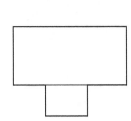

6. Two views of an object are shown.

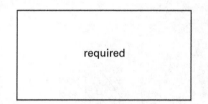 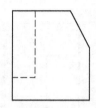

Which option represents the missing third view?

(A)

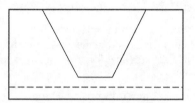

(B)

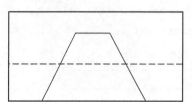

(C)

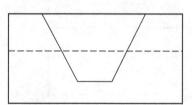

(D)

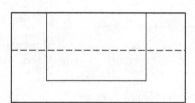

7. A parallel-scale nomograph has been prepared to solve the equation $D = 1.075\sqrt{WH}$. Using the nomograph, what is most nearly the value of D when $H = 10$ and $W = 40$?

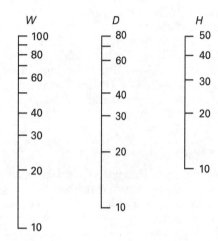

(A) 20
(B) 22
(C) 27
(D) 32

8. Which statement below is accurate for the surface finish designation shown?

(A) The waviness height limit is 0.0013 μm.
(B) The lay is perpendicular.
(C) The surface roughness height limit is 0.003 μin.
(D) 32% of the surface may exceed the specification.

9. How would the following surface finish designation be interpreted?

(A) The surface is rolled.
(B) The maximum roughness width is 0.6 μm.
(C) The piece is to be used exactly as it comes directly from the manufacturing process.
(D) The sample length is 0.6 m.

SOLUTIONS

1. Orthographic views are used in engineering drawings to depict a three-dimensional object in two dimensions by taking the projection of the object onto two-dimensional surfaces of some combination of views from the top, bottom, right, left, front, and back. Imagine taking the projection of the two lines in the x-y plane and the x-z plane. If the two lines intercept, the lines will intercept in the projections and the value of x will be the same in both projections. If the two lines do not intercept, then it is possible for the lines to intercept in the projections in the x-y plane and the x-z plane, but the value for x will always be different in the two projections.

The answer is (B).

2. Both views project the cube's front face as a square, so the views are oblique. However, neither view is isometric, principal, or orthographic. The left view uses equal-length lines for the x-, y-, and z-projectors, so it is a cavalier projection. The right view shortens the z-projectors to 50%, so it is a cabinet projection.

The answer is (D).

3. Near the top of the illustration, the top railing, the floor, and the vertical poles intersect with approximately 120° angles, so the view is isometric.

The answer is (A).

4. The complete set of views is

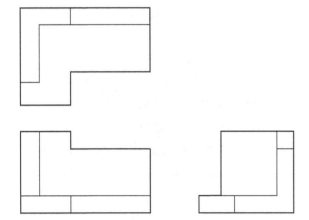

The answer is (D).

5. The front and right side views are provided, but the top view is missing. The complete set of views is

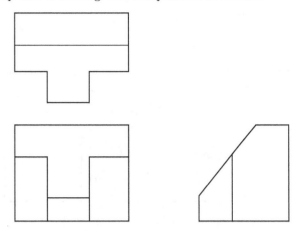

The drawing shown in option A does not show the sloped surface, so it is incorrect.

The drawing shown in option C shows both the back flat surface and back slope as seen in the side view but shows a flat surface for the nose extension, so it is incorrect.

The drawing shown in option D shows either a continuous slope with no flat on the back portion or a flat section to the nose extension, so it is incorrect.

The answer is (B).

6. The complete set of views is

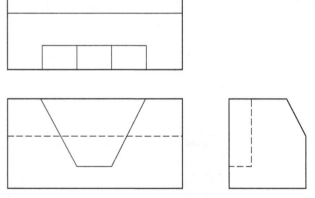

The drawing shown in option A shows the back slope as lower than the depth of the V wedge cut, so it is incorrect.

The drawing shown in option B shows the V wedge from the top view upside down, so it is incorrect.

The drawing shown in option D shows the V wedge cut from the top view as a straight square cut, so it is incorrect.

The answer is (C).

7. A straight line through the points $H = 10$ and $W = 40$ intersects the D-scale at approximately 22.

The answer is (B).

8. The values 0.0013 and 0.003 are too "fine" to be in micrometers. These values must be in microinches. The roughness height limit is 32 μin. The waviness height limit is 0.0013 μin. The roughness width limit is 0.003 μin. The lay is perpendicular.

The answer is (B).

9. The circle in the crook of the surface finish arrow means that no resurfacing is permitted (i.e., it must be used exactly as it comes from the manufacturing process). [Surface Texture Symbols and Construction]

The maximum height of the roughness is 0.6 μm. In addition, since no length of sampling is specified, the default limit of five sample lengths (i.e., between the five highest peaks and the five lowest valleys) applies. Also, since no other value is specified, a maximum of 16% (the default value) of the item may exceed the minimum surface roughness.

The answer is (C).

3 Algebra

Content in blue refers to the NCEES Handbook.

PRACTICE PROBLEMS

1. What are the roots of the quadratic equation $x^2 - 7x - 44 = 0$?

(A) −11, 4
(B) −7.5, 3.5
(C) 7.5, −3.5
(D) 11, −4

2. What is most nearly the value of x that satisfies the expression $17.3 = e^{1.1x}$?

(A) 0.17
(B) 2.6
(C) 5.8
(D) 15

3. What is most nearly the following sum?

$$\sum_{j=1}^{5}\left((j+1)^2 - 1\right)$$

(A) 15
(B) 24
(C) 35
(D) 85

4. A quantity increases by 0.1% of its current value every 0.1 sec. The increase in quantity is not necessarily a smooth function. The doubling time is most nearly

(A) 14 sec
(B) 69 sec
(C) 690 sec
(D) 69,000 sec

SOLUTIONS

1. *method 1:* Use inspection.

Assuming that the roots are integers, there are only a few ways to arrive at 44 by multiplication: 1×44, 2×22, and 4×11. Of these, only factors 4 and 11 result in a difference of 7. Therefore, the quadratic equation can be factored into $(x - 11)(x + 4) = 0$. The roots are 11 and −4.

method 2: Use the quadratic formula, where $a = 1$, $b = -7$, and $c = -44$.

The roots are

$$x_1, x_2 = \frac{-b \pm \sqrt{b^2 - 4ac}}{2a} = \frac{-(-7) \pm \sqrt{(-7)^2 - (4)(1)(-44)}}{(2)(1)}$$

$$= \frac{7 \pm 15}{2}$$

$$= 11, -4$$

method 3: Use completing the square.

$$\left(x - \frac{7}{2}\right)^2 = 44 + \left(\frac{7}{2}\right)^2$$

$$(x - 3.5)^2 = 56.25$$

$$(x - 3.5) = \pm\sqrt{56.25}$$

$$= \pm 7.5$$

The roots are $x_1, x_2 = 11, -4$.

The answer is (D).

2. Take the natural logarithm of both sides, using the identity $\log_b b^n = n$.

$$17.3 = e^{1.1x}$$

$$\ln 17.3 = \ln e^{1.1x}$$

$$= 1.1x$$

$$x = \frac{\ln 17.3}{1.1}$$

$$= 2.592 \ (2.6)$$

The answer is (B).

3. Expand the summation.

$$S = \sum_{j=1}^{5}((j+1)^2 - 1)$$
$$= ((1+1)^2 - 1) + ((2+1)^2 - 1) + ((3+1)^2 - 1)$$
$$+ ((4+1)^2 - 1) + ((5+1)^2 - 1)$$
$$= 3 + 8 + 15 + 24 + 35$$
$$= 85$$

The answer is (D).

4. Let n represent the number of elapsed periods of 0.1 sec and let y_n represent the amount present after n periods. y_0 represents the initial quantity. Express y_1 and y_2 in terms of y_0.

$$y_1 = 1.001 y_0$$
$$y_2 = 1.001 y_1 = (1.001)^2 y_0$$

A pattern emerges for subsequent terms.

$$y_n = (1.001)^n y_0$$

The initial quantity doubles in value after n periods.

$$y_n = 2y_0 = (1.001)^n y_0$$
$$2 = (1.001)^n$$

Determine n using an identity of logarithms, $\log x^n = n (\log x)$.

$$n \log 1.001 = \log 2 = 0.3010$$
$$n = \frac{0.3010}{\log 1.001}$$
$$= 693.4 \text{ periods}$$

Because the increase in quantity is not necessarily a smooth function, the doubling may not be achieved until the next period. Therefore, the result must be rounded up to 694 periods.

Since each period is 0.1 sec, the doubling time is

$$t = (694)(0.1 \text{ sec})$$
$$= 69.4 \text{ sec } (69 \text{ sec})$$

The answer is (B).

Linear Algebra

Content in blue refers to the *NCEES Handbook*.

PRACTICE PROBLEMS

1. What is most nearly the determinant of matrix **A**?

$$\mathbf{A} = \begin{bmatrix} 8 & 2 & 0 & 0 \\ 2 & 8 & 2 & 0 \\ 0 & 2 & 8 & 2 \\ 0 & 0 & 2 & 4 \end{bmatrix}$$

(A) 459

(B) 832

(C) 1552

(D) 1776

2. Use Cramer's rule to solve for the values of x, y, and z that simultaneously satisfy the following equations.

$$x + y = -4$$
$$x + z - 1 = 0$$
$$2z - y + 3x = 4$$

(A) $(x, y, z) = (3, 2, 1)$

(B) $(x, y, z) = (-3, -1, 2)$

(C) $(x, y, z) = (3, -1, -3)$

(D) $(x, y, z) = (-1, -3, 2)$

SOLUTIONS

1. Expand by cofactors of the first column since there are two zeros in that column.

$$|\mathbf{A}| = 8 \begin{vmatrix} 8 & 2 & 0 \\ 2 & 8 & 2 \\ 0 & 2 & 4 \end{vmatrix} - 2 \begin{vmatrix} 2 & 0 & 0 \\ 2 & 8 & 2 \\ 0 & 2 & 4 \end{vmatrix} + 0 - 0$$

By first column:

$$\begin{vmatrix} 8 & 2 & 0 \\ 2 & 8 & 2 \\ 0 & 2 & 4 \end{vmatrix} = (8)((8)(4) - (2)(2)) - (2)((2)(4) - (2)(0))$$

$$= (8)(28) - (2)(8)$$

$$= 208$$

By first column:

$$\begin{vmatrix} 2 & 0 & 0 \\ 2 & 8 & 2 \\ 0 & 2 & 4 \end{vmatrix} = (2)((8)(4) - (2)(2))$$

$$= 56$$

$$|\mathbf{A}| = (8)(208) - (2)(56) = 1552$$

The answer is (C).

2. Rearrange the equations.

$$x + y = -4$$
$$x + z = 1$$
$$3x - y + 2z = 4$$

Write the set of equations in matrix form: $\mathbf{AX} = \mathbf{B}$.

$$\begin{bmatrix} 1 & 1 & 0 \\ 1 & 0 & 1 \\ 3 & -1 & 2 \end{bmatrix} \begin{bmatrix} x \\ y \\ z \end{bmatrix} = \begin{bmatrix} -4 \\ 1 \\ 4 \end{bmatrix}$$

Find the determinant of the matrix **A**.

$$|\mathbf{A}| = \begin{vmatrix} 1 & 1 & 0 \\ 1 & 0 & 1 \\ 3 & -1 & 2 \end{vmatrix}$$

$$= 1\begin{vmatrix} 0 & 1 \\ -1 & 2 \end{vmatrix} - 1\begin{vmatrix} 1 & 0 \\ -1 & 2 \end{vmatrix} + 3\begin{vmatrix} 1 & 0 \\ 0 & 1 \end{vmatrix}$$

$$= (1)\big((0)(2) - (1)(-1)\big)$$
$$\quad - (1)\big((1)(2) - (-1)(0)\big)$$
$$\quad + (3)\big((1)(1) - (0)(0)\big)$$
$$= (1)(1) - (1)(2) + (3)(1)$$
$$= 1 - 2 + 3$$
$$= 2$$

Find the determinant of the substitutional matrix $\mathbf{A}_1$.

$$|\mathbf{A}_1| = \begin{vmatrix} -4 & 1 & 0 \\ 1 & 0 & 1 \\ 4 & -1 & 2 \end{vmatrix}$$

$$= -4\begin{vmatrix} 0 & 1 \\ -1 & 2 \end{vmatrix} - 1\begin{vmatrix} 1 & 0 \\ -1 & 2 \end{vmatrix} + 4\begin{vmatrix} 1 & 0 \\ 0 & 1 \end{vmatrix}$$

$$= (-4)\big((0)(2) - (1)(-1)\big)$$
$$\quad - (1)\big((1)(2) - (-1)(0)\big)$$
$$\quad + (4)\big((1)(1) - (0)(0)\big)$$
$$= (-4)(1) - (1)(2) + (4)(1)$$
$$= -4 - 2 + 4$$
$$= -2$$

Find the determinant of the substitutional matrix $\mathbf{A}_2$.

$$|\mathbf{A}_2| = \begin{vmatrix} 1 & -4 & 0 \\ 1 & 1 & 1 \\ 3 & 4 & 2 \end{vmatrix}$$

$$= 1\begin{vmatrix} 1 & 1 \\ 4 & 2 \end{vmatrix} - 1\begin{vmatrix} -4 & 0 \\ 4 & 2 \end{vmatrix} + 3\begin{vmatrix} -4 & 0 \\ 1 & 1 \end{vmatrix}$$

$$= (1)\big((1)(2) - (4)(1)\big)$$
$$\quad - (1)\big((-4)(2) - (4)(0)\big)$$
$$\quad + (3)\big((-4)(1) - (1)(0)\big)$$
$$= (1)(-2) - (1)(-8) + (3)(-4)$$
$$= -2 + 8 - 12$$
$$= -6$$

Find the determinant of the substitutional matrix $\mathbf{A}_3$.

$$|\mathbf{A}_3| = \begin{vmatrix} 1 & 1 & -4 \\ 1 & 0 & 1 \\ 3 & -1 & 4 \end{vmatrix}$$

$$= 1\begin{vmatrix} 0 & 1 \\ -1 & 4 \end{vmatrix} - 1\begin{vmatrix} 1 & -4 \\ -1 & 4 \end{vmatrix} + 3\begin{vmatrix} 1 & -4 \\ 0 & 1 \end{vmatrix}$$

$$= (1)\big((0)(4) - (-1)(1)\big)$$
$$\quad - (1)\big((1)(4) - (-1)(-4)\big)$$
$$\quad + (3)\big((1)(1) - (0)(-4)\big)$$
$$= (1)(1) - (1)(0) + (3)(1)$$
$$= 1 - 0 + 3$$
$$= 4$$

Use Cramer's rule.

$$x = \frac{|\mathbf{A}_1|}{|\mathbf{A}|} = \frac{-2}{2} = -1$$

$$y = \frac{|\mathbf{A}_2|}{|\mathbf{A}|} = \frac{-6}{2} = -3$$

$$z = \frac{|\mathbf{A}_3|}{|\mathbf{A}|} = \frac{4}{2} = 2$$

The answer is (D).

5 Vectors

Content in blue refers to the NCEES Handbook.

PRACTICE PROBLEMS

1. The dot product of the vectors $V_1 = 2i - 3j + 6k$ and $V_2 = 8i + 2j - 3k$ is

(A) -12
(B) -8
(C) 37
(D) 40

2. The dot product of the vectors $V_1 = 6i + 2j + 3k$ and $V_2 = i + k$ is

(A) -11
(B) 9
(C) 11
(D) 13

3. The angle between the vectors $V_1 = 1i + 4j$ and $V_2 = 9i - 3j$ is most nearly

(A) $70°$
(B) $85°$
(C) $90°$
(D) $95°$

4. The angle between the vectors $V_1 = 7i - 3j$ and $V_2 = 3i + 4j$ is most nearly

(A) $73°$
(B) $76°$
(C) $100°$
(D) $120°$

SOLUTIONS

1. The dot product is

$$V_1 \cdot V_2 = V_{1x}V_{2x} + V_{1y}V_{2y} + V_{1z}V_{2z}$$
$$= (2)(8) + (-3)(2) + (6)(-3)$$
$$= -8$$

The answer is (B).

2. The dot product is

$$V_1 \cdot V_2 = V_{1x}V_{2x} + V_{1y}V_{2y} + V_{1z}V_{2z}$$
$$= (6)(1) + (2)(0) + (3)(1)$$
$$= 9$$

The answer is (B).

3. The dot product of the vectors is

$$V_1 \cdot V_2 = V_{1x}V_{2x} + V_{1y}V_{2y}$$
$$= (1)(9) + (4)(-3)$$
$$= -3$$

The angle between the vectors is

$$\cos\phi = \frac{V_1 \cdot V_2}{|V_1||V_2|} = \frac{-3}{\sqrt{(1)^2 + (4)^2}\sqrt{(9)^2 + (-3)^2}}$$
$$= -0.0767$$
$$\phi = \arccos(-0.0767) = 94.4° \quad (95°)$$

The answer is (D).

4. The dot product of the vectors is

$$V_1 \cdot V_2 = V_{1x}V_{2x} + V_{1y}V_{2y}$$
$$= (7)(3) + (-3)(4)$$
$$= 9$$

The angle between the vectors is

$$\cos\phi = \frac{\mathbf{V}_1 \cdot \mathbf{V}_2}{|\mathbf{V}_1||\mathbf{V}_2|} = \frac{9}{\sqrt{(7)^2 + (-3)^2}\sqrt{(3)^2 + (4)^2}}$$
$$= 0.236$$
$$\phi = \arccos 0.236 = 76.3° \quad (76°)$$

The answer is (B).

6 Trigonometry

Content in blue refers to the *NCEES Handbook*.

PRACTICE PROBLEMS

1. A 5 lbm (5 kg) block sits on a 20° incline without slipping.

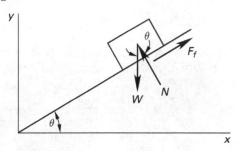

What is most nearly the magnitude of the frictional force holding the block stationary?

(A) 1.7 lbf (17 N)
(B) 3.4 lbf (33 N)
(C) 4.7 lbf (46 N)
(D) 5.0 lbf (49 N)

2. Complete the following calculation related to a catenary cable.

$$S = c\left(\cosh\frac{a}{h} - 1\right) = (245 \text{ ft})\left(\cosh\frac{50 \text{ ft}}{245 \text{ ft}} - 1\right)$$

(A) 3.9 ft
(B) 4.5 ft
(C) 5.1 ft
(D) 7.4 ft

3. Part of a turn-around area in a parking lot is shaped as a circular segment. The segment has a central angle of 120° and a radius of 75 ft. If the circular segment is to receive a special surface treatment, what area will be treated?

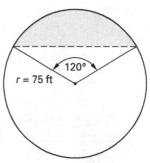

(A) 49 ft²
(B) 510 ft²
(C) 3500 ft²
(D) 5900 ft²

4. A boat is traveling across a river at 15 ft/sec. To ensure it reaches the bank on the other side of the river at the same point it leaves from, the boat travels at an upstream angle of 30°.

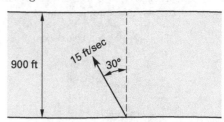

What is the speed of the current of the river, and how long will the boat take to reach the other side?

(A) 2.9 ft/sec, 38 sec
(B) 3.8 ft/sec, 47 sec
(C) 4.7 ft/sec, 60 sec
(D) 7.5 ft/sec, 69 sec

SOLUTIONS

1.

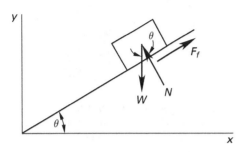

Customary U.S. Solution

The mass of the block is $m = 5$ lbm. The angle of inclination is $\theta = 20°$. The weight is

Newton's Second Law (Equations of Motion)

$$W = \frac{mg}{g_c}$$

$$= \frac{(5 \text{ lbm})\left(32.174 \, \frac{\text{ft}}{\text{sec}^2}\right)}{32.174 \, \frac{\text{lbm-ft}}{\text{lbf-sec}^2}}$$

$$= 5 \text{ lbf}$$

The frictional force is

$$F_f = W \sin\theta = (5 \text{ lbf})\sin 20°$$
$$= 1.71 \text{ lbf} \quad (1.7 \text{ lbf})$$

SI Solution

The mass of the block is $m = 5$ kg. The angle of inclination is $\theta = 20°$. The gravitational force is

Newton's Second Law (Equations of Motion)

$$W = mg$$
$$= (5 \text{ kg})\left(9.81 \, \frac{\text{m}}{\text{s}^2}\right)$$
$$= 49.1 \text{ N}$$

The frictional force is

$$F_f = W \sin\theta = (49.1 \text{ N})\sin 20°$$
$$= 16.8 \text{ N} \quad (17 \text{ N})$$

The answer is (A).

2. This calculation contains a hyperbolic cosine function.

$$S = c\left(\cosh\frac{a}{h} - 1\right)$$
$$= (245 \text{ ft})\left(\cosh\frac{50 \text{ ft}}{245 \text{ ft}} - 1\right)$$
$$= (245 \text{ ft})(1.0209 - 1)$$
$$= 5.12 \text{ ft} \quad (5.1 \text{ ft})$$

The answer is (C).

3. The area of a circular segment is

Circular Segment

$$A = \frac{r^2(\phi - \sin\phi)}{2}$$

Since ϕ appears in the expression by itself, it must be expressed in radians. [Measurement Relationships]

$$\phi = \frac{(120°)(\pi)}{180°}$$
$$= 2.094 \text{ rad}$$

$$A = \frac{r^2(\phi - \sin\phi)}{2}$$
$$= \frac{(75 \text{ ft})^2(2.094 - \sin 120°)}{2}$$
$$= 3453.7 \text{ ft}^2 \quad (3500 \text{ ft}^2)$$

The answer is (C).

4. The angle and speed of the boat are known. Find the speed of the river current.

Trigonometry: Basics

$$\sin\theta = \frac{y}{r}$$
$$y = r\sin\theta$$
$$= \left(15 \, \frac{\text{ft}}{\text{sec}}\right)(\sin 30°)$$
$$= 7.5 \text{ ft/sec}$$

The time to cross the river is

$$\cos\theta = \frac{x}{r}$$
$$x = r\cos\theta$$
$$= \left(15 \ \frac{\text{ft}}{\text{sec}}\right)(\cos 30°)$$
$$= 12.99 \ \text{ft/sec}$$

$$t = \frac{d}{x}$$
$$= \frac{900 \ \text{ft}}{12.99 \ \frac{\text{ft}}{\text{sec}}}$$
$$= 69.28 \ \text{sec} \quad (69 \ \text{sec})$$

The answer is (D).

Analytic Geometry

Content in blue refers to the *NCEES Handbook*.

PRACTICE PROBLEMS

1. The diameter of a sphere and the base of a cone are equal. What approximate percentage of that diameter must the cone's height be so that both volumes are equal?

(A) 133%
(B) 150%
(C) 166%
(D) 200%

2. The distance that vehicles travel along on a horizontal circular roadway curve is 747 ft. The radius of the curve is 400 ft. The central angle between the entrance and exit points is most nearly

(A) 0.27°
(B) 54°
(C) 110°
(D) 340°

3. A vertical parabolic roadway crest curve starts deviating from a constant grade at station 103 (i.e., 10,300 ft from an initial benchmark). At sta 103+62, the curve is 2.11 ft lower than the tangent (i.e., from the straight line extension of the constant grade). Approximately how far will the curve be from the tangent at sta 103+87?

(A) 1.1 ft
(B) 1.5 ft
(C) 3.0 ft
(D) 4.2 ft

4. A pile driving hammer emits 143 W of sound power with each driving stroke. Assume isotropic emission and disregard reflected power. What is most nearly the maximum areal sound power density at the ground when 10.7 m of pile remains to be driven?

(A) 0.030 W/m^2
(B) 0.10 W/m^2
(C) 0.53 W/m^2
(D) 1.1 W/m^2

5. A cylinder with a dome bottom has a diameter of 52 in, a wall thickness of 2 in, and a height of 120 in. Water flows into the container at a rate of 15 gpm.

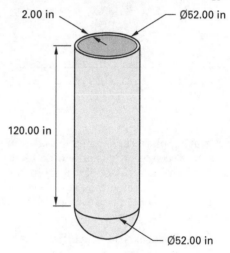

The time it will take to fill the container is most nearly

(A) 10 min
(B) 15 min
(C) 71 min
(D) 79 min

Illustration for Problem 6

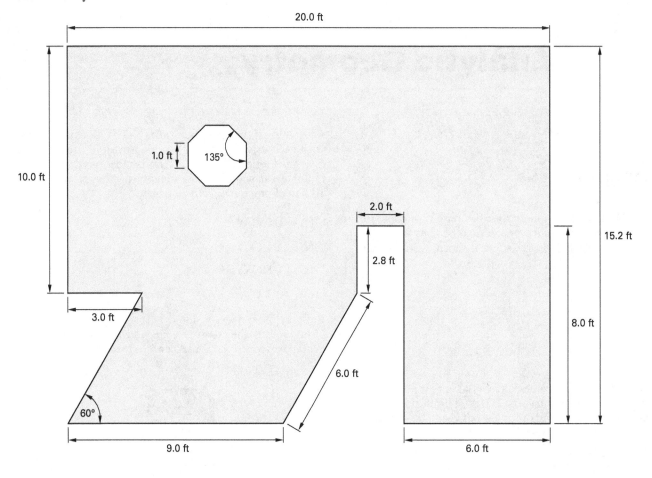

6. A steel grating platform is being installed in a workshop to create a second floor that will be used for storage. Due to the placement of large equipment in the workshop, the shape of the platform will be adjusted to fit around the equipment. The proposed platform layout is shown in *Illustration for Problem 6*.

The area of the platform is most nearly:

(A) 210 ft²

(B) 270 ft²

(C) 340 ft²

(D) 390 ft²

7. A circular channel with a diameter of 4 ft has a gate that covers 70% of the height of the diameter.

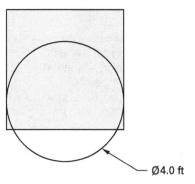

The cross-sectional area of the channel opening that still allows fluid flow is most nearly

(A) 1.87 ft²

(B) 2.56 ft²

(C) 3.16 ft²

(D) 3.74 ft²

SOLUTIONS

1. Let D be the diameter of the sphere and the base of the cone.

The volume of the sphere is

Sphere
$$V_{\text{sphere}} = \frac{4}{3}\pi r^3 = \frac{4}{3}\pi\left(\frac{D}{2}\right)^3$$
$$= \frac{1}{6}\pi D^3$$

The volume of the circular cone is

$$V_{\text{cone}} = \frac{1}{3}\pi r^2 h = \frac{1}{3}\pi\left(\frac{D}{2}\right)^2 h$$
$$= \frac{1}{12}\pi D^2 h$$

Since the volume of the sphere and cone are equal,

$$V_{\text{cone}} = V_{\text{sphere}}$$
$$\frac{1}{12}\pi D^2 h = \frac{1}{6}\pi D^3$$
$$h = 2D$$

The height of the cone must be 200% of the diameter of the sphere.

The answer is (D).

2. Horizontal roadway curves are circular arcs. The circumference (perimeter) of an entire circle with a radius of 400 ft is

$$P = 2\pi r = (2\pi)(400 \text{ ft}) = 2513.3 \text{ ft}$$

From a ratio of curve length to angles,

Circular Segment
$$\phi = \frac{s}{r} = \frac{s}{\dfrac{P}{2\pi}} = \left(\frac{s}{P}\right)(2\pi)$$
$$= \left(\frac{747 \text{ ft}}{2513.3 \text{ ft}}\right)(360°)$$
$$= 107° \quad (110°)$$

The answer is (C).

3. Vertical roadway curves are parabolic arcs. Parabolas are second-degree polynomials. Deviations, y, from a baseline are proportional to the square of the separation distance. That is, $y \propto x^2$.

$$\frac{y_1}{y_2} = \left(\frac{x_1}{x_2}\right)^2$$
$$y_2 = y_1\left(\frac{x_2}{x_1}\right)^2 = (2.11 \text{ ft})\left(\frac{87 \text{ ft}}{62 \text{ ft}}\right)^2$$
$$= 4.15 \text{ ft} \quad (4.2 \text{ ft})$$

The answer is (D).

4. The power is emitted isotropically in all directions. The maximum sound power will occur at the surface, adjacent to the pile. The surface area of a sphere with a radius of 10.7 m is

Sphere
$$A = 4\pi r^2 = (4\pi)(10.7 \text{ m})^2 = 1438.7 \text{ m}^2$$

The areal sound power density is

$$\rho_S = \frac{P}{A} = \frac{143 \text{ W}}{1438.7 \text{ m}^2}$$
$$= 0.0994 \text{ W/m}^2 \quad (0.10 \text{ W/m}^2)$$

The answer is (B).

5. The volume of the cylinder is

Right Circular Cylinder
$$V_{\text{cylinder}} = \pi r^2 h = \frac{(\pi D^2 h)}{4}$$
$$= \frac{\pi(52 \text{ in} - (2)(2 \text{ in}))^2 (120 \text{ in})}{4}$$
$$= 217{,}147 \text{ in}^3$$

Use the equation for a sphere to find the volume of the dome bottom.

Sphere
$$V = \frac{4\pi r^3}{3} = \frac{\pi D^3}{6}$$

$$V_{\text{dome}} = \frac{\dfrac{\pi D^3}{6}}{2} = \frac{\pi D^3}{12} = \frac{\pi(52 \text{ in} - (2)(2 \text{ in}))^3}{12}$$
$$= 28{,}953 \text{ in}^3$$

Convert the total volume of the container to cubic feet.

$$V_t = \frac{217{,}147 \text{ in}^3 + 28{,}953 \text{ in}^3}{(12)^3 \dfrac{\text{in}^3}{\text{ft}^3}}$$

$$= 142.4 \text{ ft}^3$$

Convert the flow rate of water into the container from gallons per minute to cubic feet per minute. [Measurement Relationships]

$$Q = \left(15 \; \frac{\text{gal}}{\text{min}}\right)\left(0.134 \; \frac{\text{ft}^3}{\text{gal}}\right)$$

$$= 2.01 \text{ ft}^3/\text{min}$$

Divide the total volume of the container by the flow rate of the water to find the fill time.

$$t = \frac{142.4 \text{ ft}^3}{2.01 \; \dfrac{\text{ft}^3}{\text{min}}}$$

$$= 70.8 \text{ min} \quad (71 \text{ min})$$

The answer is (C).

6. Break the floor plan into sections and calculate the area of each section (see *Illustration for Solution 6*).

Find the area of section 1.

$$\phi = \frac{2\pi}{n}$$

$$= \frac{2\pi}{8}$$

$$= 0.7854$$

Regular Polygon With n Equal Sides

$$s = 2r\left(\tan\frac{\phi}{2}\right)$$

$$r = \frac{s}{(2)\left(\tan\dfrac{\phi}{2}\right)}$$

$$= \frac{1 \text{ ft}}{(2)\left(\tan\dfrac{0.7854}{2}\right)}$$

$$= 1.207 \text{ ft}$$

$$A_{\text{polygon}} = \frac{nsr}{2}$$

$$= \frac{(8)(1 \text{ ft})(1.207 \text{ ft})}{2}$$

$$= 4.83 \text{ ft}^2$$

Parallelogram

$$A_1 = ah - A_{\text{polygon}}$$

$$= (20 \text{ ft} - 2 \text{ ft} - 6 \text{ ft})(10 \text{ ft}) - 4.83 \text{ ft}^2$$

$$= 115.2 \text{ ft}^2$$

Find the areas of sections 2, 3, and 4.

$$A_2 = ah$$

$$= (2 \text{ ft} + 6 \text{ ft})(15.2 \text{ ft} - 8 \text{ ft})$$

$$= 57.6 \text{ ft}^2$$

$$A_3 = ah$$

$$= (8 \text{ ft})(6 \text{ ft})$$

$$= 48 \text{ ft}^2$$

$$A_4 = ab(\sin \phi)$$

$$= (9 \text{ ft})(6 \text{ ft})(\sin 60°)$$

$$= 46.8 \text{ ft}^2$$

The total area of the platform is

$$A_T = 115.2 \text{ ft}^2 + 57.6 \text{ ft}^2 + 48 \text{ ft}^2 + 46.8 \text{ ft}^2$$

$$= 267.6 \text{ ft}^2$$

The answer is (B).

7. The gate covers 70% of the height of the opening of the channel. Find the height of the part of the channel that remains open, which is the diameter of the opening, D_{opening}.

$$D_{\text{opening}} = D \times \text{percentage open}$$

$$= (4 \text{ ft})\left(1 - \frac{70}{100}\right)$$

$$= 1.2 \text{ ft}$$

Determine the total angle of the opening from the center of the circular channel.

Circular Segment

$$\phi = 2 \arccos \frac{(r - D_{\text{opening}})}{r}$$

$$= 2 \arccos \frac{(2 \text{ ft} - 1.2 \text{ ft})}{2 \text{ ft}}$$

$$= 132.8°$$

Illustration for Solution 6

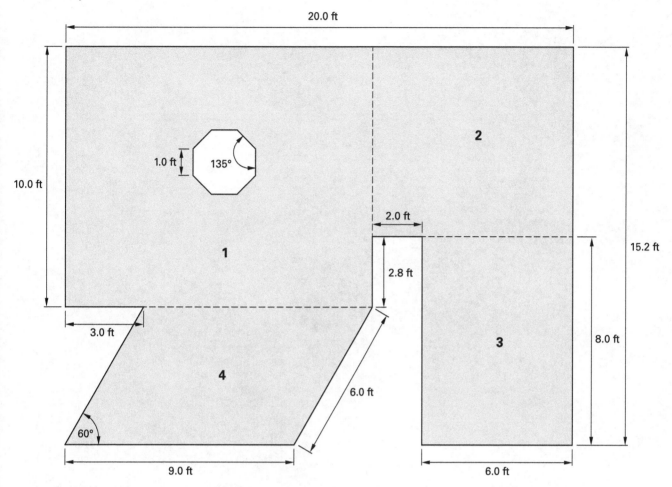

Find the cross-sectional area of the opening using the equation for the area of a circular segment.

Circular Segment

$$A = \frac{r^2\,(\phi - \sin\phi)}{2}$$

$$= \frac{(2\text{ ft})^2\left((132.8°)\left(\dfrac{\pi}{180°}\right) - \sin 132.8°\right)}{2}$$

$$= 3.16\text{ ft}^2$$

The answer is (C).

Differential Calculus

Content in blue refers to the *NCEES Handbook*.

PRACTICE PROBLEMS

1. Find all minima, maxima, and inflection points for

$$y = x^3 - 9x^2 - 3$$

(A) inflection at $x = -3$, maximum at $x = 0$, minimum at $x = -6$

(B) inflection at $x = 3$, maximum at $x = 0$, minimum at $x = 6$

(C) inflection at $x = 3$, maximum at $x = 6$, minimum at $x = 0$

(D) inflection at $x = 0$, maximum at $x = -3$, minimum at $x = 3$

2. The equation for the elevation above mean sea level of a sag vertical roadway curve is

$$y(x) = 0.56x^2 - 3.2x + 708.28$$

y is measured in feet, and x is measured in 100 ft stations past the beginning of the curve. What is most nearly the elevation of the turning point (i.e., the lowest point on the curve)?

(A) 702 ft

(B) 704 ft

(C) 705 ft

(D) 706 ft

3. A car drives on a highway with a legal speed limit of 100 km/h. The fuel usage, Q (in liters per 100 kilometers driven), of a car driven at speed v (in km/h) is

$$Q(v) = \frac{1750v}{v^2 + 6700}$$

At what approximate legal speed should the car travel in order to maximize the fuel efficiency?

(A) 82 km/h

(B) 87 km/h

(C) 93 km/h

(D) 100 km/h

4. A chemical feed storage tank is needed with a volume of 3000 ft³ (gross of fittings). The tank will be formed as a circular cylinder with barrel length, L, capped by two hemispherical ends of radius, r. The manufacturing cost per unit area of hemispherical ends is double that of the cylinder. The dimensions that will minimize the manufacturing cost are most nearly

(A) radius = 4½ ft; cylinder barrel length = 42 ft

(B) radius = 5 ft; cylinder barrel length = 31½ ft

(C) radius = 5½ ft; cylinder barrel length = 22½ ft

(D) radius = 6 ft; cylinder barrel length = 18½ ft

SOLUTIONS

1. Determine the critical points by taking the first derivative of the function and setting it equal to zero.

$$\frac{dy}{dx} = 3x^2 - 18x = 3x(x-6)$$
$$3x(x-6) = 0$$
$$x(x-6) = 0$$

The critical points are located at $x = 0$ and $x = 6$.

Determine the inflection points by setting the second derivative equal to zero. Take the second derivative.

$$\frac{d^2y}{dx^2} = \left(\frac{d}{dx}\right)\left(\frac{dy}{dx}\right)$$
$$= \frac{d}{dx}(3x^2 - 18x)$$
$$= 6x - 18$$

Set the second derivative equal to zero.

$$\frac{d^2y}{dx^2} = 0 = 6x - 18 = (6)(x-3)$$
$$(6)(x-3) = 0$$
$$x - 3 = 0$$
$$x = 3$$

The inflection point is at $x = 3$.

Determine the local maximum and minimum by substituting the critical points into the expression for the second derivative.

At the critical point $x = 0$,

$$\left.\frac{d^2y}{dx^2}\right|_{x=0} = (6)(x-3) = (6)(0-3) = -18$$

Since $-18 < 0$, $x = 0$ is a local maximum.

At the critical point $x = 6$,

$$\left.\frac{d^2y}{dx^2}\right|_{x=6} = (6)(x-3) = (6)(6-3) = 18$$

Since $18 > 0$, $x = 6$ is a local minimum.

The answer is (B).

2. Set the derivative of the curve's equation to zero.

$$\frac{dy(x)}{dx} = \frac{d}{dx}(0.56x^2 - 3.2x + 708.28) = 1.12x - 3.2$$

$$x_c = \frac{3.2}{1.12} = 2.857 \text{ sta}$$

Insert x_c into the elevation equation.

$$y_{\min} = y(x_c) = 0.56x^2 - 3.2x + 708.28$$
$$= (0.56)(2.857 \text{ sta})^2 - (3.2)(2.857 \text{ sta}) + 708.28$$
$$= 703.71 \text{ ft} \quad (704 \text{ ft})$$

The answer is (B).

3. Use the quotient rule to calculate the derivative.

$$\mathbf{D}\left(\frac{f(x)}{g(x)}\right) = \frac{g(x)\mathbf{D}f(x) - f(x)\mathbf{D}g(x)}{(g(x))^2}$$

$$\frac{dQ(v)}{dv} = \frac{d}{dv}\left(\frac{1750v}{v^2 + 6700}\right)$$
$$= \frac{(v^2 + 6700)(1750) - (1750v)(2v)}{(v^2 + 6700)^2}$$

Combining terms and simplifying,

$$\frac{dQ(v)}{dv} = \frac{-1750v^2 + 11{,}725{,}000}{v^4 + 13{,}400v^2 + 44{,}890{,}000}$$

Set the derivative of the fuel consumption equation to zero. In order for the derivative to be zero, the numerator must be zero.

$$-1750v^2 + 11{,}725{,}000 = 0$$

$$v = \sqrt{\frac{11{,}725{,}000}{1750}} = 81.85 \quad (82 \text{ km/h})$$

Maximizing the fuel efficiency is the same as minimizing the fuel usage. It is not known if setting $dQ(v)/dt = 0$ results in a minimum or maximum. While using $d^2Q(v)/dt^2$ is possible, it is easier just to plot the points.

v	Q(v)
82 km/h	10.689 L
87 km/h	10.669 L
93 km/h	10.603 L
100 km/h	10.479 L

Clearly, $Q(82 \text{ km/h})$ is a maximum, and the minimum fuel usage occurs at the endpoint of the range, at 100 km/h.

The answer is (D).

4. The volume of the tank will be the combined volume of a cylinder and a sphere. The cylinder and sphere have the same radius, r.

$$V = \pi r^2 L + \frac{4}{3}\pi r^3 = 3000 \text{ ft}^3$$

Per the problem statement, the cost of the hemispherical ends is double that of the cylinder. For any given cost per unit area (arbitrarily selected as $1/ft^2$ for the cylinder), the cost function is

$$C(r, L) = \left(1 \ \frac{\$}{\text{ft}^2}\right)(A_{\text{cylinder}} + 2A_{\text{sphere}}) = 2\pi r L + 2(4\pi r^2)$$
$$= 2\pi r L + 8\pi r^2$$

Solve the volume equation for barrel length, L.

$$L = \frac{3000 - \frac{4}{3}\pi r^3}{\pi r^2}$$

Substitute L into the cost equation to get a cost function of a single variable.

$$C(r) = 2\pi r L + 8\pi r^2 = 2\pi r \left(\frac{3000 - \frac{4}{3}\pi r^3}{\pi r^2}\right) + 8\pi r^2$$
$$= \frac{6000}{r} - \frac{8\pi r^2}{3} + 8\pi r^2$$
$$= \frac{6000}{r} + \frac{16\pi r^2}{3}$$

Find the optimal value of r by setting the first derivative of the cost function equal to zero.

$$\frac{dC(r)}{dr} = \frac{d}{dr}\left(\frac{6000}{r} + \frac{16\pi r^2}{3}\right) = \frac{-6000}{r^2} + \frac{32\pi r}{3} = 0$$

$$\frac{6000}{r^2} = \frac{32\pi r}{3}$$

Cross multiply, and solve for the optimal value of radius, r.

$$32\pi r^3 = 18{,}000$$
$$r = \sqrt[3]{\frac{18{,}000}{32\pi}}$$
$$= 5.636 \text{ ft} \quad (5\frac{1}{2} \text{ ft})$$

Calculate the optimal value of the barrel length, L.

$$L = \frac{3000 - \frac{4}{3}\pi r^3}{\pi r^2}$$
$$= \frac{3000 - \frac{4}{3}\pi (5.636 \text{ ft})^3}{\pi (5.636 \text{ ft})^2}$$
$$= 22.55 \text{ ft} \quad (22\frac{1}{2} \text{ ft})$$

The answer is (C).

9 Integral Calculus

Content in blue refers to the *NCEES Handbook*.

PRACTICE PROBLEMS

1. Find the indefinite integral of

$$\int \frac{x^2}{x^2 + x - 6}\, dx$$

(A) $\ln|(x+3)| + \frac{4}{5}\ln|(x+2)| + C$

(B) $\ln|(x-3)| + \frac{4}{5}\ln|(x-2)| + C$

(C) $x - \frac{5}{9}\ln|(x-3)| + \frac{4}{5}\ln|(x-2)| + C$

(D) $x - \frac{9}{5}\ln|(x+3)| + \frac{4}{5}\ln|(x-2)| + C$

2. Calculate the definite integral of

$$\int_1^2 (4x^3 - 3x^2)\, dx$$

(A) 8

(B) 16

(C) 24

(D) 32

3. Find the area bounded by $x = 1$, $x = 3$, $y + x + 1 = 0$, and $y = 6x - x^2$.

(A) $7\frac{1}{2}$

(B) $13\frac{1}{3}$

(C) $21\frac{1}{3}$

(D) $25\frac{1}{2}$

4. The velocity profile of a fluid experiencing laminar flow in a pipe of radius R is

$$v(r) = v_{max}\left[1 - \left(\frac{r}{R}\right)^2\right]$$

r is the distance from the centerline. v_{max} is the centerline velocity. What is the volumetric flow rate?

(A) $\dfrac{\pi v_{max} R^2}{2}$

(B) $\pi v_{max} R^2$

(C) $\dfrac{3\pi v_{max} R^2}{2}$

(D) $2\pi v_{max} R^2$

5. A waveform is shown.

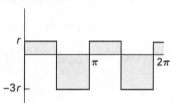

The first term in a Fourier series represents a waveform's average value. The first term of the Fourier series approximation for the waveform is

(A) $-r$

(B) $-\frac{1}{2}r$

(C) r

(D) $2r$

6. A waveform is shown.

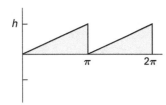

The first term in a Fourier series represents a waveform's average value. The first term of the Fourier series approximation for the waveform is

(A) $-\dfrac{h}{2}$

(B) $\dfrac{h}{2}$

(C) h

(D) $2h$

7. The volumetric flow rate of a fluid experiencing laminar flow in a pipe of radius R is

$$v(r) = v_{max}\left[1 - \left(\dfrac{r}{R}\right)^2\right]$$

$$Q = \dfrac{\pi v_{max} R^2}{2}$$

v_{max} is the centerline velocity. What is the average velocity?

(A) $\dfrac{v_{max}}{6}$

(B) $\dfrac{v_{max}}{4}$

(C) $\dfrac{v_{max}}{3}$

(D) $\dfrac{v_{max}}{2}$

SOLUTIONS

1. Solve for the indefinite integral.

$$\dfrac{x^2}{x^2+x-6} = \dfrac{x^2+x-x-6+6}{x^2+x-6}$$

$$= 1 - \dfrac{x-6}{x^2+x-6}$$

$$= 1 - \dfrac{x-6}{(x+3)(x-2)}$$

$$= 1 - \dfrac{\tfrac{9}{5}}{x+3} + \dfrac{\tfrac{4}{5}}{x-2}$$

$$\int \dfrac{x^2}{x^2+x-6}\,dx = \int \left(1 - \dfrac{\tfrac{9}{5}}{x+3} + \dfrac{\tfrac{4}{5}}{x-2}\right) dx$$

$$= \int dx - \int \dfrac{\tfrac{9}{5}}{x+3}\,dx + \int \dfrac{\tfrac{4}{5}}{x-2}\,dx$$

$$= x - \tfrac{9}{5}\ln|(x+3)| + \tfrac{4}{5}\ln|(x-2)| + C$$

The answer is (D).

2. The definite integral is

$$\int_1^2 (4x^3 - 3x^2)\,dx = (x^4 - x^3)\Big|_1^2$$

$$= (2)^4 - (2)^3 - \left((1)^4 - (1)^3\right)$$

$$= 8$$

The answer is (A).

3. The bounded area is shown.

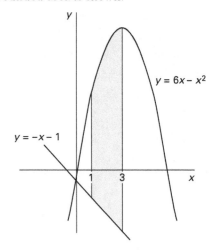

The area is

$$A = \int_1^3 \left((6x - x^2) - (-x - 1)\right) dx$$
$$= \int_1^3 (-x^2 + 7x + 1) dx$$
$$= \left(-\frac{x^3}{3} + \frac{7}{2}x^2 + x\right)\bigg|_1^3$$
$$= -\frac{(3)^3}{3} + \left(\frac{7}{2}\right)(3)^2 + 3 - \left(-\frac{(1)^3}{3} + \left(\frac{7}{2}\right)(1)^2 + 1\right)$$
$$= 21\frac{1}{3}$$

The answer is (C).

4. Divide the circular internal area of the pipe into small annular rings. The radius at the ring is r; the differential thickness of the ring is dr; the differential area of the ring is $dA = 2\pi r \, dr$; and the velocity is $v(r)$. Solve for the volumetric flow rate through the annular ring.

$$dQ = v(r) \, dA = 2\pi r v_{\max}\left[1 - \left(\frac{r}{R}\right)^2\right] dr$$
$$Q = \int_{r=0}^{r=R} 2\pi r v_{\max}\left[1 - \left(\frac{r}{R}\right)^2\right] dr$$
$$= 2\pi v_{\max} \int_{r=0}^{r=R} \left(r - \frac{r^3}{R^2}\right) dr$$
$$= 2\pi v_{\max} \left(\frac{r^2}{2} - \frac{r^4}{4R^2}\right)\bigg|_0^R$$
$$= 2\pi v_{\max} \left(\frac{R^2}{2} - \frac{R^4}{4R^2} - 0 - 0\right)$$
$$= \frac{\pi v_{\max} R^2}{2}$$

The answer is (A).

5. For the given waveform, the first term of the Fourier series approximation is

$$a_0 = \frac{1}{2\pi}\int_0^{2\pi} f(t) \, dt = \frac{1}{\pi}\int_0^{\pi} f(t) \, dt$$
$$= \frac{1}{\pi}\left(r\left(\frac{\pi}{2}\right) + (-3r)\left(\frac{\pi}{2}\right)\right)$$
$$= -r$$

The answer is (A).

6. For the given waveform, the first term of the Fourier series approximation is

$$a_0 = \frac{1}{2\pi}\int_0^{2\pi} f(t) \, dt = \frac{1}{\pi}\int_0^{\pi} f(t) \, dt$$
$$= \left(\frac{1}{\pi}\right)\left(\frac{1}{2}\pi h\right)$$
$$= \frac{h}{2}$$

The answer is (B).

7. The average velocity is

$$\bar{v} = \frac{Q}{A} = \frac{\dfrac{\pi v_{\max} R^2}{2}}{\pi R^2} = \frac{v_{\max}}{2}$$

The answer is (D).

10 Differential Equations

Content in blue refers to the *NCEES Handbook*.

PRACTICE PROBLEMS

1. Solve the following differential equation for y.

$$y' - y = 2xe^{2x} \quad y(0) = 1$$

(A) $y = 2e^{-2x}(x-1) + 3e^{-x}$

(B) $y = 2e^{2x}(x-1) + 3e^{x}$

(C) $y = -2e^{-2x}(x-1) + 3e^{-x}$

(D) $y = 2e^{2x}(x-1) + 3e^{-x}$

2. The oscillation exhibited by the top story of a certain building in free motion is given by the following differential equation.

$$x'' + 2x' + 2x = 0 \quad x(0) = 0 \quad x'(0) = 1$$

If $x(t) = e^{-t}\sin t$, the amplitude of oscillation is most nearly

(A) 0.32

(B) 0.54

(C) 1.7

(D) 6.6

3. The oscillation exhibited by the top story of a certain building in free motion is given by the following differential equation.

$$x'' + 2x' + 2x = 0 \quad x(0) = 0 \quad x'(0) = 1$$

A lateral wind load is applied with a form of $\sin t$. The general complementary solution is

$$x_c(t) = A_1 e^{-t}\cos t + A_2 e^{-t}\sin t$$

What is x as a function of time?

(A) $\frac{6}{5}e^{-t}\sin t + \frac{2}{5}e^{-t}\cos t$

(B) $\frac{6}{5}e^{t}\sin t + \frac{2}{5}e^{t}\cos t$

(C) $\frac{2}{5}e^{t}\sin t + \frac{6}{5}e^{-t}\cos t + \frac{2}{5}\sin t - \frac{1}{5}\cos t$

(D) $\frac{2}{5}e^{-t}\cos t + \frac{6}{5}e^{-t}\sin t - \frac{2}{5}\cos t + \frac{1}{5}\sin t$

4. A 90 lbm (40 kg) bag of a chemical is accidentally dropped in an aerating lagoon. The chemical is water soluble and nonreacting. The lagoon is 120 ft (35 m) in diameter and filled to a depth of 10 ft (3 m). The aerators circulate and distribute the chemical evenly throughout the lagoon.

Water enters the lagoon at a rate of 30 gal/min (115 L/min). Fully mixed water is pumped into a reservoir at a rate of 30 gal/min (115 L/min). The established safe concentration of this chemical is 1 ppb (part per billion).

The number of days it will take for the concentration of the discharge water to reach this level is most nearly

(A) 25 days

(B) 50 days

(C) 100 days

(D) 200 days

5. A tank contains 100 gal (100 L) of brine made by dissolving 60 lbm (60 kg) of salt in pure water. Salt water with a concentration of 1 lbm/gal (1 kg/L) enters the tank at a rate of 2 gal/min (2 L/min). A well-stirred mixture is drawn from the tank at a rate of 3 gal/min (3 L/min). The mass of salt in the tank after one hour is most nearly

(A) 13 lbm (13 kg)

(B) 37 lbm (37 kg)

(C) 43 lbm (43 kg)

(D) 51 lbm (51 kg)

SOLUTIONS

1. The equation is a first-order linear differential equation of the form

$$y' + p(x)y = g(x)$$
$$p(x) = -1$$
$$g(x) = 2xe^{2x}$$

The integration factor $u(x)u(x)$ is

$$u(x) = \exp\left(\int p(x)\,dx\right) = \exp\left(\int(-1)\,dx\right) = e^{-x}$$

The closed form of the solution is

$$\begin{aligned}y &= \frac{1}{u(x)}\left(\int u(x)g(x)\,dx + C\right) \\ &= \frac{1}{e^{-x}}\left(\int (e^{-x})(2xe^{2x})\,dx + C\right) \\ &= e^{x}\bigl(2(xe^{x} - e^{x}) + C\bigr) \\ &= e^{x}\bigl(2e^{x}(x-1) + C\bigr) \\ &= 2e^{2x}(x-1) + Ce^{x}\end{aligned}$$

Apply the initial condition $y(0) = 1$ to obtain the integration constant, C.

$$\begin{aligned}y(0) = 2e^{(2)(0)}(0-1) + Ce^{0} &= 1 \\ (2)(1)(-1) + C(1) &= 1 \\ -2 + C &= 1 \\ C &= 3\end{aligned}$$

Substituting in the value for the integration constant, C, the solution is

$$y = 2e^{2x}(x-1) + 3e^{x}$$

The answer is (B).

2. The differential equation is a homogeneous second-order linear differential equation with constant coefficients. Write the characteristic equation.

$$r^2 + 2r + 2 = 0$$

This is a quadratic equation of the form $ar^2 + br + c = 0$, where $a = 1$, $b = 2$, and $c = 2$.

Solve for r.

$$\begin{aligned}r &= \frac{-b \pm \sqrt{b^2 - 4ac}}{2a} = \frac{-2 \pm \sqrt{(2)^2 - (4)(1)(2)}}{(2)(1)} \\ &= \frac{-2 \pm \sqrt{4-8}}{2} \\ &= -1 \pm \sqrt{-1} \\ &= -1 \pm i \\ r_1 &= -1 + i, \text{ and } r_2 = -1 - i\end{aligned}$$

Since the roots are imaginary and of the form $\alpha + i\omega$ and $\alpha + i\omega$, where $\alpha = -1$ and $\omega = 1$, the general form of the solution is

$$\begin{aligned}x(t) &= A_1 e^{\alpha t} = \cos\omega t + A_2 e^{\alpha t}\sin\omega t \\ &= A_1 e^{-1t} = \cos t + A_2 e^{-1t}\sin t \\ &= A_1 e^{-t} = \cos t + A_2 e^{-t}\sin t\end{aligned}$$

Apply the initial conditions, $x(0) = 0$ and $x'(0) = 1$, to solve for A_1 and A_2.

First, apply the initial condition, $x(0) = 0$.

$$\begin{aligned}x(t) = A_1 e^0 = \cos 0 + A_2 e^0 \sin 0 &= 0 \\ A_1(1)(1) + A_2(1)(0) &= 0 \\ A_1 &= 0\end{aligned}$$

Substituting, the solution of the differential equation becomes

$$x(t) = A_2 e^{-t}\sin t$$

To apply the second initial condition, take the first derivative.

$$\begin{aligned}x'(t) &= \frac{d}{dt}(A_2 e^{-t}\sin t) = A_2\frac{d}{dt}(e^{-t}\sin t) \\ &= A_2\left(\sin t\frac{d}{dt}(e^{-t}) e^{-t}\frac{d}{dt}\sin t\right) \\ &= A_2(\sin t(-e^{-t}) + e^{-t}(\cos t)) \\ &= A_2(e^{-t})(-\sin t + \cos t)\end{aligned}$$

Apply the initial condition, $x'(0) = 1$.

$$\begin{aligned}x'(0) = A_2 e^0(-\sin 0 + \cos 0) &= 1 \\ A_2(1)(0+1) &= 1 \\ A_2 &= 1\end{aligned}$$

The solution is

$$\begin{aligned}x(t) = A_2 e^{-t}\sin t &= (1)e^{-t}\sin t \\ &= e^{-t}\sin t\end{aligned}$$

The amplitude of the oscillation is the maximum displacement.

Take the derivative of $x(t) = e^{-t} \sin t$.

$$x'(t) = \frac{d}{dt}(e^{-t} \sin t) = \sin t \frac{d}{dt}(e^{-t}) + e^{-t} \frac{d}{dt} \sin t$$
$$= \sin t(-e^{-t}) + e^{-t} \cos t$$
$$= e^{-t}(\cos t - \sin t)$$

The maximum displacement occurs at $x'(t) = 0$.
Since $e^{-t} \neq 0$ except as t approaches infinity,

$$\cos t - \sin t = 0$$
$$\tan t = 1$$
$$t = \tan^{-1}(1)$$
$$= 0.785 \text{ rad}$$

At $t = 0.785$ rad, the displacement is maximum. Substitute into the original equation for $x(t)$ to obtain a value for the maximum displacement.

$$x(0.785) = e^{-0.785} \sin 0.785 = 0.322$$

The amplitude is 0.322 (0.32).

The answer is (A).

3. (An alternative solution using Laplace transforms follows this solution.) The application of a lateral wind load with the form $\sin t$ revises the differential equation to the form

$$x'' + 2x' + 2x = \sin t$$

Express the solution as the sum of the complementary x_c and particular x_p solutions.

$$x(t) = x_c(t) + x_p(t)$$

The general form of the particular solution is given by

$$x_p(t) = x^s(A_3 \cos t + A_4 \sin t)$$

Determine the value of s; check to see if the terms of the particular solution solve the homogeneous equation.

Examine the term $A_3 \cos t$.

Take the first derivative.

$$\frac{d}{dx}(A_3 \cos t) = -A_3 \sin t$$

Take the second derivative.

$$\frac{d}{dx}\left(\frac{d}{dx}(A_3 \cos t)\right) = \frac{d}{dx}(-A_3 \sin t)$$
$$= -A_3 \cos t$$

Substitute the terms into the homogeneous equation.

$$x'' + 2x' + 2x = -A_3 \cos t + (2)(-A_3 \sin t)$$
$$+ (2)(-A_3 \cos t)$$
$$= A_3 \cos t - 2A_3 \sin t$$
$$\neq 0$$

Except for the trivial solution $A_3 = 0$, the term $A_3 \cos t$ does not solve the homogeneous equation.

Examine the second term, $A_4 \sin t$.

Take the first derivative.

$$\frac{d}{dx}(A_4 \sin t) = A_4 \cos t$$

Take the second derivative.

$$\frac{d}{dx}\left(\frac{d}{dx}(A_4 \sin t)\right) = \frac{d}{dx}(A_4 \cos t) = -A_4 \sin t$$

Substitute the terms into the homogeneous equation.

$$x'' + 2x' + 2x = -A_4 \sin t + (2)(A_4 \cos t)$$
$$+ (2)(A_4 \sin t)$$
$$= A_4 \sin t + 2A_4 \cos t$$
$$\neq 0$$

Except for the trivial solution $A_4 = 0$, the term $A_4 \sin t$ does not solve the homogeneous equation.

Neither of the terms satisfies the homogeneous equation, so $s = 0$. Therefore, the particular solution is of the form

$$x_p(t) = A_3 \cos t + A_4 \sin t$$

Use the method of undetermined coefficients to solve for A_3 and A_4. Take the first derivative.

$$x'_p(t) = \frac{d}{dx}(A_3 \cos t + A_4 \sin t)$$
$$= -A_3 \sin t + A_4 \cos t$$

Take the second derivative.

$$x_p''(t) = \frac{d}{dx}\left[\frac{d}{dx}(A_3 \cos t + A_4 \sin t)\right]$$
$$= \frac{d}{dx}(-A_3 \sin t + A_4 \cos t)$$
$$= -A_3 \cos t - A_4 \sin t$$

Substitute the expressions for the derivatives into the differential equation.

$$x'' + 2x' + 2x = (-A_3 \cos t - A_4 \sin t)$$
$$+ (2)(-A_3 \sin t + A_4 \cos t)$$
$$+ (2)(A_3 \cos t + A_4 \sin t)$$
$$= \sin t$$

Rearranging terms gives

$$(-A_3 + 2A_4 + 2A_3)\cos t$$
$$+ (-A_4 - 2A_3 + 2A_4)\sin t = \sin t$$
$$(A_3 + 2A_4)\cos t + (-2A_3 + A_4)\sin t = \sin t$$

Equating coefficients gives

$$A_3 + 2A_4 = 0$$
$$-2A_3 + A_4 = 1$$

Multiplying the first equation by 2 and adding equations gives

$$2A_3 + 4A_4 = 0$$
$$+(-2A_3 + A_4) = 1$$
$$5A_4 = 1 \text{ or } A_4 = \frac{1}{5}$$

From the first equation for $A_4 = 1/5$, $A_3 + (2)(1/5) = 0$, and $A_3 = -2/5$.

Substituting for the coefficients, the particular solution becomes

$$x_p(t) = -\tfrac{2}{5}\cos t + \tfrac{1}{5}\sin t$$

Combining the complementary and particular solutions gives

$$x(t) = x_c(t) + x_p(t)$$
$$= A_1 e^{-t}\cos t + A_2 e^{-t}\sin t - \tfrac{2}{5}\cos t + \tfrac{1}{5}\sin t$$

Apply the initial conditions to solve for the coefficients A_1 and A_2, then apply the first initial condition, $x(0) = 0$.

$$x(t) = A_1 e^0 \cos 0 + A_2 e^0 \sin 0$$
$$\tfrac{2}{5}\cos 0 + \tfrac{1}{5}\sin 0 = 0$$
$$A_1(1)(1) + A_2(1)(0) + \left(-\tfrac{2}{5}\right)(1) + \left(\tfrac{1}{5}\right)(0) = 0$$
$$A_1 - \tfrac{2}{5} = 0$$
$$A_1 = \tfrac{2}{5}$$

Substituting for A_1, the solution becomes

$$x(t) = \tfrac{2}{5}e^{-t}\cos t + A_2 e^{-t}\sin t - \tfrac{2}{5}\cos t + \tfrac{1}{5}\sin t$$

Take the first derivative.

$$x'(t) = \frac{d}{dx}\left[\begin{array}{l}\left(\tfrac{2}{5}e^{-t}\cos t + A_2 e^{-t}\sin t\right) \\ +\left(-\tfrac{2}{5}\cos t + \tfrac{1}{5}\sin t\right)\end{array}\right]$$
$$= \left(\tfrac{2}{5}\right)(-e^{-t}\cos t - e^{-t}\sin t)$$
$$+ A_2(-e^{-t}\sin t + e^{-t}\cos t)$$
$$+ \left(-\tfrac{2}{5}\right)(-\sin t) + \tfrac{1}{5}\cos t$$

Apply the second initial condition, $x'(0) = 1$.

$$x'(0) = \left(\tfrac{2}{5}\right)(-e^0 \cos 0 - e^0 \sin 0)$$
$$+ A_2(-e^0 \sin 0 + e^0 \cos 0)$$
$$+ \left(-\tfrac{2}{5}\right)(-\sin 0) + \tfrac{1}{5}\cos 0$$
$$= 1$$

$$\left(\tfrac{2}{5}\right)(-(1)(1) - (1)(0)) + A_2(-(1)(0) + (1)(1))$$
$$+ \left(-\tfrac{2}{5}\right)(0) + \left(\tfrac{1}{5}\right)(1) = 1$$
$$\left(\tfrac{2}{5}\right)(-1) + A_2(1) + \left(\tfrac{1}{5}\right) = 1$$
$$A_2 = \tfrac{6}{5}$$

Substituting for A_2, the solution becomes

$$x(t) = \tfrac{2}{5}e^{-t}\cos t + \tfrac{6}{5}e^{-t}\sin t - \tfrac{2}{5}\cos t + \tfrac{1}{5}\sin t$$

Alternate solution:

Use the Laplace transform method.

$$x'' + 2x' + 2x = \sin t$$
$$\mathcal{L}(x'') + 2\mathcal{L}(x') + 2\mathcal{L}(x) = \mathcal{L}(\sin t)$$
$$s^2\mathcal{L}(x) - 1 + 2s\mathcal{L}(x) + 2\mathcal{L}(x) = \frac{1}{s^2+1}$$
$$\mathcal{L}(x)(s^2+2s+2) - 1 = \frac{1}{s^2+1}$$

$$\mathcal{L}(x) = \frac{1}{s^2+2s+2} + \frac{1}{(s^2+1)(s^2+2s+2)}$$
$$= \frac{1}{(s+1)^2+1} + \frac{1}{(s^2+1)(s^2+2s+2)}$$

Use partial fractions to expand the second term, then cross multiply.

$$\frac{1}{(s^2+1)(s^2+2s+2)} = \frac{A_1 + B_1 s}{s^2+1} + \frac{A_2 + B_2 s}{s^2+2s+2}$$

$$= \frac{\begin{array}{l}A_1 s^2 + 2A_1 s + 2A_1 + B_1 s^3 + 2B_1 s^2 + 2B_1 s \\ + A_2 s^2 + A_2 + B_2 s^3 + B_2 s\end{array}}{(s^2+1)(s^2+2s+2)}$$

$$= \frac{s^3(B_1+B_2) + s^2(A_1+A_2+2B_1) + s(2A_1+2B_1+B_2) + 2A_1+A_2}{(s^2+1)(s^2+2s+2)}$$

Compare numerators to obtain the following four simultaneous equations.

$$\begin{aligned} B_1 + B_2 &= 0 \\ A_1 + A_2 + 2B_1 &= 0 \\ 2A_1 + 2B_1 + B_2 &= 0 \\ 2A_1 + A_2 &= 1 \end{aligned}$$

Use Cramer's rule to find A_1.

$$A_1 = \frac{\begin{vmatrix} 0 & 0 & 1 & 1 \\ 0 & 1 & 2 & 0 \\ 0 & 0 & 2 & 1 \\ 1 & 1 & 0 & 0 \end{vmatrix}}{\begin{vmatrix} 0 & 0 & 1 & 1 \\ 1 & 1 & 2 & 0 \\ 2 & 0 & 2 & 1 \\ 2 & 1 & 0 & 0 \end{vmatrix}} = \frac{-1}{-5} = \frac{1}{5}$$

The rest of the coefficients are found similarly.

$$A_1 = \frac{1}{5}$$
$$A_2 = \frac{3}{5}$$
$$B_1 = -\frac{2}{5}$$
$$B_2 = \frac{2}{5}$$

Then,

$$\mathcal{L}(x) = \frac{1}{(s+1)^2+1} + \frac{\frac{1}{5}}{s^2+1} + \frac{-\frac{2}{5}s}{s^2+1}$$
$$+ \frac{\frac{3}{5}}{s^2+2s+2} + \frac{\frac{2}{5}s}{s^2+2s+2}$$

Take the inverse transform.

$$\begin{aligned} x(t) &= \mathcal{L}^{-1}\{\mathcal{L}(x)\} \\ &= e^{-t}\sin t + \tfrac{1}{5}\sin t - \tfrac{2}{5}\cos t + \tfrac{3}{5}e^{-t}\sin t \\ &\quad + \tfrac{2}{5}(e^{-t}\cos t - e^{-t}\sin t) \\ &= \tfrac{2}{5}e^{-t}\cos t + \tfrac{6}{5}e^{-t}\sin t + \tfrac{2}{5}\cos t - \tfrac{1}{5}\sin t \end{aligned}$$

The answer is (D).

4. *Customary U.S. Solution*

First, find the mass of chemicals at a concentration of 1 ppb.

The initial volume of water in the lagoon is

$$\begin{aligned} V(t) &= \left(\frac{\pi}{4}\right)(\text{diameter of lagoon})^2(\text{depth of lagoon}) \\ &= \left(\frac{\pi}{4}\right)(120 \text{ ft})^2(10 \text{ ft}) \\ &= 113{,}097 \text{ ft}^3 \end{aligned}$$

The initial mass of the water in the lagoon is

$$\begin{aligned} m_i &= V\rho = (113{,}097 \text{ ft}^3)\left(62.4 \ \frac{\text{lbm}}{\text{ft}^3}\right) \\ &= 7.06 \times 10^6 \text{ lbm} \quad (7.06 \times 10^6 \text{ lbm}) \end{aligned}$$

The final mass of chemicals at a concentration of 1 ppb is

$$m_f = \frac{7.06 \times 10^6 \text{ lbm}}{1 \times 10^9}$$
$$= 7.06 \times 10^{-3} \text{ lbm} \quad (0.007 \text{ lbm})$$

Substituting into the general form of the differential equation gives

$$m'(t) = a(t) - \frac{m(t)o(t)}{V(t)}$$
$$= 0 - m(t) \left(\frac{4.01 \frac{\text{ft}^3}{\text{min}}}{113{,}097 \text{ ft}^3} \right)$$
$$= -\left(\frac{3.55 \times 10^{-5}}{\text{min}} \right) m(t)$$

$$m'(t) + \left(\frac{3.55 \times 10^{-5}}{\text{min}} \right) m(t) = 0$$

The differential equation of the problem has the following characteristic equation.

$$r + \left(\frac{3.55 \times 10^{-5}}{\text{min}} \right) = 0$$
$$r = -3.55 \times 10^{-5} \, / \, \text{min}$$

The general form of the solution is

$$m(t) = A e^{rt}$$

Substituting the root, r, gives

$$m(t) = A e^{(-3.55 \times 10^{-5} / \text{min})t}$$

Apply the initial condition $m(0) = 90$ lbm at time $t = 0$.

$$m(0) = A e^{(-3.55 \times 10^{-5} / \text{min})(0)} = 90 \text{ lbm}$$
$$A e^0 = 90 \text{ lbm}$$
$$A = 90 \text{ lbm}$$

Therefore,

$$m(t) = (90 \text{ lbm}) e^{(-3.55 \times 10^{-5} / \text{min})t}$$

Solve for t.

$$\frac{m(t)}{90 \text{ lbm}} = e^{(-3.55 \times 10^{-5} / \text{min})t}$$
$$\ln\left(\frac{m(t)}{90 \text{ lbm}} \right) = \ln\left(e^{(-3.55 \times 10^{-5} / \text{min})t} \right)$$
$$= \left(\frac{-3.55 \times 10^{-5}}{\text{min}} \right) t$$
$$t = \frac{\ln\left(\frac{m(t)}{90 \text{ lbm}} \right)}{\frac{3.55 \times 10^{-5}}{\text{min}}}$$

$$t = \left(\frac{\ln\left(\frac{m(t)}{90 \text{ lbm}} \right)}{\frac{-3.55 \times 10^{-5}}{\text{min}}} \right) \left(\frac{1 \text{ hr}}{60 \text{ min}} \right) \left(\frac{1 \text{ day}}{24 \text{ hr}} \right)$$

$$= \left(\frac{\ln\left(\frac{7.06 \times 10^{-3} \text{ lbm}}{90 \text{ lbm}} \right)}{\frac{-3.55 \times 10^{-5}}{\text{min}}} \right) \left(\frac{1 \text{ hr}}{60 \text{ min}} \right) \left(\frac{1 \text{ day}}{24 \text{ hr}} \right)$$

$$= 185 \text{ days} \quad (200 \text{ days})$$

SI Solution

First, find the mass of chemicals at a concentration of 1 ppb.

The initial volume of water in the lagoon is

$$V(t) = \left(\frac{\pi}{4} \right) (\text{diamerter of lagoon})^2 (\text{depth of lagoon})$$
$$= \left(\frac{\pi}{4} \right) (35 \text{ m})^2 (3 \text{ m})$$
$$= 2886 \text{ m}^3$$

The initial mass of water in the lagoon is

$$m_i = V\rho = (2886 \text{ m}^3)\left(1000 \, \frac{\text{kg}}{\text{m}^3} \right)$$
$$= 2.886 \times 10^6 \text{ kg} \quad (2.9 \times 10^6 \text{ kg})$$

The final mass of chemicals at a concentration of 1 ppb is

$$m_f = \frac{2.886 \times 10^6 \text{ kg}}{1 \times 10^9}$$
$$= 2.886 \times 10^{-3} \text{ kg} \quad (0.003 \text{ kg})$$

Substituting into the general form of the differential equation gives

$$m'(t) = a(t) - \frac{m(t)\,o(t)}{V(t)}$$

$$= 0 - m(t)\left(\frac{0.115\,\dfrac{\text{ft}^3}{\text{min}}}{2886\,\text{m}^3}\right)$$

$$= -\left(\frac{3.985 \times 10^{-5}}{\text{min}}\right) m(t)$$

$$m'(t) + \left(\frac{3.985 \times 10^{-5}}{\text{min}}\right) m(t) = 0$$

The differential equation of the problem has the following characteristic equation.

$$r + \left(\frac{3.985 \times 10^{-5}}{\text{min}}\right) = 0$$

$$r = -3.985 \times 10^{-5}/\text{min}$$

The general form of the solution is

$$m(t) = A e^{rt}$$

Substituting in for the root, r, gives

$$m(t) = A e^{(-3.985 \times 10^{-5}/\text{min})t}$$

Apply the initial condition $m(0) = 40$ kg at time $t = 0$.

$$m(0) = A e^{(-3.985 \times 10^{-5}/\text{min})(0)} = 40\,\text{kg}$$
$$A e^0 = 40\,\text{kg}$$
$$A = 40\,\text{kg}$$

Therefore,

$$m(t) = (40\,\text{kg}) e^{(-3.985 \times 10^{-5}/\text{min})t}$$

Solve for t.

$$\frac{m(t)}{40\,\text{kg}} = e^{(-3.985 \times 10^{-5}/\text{min})t}$$

$$\ln\left(\frac{m(t)}{40\,\text{kg}}\right) = \ln\left(e^{(-3.985 \times 10^{-5}/\text{min})t}\right)$$

$$= \left(\frac{-3.985 \times 10^{-5}}{\text{min}}\right) t$$

$$t = \frac{\ln\left(\dfrac{m(t)}{40\,\text{kg}}\right)}{\dfrac{-3.985 \times 10^{-5}}{\text{min}}}$$

Find the time required to achieve a mass of 2.886×10^{-3} kg.

$$t = \left(\frac{\ln\left(\dfrac{m(t)}{40\,\text{kg}}\right)}{\dfrac{-3.985 \times 10^{-5}}{\text{min}}}\right)\left(\frac{1\,\text{hr}}{60\,\text{min}}\right)\left(\frac{1\,\text{day}}{24\,\text{hr}}\right)$$

$$= \left(\frac{\ln\left(\dfrac{2.886 \times 10^{-3}\,\text{kg}}{40\,\text{kg}}\right)}{\dfrac{-3.985 \times 10^{-5}}{\text{min}}}\right)\left(\frac{1\,\text{hr}}{60\,\text{min}}\right)\left(\frac{1\,\text{day}}{24\,\text{hr}}\right)$$

$$= 166\,\text{days}\quad(200\,\text{days})$$

The answer is (D).

5. Let

$$m(t) = \text{mass of salt in tank at time } t$$
$$m_0 = 60\text{ mass units (lbm or kg)}$$
$$m'(t) = \text{rate at which salt content is changing}$$

Two mass units of salt enter each minute, and three volumes leave each minute. The amount of salt leaving each minute is

$$\left(3\,\frac{\text{vol}}{\text{min}}\right)\left(\text{concentration in }\frac{\text{mass}}{\text{vol}}\right)$$

$$= \left(3\,\frac{\text{vol}}{\text{min}}\right)\left(\frac{\text{salt mass}}{\text{volume}}\right)$$

$$= \left(3\,\frac{\text{vol}}{\text{min}}\right)\left(\frac{m(t)}{100 - t}\right)$$

$$m'(t) = 2 - (3)\left[\frac{m(t)}{100-t}\right] \text{ or } m'(t) + \frac{3m(t)}{100-t}$$
$$= 2 \text{ mass units/min}$$

This is a first-order linear differential equation. The integrating factor is

$$m = \exp\left(3\int \frac{dt}{100-t}\right)$$
$$= \exp\big((3)(-\ln(100-t))\big)$$
$$= (100-t)^{-3}$$
$$m(t) = (100-t)^3\left[2\int \frac{dt}{(100-t)^3} + k\right]$$
$$= 100 - t + k(100-t)^3$$

$m = 60$ mass units at $t = 0$, so $k = -0.00004$.

$$m(t) = 100 - t - (0.00004)(100-t)^3$$

At $t = 60$ min,

$$m = 100 - 60 \text{ min} - (0.00004)(100 - 60 \text{ min})^3$$
$$= 37.44 \text{ mass units} \quad (37 \text{ lbm per kg})$$

The answer is (B).

11 Probability and Statistical Analysis of Data

Content in blue refers to the *NCEES Handbook*.

PRACTICE PROBLEMS

1. Four military recruits whose respective shoe sizes are 7, 8, 9, and 10 report to the supply clerk to be issued boots. The supply clerk selects one pair of boots in each of the four required sizes and hands them at random to the recruits. Using exhaustive enumeration, the probability that all recruits will receive boots of an incorrect size is most nearly

(A) 0.25
(B) 0.38
(C) 0.45
(D) 0.61

2. Four military recruits whose respective shoe sizes are 7, 8, 9, and 10 report to the supply clerk to be issued boots. The supply clerk selects one pair of boots in each of the four required sizes and hands them at random to the recruits. The probability that exactly three recruits will receive boots of the correct size is most nearly

(A) 0
(B) 0.063
(C) 0.17
(D) 0.25

3. The time taken by a toll taker to collect the toll from vehicles crossing a bridge is an exponential distribution with a mean of 23 sec. The probability that a random vehicle will be processed in more than 25 sec is most nearly

(A) 0.17
(B) 0.25
(C) 0.34
(D) 0.52

4. The number of cars entering a toll plaza on a bridge during the hour after midnight follows a Poisson distribution with a mean of 20. The percent probability that exactly 17 cars will pass through the toll plaza during that hour on any given night is most nearly

(A) 0.08
(B) 0.12
(C) 0.16
(D) 0.23

5. The number of cars entering a toll plaza on a bridge during the hour after midnight follows a Poisson distribution with a mean of 20. The percent probability that three or fewer cars will pass through the toll plaza at that hour on any given night is most nearly

(A) 0.00032%
(B) 0.0019%
(C) 0.079%
(D) 0.11%

6. A survey field crew measures one leg of a traverse four times. The following results are obtained.

repetition	measurement	direction
1	1249.529	forward
2	1249.494	backward
3	1249.384	forward
4	1249.348	backward

What is the standard deviation?

(A) 0.001
(B) 0.012
(C) 0.055
(D) 0.086

7. A survey field crew measures one leg of a traverse four times. The crew chief is under orders to obtain readings with upper one-tailed confidence limits of 90%. Which readings are unacceptable?

(A) Readings inside the 90% confidence limits are unacceptable.

(B) Readings outside the 90% confidence limits are acceptable.

(C) Readings outside both 90% confidence limits are unacceptable.

(D) Readings outside the lower 90% confidence limit are unacceptable.

8. A survey field crew measures one leg of a traverse four times. The following results are obtained.

repetition	measurement	direction
1	1249.529	forward
2	1249.494	backward
3	1249.384	forward
4	1249.348	backward

The crew chief is under orders to obtain readings with confidence limits of 90%. The most probable value of the distance is most nearly

(A) 1249.399

(B) 1249.410

(C) 1249.439

(D) 1249.452

9. A survey field crew measures one leg of a traverse four times. The following results are obtained.

repetition	measurement	direction
1	1249.529	forward
2	1249.494	backward
3	1249.384	forward
4	1249.348	backward

The crew chief is under orders to obtain readings with confidence limits of 90%. The error in the most probable value (at 90% confidence) is most nearly

(A) 0.08

(B) 0.10

(C) 0.14

(D) 0.19

10. A survey field crew measures one leg of a traverse four times. The following results are obtained.

repetition	measurement	direction
1	1249.529	forward
2	1249.494	backward
3	1249.384	forward
4	1249.348	backward

The crew chief is under orders to obtain readings with confidence limits of 90%. If the distance is one side of a square traverse whose sides are all equal, the most probable closure error is most nearly

(A) 0.14

(B) 0.20

(C) 0.28

(D) 0.35

11. A survey field crew measures one leg of a traverse four times. The following results are obtained.

repetition	measurement	direction
1	1249.529	forward
2	1249.494	backward
3	1249.384	forward
4	1249.348	backward

The crew chief is under orders to obtain readings with confidence limits of 90%. If the distance is one side of a square traverse whose sides are all equal, the most probable error expressed as a fraction is most nearly

(A) 1:17,600

(B) 1:14,200

(C) 1:12,500

(D) 1:10,900

12. Which of the following statements indicates a key difference between accuracy and precision?

(A) If an experiment can be repeated with identical results, the results are considered accurate.

(B) If an experiment has a small bias, the results are considered precise.

(C) If an experiment is precise, it cannot also be accurate.

(D) If an experiment is unaffected by experimental error, the results are accurate.

13. Which of the following is an example of a systematic error in a field survey?

(A) measuring river depth as a motorized ski boat passes by

(B) using a steel tape that is too short to measure consecutive distances

(C) locating magnetic north while near a large iron ore deposit along an overland route

(D) determining local wastewater biological oxygen demand after a toxic spill

14. Two resistances, a meter resistor and a shunt resistor, are connected in parallel in an ammeter. Most of the current passing through the meter goes through the shunt resistor. In order to determine the accuracy of the resistance of shunt resistors being manufactured for a line of ammeters, a manufacturer tests a sample of 100 shunt resistors. The numbers of shunt resistors with the resistance indicated (to the nearest hundredth of an ohm) are as follows.

0.200 Ω, 1; 0.210 Ω, 3; 0.220 Ω, 5; 0.230 Ω, 10; 0.240 Ω, 17; 0.250 Ω, 40; 0.260 Ω, 13; 0.270 Ω, 6; 0.280 Ω, 3; 0.290 Ω, 2.

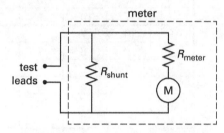

The mean resistance is most nearly

(A) 0.235 Ω

(B) 0.247 Ω

(C) 0.251 Ω

(D) 0.259 Ω

15. Two resistances, a meter resistor and a shunt resistor, are connected in parallel in an ammeter. Most of the current passing through the meter goes through the shunt resistor. In order to determine the accuracy of the resistance of shunt resistors being manufactured for a line of ammeters, a manufacturer tests a sample of 100 shunt resistors. The numbers of shunt resistors with the resistance indicated (to the nearest hundredth of an ohm) are as follows.

0.200 Ω, 1; 0.210 Ω, 3; 0.220 Ω, 5; 0.230 Ω, 10; 0.240 Ω, 17; 0.250 Ω, 40; 0.260 Ω, 13; 0.270 Ω, 6; 0.280 Ω, 3; 0.290 Ω, 2.

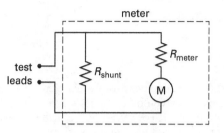

The median resistance is most nearly

(A) 0.22 Ω

(B) 0.24 Ω

(C) 0.25 Ω

(D) 0.26 Ω

16. Two resistances, a meter resistor and a shunt resistor, are connected in parallel in an ammeter. Most of the current passing through the meter goes through the shunt resistor. In order to determine the accuracy of the resistance of shunt resistors being manufactured for a line of ammeters, a manufacturer tests a sample of 100 shunt resistors. The numbers of shunt resistors with the resistance indicated (to the nearest hundredth of an ohm) are as follows.

0.200 Ω, 1; 0.210 Ω, 3; 0.220 Ω, 5; 0.230 Ω, 10; 0.240 Ω, 17; 0.250 Ω, 40; 0.260 Ω, 13; 0.270 Ω, 6; 0.280 Ω, 3; 0.290 Ω, 2.

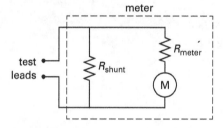

The sample variance is most nearly

(A) 0.00027 Ω^2

(B) 0.0083 Ω^2

(C) 0.0114 Ω^2

(D) 0.0163 Ω^2

17. California law requires a statistical analysis of the average speed driven by motorists on a road prior to the use of radar speed control. The following speeds (all in mph) were observed in a random sample of 40 cars.

44, 48, 26, 25, 20, 43, 40, 42, 29, 39, 23, 26, 24, 47, 45, 28, 29, 41, 38, 36, 27, 44, 42, 43, 29, 37, 34, 31, 33, 30, 42, 43, 28, 41, 29, 36, 35, 30, 32, 31

The upper quartile speed is most nearly

(A) 30 mph
(B) 35 mph
(C) 40 mph
(D) 45 mph

18. California law requires a statistical analysis of the average speed driven by motorists on a road prior to the use of radar speed control. The following speeds (all in mph) were observed in a random sample of 40 cars.

44, 48, 26, 25, 20, 43, 40, 42, 29, 39, 23, 26, 24, 47, 45, 28, 29, 41, 38, 36, 27, 44, 42, 43, 29, 37, 34, 31, 33, 30, 42, 43, 28, 41, 29, 36, 35, 30, 32, 31

The arithmetic mean speed is most nearly

(A) 31 mph
(B) 33 mph
(C) 35 mph
(D) 37 mph

19. California law requires a statistical analysis of the average speed driven by motorists on a road prior to the use of radar speed control. The following speeds (all in mph) were observed in a random sample of 40 cars.

44, 48, 26, 25, 20, 43, 40, 42, 29, 39, 23, 26, 24, 47, 45, 28, 29, 41, 38, 36, 27, 44, 42, 43, 29, 37, 34, 31, 33, 30, 42, 43, 28, 41, 29, 36, 35, 30, 32, 31

The standard deviation of the sample data is most nearly

(A) 2.1 mph
(B) 6.1 mph
(C) 6.8 mph
(D) 7.4 mph

20. California law requires a statistical analysis of the average speed driven by motorists on a road prior to the use of radar speed control. The following speeds (all in mph) were observed in a random sample of 40 cars.

44, 48, 26, 25, 20, 43, 40, 42, 29, 39, 23, 26, 24, 47, 45, 28, 29, 41, 38, 36, 27, 44, 42, 43, 29, 37, 34, 31, 33, 30, 42, 43, 28, 41, 29, 36, 35, 30, 32, 31

The sample variance is most nearly

(A) 60 mi²/hr²
(B) 320 mi²/hr²
(C) 1200 mi²/hr²
(D) 3100 mi²/hr²

SOLUTIONS

1. There are $4! = 24$ different possible outcomes. By enumeration, there are 9 completely wrong combinations. So,

$$p\{\text{all wrong}\} = \frac{9}{24} = 0.375 \quad (0.38)$$

	sizes				
correct →	7	8	9	10	all wrong
sizes issued	7	8	9	10	
	7	8	10	9	
	7	9	8	10	
	7	9	10	8	
	7	10	8	9	
	7	10	9	8	
	8	9	10	7	X
	8	9	7	10	
	8	10	9	7	
	8	10	7	9	X
	8	7	9	10	
	8	7	10	9	X
	9	10	7	8	X
	9	10	8	7	X
	9	7	10	8	X
	9	7	8	10	
	9	8	7	10	
	9	8	10	7	
	10	7	8	9	X
	10	7	9	8	
	10	8	7	9	
	10	8	9	7	
	10	9	8	7	X
	10	9	7	8	X

The answer is (B).

2. If three recruits receive the correct size, the fourth recruit will also receive the correct size since only one pair remains.

$$p\{\text{exactly 3}\} = 0$$

The answer is (A).

3. For an exponential distribution function, the mean is

$$\mu = \frac{1}{\lambda}$$

Using the exponential distribution function with a mean of 23, the average times per toll taken per period is

$$\mu = 23 = \frac{1}{\lambda}$$
$$\lambda = \frac{1}{23} = 0.0435$$

Using the continuous distribution for an exponential distribution function, the function is

$$p = F(x) = 1 - e^{-\lambda x}$$
$$p\{X < x\} = 1 - p$$
$$p\{X > x\} = 1 - F(x)$$
$$= 1 - (1 - e^{-\lambda x})$$
$$= e^{-\lambda x}$$

The probability of a random vehicle being processed in more than 25 sec is

$$p\{x > 25\} = e^{-(0.0435)(25)}$$
$$= 0.337 \quad (0.34)$$

The answer is (C).

4. The distribution is a Poisson distribution with an average of $\lambda = 20$.

The probability for a Poisson distribution is

$$p\{x\} = f(x)$$
$$= \frac{e^{-\lambda}\lambda^x}{x!}$$

Therefore, the probability of 17 cars is

$$p\{x = 17\} = f(17)$$
$$= \frac{e^{-20}20^{17}}{17!}$$
$$= 0.076 \quad (0.08)$$

The answer is (A).

5. The probability for a Poisson distribution is

$$p\{x\} = f(x)$$
$$= \frac{e^{-\lambda}\lambda^x}{x!}$$

The probability of three or fewer cars is

$$p\{x \leq 3\} = p\{x = 0\} + p\{x = 1\} + p\{x = 2\}$$
$$+ p\{x = 3\}$$
$$= f(0) + f(1) + f(2) + f(3)$$
$$= \frac{e^{-20}20^0}{0!} + \frac{e^{-20}20^1}{1!}$$
$$+ \frac{e^{-20}20^2}{2!} + \frac{e^{-20}20^3}{3!}$$
$$= 2 \times 10^{-9} + 4.1 \times 10^{-8}$$
$$+ 4.12 \times 10^{-7} + 2.75 \times 10^{-6}$$
$$= 3.2 \times 10^{-6}$$
$$= 0.0000032 \quad (0.00032\%)$$

The answer is (A).

6. Since the sample population is small, the sample standard deviation is

Dispersion, Mean, Median, and Mode Values

$$s = \sqrt{\left(\frac{1}{n-1}\right)\sum_{i=1}^{n}(X_i - \overline{X})^2}$$

$$= \sqrt{\left(\frac{1}{4-1}\right)\begin{pmatrix}(1249.529 - 1249.439)^2 \\ +(1249.494 - 1249.439)^2 \\ +(1249.384 - 1249.439)^2 \\ +(1249.348 - 1249.439)^2\end{pmatrix}}$$

$$= 0.08647 \quad (0.086)$$

The answer is (D).

7.

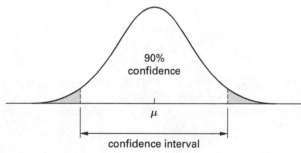

The readings outside any of the two limits aren't acceptable to obtain the upper one-tailed confidence limits of 90%.

The answer is (C).

8. Using the equation for standard deviation, the arithmetic mean is

Dispersion, Mean, Median, and Mode Values

$$\overline{X} = \left(\frac{1}{n}\right)(X_1 + X_2 + \cdots + X_n) = \frac{1}{n}\sum_{i=1}^{n} X_i$$

$$= \left(\frac{1}{4}\right)(1249.529 + 1249.494 + 1249.384 + 1249.348)$$

$$= 1249.439$$

The unbiased estimate of the most probable distance is 1249.439.

The answer is (C).

9. Using the equation for standard deviation, the arithmetic mean is

Dispersion, Mean, Median, and Mode Values

$$\overline{X} = \left(\frac{1}{n}\right)(X_1 + X_2 + \cdots + X_n) = \frac{1}{n}\sum_{i=1}^{n} X_i$$

$$= \left(\frac{1}{4}\right)(1249.529 + 1249.494 + 1249.384 + 1249.348)$$

$$= 1249.439$$

Since the sample population is small, the standard deviation is

Dispersion, Mean, Median, and Mode Values

$$s = \sqrt{\left(\frac{1}{n-1}\right)\sum_{i=1}^{n}(X_i - \overline{X})^2}$$

$$= \sqrt{\left(\frac{1}{4-1}\right)\begin{pmatrix}(1249.529 - 1249.439)^2 \\ +(1249.494 - 1249.439)^2 \\ +(1249.384 - 1249.439)^2 \\ +(1249.348 - 1249.439)^2\end{pmatrix}}$$

$$= 0.08647$$

Using a table for area under the standard normal curve, two-tail 90% confidence limits are located $1.645s$ from $\overline{X}$. Therefore, the error in the most probable value for the 90% confidence range is

$$\text{error} = \pm 1.645s$$
$$= \pm(1.645)(0.08647)$$
$$= \pm 0.142 \quad (0.14)$$

The answer is (C).

10. If the surveying crew places a marker, measures a distance x, places a second marker, and then measures the same distance x back to the original marker, the ending point should coincide with the original marker. If, due to measurement errors, the ending and starting points do not coincide, the difference is the closure error.

In this example, the survey crew moves around the four sides of a square, so there are two measurements in the x-direction and two measurements in the y-direction. If the errors E_1 and E_2 are known for two measurements, x_1 and x_2, the error associated with the sum or difference, $x_1 \pm x_2$, is

$$E\{x_1 \pm x_2\} = \sqrt{E_1^2 + E_2^2}$$

To calculate the total error, the error associated with the confidence interval must be calculated. Using the equation for standard deviation, the arithmetic mean is

Dispersion, Mean, Median, and Mode Values

$$\overline{X} = \left(\frac{1}{n}\right)(X_1 + X_2 + \cdots + X_n) = \frac{1}{n}\sum_{i=1}^{n} X_i$$

$$= \left(\frac{1}{4}\right)(1249.529 + 1249.494 + 1249.384 + 1249.348)$$

$$= 1249.439$$

Since the sample population is small, the standard deviation is

Dispersion, Mean, Median, and Mode Values

$$s = \sqrt{\left(\frac{1}{n-1}\right)\sum_{i=1}^{n}(X_i - \overline{X})^2}$$

$$= \sqrt{\left(\frac{1}{4-1}\right)\begin{pmatrix}(1249.529 - 1249.439)^2 \\ +(1249.494 - 1249.439)^2 \\ +(1249.384 - 1249.439)^2 \\ +(1249.348 - 1249.439)^2\end{pmatrix}}$$

$$= 0.08647$$

Using the table for area under the standard normal curve, two-tail 90% confidence limits are located $1.645s$ from $\overline{X}$. Therefore, the error associated with the confidence interval is

$$\text{error} = \pm 1.645s$$
$$= \pm(1.645)(0.08647)$$
$$= \pm 0.1422$$

The total error in the x-direction is

$$E_x = \sqrt{(0.1422)^2 + (0.1422)^2}$$
$$= 0.2011$$

The error in the y-direction is calculated the same way

and is also 0.2011. E_x and E_y are combined by the Pythagorean theorem to yield the probable closure error.

$$E_{\text{closure}} = \sqrt{(0.2011)^2 + (0.2011)^2}$$
$$= 0.2844 \quad (0.28)$$

The answer is (C).

11. If the surveying crew places a marker, measures a distance x, places a second marker, and then measures the same distance x back to the original marker, the ending point should coincide with the original marker. If, due to measurement errors, the ending and starting points do not coincide, the difference is the closure error.

In this example, the survey crew moves around the four sides of a square, so there are two measurements in the x-direction and two measurements in the y-direction. If the errors E_1 and E_2 are known for two measurements, x_1 and x_2, the error associated with the sum or difference, $x_1 \pm x_2$, is

$$E\{x_1 \pm x_2\} = \sqrt{E_1^2 + E_2^2}$$

To calculate the total error, the error associated with the confidence interval must be calculated. Using the equation for standard deviation, the arithmetic mean is

Dispersion, Mean, Median, and Mode Values

$$\overline{X} = \left(\frac{1}{n}\right)(X_1 + X_2 + \cdots + X_n) = \frac{1}{n}\sum_{i=1}^{n} X_i$$
$$= \left(\frac{1}{4}\right)(1249.529 + 1249.494 + 1249.384 + 1249.348)$$
$$= 1249.439$$

Since the sample population is small, the sample standard deviation is

Dispersion, Mean, Median, and Mode Values

$$s = \sqrt{\left(\frac{1}{n-1}\right)\sum_{i=1}^{n}(X_i - \overline{X})^2}$$

$$= \sqrt{\left(\frac{1}{4-1}\right)\begin{pmatrix}(1249.529 - 1249.439)^2 \\ +(1249.494 - 1249.439)^2 \\ +(1249.384 - 1249.439)^2 \\ +(1249.348 - 1249.439)^2\end{pmatrix}}$$

$$= 0.08647$$

Using the table for area under the standard normal curve table, two-tail 90% confidence limits are located $1.645s$ from $\overline{X}$. Therefore, the error associated with the confidence interval is

$$\text{error} = \pm 1.645s$$
$$= \pm(1.645)(0.08647)$$
$$= \pm 0.1422$$

The total error in the x-direction is

$$E_x = \sqrt{(0.1422)^2 + (0.1422)^2}$$
$$= 0.2011$$

The error in the y-direction is calculated the same way and is also 0.2011. E_x and E_y are combined by the Pythagorean theorem to yield the probable closure error.

$$E_{\text{closure}} = \sqrt{(0.2011)^2 + (0.2011)^2}$$
$$= 0.2844$$

In surveying, error may be expressed as a fraction of one or more legs of the traverse. Assume that the total of all four legs is to be used as the basis.

$$\frac{0.2844}{(4)(1249)} = 1/17{,}567 \quad (1{:}17{,}600)$$

The answer is (A).

12. An experiment is accurate if it is unchanged by experimental error. Precision is concerned with the repeatability of experimental results. If an experiment is repeated with identical results, the experiment is said to be precise. However, it is possible to have a highly precise experiment with a large bias.

The answer is (D).

13. A systematic error is one that is always present and is unchanged from sample to sample. For example, a steel tape that is 0.02 ft too short to measure consecutive distances introduces a systematic error.

The answer is (B).

14. For convenience, tabulate the frequency-weighted values of R and R_2.

R	f	fR	fR^2
0.200	1	0.200	0.0400
0.210	3	0.630	0.1323
0.220	5	1.100	0.2420
0.230	10	2.300	0.5290
0.240	17	4.080	0.9792
0.250	40	10.000	2.5000
0.260	13	3.380	0.8788
0.270	6	1.620	0.4374
0.280	3	0.840	0.2352
0.290	2	0.580	0.1682
	100	24.730	6.1421

The mean resistance is a weighted arithmetic mean.

Dispersion, Mean, Median, and Mode Values

$$\overline{X}_w = \frac{\sum w_i X_i}{\sum w_i} = \overline{R}$$

$$\overline{R} = \frac{24.730\ \Omega}{100}$$

$$= 0.2473\ \Omega \quad (0.247\ \Omega)$$

The answer is (B).

15. The 50th and 51st values are both 0.25 Ω. The median is 0.25 Ω. [Dispersion, Mean, Median, and Mode Values]

The answer is (C).

16. The sample variance is

Dispersion, Mean, Median, and Mode Values

$$s = \sqrt{\frac{\sum fR^2 - \frac{\left(\sum fR^2\right)^2}{n}}{n-1}}$$

$$= \sqrt{\frac{6.1421\ \Omega - \frac{(24.73\ \Omega)^2}{100}}{99}}$$

$$= 0.0163\ \Omega$$

$$s^2 = (0.0163\ \Omega)^2$$

$$= 0.0002658\ \Omega^2 \quad (0.00027\ \Omega^2)$$

The answer is (A).

17. Tabulate the frequency distribution and the cumulative frequency distribution of the data. The lowest speed is 20 mph and the highest speed is 48 mph; therefore, the range is 28 mph. Choose 10 cells with a width of 3 mph.

mid-point	interval (mph)	frequency	cumulative frequency	cumulative percent
21	20–22	1	1	3
24	23–25	3	4	10
27	26–28	5	9	23
30	29–31	8	17	43
33	32–34	3	20	50
36	35–37	4	24	60
39	38–40	3	27	68
42	41–43	8	35	88
45	44–46	3	38	95
48	47–49	2	40	100

To find the upper quartile speed, draw a cumulative frequency graph for the data given.

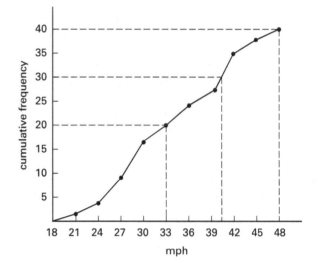

From the cumulative frequency graph, the upper quartile speed occurs at 30 cars or 75%, which corresponds to approximately 40 mph.

The answer is (C).

18. Calculate the following quantities.

$$\sum X_i = 1390\ \text{mi/hr}$$
$$n = 40$$

The arithmetic mean is

Dispersion, Mean, Median, and Mode Values

$$\overline{X} = \frac{1}{n} \sum_{i=1}^{n} X_i$$

$$= \frac{1390\ \frac{\text{mi}}{\text{hr}}}{40}$$

$$= 34.75\ \text{mi/hr} \quad (35\ \text{mph})$$

The answer is (C).

19. Sum the speeds of all 40 cars.

$$\sum X_i = 1390 \text{ mi/hr}$$
$$n = 40$$

The arithmetic mean is

Dispersion, Mean, Median, and Mode Values

$$\overline{X} = \frac{1}{n}\sum_{i=1}^{n} X_i$$

$$\overline{X} = \mu \quad [\text{for sufficiently large values of } n]$$

$$\mu = \frac{1390 \frac{\text{mi}}{\text{hr}}}{40}$$

$$= 34.75 \text{ mi/hr}$$

Using the standard deviation formula, the sample mean can be found. [Dispersion, Mean, Median, and Mode Values]

$$\sigma = \sqrt{\frac{\sum x^2}{N} - u^2}$$

$$\sum x^2 = 50{,}496 \text{ mi}^2/\text{hr}^2$$

Use the sample mean as an unbiased estimator of the population mean, μ.

$$\sigma = \sqrt{\frac{\sum x^2}{N} - \mu^2} = \sqrt{\frac{50{,}496 \frac{\text{mi}^2}{\text{hr}^2}}{40} - \left(34.75 \frac{\text{mi}}{\text{hr}}\right)^2}$$

$$= 7.405 \text{ mi/hr} \quad (7.4 \text{ mph})$$

The answer is (D).

20. The sample standard deviation is

Dispersion, Mean, Median, and Mode Values

$$s = \sqrt{\left(\frac{1}{n-1}\right)\left(\sum_{i=1}^{n}(X_i - \overline{X})^2\right)}$$

$$= \sqrt{\frac{50{,}496 \frac{\text{mi}^2}{\text{hr}^2} - \frac{\left(1390 \frac{\text{mi}}{\text{hr}}\right)^2}{40}}{40 - 1}}$$

$$= 7.5 \text{ mi/hr} \quad (7.5 \text{ mph})$$

The sample variance is given by the square of the sample standard deviation.

[Dispersion, Mean, Median, and Mode Values]

$$s^2 = \left(7.5 \frac{\text{mi}}{\text{hr}}\right)^2$$

$$= 56.25 \text{ mi}^2/\text{hr}^2 \quad (60 \text{ mi}^2/\text{hr}^2)$$

The answer is (A).

12 Numbering Systems

Content in blue refers to the *NCEES Handbook*.

PRACTICE PROBLEMS

1. The binary (base-2) calculation of 100×11 is

(A) 0011
(B) 1000
(C) 1100
(D) 1110

2. The binary (base-2) calculation of 1011×1101 is

(A) 10001111
(B) 11100111
(C) 11110000
(D) 11111111

3. The octal (base-8) calculation of 325×36 is

(A) 11700
(B) 12346
(C) 14366
(D) 14936

4. The octal (base-8) calculation of 3251×161 is

(A) 523411
(B) 550111
(C) 566011
(D) 570231

5. The hexadecimal (base-16) calculation of $FE \times EF$ is

(A) 22FE
(B) ED22
(C) EE11
(D) EF22

6. The hexadecimal (base-16) calculation of $17 \times 7A$ is

(A) 8A
(B) 6AE
(C) AF6
(D) FE8

7. Converting $(734.262)_8$ to a decimal (base-10) number gives

(A) 262.734
(B) 349.674
(C) 448.763
(D) 476.348

8. Converting $(1011.11)_2$ to a decimal (base-10) number gives

(A) 11.75
(B) 14.75
(C) 21.75
(D) 38.75

9. Converting $(0.97)_{10}$ to a binary (base-2) number gives

(A) $(0.101110\cdots)_2$
(B) $(0.111110\cdots)_2$
(C) $(110111.0)_2$
(D) $(111011.1)_2$

10. Converting $(321.422)_8$ to a binary (base-2) number gives

(A) $(001100011.100001000)_2$
(B) $(010010100.001001100)_2$
(C) $(011000101.010010001)_2$
(D) $(011010001.100010010)_2$

SOLUTIONS

1. The binary calculation is

$$\begin{array}{r} 100 \\ \times\ 11 \\ \hline 100 \\ 100 \\ \hline 1100 \end{array}$$

The answer is (C).

2. The binary calculation is

$$\begin{array}{r} 1011 \\ \times\ 1101 \\ \hline 1011 \\ 0000 \\ 1011 \\ 1011 \\ \hline 10001111 \end{array}$$

The answer is (A).

3. The octal calculation is

$$\begin{array}{r} 325 \\ \times\ \ \ 36 \\ \hline 2376 \\ 1177 \\ \hline 14366 \end{array}$$

The answer is (C).

4. The octal calculation is

$$\begin{array}{r} 3251 \\ \times\ \ \ 161 \\ \hline 3251 \\ 23766 \\ 3251 \\ \hline 570231 \end{array}$$

The answer is (D).

5. The hexadecimal calculation is

$$\begin{array}{r} FE \\ \times\ EF \\ \hline EE2 \\ DE4 \\ \hline ED22 \end{array}$$

The answer is (B).

6. The hexadecimal calculation is

$$\begin{array}{r} 17 \\ \times\ 7A \\ \hline E6 \\ A1 \\ \hline AF6 \end{array}$$

The answer is (C).

7. Converting,

$$(734.262)_8 = (7)(8)^2 + (3)(8)^1 + (4)(8)^0$$
$$+ (2)(8)^{-1} + (6)(8)^{-2} + (2)(8)^{-3}$$
$$= 476.348$$

The answer is (D).

8. Converting,

$$(1011.11)_2 = (1)(2)^3 + (0)(2)^2 + (1)(2)^1$$
$$+ (1)(2)^0 + (1)(2)^{-1} + (1)(2)^{-2}$$
$$= 11.75$$

The answer is (A).

9. Convert to a binary number.

$$\begin{array}{ll} 0.97 \times 2 = 1 & \text{remainder } 0.94 \\ 0.94 \times 2 = 1 & \text{remainder } 0.88 \\ 0.88 \times 2 = 1 & \text{remainder } 0.76 \\ 0.76 \times 2 = 1 & \text{remainder } 0.52 \\ 0.52 \times 2 = 1 & \text{remainder } 0.04 \\ 0.04 \times 2 = 0 & \text{remainder } 0.08 \\ & \vdots \end{array}$$

$$(0.97)_{10} = (0.111110\cdots)_2$$

The answer is (B).

10. Since $8 = (2)^3$, convert each octal digit into its binary equivalent.

$$321.422:\quad \begin{array}{l} 3 = 011 \\ 2 = 010 \\ 1 = 001 \\ 4 = 100 \\ 2 = 010 \\ 2 = 010 \end{array}$$

$$(321.422)_8 = (011010001.100010010)_2$$

The answer is (D).

13 Numerical Analysis

Content in blue refers to the *NCEES Handbook*.

PRACTICE PROBLEMS

1. A function is given as $y = 3x^{0.93} + 4.2$. What is most nearly the percent relative error if the value of y at $x = 2.7$ is found by using straight-line interpolation between $x = 2$ and $x = 3$?

(A) 0.06%
(B) 0.18%
(C) 2.5%
(D) 5.4%

2. Given the following data points, estimate y by straight-line interpolation for $x = 2.75$.

x	y
1	4
2	6
3	2
4	−14

(A) 2.1
(B) 2.4
(C) 2.7
(D) 3.0

3. Using the bisection method, find all of the roots of $f(x) = 0$ to the nearest 0.000005.

$$f(x) = x^3 + 2x^2 + 8x - 2$$

4. The increase in concentration of mixed-liquor suspended solids (MLSS) in an activated sludge aeration tank as a function of time is given in the table. Use a second-order Lagrangian interpolation to estimate the MLSS after 16 min of aeration.

t (min)	MLSS (mg/L)
0	0
10	227
15	362
20	517
22.5	602
30	901

(A) 350 mg/L
(B) 390 mg/L
(C) 540 mg/L
(D) 640 mg/L

SOLUTIONS

1. The actual value at $x = 2.7$ is

$$y(x) = 3x^{0.93} + 4.2$$
$$y(2.7) = (3)(2.7)^{0.93} + 4.2$$
$$= 11.756$$

At $x = 3$,

$$y(3) = (3)(3)^{0.93} + 4.2 = 12.534$$

At $x = 2$,

$$y(2) = (3)(2)^{0.93} + 4.2 = 9.916$$

Use straight-line interpolation.

$$\frac{x_2 - x}{x_2 - x_1} = \frac{y_2 - y}{y_2 - y_1}$$
$$\frac{3 - 2.7}{3 - 2} = \frac{12.534 - y}{12.534 - 9.916}$$
$$y = 11.749$$

The relative error is

$$\frac{\text{actual value} - \text{predicted value}}{\text{actual value}} = \frac{11.756 - 11.749}{11.756}$$
$$= 0.0006 \quad (0.06\%)$$

The answer is (A).

2. Let $x_1 = 2$; therefore, from the table of data points, $y_1 = 6$. Let $x_2 = 3$; therefore, from the table of data points, $y_2 = 2$.

Let $x = 2.75$. By straight-line interpolation,

$$\frac{x_2 - x}{x_2 - x_1} = \frac{y_2 - y}{y_2 - y_1}$$
$$\frac{3 - 2.75}{3 - 2} = \frac{2 - y}{2 - 6}$$
$$y = 3.0$$

The answer is (D).

3. Use the equation $f(x) = x^3 + 2x^2 + 8x - 2$ to try to find an interval in which there is a root.

x	$f(x)$
0	-2
1	9

A root exists in the interval $[0, 1]$.

Try $x = \left(\frac{1}{2}\right)(0 + 1) = 0.5$.

$$f(0.5) = (0.5)^3 + (2)(0.5)^2 + (8)(0.5) - 2 = 2.625$$

A root exists in $[0, 0.5]$.

Try $x = 0.25$.

$$f(0.25) = (0.25)^3 + (2)(0.25)^2 + (8)(0.25) - 2 = 0.1406$$

A root exists in $[0, 0.25]$.

Try $x = 0.125$.

$$f(0.125) = (0.125)^3 + (2)(0.125)^2 + (8)(0.125) - 2$$
$$= -0.967$$

A root exists in $[0.125, 0.25]$.

Try $x = \left(\frac{1}{2}\right)(0.125 + 0.25) = 0.1875$.

Continuing,

$$f(0.1875) = -0.42 \quad [0.1875, 0.25]$$
$$f(0.21875) = -0.144 \quad [0.21875, 0.25]$$
$$f(0.234375) = -0.002 \quad [\text{This is close enough.}]$$

One root is $x_1 \approx 0.234375 \ (0.234)$.

Try to find the other two roots. Use long division to factor the polynomial.

$$\begin{array}{r}
x^2 + 2.234375x + 8.52368 \\
x - 0.234375 \overline{\smash{\big)}\ x^3 + 2x^2 + 8x - 2} \\
\underline{-(x^3 - 0.234375x^2)} \\
2.234375x^2 + 8x - 2 \\
\underline{-(2.234375x^2 - 0.52368x)} \\
8.52368x - 2 \\
\underline{-(8.52368x - 1.9977)} \\
\approx 0
\end{array}$$

Use the quadratic equation to find the roots of $x^2 + 2.234375x + 8.52368$.

$$x_2, x_3 = \frac{-2.234375 \pm \sqrt{(2.234375)^2 - (4)(1)(8.52368)}}{(2)(1)}$$
$$= -1.117189 \pm i2.697327 \quad [\text{both imaginary}]$$

4. Choose the three data points that bracket $t=16$ as closely as possible. These three points are $t_0 = 10, t_1 = 15,$ and $t_2 = 20$.

	$i = 0$	$i = 1$	$i = 2$
$k = 0$: $S_0(16) = -0.08$	~~$\left(\dfrac{16-10}{10-10}\right)$~~	$\left(\dfrac{16-15}{10-15}\right)$	$\left(\dfrac{16-20}{10-20}\right)$
$k = 1$: $S_1(16) = 0.96$	$\left(\dfrac{16-10}{15-10}\right)$	~~$\left(\dfrac{16-15}{15-15}\right)$~~	$\left(\dfrac{16-20}{15-20}\right)$
$k = 2$: $S_2(16) = 0.12$	$\left(\dfrac{16-10}{20-10}\right)$	$\left(\dfrac{16-15}{20-15}\right)$	~~$\left(\dfrac{16-20}{20-20}\right)$~~

Use Eq. 13.3.

$$\begin{aligned}
S &= S(10)S_0(16) + S(15)S_1(16) + S(20)S_2(16) \\
&= \left(227\,\frac{\text{mg}}{\text{L}}\right)(-0.08) + \left(362\,\frac{\text{mg}}{\text{L}}\right)(0.96) \\
&\quad + \left(517\,\frac{\text{mg}}{\text{L}}\right)(0.12) \\
&= 391 \text{ mg/L} \quad (390 \text{ mg/L})
\end{aligned}$$

The answer is (B).

14 Fluid Properties

Content in blue refers to the *NCEES Handbook*.

PRACTICE PROBLEMS

(Use $g = 32.2$ ft/sec² or 9.81 m/s² unless told to do otherwise in the problem.)

1. Atmospheric pressure is 14.7 lbf/in² (101.3 kPa). What is most nearly the absolute pressure in a tank if a gauge on the tank reads 8.7 lbf/in² (60 kPa) vacuum?

(A) 4 psia (27 kPa)

(B) 6 psia (41 kPa)

(C) 8 psia (55 kPa)

(D) 10 psia (68 kPa)

2. The absolute viscosity of nitrogen at 80°F (27°C) and 70 psia (480 kPa) is 1.206×10^{-5} lbm/ft-s (1.795×10^{-5} Pa·s). The kinematic viscosity of nitrogen at the same temperature and pressure is most nearly

(A) 3.6×10^{-5} ft²/sec (3.0×10^{-6} m²/s)

(B) 4.0×10^{-5} ft²/sec (4.0×10^{-6} m²/s)

(C) 5.0×10^{-5} ft²/sec (5.0×10^{-6} m²/s)

(D) 6.0×10^{-5} ft²/sec (6.0×10^{-6} m²/s)

3. Three solutions of nitric acid are combined: one with 8% nitric acid by volume, one with 10% nitric acid by volume, and one with 20% nitric acid by volume. The combined solutions produce 100 mL of a solution that is 12% nitric acid by volume. The 8% solution contributes half of the total volume of the nitric acid contributed by the 10% and 20% solutions. The volume of 10% solution in the 12% solution is most nearly

(A) 20 mL

(B) 30 mL

(C) 50 mL

(D) 80 mL

4. A mixture of a 30% (by volume) ethylene glycol aqueous solution and water is used in a solar heating application. The components of the mixture are nonreacting. The ethylene glycol solution is at standard atmospheric pressure and at an average temperature of 140°F. The specific gravity of the solution is most nearly

(A) 1.01

(B) 1.02

(C) 1.03

(D) 1.04

5. The absolute viscosity of an unknown fluid at atmospheric pressure is measured at three temperatures.

temperature (°C)	absolute viscosity (N·s/m²)
0	0.0018
30	0.001
60	0.0006

Of the following, the unknown fluid is most likely to be

(A) water

(B) benzene

(C) ethyl alcohol

(D) carbon tetrachloride

SOLUTIONS

1. Customary U.S. Solution

Absolute pressure is equal to atmospheric pressure minus the vacuum gauge pressure reading. [Pressure Field in a Static Liquid]

$$p_{absolute} = p_{atmospheric} - p_{vacuum}$$
$$= 14.7 \, \frac{lbf}{in^2} - 8.7 \, \frac{lbf}{in^2}$$
$$= 6.0 \, lbf/in^2 \quad (6 \, psia)$$

SI Solution

Absolute pressure is equal to atmospheric pressure minus the vacuum gauge pressure reading. [Pressure Field in a Static Liquid]

$$p_{absolute} = p_{atmospheric} - p_{vacuum}$$
$$= 101.3 \, kPa - 60 \, kPa$$
$$= 41.3 \, kPa \quad (41 \, kPa)$$

The answer is (B).

2. Customary U.S. Solution

Find the absolute temperature of the nitrogen. [Temperature Conversions]

$$T = 80°F + 460° = 540°R$$

Use the ideal gas law and solve for density, ρ, which is the reciprocal of specific volume, v.

Ideal Gas

$$pv = RT$$
$$\rho = \frac{1}{v} = \frac{p}{RT}$$

Substitute this expression for density into the equation for kinematic viscosity. The specific gas constant for nitrogen is 55.16 ft-lbf/lbm-°R. [Thermal and Physical Properties of Ideal Gases (at Room Temperature)]

Stress, Pressure, and Viscosity

$$\nu = \frac{\mu}{\rho}$$
$$= \frac{\mu R T}{p}$$
$$= \frac{\left(1.206 \times 10^{-5} \, \frac{lbm}{ft\text{-}sec}\right) \times \left(55.16 \, \frac{ft\text{-}lbf}{lbm\text{-}°R}\right)(540°R)}{\left(70 \, \frac{lbf}{in^2}\right)\left(12 \, \frac{in}{ft}\right)^2}$$
$$= 3.56 \times 10^{-5} \, ft^2/sec \quad (3.6 \times 10^{-5} \, ft^2/sec)$$

SI Solution

Find the absolute temperature of the nitrogen. [Temperature Conversions]

$$T = 27°C + 273° = 300K$$

Use the ideal gas law and solve for density, ρ, which is the reciprocal of specific volume, v.

Ideal Gas

$$pv = RT$$
$$\rho = \frac{1}{v} = \frac{p}{RT}$$

Substitute this expression for density into the equation for kinematic viscosity. The specific gas constant for nitrogen is 0.2968 kJ/kg·K. [Thermal and Physical Properties of Ideal Gases (at Room Temperature)]

Stress, Pressure, and Viscosity

$$\nu = \frac{\mu}{\rho}$$
$$= \frac{\mu R T}{p}$$
$$= \frac{(1.795 \times 10^{-5} \, Pa·s) \times \left(0.2968 \, \frac{kJ}{kg·K}\right)\left(1000 \, \frac{J}{kJ}\right) \times (300K)}{(480 \, kPa)\left(1000 \, \frac{Pa}{kPa}\right)}$$
$$= 3.33 \times 10^{-6} \, m^2/s \quad (3.0 \times 10^{-6} \, m^2/s)$$

The answer is (A).

3. Let

x = volume of 8% solution
$0.08x$ = volume of nitric acid contributed by 8% solution
y = volume of 10% solution
$0.10y$ = volume of nitric acid contributed by 10% solution
z = volume of 20% solution
$0.20z$ = volume of nitric acid contributed by 20% solution

The three conditions that must be satisfied are

$$x + y + z = 100 \text{ mL}$$
$$0.08x + 0.10y + 0.20z = (0.12)(100 \text{ mL}) = 12 \text{ mL}$$
$$0.08x = \frac{1}{2}(0.10y + 0.20z)$$

Simplifying these equations,

$$x + y + z = 100$$
$$4x + 5y + 10z = 600$$
$$8x - 5y - 10z = 0$$

Adding the second and third equations gives

$$12x = 600$$
$$x = 50 \text{ mL}$$

Work with the first two equations to get

$$y + z = 100 - 50 = 50$$
$$5y + 10z = 600 - (4)(50) = 400$$

Multiplying the top equation by -5 and adding to the bottom equation,

$$5z = 150$$
$$z = 30 \text{ mL}$$

From the first equation,

$$y = 20 \text{ mL}$$

The answer is (A).

4. Find the density of the ethylene glycol aqueous solution at 140°F. [Density of Aqueous Solutions of Ethylene Glycol]

$$\rho_{\text{sol}} = 64.15 \text{ lbm/ft}^3$$

Find the density of water at standard conditions. [Properties of Water at Standard Conditions]

$$\rho_{\text{water}} = 62.4 \text{ lbm/ft}^3$$

The specific gravity of the solution is

$$\text{SG} = \frac{\rho_{\text{sol}}}{\rho_{\text{water}}} = \frac{64.15 \frac{\text{lbm}}{\text{ft}^3}}{62.4 \frac{\text{lbm}}{\text{ft}^3}}$$
$$= 1.028 \quad (1.03)$$

The answer is (C).

5. Use a graph of the absolute viscosities of fluids at various temperatures. [Absolute Viscosity (Left) and Kinematic Viscosity (Right) of Common Fluids at 1 atm]

At 0°C, benzene has an absolute viscosity of about 0.0009 N·s/m², and carbon tetrachloride has an absolute viscosity of about 0.0014 N·s/m². Both water and ethyl alcohol have absolute viscosities of about 0.0018 N·s/m², so the unknown fluid is more likely to be water or ethyl alcohol.

At 60°C, water has an absolute viscosity of about 0.0005 N·s/m², and ethyl alcohol has an absolute viscosity of about 0.0006 N·s/m². The unknown fluid is most likely to be ethyl alcohol.

The answer is (C).

15 Fluid Statics

Content in blue refers to the *NCEES Handbook*.

PRACTICE PROBLEMS

(Use $g = 32.2$ ft/sec^2 or 9.81 m/s^2 unless told to do otherwise in the problem.)

1. A 4000 lbm (1800 kg) blimp contains 10,000 lbm (4500 kg) of hydrogen (specific gas constant = 766.5 ft-lbf/lbm-°R (4124 J/kg·K)) at 56°F (13°C) and 30.2 in Hg (770 mm Hg). If the hydrogen and air are in thermal and pressure equilibrium, what is most nearly the blimp's lift (net upward force)?

(A) 7.6×10^3 lbf (3.4×10^4 N)

(B) 1.2×10^4 lbf (5.3×10^4 N)

(C) 1.3×10^5 lbf (5.7×10^5 N)

(D) 1.7×10^5 lbf (7.7×10^5 N)

2. A hollow 6 ft (1.8 m) diameter sphere floats half-submerged in seawater (the density of the seawater is 64.0 lbm/ft^3 (1025 kg/m^3)). The mass of concrete (density = 150 lbm/ft^3 (2400 kg/m^3)) that is required as an external anchor to just submerge the sphere completely is most nearly

(A) 2700 lbm (1200 kg)

(B) 4200 lbm (1900 kg)

(C) 5500 lbm (2500 kg)

(D) 6300 lbm (2700 kg)

3. Water removed from Lake Superior (elevation, 601 ft above mean sea level; water density, 62.4 lbm/ft^3) is transported by tanker ship to the Atlantic Ocean (elevation, 0 ft) through 16 Seaway locks. A tanker's displacement is 32,000 tonnes when loaded, and 5100 tonnes when empty. Each lock is 766 ft long and 80 ft wide. Water pumped from each lock flows to the Atlantic Ocean. Compared to a passage from Lake Superior to the Atlantic Ocean when empty, what is most nearly the change in water loss from Lake Superior when a ship passes through the locks fully loaded?

(A) 27,000 tonnes less loss

(B) no change in loss

(C) 27,000 tonnes additional loss

(D) 54,000 tonnes additional loss

4. A manometer is set up to measure the pressure of 70°F (21°C) water in a tank. The setup and readings for the manometer are shown in the illustration. The manometer uses mercury and SAE 50W oil. The SAE 50W oil has a specific weight of 56.25 lbf/ft^3 (8.843 kN/m^3). The gage pressure at the oil level line is 25 psi (172.37 kN/m^2). What is the pressure of the water in the tank?

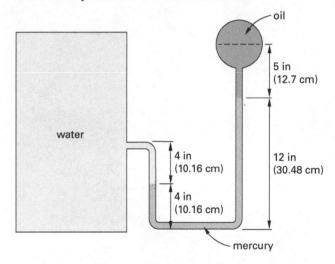

(A) 25 psi (175 kN/m^2)

(B) 29 psi (199 kN/m^2)

(C) 40 psi (281 kN/m^2)

(D) 47 psi (327 kN/m^2)

5. A small gate with a thickness of 12 in (30.5 cm) is placed on a dam with the gate bottom so that the bottom of the gate is at a depth of 32.8 ft (10 m) below the surface of a reservoir, at 50°F (10°C). The reservoir is at sea level. The gate is vertical in the water (i.e., it does not have a slope). The pressure of air at sea level is 1 atm (101.3 kPa). The maximum pressure exerted on any part of the gate is most nearly

(A) 26 psi (180 kPa)

(B) 29 psi (200 kPa)

(C) 32 psi (220 kPa)

(D) 35 psi (240 kPa)

SOLUTIONS

1. *Customary U.S. Solution*

The lift (lifting force) of the hydrogen-filled blimp, F_{lift}, is equal to the difference between the buoyant force, F_b, and the weight of the hydrogen contained in the blimp, W_{H}.

$$F_{\text{lift}} = F_b - W_{\text{H}} - W_{\text{blimp}}$$

The weight of the hydrogen is calculated from the mass of hydrogen.

Newton's Second Law (Equations of Motion)

$$W_{\text{H}} = \frac{m_{\text{H}} g}{g_c} = \frac{(10{,}000 \text{ lbm})\left(32.2 \, \dfrac{\text{ft}}{\text{sec}^2}\right)}{32.2 \, \dfrac{\text{lbm-ft}}{\text{lbf-sec}^2}}$$

$$= 10{,}000 \text{ lbf}$$

The buoyant force is equal to the weight of the displaced air. The volume of the air displaced is equal to the volume of hydrogen enclosed in the blimp.

The absolute temperature of the hydrogen is

Temperature Conversions

$$°R = °F + 459.69° = T$$
$$T = 56°F + 459.69°$$
$$= 515.69°R$$

Find the pressure of the hydrogen. [Measurement Relationships]

$$p = \frac{(30.2 \text{ in Hg})\left(12 \, \dfrac{\text{in}}{\text{ft}}\right)^2}{2.036 \, \dfrac{\text{in Hg}}{\dfrac{\text{lbf}}{\text{in}^2}}} = 2136 \text{ lbf/ft}^2$$

Compute the volume of hydrogen from the ideal gas law equation. [Thermal and Physical Properties of Ideal Gases (at Room Temperature)]

Ideal Gas

$$pV_{\text{H}} = m_{\text{H}} RT$$
$$V_{\text{H}} = \frac{m_{\text{H}} RT}{p}$$

$$= \frac{(10{,}000 \text{ lbm})\left(766.5 \, \dfrac{\text{ft-lbf}}{\text{lbm-°R}}\right)(515.69°R)}{2136 \, \dfrac{\text{lbf}}{\text{ft}^2}}$$

$$= 1.85 \times 10^6 \text{ ft}^3$$

Since the volume of the hydrogen contained in the blimp is equal to the air displaced, the air displaced can be computed from the ideal gas equation. Since the air and hydrogen are in thermal and pressure equilibrium, the temperature and pressure are equal to the values given for the hydrogen.

For air, $R = 53.35$ ft-lbf/lbm-°R. [Thermal and Physical Properties of Ideal Gases (at Room Temperature)]

$$m_{air} = \frac{pV_H}{RT} = \frac{\left(2136\,\dfrac{\text{lbf}}{\text{ft}^2}\right)(1.85 \times 10^6\text{ ft}^3)}{\left(53.35\,\dfrac{\text{ft-lbf}}{\text{lbm-°R}}\right)(515.69\text{°R})}$$

$$= 1.436 \times 10^5 \text{ lbm}$$

The buoyant force is equal to the weight of the air.

Newton's Second Law (Equations of Motion)

$$F_b = W_{air} = \frac{m_{air}g}{g_c}$$

$$= \frac{(1.436 \times 10^5\text{ lbm})\left(32.2\,\dfrac{\text{ft}}{\text{sec}^2}\right)}{32.2\,\dfrac{\text{lbm-ft}}{\text{lbf-sec}^2}}$$

$$= 1.436 \times 10^5 \text{ lbf}$$

The lift (lifting force) is

$$F_{lift} = F_b - W_H - W_{blimp}$$
$$= 1.436 \times 10^5 \text{ lbf} - 10{,}000 \text{ lbf} - 4000 \text{ lbf}$$
$$= 1.296 \times 10^5 \text{ lbf} \quad (1.3 \times 10^5 \text{ lbf})$$

SI Solution

The lift (lifting force) of the hydrogen-filled blimp, F_{lift}, is equal to the difference between the buoyant force, F_b, and the weight of the hydrogen contained in the blimp, W_H.

$$F_{lift} = F_b - W_H - W_{blimp}$$

The weight of the hydrogen is calculated from the mass of hydrogen.

Newton's Second Law (Equations of Motion)

$$W_H = m_H g$$
$$= (4500 \text{ kg})\left(9.81\,\dfrac{\text{m}}{\text{s}^2}\right)$$
$$= 44\,145 \text{ N}$$

The buoyant force is equal to the weight of the displaced air. The volume of the air displaced is equal to the volume of hydrogen enclosed in the blimp.

The absolute temperature of the hydrogen is

Temperature Conversions

$$K = °C + 273.15° = T$$
$$T = 13°C + 273.15°$$
$$= 286.15 K$$

The absolute pressure of the hydrogen is

$$p = \frac{(770 \text{ mm Hg})\left(133.4\,\dfrac{\text{kPa}}{\text{m}}\right)}{1000\,\dfrac{\text{mm}}{\text{m}}}$$

$$= 102.7 \text{ kPa}$$

Compute the volume of hydrogen from the ideal gas law.

Ideal Gas

$$pV_H = m_H RT$$

$$V_H = \frac{m_H RT}{p}$$

$$= \frac{(4500 \text{ kg})\left(4124\,\dfrac{\text{J}}{\text{kg}\cdot\text{K}}\right)(286.15\text{K})}{(102.7 \text{ kPa})\left(1000\,\dfrac{\text{Pa}}{\text{kPa}}\right)}$$

$$= 5.17 \times 10^4 \text{ m}^3$$

Since the volume of the hydrogen contained in the blimp is equal to the air displaced, the air displaced can be computed from the ideal gas equation. Since the air and hydrogen are assumed to be in thermal and pressure equilibrium, the temperature and pressure are equal to the values given for the hydrogen.

For air, $R = 287.03$ J/kg·K. [Thermal and Physical Properties of Ideal Gases (at Room Temperature)]

$$m_{air} = \frac{pV_H}{RT}$$

$$= \frac{(102.7 \text{ kPa})\left(1000\,\dfrac{\text{Pa}}{\text{kPa}}\right)(5.17 \times 10^4 \text{ m}^3)}{\left(287.03\,\dfrac{\text{J}}{\text{kg}\cdot\text{K}}\right)(286.15\text{K})}$$

$$= 6.464 \times 10^4 \text{ kg}$$

The buoyant force is equal to the weight of the air.

Newton's Second Law (Equations of Motion)

$$F_b = W_{air} = m_{air} g$$

$$= (6.464 \times 10^4 \text{ kg})\left(9.81\,\dfrac{\text{m}}{\text{s}^2}\right)$$

$$= 6.34 \times 10^5 \text{ N}$$

The lift (lifting force) is

$$F_{\text{lift}} = F_b - W_H - W_{\text{blimp}}$$
$$= 6.34 \times 10^5 \text{ N} - 44\,145 \text{ N} - (1800 \text{ kg})\left(9.81 \frac{\text{m}}{\text{s}^2}\right)$$
$$= 5.7 \times 10^5 \text{ N}$$

The answer is (C).

2. *Customary U.S. Solution*

The weight of the sphere is equal to the weight of the displaced volume of water when floating.

The buoyant force is given by

Archimedes' Principle and Buoyancy

$$F_{\text{buoyant}} = \gamma V_{\text{displaced}}$$
$$\gamma = \rho g$$
$$F_b = \frac{\rho g V_{\text{displaced}}}{g_c}$$

Since the sphere is half submerged, the weight of the sphere is

$$W_{\text{sphere}} = \frac{1}{2}\left(\frac{\rho g V_{\text{sphere}}}{g_c}\right)$$

The volume of the sphere is

Mensuration of Areas and Volumes: Nomenclature

$$V_{\text{sphere}} = \frac{\pi}{6}D^3 = \left(\frac{\pi}{6}\right)(6 \text{ ft})^3$$
$$= 113.1 \text{ ft}^3$$

The weight of the sphere is

$$W_{\text{sphere}} = \frac{1}{2}\left(\frac{\rho g V_{\text{sphere}}}{g_c}\right)$$
$$= \left(\frac{1}{2}\right)\left(\frac{\left(64.0 \frac{\text{lbm}}{\text{ft}^3}\right)\left(32.2 \frac{\text{ft}}{\text{sec}^2}\right)(113.1 \text{ ft}^3)}{32.2 \frac{\text{lbm-ft}}{\text{lbf-sec}^2}}\right)$$
$$= 3619 \text{ lbf}$$

The equilibrium equation for a fully submerged sphere and anchor can be solved for the concrete volume.

$$W_{\text{sphere}} + W_{\text{concrete}} = (V_{\text{sphere}} + V_{\text{concrete}})\rho_{\text{water}}$$
$$W_{\text{sphere}} + \rho_{\text{concrete}} V_{\text{concrete}}\left(\frac{g}{g_c}\right)$$
$$= (V_{\text{sphere}} + V_{\text{concrete}})\rho_{\text{water}}\left(\frac{g}{g_c}\right)$$

$$3619 \text{ lbf} + \left(150 \frac{\text{lbm}}{\text{ft}^3}\right)V_{\text{concrete}}\left(\frac{32.2 \frac{\text{ft}}{\text{sec}^2}}{32.2 \frac{\text{lbm-ft}}{\text{lbf-sec}^2}}\right)$$
$$= (113.1 \text{ ft}^3 + V_{\text{concrete}})\left(64.0 \frac{\text{lbm}}{\text{ft}^3}\right)\left(\frac{32.2 \frac{\text{ft}}{\text{sec}^2}}{32.2 \frac{\text{lbm-ft}}{\text{lbf-sec}^2}}\right)$$

$$V_{\text{concrete}} = 42.09 \text{ ft}^3$$
$$m_{\text{concrete}} = \rho_{\text{concrete}} V_{\text{concrete}}$$
$$= \left(150 \frac{\text{lbm}}{\text{ft}^3}\right)(42.09 \text{ ft}^3)$$
$$= 6314 \text{ lbm} \quad (6300 \text{ lbm})$$

SI Solution

The weight of the sphere is equal to the weight of the displaced volume of water when floating.

The buoyant force is given by

Archimedes' Principle and Buoyancy

$$F_{\text{buoyant}} = \gamma V_{\text{displaced}}$$
$$\gamma = \rho g$$
$$F_b = \rho g V_{\text{displaced}}$$

Since the sphere is half submerged, the weight of the sphere is

$$W_{\text{sphere}} = \frac{1}{2}\rho g V_{\text{sphere}}$$

The volume of the sphere is

Mensuration of Areas and Volumes: Nomenclature

$$V_{\text{sphere}} = \frac{\pi}{6}D^3$$
$$= \left(\frac{\pi}{6}\right)(1.8 \text{ m})^3$$
$$= 3.054 \text{ m}^3$$

The weight of the sphere is

$$W_{sphere} = \frac{1}{2}\rho g V_{sphere}$$
$$= \left(\frac{1}{2}\right)\left(1025 \, \frac{kg}{m^3}\right)\left(9.81 \, \frac{m}{s^2}\right)(3.054 \, m^3)$$
$$= 15\,354 \, N$$

The equilibrium equation for a fully submerged sphere and anchor can be solved for the concrete volume.

$$W_{sphere} + W_{concrete} = g(V_{sphere} + V_{concrete})\rho_{water}$$
$$W_{sphere} + \rho_{concrete}gV_{concrete} = g(V_{sphere} + V_{concrete})\rho_{water}$$
$$15\,354 \, N + \left(2400 \, \frac{kg}{m^3}\right)\left(9.81 \, \frac{m}{s^2}\right)V_{concrete}$$
$$= \left(9.81 \, \frac{m}{s^2}\right)(3.054 \, m^3 + V_{concrete})\left(1025 \, \frac{kg}{m^3}\right)$$
$$V_{concrete} = 1.138 \, m^3$$
$$m_{concrete} = \rho_{concrete}V_{concrete}$$
$$= \left(2400 \, \frac{kg}{m^3}\right)(1.138 \, m^3)$$
$$= 2731 \, kg \quad (2700 \, kg)$$

The answer is (D).

3. From Archimedes' principle, each tonne of water carried in the tanker displaces a tonne of water in the lock. So, each tonne of water transported out of the lake results in a tonne less of lock loss. Compared to an empty tanker, the net result is zero (no change in loss). [Archimedes' Principle and Buoyancy]

The answer is (B).

4. The specific weight of water at 70°F is 62.3 lbf/ft³. The density of mercury is 845 lbm/ft³. [Properties of Water (I-P Units)] [Properties of Metals - I-P Units]

Calculate the specific weight of mercury.

$$\gamma_{mercury} = \rho_{mercury} \frac{g}{g_c}$$
$$= \frac{\left(845 \, \frac{lbm}{ft^3}\right)\left(32.2 \, \frac{ft}{sec^2}\right)}{\left(32.2 \, \frac{lbm\text{-}ft}{lbf\text{-}sec^2}\right)}$$
$$= 845 \, lbf/ft^3$$

Calculate the pressure of the flowing water.

Manometers

$$p_{gage} = p_{water} + \gamma_{water}h_{water} - \gamma_{mercury}h_{mercury} - \gamma_{oil}h_{oil}$$
$$p_{water} = p_{gage} - \gamma_{water}h_{water} + \gamma_{mercury}h_{mercury} + \gamma_{oil}h_{oil}$$
$$= 25 \, \frac{lbf}{in^2} - \left(62.3 \, \frac{lbf}{ft^3}\right)(4 \, in)\left(\frac{1 \, ft}{12 \, in}\right) + \left(845 \, \frac{lbf}{ft^3}\right)$$
$$\times (8 \, in)\left(\frac{1 \, ft}{12 \, in}\right) + \left(56.25 \, \frac{lbf}{ft^3}\right)(5 \, in)\left(\frac{1 \, ft}{12 \, in}\right)$$
$$= 25 \, \frac{lbf}{in^2} + \left(566.0 \, \frac{lbf}{ft^2}\right)\left(\frac{1 \, \frac{lbf}{in^2}}{144 \, \frac{lbf}{ft^2}}\right)$$
$$= 28.93 \, lbf/in^2 \quad (29 \, psi)$$

SI Solution

The specific weight of water at 21°C is 9.787 kN/m³. The density of mercury is 13 547 kg/m³. [Properties of Water (SI Units)] [Properties of Metals - SI Units]

Calculate the specific weight of mercury.

$$\gamma_{mercury} = \rho_{mercury} \, g$$
$$= \left(13547 \, \frac{kg}{m^3}\right)\left(9.81 \, \frac{m}{s^2}\right)$$
$$= 132.896 \times 10^3 \, N/m^3$$
$$= 132.896 \, kN/m^3$$

Calculate the pressure of the fluid flow.

Manometers

$$p_{gage} = p_{water} + \gamma_{water}h_{water} - \gamma_{mercury}h_{mercury} - \gamma_{oil}h_{oil}$$
$$p_{water} = p_{gage} - \gamma_{water}h_{water} + \gamma_{mercury}h_{mercury} + \gamma_{oil}h_{oil}$$
$$= 172.37 \, \frac{kN}{m^2} - \left(9.787 \, \frac{kN}{m^3}\right)(0.1016 \, m) + \left(132.896 \, \frac{kN}{m^3}\right)(0.2032 \, m)$$
$$+ \left(8.843 \, \frac{kN}{m^3}\right)(0.127 \, m)$$
$$= 199.50 \, kN/m^2 \quad (199 \, kN/m^2)$$

The answer is (B).

5. *U.S. Customary Solution*

The density of water at atmospheric pressure at 50°F is 1.940 lbf-sec²/ft⁴. The pressure of air at sea level is 1 atm. [Properties of Water (I-P Units)]

The pressure on the gate is greatest at the bottom edge. The pressure at this depth is

Forces on Submerged Surfaces and the Center of Pressure

$$p = p_o + \rho g h$$

Since the gate is vertical, the angle θ is 90°.

$$h = 32.8 \text{ ft}$$

$$p = 14.7 \, \frac{\text{lbf}}{\text{in}^2} + \left(1.940 \, \frac{\text{lbf-sec}^2}{\text{ft}^4}\right)\left(32.17 \, \frac{\text{ft}}{\text{sec}^2}\right)$$

$$\times (32.8 \text{ ft}) \left(\frac{1 \text{ ft}^2}{\left(12 \, \frac{\text{in}}{\text{ft}}\right)^2}\right)$$

$$= 28.92 \text{ lbf/in}^2 \quad (29 \text{ psi})$$

SI Solution

The specific weight of water at atmospheric pressure at 10°C is 999.7 kg/m³. The pressure of air at sea level is 101.325 kPa. [Properties of Water (SI Units)]

The pressure on the gate is greatest at the bottom edge. The pressure at this depth is

Forces on Submerged Surfaces and the Center of Pressure

$$p = p_o + \rho g h = p_o + \gamma \cdot h$$

Since the gate is vertical, the angle θ is 90°.

$$h = 10 \text{ m}$$

$$p = 101.325 \text{ kPa} + \left(999.7 \, \frac{\text{kg}}{\text{m}^3}\right)\left(9.81 \, \frac{\text{m}}{\text{sec}^2}\right)(10 \text{ m})\left(\frac{1}{1000 \, \frac{\text{Pa}}{\text{kPa}}}\right)$$

$$= 199.4 \text{ kPa} \quad (200 \text{ kPa})$$

The answer is (B).

16 Fluid Flow Parameters

Content in blue refers to the NCEES Handbook.

PRACTICE PROBLEMS

Use the following values unless told to do otherwise in the problem:

$$g = 32.2 \text{ ft/sec}^2 (9.81 \text{ m/s}^2)$$
$$\rho_{\text{water}} = 62.4 \text{ lbm/ft}^3 (1000 \text{ kg/m}^3)$$
$$p_{\text{atmospheric}} = 14.7 \text{ psia} (101.3 \text{ kPa})$$

1. A 10 in (25 cm) diameter composition pipe is compressed by a tree root into an elliptical cross section until its inside height is only 7.2 in (18 cm). When the pipe is flowing full, its hydraulic diameter is most nearly

(A) 4.4 in (11.0 cm)
(B) 5.4 in (13.8 cm)
(C) 6.4 in (16.2 cm)
(D) 8.8 in (22.0 cm)

2. A pipe with an inside diameter of 18.8 in contains water to a depth of 15.7 in. What is most nearly the hydraulic diameter?

(A) 8.8 in
(B) 10.2 in
(C) 23 in
(D) 30.5 in

3. Water at 6 gpm flows through a pipe with a 1/2 in diameter. At a point 1.5 ft above ground level, a pressure gage reads 40 psi. At a point further downstream and 5.0 ft above ground, a second pressure gage reads 1 psi. The head loss between the two pressure gages is most nearly

(A) 22 ft
(B) 54 ft
(C) 87 ft
(D) 97 ft

4. Outside air is drawn through a round duct with a 12 in diameter at 600 cfm. The duct entrance is centered 15 ft above ground level. The air exits into a room through a round duct with an 8 in diameter that is centered 2 ft above ground level. Because of the pressure in the room, the velocity through the 8 in duct is limited to 10 ft/sec. The pressure in the room is most nearly

(A) 14.7 psia
(B) 15.4 psia
(C) 16.6 psia
(D) 17.1 psia

5. Water at 70°F and atmospheric pressure flows at 30 gpm through a rough pipe that has a diameter of 6 in. According to the Moody diagram, the flow is

(A) turbulent
(B) laminar
(C) critical
(D) transitional

SOLUTIONS

1. *Customary U.S. Solution*

The inside perimeter of the pipe before compression was

$$p = \pi D = \pi(10 \text{ in}) = 31.42 \text{ in}$$

The perimeter of the pipe is the same after compression into an ellipse. The minor axis of the ellipse is one-half its height.

$$b = \frac{7.2 \text{ in}}{2} = 3.6 \text{ in}$$

Use the formula for the perimeter of an ellipse, and solve for the major axis.

Ellipse

$$p_{\text{approx}} = 2\pi\sqrt{\frac{a^2 + b^2}{2}}$$

$$a = \sqrt{2\left(\frac{p}{2\pi}\right)^2 - b^2}$$

$$= \sqrt{(2)\left(\frac{31.42 \text{ in}}{2\pi}\right)^2 - (3.6 \text{ in})^2}$$

$$= 6.09 \text{ in}$$

Use the formula for the area of an ellipse to find the cross-sectional area inside the pipe.

Ellipse

$$A_{\text{ellipse}} = \pi ab$$

$$= \pi(6.09 \text{ in})(3.6 \text{ in})$$

$$= 68.88 \text{ in}^2$$

Find the hydraulic diameter when the pipe is flowing full.

Flow in Noncircular Conduits

$$D_H = \frac{4 \times \text{cross-sectional area of flowing fluid}}{\text{wetting perimeter}}$$

$$= \frac{(4)(68.88 \text{ in}^2)}{31.42 \text{ in}}$$

$$= 8.769 \text{ in} \quad (8.8 \text{ in})$$

SI Solution

The inside perimeter of the pipe before compression was

$$p = \pi D = \pi(25 \text{ cm}) = 78.54 \text{ cm}$$

The perimeter of the pipe is the same after compression into an ellipse. The minor axis of the ellipse is one-half its height.

$$b = \frac{18 \text{ cm}}{2} = 9 \text{ cm}$$

Use the formula for the perimeter of an ellipse, and solve for the major axis.

Ellipse

$$p_{\text{approx}} = 2\pi\sqrt{\frac{a^2 + b^2}{2}}$$

$$a = \sqrt{2\left(\frac{p}{2\pi}\right)^2 - b^2}$$

$$= \sqrt{(2)\left(\frac{78.54 \text{ cm}}{2\pi}\right)^2 - (9 \text{ cm})^2}$$

$$= 15.22 \text{ cm}$$

Use the formula for the area of an ellipse to find the cross-sectional area inside the pipe.

$$A_{\text{ellipse}} = \pi ab$$

$$= \pi(15.22 \text{ cm})(9 \text{ cm})$$

$$= 430.3 \text{ cm}^2$$

Find the hydraulic diameter when the pipe is flowing full.

Flow in Noncircular Conduits

$$D_H = \frac{4 \times \text{cross-sectional area of flowing fluid}}{\text{wetting perimeter}}$$

$$= \frac{(4)(430.3 \text{ cm}^2)}{78.54 \text{ cm}}$$

$$= 21.91 \text{ cm} \quad (22.0 \text{ cm})$$

The answer is (D).

2. The area of flow in the pipe is the entire cross-sectional area minus the area of the circular segment that is empty. The entire cross-sectional area is

$$A_{\text{pipe}} = \frac{\pi D^2}{4} = \frac{\pi(18.8 \text{ in})^2}{4} = 277.6 \text{ in}^2$$

The empty circular segment has a depth and radius of

$$d = 18.8 \text{ in} - 15.7 \text{ in} = 3.1 \text{ in}$$

$$r = \frac{18.8 \text{ in}}{2} = 9.4 \text{ in}$$

Find the angle of the circular segment.

$$\phi = 2\arccos\frac{r-d}{r}$$

Circular Segment

$$= 2\arccos\frac{9.4\text{ in} - 3.1\text{ in}}{9.4\text{ in}}$$

$$= 1.673\text{ rad}$$

Find the area of the circular segment.

Circular Segment

$$A_{\text{empty}} = \frac{r^2(\phi - \sin\phi)}{2}$$

$$= \frac{(9.4\text{ in})^2\left(\begin{array}{c}1.673\text{ rad}\\-\sin(1.673\text{ rad})\end{array}\right)}{2}$$

$$= 29.96\text{ in}^2$$

The area of flow is

$$A_{\text{flow}} = A_{\text{pipe}} - A_{\text{empty}}$$

$$= 277.6\text{ in}^2 - 29.96\text{ in}^2$$

$$= 247.6\text{ in}^2$$

The wetted perimeter is the entire inner perimeter of the pipe minus the arc of the empty circular segment. The inner perimeter of the pipe is

$$P_{\text{total}} = \pi D = \pi(18.8\text{ in}) = 59.06\text{ in}$$

Find the arc of the empty circular segment, s.

Circular Segment

$$\phi = \frac{s}{r}$$

$$s = \phi r = (1.673\text{ rad})(9.4\text{ in}) = 15.73\text{ in}$$

The wetted perimeter is

$$P_{\text{wetted}} = P_{\text{total}} - s$$

$$= 59.06\text{ in} - 15.73\text{ in}$$

$$= 43.33\text{ in}$$

Find the hydraulic diameter of the pipe when flowing to a depth of 15.7 in.

Flow in Noncircular Conduits

$$D_H = \frac{4 \times \text{cross-sectional area of flowing fluid}}{\text{wetted perimeter}}$$

$$= \frac{4A_{\text{flow}}}{P_{\text{wetted}}} = \frac{(4)(247.6\text{ in}^2)}{43.33\text{ in}} = 22.86\text{ in}\quad(23\text{ in})$$

The answer is (C).

3. Use the Bernoulli equation.

Bernoulli Equation

$$\frac{p_1 g_c}{\rho g} + z_1 + \frac{v_1^2}{2g} = \frac{p_2 g_c}{\rho g} + z_2 + \frac{v_2^2}{2g} + h_f$$

The flow rate and pipe diameter do not change, so the velocity is constant and the velocity terms cancel out. Solve for the head loss. The density of water at standard conditions is 62.4 lbm/ft³. [Properties of Water at Standard Conditions]

$$h_f = \frac{(p_1 - p_2)g_c}{\rho g} + z_1 - z_2$$

$$= \frac{\left(\left(40\,\frac{\text{lbf}}{\text{in}^2} - 1\,\frac{\text{lbf}}{\text{in}^2}\right)\left(12\,\frac{\text{in}}{\text{ft}}\right)^2\right)}{\left(62.4\,\frac{\text{lbm}}{\text{ft}^3}\right)\left(32.17\,\frac{\text{ft}}{\text{sec}^2}\right)}$$

$$+ 1.5\text{ ft} - 5.0\text{ ft}$$

$$= 86.5\text{ ft}\quad(87\text{ ft})$$

The answer is (C).

4. The cross-sectional areas of the duct entrance (point 1) and exit (point 2) are

$$A_1 = \frac{\pi D_1^2}{4} = \frac{\pi\left(\frac{12\text{ in}}{12\,\frac{\text{in}}{\text{ft}}}\right)^2}{4} = 0.7854\text{ ft}^2$$

$$A_2 = \frac{\pi D_2^2}{4} = \frac{\pi\left(\frac{8\text{ in}}{12\,\frac{\text{in}}{\text{ft}}}\right)^2}{4} = 0.3491\text{ ft}^2$$

The flow rate in cubic feet per second is

$$\frac{600\,\frac{\text{ft}^3}{\text{min}}}{60\,\frac{\text{sec}}{\text{min}}} = 10\text{ ft}^3/\text{sec}$$

Use the continuity equation to calculate the velocity of the air at the entrance.

Continuity Equation

$$Q = Av$$

$$v_1 = \frac{Q}{A_1} = \frac{10 \frac{\text{ft}^3}{\text{sec}}}{0.7854 \text{ ft}^2} = 12.73 \text{ ft/sec}$$

From the problem statement, the velocity of the air at the exit is 10 ft/sec. Use the Bernoulli equation.

Continuity Equation

$$\frac{v^2}{2g_c} + \frac{p}{\rho} + \frac{gz}{g_c} = \text{constant}$$

With the entrance as point 1 and the exit as point 2, this can be expressed as

$$\frac{v_1^2}{2g_c} + \frac{p_1}{\rho} + \frac{gz_1}{g_c} = \frac{v_2^2}{2g_c} + \frac{p_2}{\rho} + \frac{gz_2}{g_c}$$

Solve for the pressure at the exit. The density of air at standard conditions is 0.075 lbm/ft³. [Standard Dry Air Conditions at Sea Level]

$$p_2 = \frac{\rho(v_1^2 - v_2^2)}{2g_c} + \frac{\rho g(z_1 - z_2)}{g_c} + p_1$$

$$= \frac{\left(0.075 \frac{\text{lbm}}{\text{ft}^3}\right)\left[\left(12.73 \frac{\text{ft}}{\text{sec}}\right)^2 - \left(10 \frac{\text{ft}}{\text{sec}}\right)^2\right]}{(2)\left(32.17 \frac{\text{lbm-ft}}{\text{lbf-sec}^2}\right)\left(12 \frac{\text{in}}{\text{ft}}\right)^2}$$

$$+ \frac{\left(0.075 \frac{\text{lbm}}{\text{ft}^3}\right)\left(32.17 \frac{\text{ft}}{\text{sec}^2}\right) \times (15 \text{ ft} - 2 \text{ ft})}{\left(32.17 \frac{\text{lbm-ft}}{\text{lbf-sec}^2}\right)\left(12 \frac{\text{in}}{\text{ft}}\right)^2}$$

$$+ 14.7 \frac{\text{lbf}}{\text{in}^2}$$

$$= 14.71 \text{ lbf/in}^2 \quad (14.7 \text{ psia})$$

The answer is (A).

5. The flow is characterized by identifying the Reynolds number. To find the Reynolds number, the fluid's velocity and kinematic viscosity must be known.

The cross-sectional area of the pipe is

$$A = \frac{\pi D^2}{4} = \frac{\pi (0.5 \text{ ft})^2}{4} = 0.1963 \text{ ft}^2$$

The flow rate converted to cubic feet per second is

$$Q = \frac{30 \frac{\text{gal}}{\text{min}}}{\left(7.481 \frac{\text{gal}}{\text{ft}^3}\right)\left(60 \frac{\text{sec}}{\text{min}}\right)}$$

$$= 0.06684 \text{ ft}^3/\text{sec}$$

Use the continuity equation, and solve for the velocity.

Continuity Equation

$$Q = Av$$

$$v = \frac{Q}{A} = \frac{0.06684 \frac{\text{ft}^3}{\text{sec}}}{0.1963 \text{ ft}^2} = 0.3405 \text{ ft/sec}$$

The kinematic viscosity of 70°F at atmospheric pressure is 1.059×10^{-5} ft²/sec. [Properties of Water (I-P Units)]

Calculate the Reynolds number.

Reynolds Number

$$\text{Re} = \frac{vD}{\nu}$$

$$= \frac{\left(0.3405 \frac{\text{ft}}{\text{sec}}\right)(0.5 \text{ ft})}{1.059 \times 10^{-5} \frac{\text{ft}^2}{\text{sec}}}$$

$$= 16,076$$

From the Moody diagram, flow in a rough pipe with a Reynolds number between 4000 and 20,000 is transitional. [Moody Diagram (Stanton Diagram)]

The answer is (D).

17 Fluid Dynamics

Content in blue refers to the NCEES Handbook.

PRACTICE PROBLEMS

(Use $g = 32.2$ ft/sec² (9.81 m/s²) and 60°F (16°C) water unless told to do otherwise in the problem.)

1. Water flows through a schedule-40 steel pipe that changes gradually in diameter from 6 in at point A to 12 in at point B. Point B is 15 ft (4.6 m) higher than point A. The water flows at a rate of 5 ft³/sec (130 L/s). The respective pressures at points A and B are 10 psia (70 kPa) and 7 psia (48.3 kPa). All minor losses are insignificant. The velocity and direction of flow at point A are most nearly

(A) 3.2 ft/sec (1 m/s); from A to B

(B) 25 ft/sec (7 m/s); from A to B

(C) 3.2 ft/sec (1 m/s); from B to A

(D) 25 ft/sec (7 m/s); from B to A

2. Water flows through a pipe network at 10 ft/sec. The pipe network consists of 50 ft of new 3 in diameter, schedule-40 iron pipe, four 45° standard elbows, 2 standard 90° elbows, and a fully open gate valve. All fittings are flanged. The element that contributes the most to specific energy loss based on equivalent pipe lengths is the

(A) four 45° elbows

(B) two 90° elbows

(C) gate valve

(D) pipe friction from the 50 ft section of pipe (excluding the fittings)

3. 300 ft of 18 in high-density polyethylene (HDPE) pipe ($C = 120$) and 400 ft of 14 in HDPE pipe ($C = 120$) are currently joined in series as part of a pipe network. The length of 16 in diameter HDPE pipe ($C = 120$) that can replace the two pipes without increasing the pumping power required is most nearly

(A) 640 ft

(B) 790 ft

(C) 840 ft

(D) 940 ft

4. A 10 in diameter steel pipe ($C = 100$) carries water flowing at 4 ft/sec. The theoretical friction loss per 100 ft of pipe length is most nearly

(A) 1.0 ft

(B) 4.2 ft

(C) 10 ft

(D) 14 ft

5. A pressurized supply line ($C = 140$) brings cold water to 600 residential connections. The line is 17,000 ft long and has a diameter of 10 in. The elevation is 1000 ft at the start of the line and 850 ft at the delivery end of the line. The average flow rate is 1.1 gal/min per residence, and the peaking factor is 2.5. The minimum required pressure at the delivery end is 60 psig. Most nearly, what must be the pressure at the start of the line during peak flow?

(A) 9.4 psig

(B) 40 psig

(C) 95 psig

(D) 110 psig

6. Three reservoirs (A, B, and C) are interconnected with a common junction (point D) at elevation 25 ft above an arbitrary reference point. The water levels for reservoirs A, B, and C are at elevations of 50 ft, 40 ft, and 22 ft, respectively. The pipe from reservoir A to the junction is 800 ft of 3 in (nominal) steel pipe. The pipe from reservoir B to the junction is 500 ft of 10 in (nominal) steel pipe. The pipe from reservoir C to the junction is 1000 ft of 4 in (nominal) steel pipe. All pipes are schedule 40 with a friction factor of 0.02. All minor losses and velocity heads can be neglected.

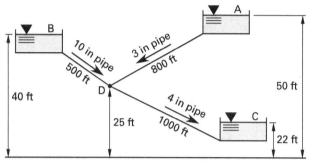

The direction of flow and the pressure at point D are most nearly

(A) out of reservoir B; 3.7 psi

(B) out of reservoir B; 6.5 psi

(C) into reservoir B; 8.1 psi

(D) into reservoir B; 9.4 psi

7. A cast-iron pipe with expansion joints throughout is 500 ft (150 m) long. 70°F (20°C) water is flowing at 6 ft/sec (2 m/s). The speed of sound in the pipe is 3972 ft/sec (1225 m/s). Over what approximate length of time must the valve be closed to create a pressure equivalent to instantaneous closure?

(A) 0.25 sec

(B) 0.68 sec

(C) 1.6 sec

(D) 2.1 sec

8. The velocity of discharge from a fire hose is 50 ft/sec (15 m/s). The hose is oriented 45° from the horizontal. Disregarding air friction, the maximum range of the discharge is most nearly

(A) 45 ft (14 m)

(B) 78 ft (23 m)

(C) 91 ft (27 m)

(D) 110 ft (33 m)

9. A full cylindrical tank that is 40 ft high has a constant diameter of 20 ft. The tank has a 4 in diameter hole in its bottom. The coefficient of discharge for the hole is 0.98. The equation for the time required to lower the fluid elevation is

$$t = \frac{2A_t(\sqrt{z_1} - \sqrt{z_2})}{C_d A_o \sqrt{2g}}$$

Where A_t and A_o are the tank and hole cross sectional areas and C_d is the coefficient of discharge. The time it will take for the water level to drop from 40 ft to 20 ft is most nearly

(A) 950 sec

(B) 1200 sec

(C) 1450 sec

(D) 1700 sec

10. A sharp-edged orifice meter with a 0.2 ft diameter opening is installed in a 1 ft diameter pipe. 70°F water approaches the orifice at 2 ft/sec. The indicated pressure drop across the orifice meter is most nearly

(A) 5.9 psi

(B) 13 psi

(C) 22 psi

(D) 47 psi

11. A mercury manometer is used to measure a pressure difference across an orifice meter in a water line. The difference in mercury levels is 7 in (17.8 cm). The pressure differential is most nearly

(A) 1.7 psi (12 kPa)

(B) 3.2 psi (22 kPa)

(C) 7.9 psi (55 kPa)

(D) 23 psi (160 kPa)

12. A pipe necks down from 24 in at point A to 12 in at point B. 8 ft³/sec of 60°F water flows from point A to point B. The pressure head at point A is 20 ft. Friction is insignificant over the distance between points A and B. The magnitude and direction of the resultant force of the water exerted on the bend are most nearly

(A) 2900 lbf, toward A

(B) 3500 lbf, toward A

(C) 2900 lbf, toward B

(D) 3500 lbf, toward B

13. A 2 in (0.05 m) diameter horizontal water jet has an absolute velocity (with respect to a stationary point) of 40 ft/sec (12 m/s) as it strikes a curved blade. The blade is moving horizontally away with an absolute velocity of 15 ft/sec. Water is deflected from the horizontal. The mass flow rate of the water is most nearly

(A) 47 lbm/sec (20 kg/s)

(B) 55 lbm/sec (25 kg/s)

(C) 100 lbm/sec (45 kg/s)

(D) 150 lbm/sec (70 kg/s)

14. 100 ft³/sec of water passes through a horizontal turbine. The water pressure is reduced from 30 psig to 5 psig vacuum. The mass flow rate is most nearly

(A) 1100 lbm/sec

(B) 3800 lbm/sec

(C) 5000 lbm/sec

(D) 6200 lbm/sec

15. A dish-shaped antenna faces directly into a 60 mi/hr wind. The projected area of the antenna is 0.8 ft², and the coefficient of drag is 1.2. The density of the air is 0.076 lbm/ft³. The total amount of drag force experienced by the antenna is most nearly

(A) 9.0 lbf

(B) 10 lbf

(C) 14 lbf

(D) 16 lbf

16. 68°F (20°C) castor oil flows through a pump whose impeller turns at 1000 rpm. A similar pump twice the first pump's size is tested with 68°F (20°C) air. The kinematic viscosity of the air is 15.72×10^{-5} ft²/sec (1.512×10^{-5} m²/s). The kinematic viscosity of the castor oil is 1110×10^{-5} ft²/sec (103×10^{-5} m²/s). To ensure similarity, the speed of the second pump's impeller should be most nearly

(A) 3.6 rpm

(B) 88 rpm

(C) 250 rpm

(D) 1600 rpm

17. Water flows through a line at a velocity of 6.5 ft/sec under a line pressure of 40 psi. If a valve in the line is closed suddenly, the resultant pressure rise is most nearly

(A) 205 psi

(B) 308 psi

(C) 350 psi

(D) 413 psi

18. Water flowing through a valve causes a valve pressure drop of 10 psi. The flow coefficient is 50. The flow rate through the valve is most nearly

(A) 50 gpm

(B) 80 gpm

(C) 158 gpm

(D) 205 gpm

SOLUTIONS

1. *Customary U.S. Solution*

For schedule-40 pipe, the inner diameter is 6.065 in for a nominal diameter of 6 in, and 11.938 in for a nominal diameter of 12 in. [Schedule 40 Steel Pipe]

Convert the units for the inner diameters to feet.

$$D_A = \frac{6.065 \text{ in}}{12 \frac{\text{in}}{\text{ft}}} = 0.5054 \text{ ft}$$

$$D_B = \frac{11.938 \text{ in}}{12 \frac{\text{in}}{\text{ft}}} = 0.9948 \text{ ft}$$

The total energy at a given point is found using the Bernoulli equation.

Duct Design: Bernoulli Equation

$$E_t = \frac{p}{\rho} + \frac{v^2}{2g_c} + \frac{zg}{g_c} = \text{constant}$$

Using the continuity equation, find the velocity at point A.

Continuity Equation

$$Q = A v = \left(\frac{\pi}{4}\right) D_A^2 v_A$$

$$v_A = \left(\frac{4}{\pi}\right)\left(\frac{Q}{D_A^2}\right) = \left(\frac{4}{\pi}\right)\left(\frac{5 \frac{\text{ft}^3}{\text{sec}}}{(0.5054 \text{ ft})^2}\right)$$

$$= 24.9 \text{ ft/sec} \quad (25 \text{ ft/sec})$$

Let point A be at zero elevation.

The pressure at point A is

$$p_A = \left(10 \frac{\text{lbf}}{\text{in}^2}\right)\left(12 \frac{\text{in}}{\text{ft}}\right)^2 = 1440 \text{ lbf/ft}^2$$

For water, $\rho \approx 62.4$ lbm/ft^3. [Properties of Water at Standard Conditions]

The total energy at point A is

Duct Design: Bernoulli Equation

$$E_t = \frac{p}{\rho} + \frac{v^2}{2g_c} + \frac{zg}{g_c} = \text{constant}$$

$$= \frac{1440 \frac{\text{lbf}}{\text{ft}^2}}{62.4 \frac{\text{lbm}}{\text{ft}^3}} + \frac{\left(24.9 \frac{\text{ft}}{\text{sec}}\right)^2}{(2)\left(32.2 \frac{\text{lbm-ft}}{\text{lbf-sec}^2}\right)} + 0$$

$$= 32.7 \text{ ft-lbf/lbm}$$

To determine the direction of the flow, find the total energy at point B. The velocity at point B is

Continuity Equation

$$Q = A v = \left(\frac{\pi}{4}\right) D_B^2 v_B$$

$$v_B = \left(\frac{4}{\pi}\right)\left(\frac{Q}{D_B^2}\right) = \left(\frac{4}{\pi}\right)\left(\frac{5 \frac{\text{ft}^3}{\text{sec}}}{(0.9948 \text{ ft})^2}\right)$$

$$= 6.436 \text{ ft/sec}$$

The pressure at point B is

$$p_B = \left(7 \frac{\text{lbf}}{\text{in}^2}\right)\left(12 \frac{\text{in}}{\text{ft}}\right)^2 = 1008 \text{ lbf/ft}^2$$

$$z_B = 15 \text{ ft}$$

The total energy at point B is

Duct Design: Bernoulli Equation

$$E_{t,B} = \frac{p_B}{\rho} + \frac{v_B^2}{2g_c} + \frac{z_B g}{g_c}$$

$$= \frac{1008 \frac{\text{lbf}}{\text{ft}^2}}{62.4 \frac{\text{lbm}}{\text{ft}^3}} + \frac{\left(6.436 \frac{\text{ft}}{\text{sec}}\right)^2}{(2)\left(32.2 \frac{\text{lbm-ft}}{\text{lbf-sec}^2}\right)}$$

$$+ \frac{(15 \text{ ft})\left(32.2 \frac{\text{ft}}{\text{sec}^2}\right)}{\left(32.2 \frac{\text{lbm-ft}}{\text{lbf-sec}^2}\right)}$$

$$= 31.8 \text{ ft-lbf/lbm}$$

Since $E_{t,A} > E_{t,B}$, the flow is from point A to point B.

SI Solution

For schedule-40 pipe, the inner diameter is 6.065 in for a nominal diameter of 6 in, and 11.938 in for a nominal diameter of 12 in. [Schedule 40 Steel Pipe]

Convert the units for the inner diameters to meters.

$$D_A = \left(\frac{6.065 \text{ in}}{12 \frac{\text{in}}{\text{ft}}}\right)\left(0.3048 \frac{\text{m}}{\text{ft}}\right) = 0.1540 \text{ m}$$

$$D_B = \left(\frac{11.938 \text{ in}}{12 \frac{\text{in}}{\text{ft}}}\right)\left(0.3048 \frac{\text{m}}{\text{ft}}\right) = 0.3032 \text{ m}$$

The total energy at a given point is found using the Bernoulli equation. [Duct Design: Bernoulli Equation]

$$E_t = \frac{p}{\rho} + \frac{\text{v}^2}{2} + zg = \text{constant}$$

At point A, the diameter is 154 mm (0.154 m). Using the continuity equation, find the velocity at point A.

Continuity Equation

$$Q = A\text{v} = \left(\frac{\pi}{4}\right)D_A^2 \text{v}_A$$

$$\text{v}_A = \left(\frac{4}{\pi}\right)\left(\frac{Q}{D_A^2}\right) = \left(\frac{4}{\pi}\right)\left(\frac{130 \frac{\text{L}}{\text{s}}}{(0.154 \text{ m})^2\left(1000 \frac{\text{L}}{\text{m}^3}\right)}\right)$$

$$= 6.98 \text{ m/s} \quad (7 \text{ m/s})$$

Let point A be at zero elevation.

The pressure at point A is

$$p_A = (70 \text{ kPa})\left(1000 \frac{\text{Pa}}{\text{kPa}}\right) = 70\,000 \text{ Pa}$$

For water, $\rho = 1000 \text{ kg/m}^3$. [Properties of Water at Standard Conditions]

The total energy at point A is

Duct Design: Bernoulli Equation

$$E_{t,A} = \frac{\text{v}_A^2}{2} + \frac{p_A}{\rho} + z_A g$$

$$= \frac{\left(6.98 \frac{\text{m}}{\text{s}}\right)^2}{2} + \frac{70\,000 \text{ Pa}}{1000 \frac{\text{kg}}{\text{m}^3}} + 0$$

$$= 94.36 \text{ J/kg}$$

To determine the direction of the flow, find the total energy at point B. The velocity at point B is

Continuity Equation

$$Q = A\text{v} = \left(\frac{\pi}{4}\right)D_B^2 \text{v}_B$$

$$\text{v}_B = \left(\frac{4}{\pi}\right)\left(\frac{Q}{D_A^2}\right) = \left(\frac{4}{\pi}\right)\left(\frac{130 \frac{\text{L}}{\text{s}}}{(0.3032 \text{ m})^2\left(1000 \frac{\text{L}}{\text{m}^3}\right)}\right)$$

$$= 1.8 \text{ m/s}$$

The pressure at point B is

$$p_B = (48.3 \text{ kPa})\left(1000 \frac{\text{Pa}}{\text{kPa}}\right) = 48\,300 \text{ Pa}$$

The total energy at point B is

Duct Design: Bernoulli Equation

$$E_{t,B} = \frac{p_B}{\rho} + \frac{\text{v}_B^2}{2} + z_B g$$

$$= \frac{48\,300 \text{ Pa}}{1000 \frac{\text{kg}}{\text{m}^3}} + \frac{\left(1.8 \frac{\text{m}}{\text{s}}\right)^2}{2} + (4.6 \text{ m})\left(9.81 \frac{\text{m}}{\text{s}^2}\right)$$

$$= 50.4 \text{ J/kg}$$

Since $E_{t,A} > E_{t,B}$, the flow is from point A to point B.

The answer is (B).

2. The factor that contributes most to the specific energy loss is determined by analysing the following factors. The equivalent length of four flanged 45° 3 in elbows is

Equivalent Length in Feet of Pipe for 90° Elbows

$$L_{e,\text{elbows}} = (4)(0.7)(9.7 \text{ ft}) = 27.16 \text{ ft}$$

Following the same process:

The equivalent length of the open gate valve is $(0.5)(9.7 \text{ ft}) = 4.85 \text{ ft}$.

The equivalent length of the 90° elbows is $(2)(9.7 \text{ ft}) = 19.4 \text{ ft}$.

The equivalent length of the pipe without fittings is 50 ft.

Comparing equivalent lengths, the 50 ft section of piping contributes the most to the specific energy loss in the piping system. Option (D) is the correct answer.

The answer is (D).

3. The pumping power will be the same if the head loss due to friction is the same. From the Hazen-Williams equation, the friction loss per foot length of pipe is

Pressure Drop of Water Flowing in Circular Pipe (Hazen-Williams)

$$h_f = \frac{10.44 Q^{1.85}}{C^{1.85} D^{4.8655}}$$

The head loss in feet of water for a pipe with length, L is

$$h_f = \frac{(10.44L)\left(\dfrac{Q}{C}\right)^{1.85}}{D^{4.8655}}$$

The 16 in diameter HDPE pipe is replacing two pipes (18 in pipe and 14 in pipe) without increasing the pumping power. The flow rate is the same in all pipe sections, as is the Hazen-Williams roughness coefficient. These terms and the constant term cancel out.

$$\frac{300 \text{ ft}}{(18 \text{ in})^{4.8655}} + \frac{400 \text{ ft}}{(14 \text{ in})^{4.8655}} = \frac{L}{(16 \text{ in})^{4.8655}}$$

$$L = 935 \text{ ft} \quad (940 \text{ ft})$$

The answer is (D).

4. From steel pipe friction tables, the friction loss for a 10 in diameter pipe is approximately 1.00 ft per 100 ft of pipe length. [Steel Pipe Friction Tables - Water]

The answer is (A).

5. The total peak flow rate is

$$Q = (600 \text{ res})(2.5)\left(1.1 \dfrac{\text{gal}}{\dfrac{\text{min}}{\text{res}}}\right) = 1650 \text{ gal/min}$$

The head loss in feet due to friction per foot of supply line is

Pressure Drop of Water Flowing in Circular Pipe (Hazen-Williams)

$$h_{f,\text{ft/ft}} = \frac{10.44 Q^{1.85}}{C^{1.85} D^{4.87}}$$

Adding the length of the supply line, L, to the previous equation gives

$$h_f = \frac{10.44 L Q^{1.85}}{C^{1.85} D^{4.87}}$$

Combine with the Bernoulli equation.

Bernoulli Equation

$$\frac{p_1}{\gamma} + z_1 = \frac{p_2}{\gamma} + z_2 + h_f$$

$$\frac{p_1}{\gamma} + z_1 = \frac{p_2}{\gamma} + z_2 + \frac{10.44 L Q^{1.85}}{C^{1.85} D^{4.8655}}$$

The specific weight of water at standard conditions is 62.4 lbf/ft³. [Properties of Water at Standard Conditions]

Solve for the pressure at the start of the line during peak flow.

$$\frac{p_{1,\text{lbf/in}^2}\left(12\,\dfrac{\text{in}}{\text{ft}}\right)^2}{62.4\,\dfrac{\text{lbf}}{\text{ft}^3}} + 1000 \text{ ft} = \frac{\left(60\,\dfrac{\text{lbf}}{\text{in}^2}\right)\left(12\,\dfrac{\text{in}}{\text{ft}}\right)^2}{62.4\,\dfrac{\text{lbf}}{\text{ft}^3}} + 850 \text{ ft}$$

$$+ \frac{(10.44)(17{,}000 \text{ ft}) \times \left(1650\,\dfrac{\text{gal}}{\text{min}}\right)^{1.85}}{(140)^{1.85}(10 \text{ in})^{4.8655}}$$

$$p_1 = 94.5 \text{ lbf/in}^2 \quad (95 \text{ psig})$$

The answer is (C).

6. The flows from reservoirs A and B are toward D and then toward C. From the continuity equation,

Continuity Equation

$$Q = Av$$

$$Q_{\text{A-D}} + Q_{\text{B-D}} = Q_{\text{D-C}}$$

$$A_A v_{\text{A-D}} + A_B v_{\text{B-D}} - A_C v_{\text{D-C}} = 0$$

From pipe and tube data tables, the area of a schedule-40 steel pipe with a 3 in nominal diameter is 7.389 in², the area of a schedule-40 steel pipe with a 10 in nominal diameter is 78.814 in², and the area of a schedule-40 steel pipe with a 4 in nominal diameter is 12.724 in². [Schedule 40 Steel Pipe]

Converting to feet,

$$A_A = \frac{7.389 \text{ in}^2}{\left(12 \frac{\text{in}}{\text{ft}}\right)^2} = 0.05134 \text{ ft}^2$$

$$A_B = \frac{78.814 \text{ in}^2}{\left(12 \frac{\text{in}}{\text{ft}}\right)^2} = 0.5473 \text{ ft}^2$$

$$A_C = \frac{12.724 \text{ in}^2}{\left(12 \frac{\text{in}}{\text{ft}}\right)^2} = 0.08836 \text{ ft}^2$$

From pipe and tube data tables, the inside diameter of a schedule-40 steel pipe with a 3 in nominal diameter is 3.068 in, the inside diameter of a schedule-40 steel pipe with a 10 in nominal diameter is 10.020 in, and the inside diameter of a schedule-40 steel pipe with a 4 in nominal diameter is 4.026 in. [Schedule 40 Steel Pipe]

Converting to feet,

$$D_A = \frac{3.068 \text{ in}}{12 \frac{\text{in}}{\text{ft}}} = 0.2557 \text{ ft}$$

$$D_B = \frac{10.020 \text{ in}}{12 \frac{\text{in}}{\text{ft}}} = 0.8350 \text{ ft}$$

$$D_C = \frac{4.026 \text{ in}}{12 \frac{\text{in}}{\text{ft}}} = 0.3355 \text{ ft}$$

The area values are substituted into the derived continuity equation.

$$0.05134 v_{\text{A-D}} + 0.5473 v_{\text{B-D}} - 0.08836 v_{\text{D-C}} = 0 \quad [\text{Eq. I}]$$

Ignoring the velocity heads, the conservation of energy equation between A and D is

Bernoulli Equation

$$\frac{p_1}{\gamma} + z_1 + \frac{v_1^2}{2g} = \frac{p_2}{\gamma} + z_2 + \frac{v_2^2}{2g} + h_f$$

$$z_A = \frac{p_D}{\gamma} + z_D + h_{f,\text{A-D}}$$

Substitute the Darcy-Weisbach equation into the conservation of energy equation to find the equation for the velocity.

Darcy-Weisbach Equation

$$h_f = \frac{fLv^2}{2Dg}$$

$$z_A = \frac{p_D}{\gamma} + z_D + h_{f,\text{A-D}} = \frac{p_D}{\gamma} + z_D + \frac{fLv^2}{2Dg}$$

$$50 \text{ ft} = \frac{p_D}{62.4 \frac{\text{lbm}}{\text{ft}^3}} + 25 \text{ ft} + \frac{(0.02)(800 \text{ ft}) v_{\text{A-D}}^2}{(2)(0.2557 \text{ ft})\left(32.2 \frac{\text{ft}}{\text{sec}^2}\right)}$$

$$v_{\text{A-D}} = \sqrt{25.73 - 0.0165 p_D} \quad [\text{Eq. II}]$$

Similarly, for B–D,

$$40 \text{ ft} = \frac{p_D}{62.4 \frac{\text{lbm}}{\text{ft}^3}} + 25 \text{ ft} + \frac{(0.02)(500 \text{ ft}) v_{\text{B-D}}^2}{(2)(0.8350 \text{ ft})\left(32.2 \frac{\text{ft}}{\text{sec}^2}\right)}$$

$$v_{\text{B-D}} = \sqrt{80.66 - 0.0862 p_D} \quad [\text{Eq. III}]$$

For D–C,

$$22 \text{ ft} = \frac{p_D}{62.4 \frac{\text{lbm}}{\text{ft}^3}} + 25 \text{ ft} - \frac{(0.02)(1000 \text{ ft}) v_{\text{D-C}}^2}{(2)(0.3355 \text{ ft})\left(32.2 \frac{\text{ft}}{\text{sec}^2}\right)}$$

$$v_{\text{D-C}} = \sqrt{3.24 + 0.0173 p_D} \quad [\text{Eq. IV}]$$

Equations I, II, III, and IV must be solved simultaneously. To do this, assume a value for p_D. This value then determines all three velocities in Eqs. II, III, and IV. These velocities are substituted into Eq. I. A trial and error solution yields

$$v_{\text{A-D}} = 3.21 \text{ ft/sec}$$
$$v_{\text{B-D}} = 0.408 \text{ ft/sec}$$
$$v_{\text{D-C}} = 4.40 \text{ ft/sec}$$
$$p_D = 933.8 \text{ lbf/ft}^2$$
$$= \left(933.8 \frac{\text{lbf}}{\text{ft}^2}\right)\left(\frac{1 \text{ ft}^2}{144 \text{ in}^2}\right)$$
$$= 6.49 \text{ psi} \quad (6.5 \text{ psi})$$

The direction of flow is from reservoir B to reservoir D.

The answer is (B).

7. *Customary U.S. Solution*

For a cast-iron (inelastic) pipe, the water hammer shock wave travels at a distance, L, from the closed valve to the source, and a rarefaction wave will return back to the valve. The total distance is $2L$.

The length of time the pressure is constant at the valve is

$$t = \frac{2L}{c} = \frac{(2)(500 \text{ ft})}{3972 \frac{\text{ft}}{\text{sec}}} = 0.25 \text{ sec}$$

SI Solution

For a cast-iron (inelastic) pipe, the water hammer shock wave travels at a distance, L, from the closed valve to the source and a rarefaction wave will return back to the valve. The total distance is $2L$.

The length of time the pressure is constant at the valve is

$$t = \frac{2L}{c} = \frac{(2)(150 \text{ m})}{1225 \dfrac{\text{m}}{\text{s}}} = 0.25 \text{ s}$$

The answer is (A).

8. *Customary U.S. Solution*

The maximum range of the discharge is found from the projectile motion equations.

The initial velocity in the y-direction is:

The discharge decelerates and reaches a velocity of zero when it reaches the highest point. After this, the discharge will accelerate and impact the ground with a velocity magnitude equal to the initial velocity but in the negative y-direction. Therefore, the final velocity in the y-direction is: $v_y = -35.35$ ft / sec

Calculate the time taken by the discharge to hit the ground.

$$v_y = -gt + v_0 \sin\theta = -gt + v_{0y}$$

Solve for t,

$$t = \frac{v_{0y} - v_y}{g} = \frac{35.35 \dfrac{\text{ft}}{\text{sec}} - \left(-35.35 \dfrac{\text{ft}}{\text{sec}}\right)}{32.2 \dfrac{\text{ft}}{\text{sec}^2}} = 2.20 \text{ sec}$$

Calculate the maximum range of the discharge.

$$x = (v_0 \cos\theta)t + x_0 = \left(50 \dfrac{\text{ft}}{\text{sec}} \times \cos 45°\right)(2.20 \text{ sec}) + 0 \text{ ft}$$
$$= 77.78 \text{ ft} \quad (78 \text{ ft})$$

SI Solution

The maximum range of the discharge is found from the projectile motion equations.

The initial velocity in the y-direction is:

$$v_{0y} = v_0 \sin\theta = \left(15 \dfrac{\text{m}}{\text{s}}\right)\sin 45° = 10.61 \text{ m / s}$$

The discharge decelerates and reaches a velocity of zero when it reaches the highest point. After this, the discharge will accelerate and impact the ground with a velocity magnitude equal to the initial velocity but in the negative y-direction. Therefore, the final velocity in the y-direction is: $v_y = -10.61$ m / s

Calculate the time taken by the discharge to hit the ground.

$$v_y = -gt + v_0 \sin\theta = -gt + v_{0y}$$

Solve for t,

$$t = \frac{v_{0y} - v_y}{g} = \frac{10.61 \dfrac{\text{m}}{\text{s}} - \left(-10.61 \dfrac{\text{m}}{\text{s}}\right)}{9.81 \dfrac{\text{m}}{\text{s}^2}} = 2.16 \text{ s}$$

Calculate the maximum range of the discharge.

$$x = (v_0 \cos\theta)t + x_0 = \left(15 \dfrac{\text{m}}{\text{s}} \times \cos 45°\right)(2.16 \text{ s}) + 0 \text{ m}$$
$$= 22.92 \text{ m} \quad (23 \text{ m})$$

The answer is (B).

9. The area of the opening and the area of the tank are

$$A_o = \left(\frac{\pi}{4}\right)\left(\frac{4 \text{ in}}{12 \dfrac{\text{in}}{\text{ft}}}\right)^2 = 0.08727 \text{ ft}^2$$

$$A_t = \left(\frac{\pi}{4}\right)(20 \text{ ft})^2 = 314.16 \text{ ft}^2$$

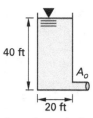

The time it takes to drop from 40 ft to 20 ft is

$$t = \frac{2A_t(\sqrt{z_1} - \sqrt{z_2})}{C_d A_o \sqrt{2g}}$$

$$= \frac{(2)(314.16 \text{ ft}^2)(\sqrt{40 \text{ ft}} - \sqrt{20 \text{ ft}})}{(0.98)(0.08727 \text{ ft}^2)\sqrt{(2)\left(32.2 \dfrac{\text{ft}}{\text{sec}^2}\right)}}$$

$$= 1696 \text{ sec} \quad (1700 \text{ sec})$$

The answer is (D).

10. From properties of water table, at 70°F, [Properties of Water (I-P Units)]

$$\gamma = 62.30 \text{ lbf/ft}^3$$
$$D_o = 0.2 \text{ ft}$$

Find the areas.

$$A_o = \left(\frac{\pi}{4}\right)(0.2 \text{ ft})^2 = 0.0314 \text{ ft}^2$$
$$A = \left(\frac{\pi}{4}\right)(1 \text{ ft})^2 = 0.7854 \text{ ft}^2$$

The flow rate is

Continuity Equation

$$Q = Av$$
$$= (0.7854 \text{ ft}^2)\left(2 \frac{\text{ft}}{\text{sec}}\right)$$
$$= 1.571 \text{ ft}^3/\text{sec}$$

The equation for the flow rate through an orifice is

Orifices

$$Q = CA_o\sqrt{\frac{2}{\rho}(p_1 - p_2)}$$

From the table of orifices and their nominal coefficients, $C = 0.61$ for a sharp-edged orifice. [Orifices]

Taking the pressure difference as $p_p - p_o$ and substituting the equation for density into the equation,

Density, Specific Weight, and Specific Gravity

$$\rho = \frac{\gamma}{g}$$

$$p_p - p_o = \left(\frac{\gamma}{2g}\right)\left(\frac{Q}{CA_o}\right)^2$$

$$= \frac{\left(\frac{62.30 \frac{\text{lbf}}{\text{ft}^3}}{(2)\left(32.2 \frac{\text{ft}}{\text{sec}^2}\right)}\right)\left(\frac{1.571 \frac{\text{ft}^3}{\text{sec}}}{(0.61)(0.0314 \text{ ft}^2)}\right)^2}{\left(12 \frac{\text{in}}{\text{ft}}\right)^2}$$

$$= 46.09 \text{ lbf/in}^2 \quad (47 \text{ psi})$$

The answer is (D).

11. *Customary U.S. Solution*

Find the densities of mercury and water. [Properties of Metals - I-P Units] [Properties of Water at Standard Conditions]

$$\rho_{\text{mercury}} = 845 \text{ lbm/ft}^3$$
$$\rho_{\text{water}} = 62.4 \text{ lbm/ft}^3$$

The manometer tube is filled with water above the mercury column. The pressure differential across the orifice meter is [Manometers]

$$\Delta p = p_1 - p_2 = (\rho_{\text{mercury}} - \rho_{\text{water}})\frac{hg}{g_c}$$

$$= \left(845 \frac{\text{lbm}}{\text{ft}^3} - 62.4 \frac{\text{lbm}}{\text{ft}^3}\right)\left(\frac{(7 \text{ in})\left(32.2 \frac{\text{ft}}{\text{sec}^2}\right)}{\left(12 \frac{\text{in}}{\text{ft}}\right)\left(32.2 \frac{\text{lbm-ft}}{\text{lbf-sec}^2}\right)}\right)$$

$$= 456.5 \text{ lbf/ft}^2 \quad (3.2 \text{ psi})$$

SI Solution

Find the densities of mercury and water. [Properties of Metals - SI Units] [Properties of Water at Standard Conditions]

$$\rho_{\text{mercury}} = 13\,547 \text{ kg/m}^3$$
$$\rho_{\text{water}} = 1000 \text{ kg/m}^3$$

The manometer tube is filled with water above the mercury column. The pressure differential across the orifice meter is [Manometers]

$$\Delta p = p_1 - p_2 = (\rho_{\text{mercury}} - \rho_{\text{water}})hg$$

$$= \left(13\,547 \frac{\text{kg}}{\text{m}^3} - 1000 \frac{\text{kg}}{\text{m}^3}\right)(0.178 \text{ m})\left(9.81 \frac{\text{m}}{\text{s}^2}\right)$$

$$= 21\,909 \text{ Pa} \quad (22 \text{ kPa})$$

The answer is (B).

12. The areas at point A and point B are

$$A_A = \left(\frac{\pi}{4}\right)\left(\frac{24 \text{ in}}{12 \frac{\text{in}}{\text{ft}}}\right)^2 = 3.142 \text{ ft}^2$$

$$A_B = \left(\frac{\pi}{4}\right)\left(\frac{12 \text{ in}}{12 \frac{\text{in}}{\text{ft}}}\right)^2 = 0.7854 \text{ ft}^2$$

The specific density of water at standard conditions is 62.4 lbm/ft³. [Properties of Water at Standard Conditions]

From the continuity equation, the velocities at point A and point B and the pressure at point A are

Continuity Equation

$$Q = Av$$

$$v_A = \frac{Q}{A_A} = \frac{8 \frac{ft^3}{sec}}{3.142 \, ft^2} = 2.546 \, ft/sec$$

$$p_A = \gamma h = \left(62.4 \frac{lbf}{ft^3}\right)(20 \, ft) = 1248 \, lbf/ft^2$$

$$v_B = \frac{Q}{A_B} = \frac{8 \frac{ft^3}{sec}}{0.7854 \, ft^2} = 10.19 \, ft/sec$$

Considering no change in elevation and frictionless flow, the Bernoulli equation simplifies to

Bernoulli Equation

$$\frac{p_1}{\rho g} + z_1 + \frac{v_1^2}{2g} = \frac{p_2}{\rho g} + z_2 + \frac{v_2^2}{2g} + h_f$$

$$\frac{p_A}{\rho} + \frac{v_A^2}{2} = \frac{p_B}{\rho} + \frac{v_B^2}{2}$$

Substitute the equation for density into the simplified equation to find the pressure at point B.

Density, Specific Weight, and Specific Gravity

$$\rho = \frac{\gamma}{g}$$

$$p_B = p_A + \left(\frac{v_A^2}{2g} - \frac{v_B^2}{2g}\right)\gamma$$

$$= 1248 \frac{lbf}{ft^2} + \frac{\left(2.546 \frac{ft}{sec}\right)^2 - \left(10.19 \frac{ft}{sec}\right)^2}{(2)\left(32.2 \frac{ft}{sec^2}\right)}$$

$$\times \left(62.4 \frac{lbf}{ft^3}\right)$$

$$= 1153.7 \, lbf/ft^2$$

Substitute the equation for the impulse-momentum principle into the continuity equation for mass flow rate to find the magnitudes of the resultant forces. The angle $\alpha = 0°$, so the cosine of α is 1.

Impulse-Momentum Principle

$$p_A A_A - p_B A_B \cos\alpha - F_x = Q\rho(v_B \cos\alpha - v_A)$$

From a diagram of forces exerted by a flowing fluid on a bend, enlargement, or contraction, a positive resultant force (in the x direction) acts towards the outlet. [Impulse-Momentum Principle]

Continuity Equation

$$\dot{m} = \rho Q$$

$$F_x = -p_B A_B + p_A A_A - \frac{\dot{m}(v_B - v_A)}{g_c}$$

$$= -\left(1153.7 \frac{lbf}{ft^2}\right)(0.7854 \, ft^2) + \left(1248 \frac{lbf}{ft^2}\right)(3.142 \, ft^2)$$

$$- \frac{\left(8 \frac{ft^3}{sec}\right)\left(62.4 \frac{lbm}{ft^3}\right)\left(10.19 \frac{ft}{sec} - 2.546 \frac{ft}{sec}\right)}{32.2 \frac{lbm\text{-}ft}{lbf\text{-}sec^2}}$$

$$= 2897 \, lbf \quad (2900 \, lbf \text{ on the fluid toward B})$$

The answer is (C).

13. *Customary U.S. Solution*

The density of the water at standard conditions is 62.4 lbm/ft³. [Properties of Water at Standard Conditions]

The mass flow rate of the water is

Continuity Equation

$$\dot{m} = \rho Q = \rho v A = \frac{\rho v \pi D^2}{4}$$

$$= \frac{\left(62.4 \frac{lbm}{ft^3}\right)\left(40 \frac{ft}{sec}\right)\pi \left(\frac{2 \, in}{12 \frac{in}{ft}}\right)^2}{4}$$

$$= 54.45 \, lbm/sec \quad (55 \, lbm/sec)$$

SI Solution

The density of the water at standard conditions is 1000 kg/m³. [Properties of Water at Standard Conditions]

The mass flow rate of the water is

Continuity Equation

$$\dot{m} = \rho Q = \rho v A = \frac{\rho v \pi D^2}{4}$$

$$= \frac{\left(1000 \frac{kg}{m^3}\right)\left(12 \frac{m}{s}\right)\pi (0.05 \, m)^2}{4}$$

$$= 23.56 \, kg/s \quad (25 \, kg/s)$$

The answer is (B).

14. The density of the water at standard conditions is 62.4 lbm/ft^3. [Properties of Water at Standard Conditions]

The mass flow rate is

Continuity Equation

$$\dot{m} = Q\rho = \left(100 \ \frac{\text{ft}^3}{\text{sec}}\right)\left(62.4 \ \frac{\text{lbm}}{\text{ft}^3}\right)$$

$$= 6240 \ \text{lbm/sec} \quad (6200 \ \text{lbm/sec})$$

The answer is (D).

15. The drag force on the antenna is

Drag Force

$$F_D = \frac{C_D A \rho \text{v}^2}{2g_c}$$

$$= \frac{(1.2)(0.8 \ \text{ft}^2)\left(0.076 \ \frac{\text{lbm}}{\text{ft}^3}\right) \times \left(\frac{\left(60 \ \frac{\text{mi}}{\text{hr}}\right)\left(5280 \ \frac{\text{ft}}{\text{mi}}\right)}{3600 \ \frac{\text{sec}}{\text{hr}}}\right)^2}{(2)\left(32.2 \ \frac{\text{lbm-ft}}{\text{lbf-sec}^2}\right)}$$

$$= 8.77 \ \text{lbf} \quad (9.0 \ \text{lbf})$$

The answer is (A).

16. *Customary U.S. Solution*

To ensure similarity between the two impellers, the Reynolds numbers for the air and castor oil must be equal. Setting the equations for both Reynolds numbers equal to each other,

Reynolds Number

$$\text{Re} = \frac{\text{v}D}{\nu}$$

$$\text{Re}_{\text{oil}} = \text{Re}_{\text{air}}$$

$$\frac{\text{v}_{\text{oil}} D_{\text{oil}}}{\nu_{\text{oil}}} = \frac{\text{v}_{\text{air}} D_{\text{air}}}{\nu_{\text{air}}}$$

$$\frac{\text{v}_{\text{oil}}}{\text{v}_{\text{air}}} = \left(\frac{\nu_{\text{oil}}}{\nu_{\text{air}}}\right)\left(\frac{D_{\text{air}}}{D_{\text{oil}}}\right)$$

v is the tangential velocity, and D is the impeller diameter. The equation for the velocity is

$$\text{v} \propto N_{\text{rpm}} D$$

$$\frac{\text{v}_{\text{oil}}}{\text{v}_{\text{air}}} = \left(\frac{N_{\text{oil}}}{N_{\text{air}}}\right)\left(\frac{D_{\text{oil}}}{D_{\text{air}}}\right)$$

Therefore,

$$\left(\frac{\nu_{\text{oil}}}{\nu_{\text{air}}}\right)\left(\frac{D_{\text{air}}}{D_{\text{oil}}}\right) = \left(\frac{N_{\text{oil}}}{N_{\text{air}}}\right)\left(\frac{D_{\text{oil}}}{D_{\text{air}}}\right)$$

$$N_{\text{air}} = N_{\text{oil}}\left(\frac{\nu_{\text{air}}}{\nu_{\text{oil}}}\right)\left(\frac{D_{\text{oil}}}{D_{\text{air}}}\right)^2$$

Since the air impeller is twice the size of the oil impeller, $D_{\text{air}} = 2D_{\text{oil}}$.

$$N_{\text{air}} = N_{\text{oil}}\left(\frac{\nu_{\text{air}}}{\nu_{\text{oil}}}\right)\left(\frac{D_{\text{oil}}}{D_{\text{air}}}\right)^2$$

$$= N_{\text{oil}}\left(\frac{\nu_{\text{air}}}{\nu_{\text{oil}}}\right)\left(\frac{D_{\text{oil}}}{2D_{\text{oil}}}\right)^2$$

$$= \tfrac{1}{4} N_{\text{oil}}\left(\frac{\nu_{\text{air}}}{\nu_{\text{oil}}}\right)$$

The necessary speed of the second pump's impeller is

$$N_{\text{air}} = \tfrac{1}{4} N_{\text{oil}}\left(\frac{\nu_{\text{air}}}{\nu_{\text{oil}}}\right)$$

$$= \left(\tfrac{1}{4}\right)\left(1000 \ \frac{\text{rev}}{\text{min}}\right)\left(\frac{15.72 \times 10^{-5} \ \frac{\text{ft}^2}{\text{sec}}}{1110 \times 10^{-5} \ \frac{\text{ft}^2}{\text{sec}}}\right)$$

$$= 3.54 \ \text{rev/min} \quad (3.6 \ \text{rpm})$$

SI Solution

The equation for the second pump's impeller is found in the same manner as for the customary U.S. solution. The necessary speed of the second pump's impeller is

$$N_{\text{air}} = \tfrac{1}{4} N_{\text{oil}} \left(\frac{\nu_{\text{air}}}{\nu_{\text{oil}}} \right)$$

$$= \left(\frac{1}{4}\right)\left(1000 \ \frac{\text{rev}}{\text{min}}\right) \left(\frac{1.512 \times 10^{-5} \ \frac{\text{m}^2}{\text{s}}}{103 \times 10^{-5} \ \frac{\text{m}^2}{\text{s}}} \right)$$

$$= 3.6 \ \text{rev/min} \quad (3.6 \ \text{rpm})$$

The answer is (A).

17. The density of water at standard conditions is 62.4 lbm/ft^3. [Properties of Water at Standard Conditions]

The velocity of sound in water is 4720 ft/sec. [Water Hammer]

Using the formula for water hammer, the pressure rise is

Water Hammer

$$\Delta p_h = \frac{\rho c_s \text{V}}{g_c}$$

$$= \frac{\left(62.4 \ \frac{\text{lbm}}{\text{ft}^3}\right)\left(4720 \ \frac{\text{ft}}{\text{sec}}\right)\left(6.5 \ \frac{\text{ft}}{\text{sec}}\right)}{\left(32.2 \ \frac{\text{lbm-ft}}{\text{lbf-sec}^2}\right)\left(12 \ \frac{\text{in}}{\text{ft}}\right)^2}$$

$$= 412.87 \ \text{lbf/in}^2 \quad (413 \ \text{psi})$$

The answer is (D).

18. From the formula for valve flow coefficient,

Valve Flow Coefficient

$$C_v = \frac{Q}{\sqrt{\Delta P}}$$

$$Q = C_v \sqrt{\Delta P}$$

$$= (50)\sqrt{10 \ \frac{\text{lbf}}{\text{in}^2}}$$

$$= 158.11 \ \text{gal/min} \quad (158 \ \text{gpm})$$

The answer is (C).

18 Hydraulic Machines and Fluid Distribution

Content in blue refers to the *NCEES Handbook*.

PRACTICE PROBLEMS

1. Two centrifugal pumps used in a water pumping application have the characteristic curves shown. The pumps operate in parallel and discharge into a common header against a head of 40 ft.

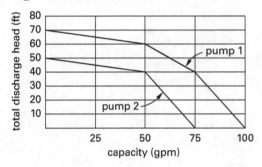

What is most nearly the discharge rate of the pumps operating in parallel?

(A) 30 gpm

(B) 50 gpm

(C) 75 gpm

(D) 130 gpm

2. A pump intended for occasional use in normal ambient conditions has an overall hydraulic efficiency of 85%. The pump will be used to develop a fluid power of 4.5 hp. The pump is driven by an electric motor with a service factor of 1.80. The smallest NEMA standard motor size suitable for this application is

(A) 3 hp

(B) 5 hp

(C) 8 hp

(D) 10 hp

3. 2000 gal/min of 60°F thickened sludge with a specific gravity of 1.2 flows through a pump with a nominal inlet diameter of 12 in and a nominal outlet diameter of 8 in. The centerlines of the inlet and outlet are at the same elevation. The inlet pressure is 8 in of mercury (vacuum). A discharge pressure gauge located 4 ft above the pump discharge centerline reads 20 psig. The pump efficiency is 85%. All pipes are schedule-40 steel. The input power of the pump is most nearly

(A) 26 hp

(B) 31 hp

(C) 37 hp

(D) 53 hp

4. 1.25 ft³/sec of 70°F water is pumped from the bottom of a tank through 700 ft of nominal 4 in schedule-40 steel pipe. The pipe has a surface roughness of 0.0002 ft. The pipeline includes a 50 ft rise in elevation, two right-angle elbows, a wide-open gate valve, and a swing check valve. All fittings and valves are regular screwed and have the equivalent lengths shown.

90° elbow: 13 ft

gate valve: 2.5 ft

check valve: 38 ft

The inlet pressure is 50 psig, and a working pressure of 20 psig is needed at the end of the pipe. The fluid power needed for this pumping application is most nearly

(A) 16 hp

(B) 23 hp

(C) 49 hp

(D) 66 hp

5. A 20 hp motor drives a centrifugal pump. The pump discharges 60°F water at a velocity of 12 ft/sec into a nominal 6 in schedule-40 steel pipe. The pump inlet is nominal 8 in schedule-40 steel pipe. The pump suction is 5 psig below standard atmospheric pressure. The friction and fitting head loss in the system is 10 ft. The pump efficiency is 70%. The suction and discharge lines are at the same elevation. Assuming standard atmospheric pressure, the height difference between the pump and the pipe discharge is most nearly

(A) 28 ft

(B) 37 ft

(C) 49 ft

(D) 81 ft

6. The pressure of 37 gal/min (65 L/s) of 80°F (27°C) water is increased from 1 atm to 40 psig (275 kPa). The additional fluid power needed at 100% efficiency is most nearly

(A) 0.45 hp (8.6 kW)

(B) 0.86 hp (18 kW)

(C) 1.8 hp (36 kW)

(D) 3.6 hp (72 kW)

7. 100 gal/min (6.3 L/s) of pressurized hot water at 280°F and 80 psia (140°C and 550 kPa) is drawn through 30 ft (10 m) of pipe into a tank pressurized to a constant 2 psig (14 kPa), as shown.

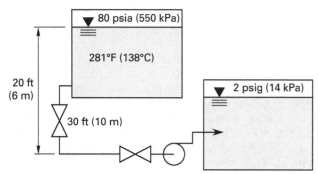

The pipe is nominal 1.5 in schedule-40 steel pipe with a surface roughness of 0.0002 ft (6.0 × 10^{-5} m). The pump's inlet and outlet are both 20 ft (6 m) below the surface of the water when the tank is full. The water's kinematic viscosity is 0.239 × 10^{-5} ft^2/sec (0.222 × 10^{-6} m^2/s), and its vapor pressure is 50.02 psia (3.431 bar). The inlet line contains a square mouth inlet, two wide-open gate valves, and two long-radius elbows. All components are regular screwed with the equivalent lengths shown.

inlet (square mouth): 3.1 ft

long radius 90° elbow: 3.4 ft

wide-open gate valves: 1.2 ft

The net positive suction head required for this application is 10 ft (3 m). What most nearly is the net positive suction head available, and will the pump cavitate?

(A) 4.0 ft (1.2 m), yes

(B) 8.9 ft (2.7 m), yes

(C) 24 ft (7.2 m), no

(D) 68 ft (21 m), no

8. The velocity of the tip of a marine propeller is 4.2 times the velocity of the boat. The propeller is located 8 ft (3 m) below the surface. The temperature of the seawater is 70°F (20°C). The density of the seawater is 64.0 lbm/ft^3 (1024 kg/m^3), and the salt content is 2.5% by weight. The practical maximum boat velocity, as limited strictly by cavitation, is most nearly

(A) 9.1 ft/sec (2.7 m/s)

(B) 12 ft/sec (3.8 m/s)

(C) 15 ft/sec (4.5 m/s)

(D) 22 ft/sec (6.6 m/s)

9. A pump is intended to run at 1750 rpm when driven by a 0.5 hp (0.37 kW) motor. The required power rating of a motor that will turn the pump at 2000 rpm is most nearly

(A) 0.25 hp (0.18 kW)

(B) 0.36 hp (0.27 kW)

(C) 0.52 hp (0.38 kW)

(D) 0.75 hp (0.55 kW)

10. A horizontal turbine reduces 100 ft^3/sec of water from 30 psia to 5 psia. Friction is negligible. The power developed is most nearly

(A) 350 hp

(B) 500 hp

(C) 650 hp

(D) 800 hp

11. 1000 ft^3/sec of 60°F water flows from a high reservoir through a hydroelectric turbine installation, exiting 625 ft lower. The head loss due to friction is 58 ft. The turbine efficiency is 89%. The power developed in the turbines is most nearly

(A) 40 kW

(B) 18 MW

(C) 43 MW

(D) 71 MW

HYDRAULIC MACHINES AND FLUID DISTRIBUTION

12. Water at 500 psig and 60°F (3.5 MPa and 15°C) drives a 250 hp (185 kW) turbine at 1750 rpm against a back pressure of 30 psig (210 kPa). The specific speed of a turbine is given by

$$N_{\text{specific}} = \frac{N_{\text{turbine}}\sqrt{P}}{h_{\text{turbine}}^{1.25}}$$

In this equation, N is rotational speed (in rpm), P is the turbine power (in horsepower or kilowatts), and h_{turbine} is the head extracted by the turbine (in feet or meters). The specific speed of this turbine is most nearly

(A) 4.4 (17)

(B) 22 (86)

(C) 75 (260)

(D) 230 (770)

13. Water at 60°F (15°C) drives a turbine. The water discharges through a 4 in (100 mm) diameter nozzle at 35 ft/sec (10.5 m/s). The water is deflected 80° by a single blade moving directly away at 10 ft/sec (3 m/s). The total force acting on a single blade is most nearly

(A) 100 lbf (450 N)

(B) 140 lbf (570 N)

(C) 160 lbf (720 N)

(D) 280 lbf (1300 N)

SOLUTIONS

1. When two pumps operate in parallel, each pump performs as it would if the other pump was not present. The total capacity is the sum of the capacities of the individual pumps. [Pump Curve Construction for Parallel Operation]

From the characteristic curves given, at a 40 ft discharge head, the capacity of pump 1 is 75 gpm, and the capacity of pump 2 is 50 gpm.

$$Q_{\text{parallel}} = Q_2 + Q_1$$
$$= 50\,\frac{\text{gal}}{\text{min}} + 75\,\frac{\text{gal}}{\text{min}}$$
$$= 125\text{ gal/min}\quad(130\text{ gpm})$$

The answer is (D).

2. The equation for the brake horsepower needed is

Centrifugal Pump Characteristics

$$P_{\text{brake}} = \frac{\rho g H Q}{\eta_{\text{pump}}} = \frac{P_{\text{fluid}}}{\eta_{\text{pump}}}$$

The motor is intended for occasional use, however, so the service factor should be included in the calculation.

$$P_{\text{brake}} = \frac{P_{\text{fluid}}}{\eta_{\text{pump}}(\text{SF})}$$
$$= \frac{4.5\text{ hp}}{(0.85)(1.80)}$$
$$= 2.94\text{ hp}$$

The smallest NEMA standard motor size with a rating greater than 2.94 hp is 3 hp.

The answer is (A).

3. Convert the flow rate to cubic feet per second. [Measurement Relationships]

$$Q = \frac{2000\,\dfrac{\text{gal}}{\text{min}}}{\left(7.481\,\dfrac{\text{gal}}{\text{ft}^3}\right)\left(60\,\dfrac{\text{sec}}{\text{min}}\right)}$$
$$= 4.456\text{ ft}^3/\text{sec}$$

From tables of pipe data, find the area of flow for schedule-40 nominal 12 in inlet pipe and the nominal 8 in outlet pipe. [Schedule 40 Steel Pipe]

$$A_{\text{in}} = \frac{111.875 \text{ in}^2}{\left(12 \frac{\text{in}}{\text{ft}}\right)^2} = 0.7769 \text{ ft}^2$$

$$A_{\text{out}} = \frac{50.002 \text{ in}^2}{\left(12 \frac{\text{in}}{\text{ft}}\right)^2} = 0.3472 \text{ ft}^2$$

Using the continuity equation to find the velocities at the inlet and outlet.

Continuity Equation

$$Q = A\text{v}$$

$$\text{v}_{\text{in}} = \frac{Q}{A_{\text{in}}} = \frac{4.456 \frac{\text{ft}^3}{\text{sec}}}{0.7769 \text{ ft}^2} = 5.736 \text{ ft/sec}$$

$$\text{v}_{\text{out}} = \frac{Q}{A_{\text{out}}} = \frac{4.456 \frac{\text{ft}^3}{\text{sec}}}{0.3472 \text{ ft}^2} = 12.83 \text{ ft/sec}$$

Convert the inlet pressure of 8 in Hg vacuum to absolute pressure in psf. [Measurement Relationships]

$$p_{\text{in}} = \left(14.7 \frac{\text{lbf}}{\text{in}^2} - \left(\frac{8 \text{ in Hg}}{2.036 \frac{\text{in Hg}}{\frac{\text{lbf}}{\text{in}^2}}}\right)\right) \times \left(12 \frac{\text{in}}{\text{ft}}\right)^2$$

$$= 1551 \text{ lbf/ft}^2$$

Convert the gage pressure of 20 psi to absolute pressure in psf.

$$p_{\text{at gage}} = \left(20 \frac{\text{lbf}}{\text{in}^2} + 14.7 \frac{\text{lbf}}{\text{in}^2}\right)\left(12 \frac{\text{in}}{\text{ft}}\right)^2$$

$$= 4997 \text{ lbf/ft}^2$$

This pressure was measured 4 ft above the outlet, so the additional pressure at the outlet due to the difference in elevation is

$$p_{\text{added}} = \frac{\rho_{\text{fluid}} g h}{g_c} = \frac{\rho_{\text{water}}(\text{SG}) g h}{g_c}$$

$$= \frac{\left(62.4 \frac{\text{lbm}}{\text{ft}^3}\right)(1.2)\left(32.17 \frac{\text{ft}}{\text{sec}^2}\right)(4 \text{ ft})}{32.17 \frac{\text{lbm-ft}}{\text{lbf-sec}^2}}$$

$$= 299.5 \text{ lbf/ft}^2$$

The absolute pressure at the outlet in psf is

$$p_{\text{out}} = 4997 \frac{\text{lbf}}{\text{ft}^2} + 299.5 \frac{\text{lbf}}{\text{ft}^2}$$

$$= 5297 \text{ lbf/ft}^2$$

Use the mechanical energy equation to find the head added by the pump.

Mechanical Energy Equation in Terms of Energy Per Unit Weight Involving Heads

$$\frac{p_{\text{in}}}{\gamma} + \frac{\text{v}_{\text{in}}^2}{2g} + z_{\text{in}} + h_{\text{shaft}} = \frac{p_{\text{out}}}{\gamma} + \frac{\text{v}_{\text{out}}^2}{2g} + z_{\text{out}} + h_{\text{loss}}$$

The pump inlet and outlet are at the same elevation, so z_{in} and z_{out} are equal, and head loss due to friction is negligible so h_{loss} is zero. Rearranging and simplifying the mechanical energy equation gives

$$h_{\text{pump}} = \frac{p_{\text{out}} - p_{\text{in}}}{\gamma} + \frac{\text{v}_{\text{out}}^2 - \text{v}_{\text{in}}^2}{2g}$$

$$= \frac{p_{\text{out}} - p_{\text{in}}}{\gamma_{\text{water}}(\text{SG})} + \frac{\text{v}_{\text{out}}^2 - \text{v}_{\text{in}}^2}{2g}$$

$$= \frac{5297 \frac{\text{lbf}}{\text{ft}^2} - 1551 \frac{\text{lbf}}{\text{ft}^2}}{\left(62.4 \frac{\text{lbf}}{\text{ft}^3}\right)(1.2)} +$$

$$\frac{\left(12.83 \frac{\text{ft}}{\text{sec}}\right)^2 - \left(5.736 \frac{\text{ft}}{\text{sec}}\right)^2}{(2)\left(32.17 \frac{\text{ft}}{\text{sec}^2}\right)}$$

$$= 52.07 \text{ ft}$$

Find the power needed by the pump to add 52.07 ft of head.

Pump Power Equation

$$P_{\text{fluid}} = \text{whp} = \frac{Q\Delta h}{3960}$$

$$P_{\text{pump}} = \frac{P_{\text{fluid}}(\text{SG})}{\eta} = \frac{Q\Delta h(\text{SG})}{3960\eta}$$

$$= \frac{\left(2000 \ \frac{\text{gal}}{\text{min}}\right)(52.07 \ \text{ft})(1.2)}{\left(3960 \ \frac{\text{ft-gal}}{\text{hp-min}}\right)(0.85)}$$

$$= 37.13 \ \text{hp} \quad (37 \ \text{hp})$$

The answer is (C).

4. From tables of pipe data, find the inside diameter and area of flow of nominal 4 in schedule-40 steel pipe. [Schedule 40 Steel Pipe]

$$D = \frac{4.026 \ \text{in}}{12 \ \frac{\text{in}}{\text{ft}}} = 0.3355 \ \text{ft}$$

$$A = \frac{12.724 \ \text{in}^2}{12 \ \frac{\text{in}}{\text{ft}}} = 0.08840 \ \text{ft}^2$$

Use the continuity equation to find the velocity in the pipe.

Continuity Equation

$$Q = A\text{v}$$

$$\text{v} = \frac{Q}{A} = \frac{1.25 \ \frac{\text{ft}^3}{\text{sec}}}{0.08840 \ \text{ft}^2} = 14.14 \ \text{ft/sec}$$

Calculate the Reynolds number. From a table of water properties, water at 70°F has a density of 62.30 lbf/ft³ and a kinematic viscosity of 1.059×10^{-5} ft²/sec. [Properties of Water at Standard Conditions]

Reynolds Number

$$\text{Re} = \frac{D\text{v}}{\nu}$$

$$= \frac{(0.3355 \ \text{ft})\left(14.14 \ \frac{\text{ft}}{\text{sec}}\right)}{1.059 \times 10^{-5} \ \frac{\text{ft}^2}{\text{sec}}}$$

$$= 4.480 \times 10^5$$

The relative roughness is

$$\frac{\epsilon}{D} = \frac{0.0002 \ \text{ft}}{0.3355 \ \text{ft}} = 0.0006$$

From a Moody diagram, the friction factor corresponding to this Reynolds number and relative roughness is about 0.018. [Moody Diagram (Stanton Diagram)]

Calculate the total equivalent length of all fittings.

$$(2)(13 \ \text{ft}) + (1)(2.5 \ \text{ft}) + (1)(38 \ \text{ft}) = 66.5 \ \text{ft}$$

The total equivalent length of the pipeline and fittings is 700 ft + 66.5 ft = 766.5 ft. Use the Darcy-Weisbach equation to find the head loss due to friction.

Darcy-Weisbach Equation

$$h_f = f\left(\frac{L}{D}\right)\left(\frac{\text{v}^2}{2g}\right)$$

$$= (0.018)\left(\frac{766.5 \ \text{ft}}{0.3355 \ \text{ft}}\right)\left(\frac{\left(14.14 \ \frac{\text{ft}}{\text{sec}}\right)^2}{(2)\left(32.17 \ \frac{\text{ft}}{\text{sec}^2}\right)}\right)$$

$$= 127.8 \ \text{ft}$$

Use the mechanical energy equation to find the head added by the pump. The mechanical energy equation for a pump in terms of energy per unit mass is

Mechanical Energy Equation in Terms of Energy Per Unit Mass

$$\frac{p_{\text{in}}}{\rho} + \frac{\text{v}_{\text{in}}^2}{2} + gz_{\text{in}} + w_{\text{shaft}} = \frac{p_{\text{out}}}{\rho} + \frac{\text{v}_{\text{out}}^2}{2} + gz_{\text{out}} + w_{\text{loss}}$$

Expressing the equation in units of feet of head, taking the bottom of the tank as point 1 and the end of the discharge pipe as point 2, and rearranging to solve for the head added by the pump gives

$$h_{\text{pump}} = \frac{(p_2 - p_1)g_c}{\rho g} + \frac{\text{v}_2^2 - \text{v}_1^2}{2g} + (z_2 - z_1) + h_f$$

$$= \frac{\left(\left(20 \ \frac{\text{lbf}}{\text{in}^2} - 50 \ \frac{\text{lbf}}{\text{in}^2}\right)\left(12 \ \frac{\text{in}}{\text{ft}}\right)^2\right) \times \left(32.17 \ \frac{\text{lbm-ft}}{\text{lbf-sec}^2}\right)}{\left(62.30 \ \frac{\text{lbm}}{\text{ft}^3}\right)\left(32.17 \ \frac{\text{ft}}{\text{sec}^2}\right)}$$

$$+ \frac{\left(14.14 \ \frac{\text{ft}}{\text{sec}}\right)^2 - \left(0 \ \frac{\text{ft}}{\text{sec}}\right)^2}{(2)\left(32.17 \ \frac{\text{ft}}{\text{sec}^2}\right)}$$

$$+ (50 \ \text{ft}) + 127.8 \ \text{ft}$$

$$= 111.6 \ \text{ft}$$

Use the continuity equation to find the mass flow rate. To use the water horsepower equation, the mass flow rate must be in units of lbm/min.

Continuity Equation

$$\dot{m} = \rho Q$$

$$= \left(62.3 \, \frac{\text{lbm}}{\text{ft}^3}\right)\left(1.25 \, \frac{\text{ft}^3}{\text{sec}}\right)\left(60 \, \frac{\text{sec}}{\text{min}}\right)$$

$$= 4673 \, \text{lbm/min}$$

Use the water horsepower equation to find the fluid power needed.

Pump Power Equation

$$\text{whp} = P_{\text{fluid}} = \frac{\dot{m}\Delta h}{33{,}000}$$

$$= \frac{\left(4673 \, \frac{\text{lbm}}{\text{min}}\right)(111.6 \, \text{ft})}{33{,}000 \, \frac{\text{ft-lbm}}{\text{hp-min}}}$$

$$= 15.80 \, \text{hp} \quad (16 \, \text{hp})$$

The answer is (A).

5. From tables of pipe data, find the areas of nominal 8 in and 6 in schedule-40 steel pipe. [Schedule 40 Steel Pipe]

$$A_{8\text{in}} = \frac{50.002 \, \text{in}^2}{\left(12 \, \frac{\text{in}}{\text{ft}}\right)^2} = 0.3472 \, \text{ft}^2$$

$$A_{6\text{in}} = \frac{28.876 \, \text{in}^2}{\left(12 \, \frac{\text{in}}{\text{ft}}\right)^2} = 0.2005 \, \text{ft}^2$$

From a table of water properties, the density of water at 60°F is 62.37 lbm/ft³. [Properties of Water (I-P Units)]

Let point 1 be the pump inlet, point 2 be the pump outlet, and point 3 be the point of discharge for the 6 in pipe. Use the continuity equation to find the mass flow rate at the pump outlet (point 2).

Continuity Equation

$$\dot{m} = \rho A \text{v}$$

$$= \rho A_{6\text{in}} \text{v}_2$$

$$= \left(62.37 \, \frac{\text{lbm}}{\text{ft}^3}\right)(0.2005 \, \text{ft}^2)\left(12 \, \frac{\text{ft}}{\text{sec}}\right)$$

$$= 150.1 \, \text{lbm/sec}$$

The mass flow rate is constant, so use the continuity equation to find the velocity at the pump inlet (point 1).

$$\dot{m} = \rho A \text{v}$$

$$\text{v}_1 = \frac{\dot{m}}{\rho A_{8\text{in}}}$$

$$= \frac{150.1 \, \frac{\text{lbm}}{\text{sec}}}{\left(62.37 \, \frac{\text{lbm}}{\text{ft}^3}\right)(0.3472 \, \text{ft}^2)}$$

$$= 6.931 \, \text{ft/sec}$$

Use the equation for pump power, and solve for the fluid power.

Centrifugal Pump Characteristics

$$P_{\text{pump}} = \frac{\rho g H Q}{\eta_{\text{pump}}}$$

$$= \frac{P_{\text{fluid}}}{\eta_{\text{pump}}}$$

$$P_{\text{fluid}} = P_{\text{pump}}\eta_{\text{pump}} = (20 \, \text{hp})(0.70)$$

$$= 14 \, \text{hp}$$

Use the equation for water horsepower, and solve for the head added by the pump.

Pump Power Equation

$$\text{whp} = \frac{\dot{m}\Delta h}{33{,}000}$$

$$\Delta h = \frac{33{,}000 P_{\text{fluid}}}{\dot{m}}$$

$$= \frac{\left(33{,}000 \, \frac{\text{ft-lbm}}{\text{hp-min}}\right)(14 \, \text{hp})}{\left(150.1 \, \frac{\text{lbm}}{\text{sec}}\right)\left(60 \, \frac{\text{sec}}{\text{min}}\right)}$$

$$= 51.30 \, \text{ft}$$

The inlet (suction) pressure is

$$p_1 = \left(14.7 \, \frac{\text{lbf}}{\text{in}^2} - 5 \, \frac{\text{lbf}}{\text{in}^2}\right)\left(12 \, \frac{\text{in}}{\text{ft}}\right)^2$$

$$= 1397 \, \text{lbf/ft}^2$$

To find the pressure at the pump outlet (point 2), use the mechanical energy equation. The mechanical energy equation for a pump in terms of energy per unit mass is

Mechanical Energy Equation in Terms of Energy Per Unit Mass

$$\frac{p_{\text{in}}}{\rho} + \frac{\text{v}_{\text{in}}^2}{2} + gz_{\text{in}} + w_{\text{shaft}} = \frac{p_{\text{out}}}{\rho} + \frac{\text{v}_{\text{out}}^2}{2} + gz_{\text{out}} + w_{\text{loss}}$$

The elevations at the pump inlet and outlet are the same, so those terms cancel, and friction loss is zero. Expressing the equation in units of feet of head, simplifying, and rearranging to solve for the pressure at the pump outlet gives

$$p_2 = p_1 - \frac{\rho(v_2^2 - v_1^2)}{2g_c} + \frac{\rho g h_{\text{pump}}}{g_c}$$
$$= 1397 \, \frac{\text{lbf}}{\text{ft}^2}$$

$$- \frac{\left(62.37 \, \frac{\text{lbm}}{\text{ft}^3}\right) \times \left(\left(12 \, \frac{\text{ft}}{\text{sec}}\right)^2 - \left(6.931 \, \frac{\text{ft}}{\text{sec}}\right)^2\right)}{(2)\left(32.17 \, \frac{\text{lbm-ft}}{\text{lbf-sec}^2}\right)}$$

$$+ \frac{\left(62.37 \, \frac{\text{lbm}}{\text{ft}^3}\right)\left(32.17 \, \frac{\text{ft}}{\text{sec}^2}\right) \times (51.30 \, \text{ft})}{32.17 \, \frac{\text{lbm-ft}}{\text{lbf-sec}^2}}$$

$$= 4504 \, \text{lbf/ft}^2$$

The pressure at the pipe discharge (point 3) is atmospheric.

$$p_3 = \left(14.7 \, \frac{\text{lbf}}{\text{in}^2}\right)\left(12 \, \frac{\text{in}}{\text{ft}}\right)^2 = 2117 \, \text{lbf/ft}^2$$

Use the mechanical energy equation again, this time between the pump outlet and the discharge (points 2 and 3). The velocity does not change between points 2 and 3, so these terms cancel. There is no added head, and the head loss is 10 ft. Expressing the equation in units of feet of head, simplifying, and rearranging to solve for the difference in elevation between the pump and the pipe discharge gives

$$z_3 - z_2 = \frac{(p_2 - p_3)g_c}{\rho g} - h_{\text{loss}}$$

$$= \frac{\left(4504 \, \frac{\text{lbf}}{\text{ft}^2} - 2117 \, \frac{\text{lbf}}{\text{ft}^2}\right) \times \left(32.17 \, \frac{\text{lbm-ft}}{\text{lbf-sec}^2}\right)}{\left(62.37 \, \frac{\text{lbm}}{\text{ft}^3}\right)\left(32.17 \, \frac{\text{ft}}{\text{sec}^2}\right)} - 10 \, \text{ft}$$

$$= 28.27 \, \text{ft} \quad (28 \, \text{ft})$$

The answer is (A).

6. *Customary U.S. Solution*

Calculate the absolute pressures before and after the increase. [Measurement Relationships]

$$p_1 = 1 \, \text{atm} = 14.70 \, \text{lbf/in}^2$$
$$p_2 = 40 \, \frac{\text{lbf}}{\text{in}^2} + 14.70 \, \frac{\text{lbf}}{\text{in}^2} = 54.70 \, \text{lbf/in}^2$$

The increase in pressure is

$$\Delta p = p_2 - p_1 = 54.70 \, \frac{\text{lbf}}{\text{in}^2} - 14.70 \, \frac{\text{lbf}}{\text{in}^2}$$
$$= 40 \, \text{lbf/in}^2$$

Use the equation relating fluid power to increase in pressure. [Pump Power Equation]

The equation for ideal pump power in ft-lbf/sec is

$$\dot{W}_{\text{ft-lbf/sec}} = Q\gamma h = Q\Delta P$$

In the preceding equation, Q is the volume flow rate in ft³/sec (cfs) and ΔP is the pressure difference across the pump in lbf / ft².

Re-write the equation for Pump power for US units.

$$\dot{W}_{hp} = \frac{Q\Delta P}{550}$$

Substitute the known values into the pump horsepower equation with appropriate unit conversions.

$$\dot{W}_{hp} = \frac{Q\Delta P}{550} = \frac{(37 \text{ gpm})\left(\dfrac{1 \dfrac{\text{ft}^3}{\text{sec}}}{449 \text{ gpm}}\right) \times \left(40 \dfrac{\text{lbf}}{\text{in}^2}\right)\left(\dfrac{144 \text{ in}^2}{\text{ft}^2}\right)}{550 \dfrac{\text{ft-lbf}}{\text{sec}}{\text{hp}}}$$

$$= 0.8629 \text{ hp} \quad (0.86 \text{ hp})$$

SI Solution

Calculate the absolute pressures before and after the increase. [Measurement Relationships]

$$p_1 = 1 \text{ atm} = 101.3 \text{ kPa}$$
$$p_2 = 275 \text{ kPa} + 101.3 \text{ kPa} = 376.3 \text{ kPa}$$

The increase in pressure is

$$\Delta p = p_2 - p_1 = 376.3 \text{ kPa} - 101.3 \text{ kPa}$$
$$= 275 \text{ kPa}$$

Use the pump power equation. [Pump Power Equation] The equation for ideal pump power in kW is

$$\dot{W}_{kW} = Q\gamma h = Q\Delta P$$

In the preceding equation, Q is the volume flow rate in m³/s and ΔP is the pressure difference across the pump in kPa (kN /m²).

Substitute the known values into the pump power equation in kW.

$$\dot{W}_{kW} = Q\Delta P = \left(65 \dfrac{\text{L}}{\text{s}}\right)\left(\dfrac{1 \text{ m}^3}{1000 \text{ L}}\right)(275 \text{ kPa})$$
$$= 17.87 \text{ kW} \quad (18 \text{ kW})$$

The answer is (B).

7. *Customary U.S. Solution*

From tables of pipe data, find the inside diameter and area of nominal 1.5 in schedule-40 steel pipe. [Schedule 40 Steel Pipe]

$$D = \frac{1.610 \text{ in}}{12 \dfrac{\text{in}}{\text{ft}}} = 0.1342 \text{ ft}$$

$$A = \frac{2.035 \text{ in}^2}{\left(12 \dfrac{\text{in}}{\text{ft}}\right)^2} = 0.01413 \text{ ft}^2$$

The flow rate in cubic feet per second is

$$Q = \frac{100 \dfrac{\text{gal}}{\text{min}}}{\left(7.481 \dfrac{\text{gal}}{\text{ft}^3}\right)\left(60 \dfrac{\text{sec}}{\text{min}}\right)}$$
$$= 0.2228 \text{ ft}^3/\text{sec}$$

Use the continuity equation to find the velocity in the pipe.

Continuity Equation

$$Q = Av$$

$$v = \frac{Q}{A} = \frac{0.2228 \dfrac{\text{ft}^3}{\text{sec}}}{0.01413 \text{ ft}^2} = 15.77 \text{ ft/sec}$$

The total equivalent length of the pipe and all fittings is

$$L = 30 \text{ ft} + 3.1 \text{ ft} + (2)(3.4 \text{ ft}) + (2)(1.2 \text{ ft})$$
$$= 42.3 \text{ ft}$$

The relative roughness is

$$\frac{\epsilon}{D} = \frac{0.0002 \text{ ft}}{0.1342 \text{ ft}} = 0.0015$$

Calculate the Reynolds number.

Reynolds Number

$$\mathrm{Re} = \frac{\mathrm{v}D}{\nu}$$

$$= \frac{\left(15.77\,\frac{\mathrm{ft}}{\mathrm{sec}}\right)(0.1342\,\mathrm{ft})}{0.239 \times 10^{-5}\,\frac{\mathrm{ft}^2}{\mathrm{sec}}}$$

$$= 8.85 \times 10^5$$

From a Moody diagram, the friction factor corresponding to this Reynolds number and relative roughness is approximately $f = 0.0215$.

Use the Darcy-Weisbach equation to find the total friction head loss from the supply reservoir to pump inlet.

Darcy-Weisbach Equation

$$h_f = f\left(\frac{L}{D}\right)\left(\frac{\mathrm{v}^2}{2g}\right)$$

$$= (0.0215)\left(\frac{42.3\,\mathrm{ft}}{0.1342\,\mathrm{ft}}\right)$$

$$\times \left(\frac{\left(15.77\,\frac{\mathrm{ft}}{\mathrm{sec}}\right)^2}{(2)\left(32.17\,\frac{\mathrm{ft}}{\mathrm{sec}^2}\right)}\right)$$

$$= 26.19\,\mathrm{ft}$$

From steam tables, the specific volume of water at 280°F is 0.0173 ft³/lbm. [Properties of Saturated Water and Steam (Temperature) - I-P Units]

The density of 281°F water is

$$\rho = \frac{1}{v_f} = \frac{1}{0.0173\,\frac{\mathrm{ft}^3}{\mathrm{lbm}}} = 57.80\,\mathrm{lbm/ft^3}$$

The net positive suction head available is

Centrifugal Pump Characteristics

Calculate the specific weight of water using the density value [Units].

$$SW = \gamma = \frac{\rho g}{g_c} = \frac{\left(57.80\,\frac{\mathrm{lbm}}{\mathrm{ft}^3}\right)\left(32.2\,\frac{\mathrm{ft}}{\mathrm{sec}^2}\right)}{32.2\,\frac{\mathrm{lbm\text{-}ft}}{\mathrm{lbf\text{-}sec}^2}} = 57.80\,\mathrm{lbf/ft^3}$$

$$NPSH_A = h_p + h_z - h_{vpa} - h_f$$

Calculate the equivalent pressure head at the water surface in the supply reservoir.

$$h_p = \frac{P_{surface}}{\gamma} = \frac{\left(80\,\frac{\mathrm{lbf}}{\mathrm{in}^2}\right)\left(144\,\frac{\mathrm{in}^2}{\mathrm{ft}^2}\right)}{57.80\,\frac{\mathrm{lbf}}{\mathrm{ft}^3}} = 199.31\,\mathrm{ft}$$

The elevation difference between the water surface in the supply reservoir and the center line of the pump is, $h_s = 20\,\mathrm{ft}$

Calculate the equivalent vapor pressure head of water at pumping temperature.

$$h_{vpa} = \frac{P_{vpa}}{\gamma} = \frac{\left(50.02\,\frac{\mathrm{lbf}}{\mathrm{in}^2}\right)\left(144\,\frac{\mathrm{in}^2}{\mathrm{ft}^2}\right)}{57.80\,\frac{\mathrm{lbf}}{\mathrm{ft}^3}} = 124.62\,\mathrm{ft}$$

The calculated value of the total friction head loss from the supply reservoir to the pump inlet is

$$h_f = 26.19\,\mathrm{ft}$$

Substitute the known values into Eqn.1.

$$NPSH_A = h_p + h_z - h_{vpa} - h_f$$
$$= 199.31\,\mathrm{ft} + 20\,\mathrm{ft} - 124.62\,\mathrm{ft} - 26.19\,\mathrm{ft}$$
$$= 68.5\,\mathrm{ft} \quad (68\,\mathrm{ft})$$

Since $NPSH_A$ (68 ft) is greater than the net positive suction head required (10 ft), the pump will not cavitate.

SI Solution

From tables of pipe data, find the inside diameter and area of nominal 1.5 in schedule-40 steel pipe. [Schedule 40 Steel Pipe]

$$D = \frac{(1.610 \text{ in})\left(2.540 \frac{\text{cm}}{\text{in}}\right)}{100 \frac{\text{cm}}{\text{m}}}$$
$$= 0.04089 \text{ m}$$

$$A = \frac{(2.035 \text{ in}^2)\left(2.540 \frac{\text{cm}}{\text{in}}\right)^2}{\left(100 \frac{\text{cm}}{\text{m}}\right)^2}$$
$$= 0.001313 \text{ m}$$

Use the continuity equation to find the velocity in the pipe.

Continuity Equation

$$Q = A\text{v}$$

$$\text{v} = \frac{Q}{A} = \frac{\left(\dfrac{6.3 \frac{\text{L}}{\text{s}}}{1000 \frac{\text{L}}{\text{m}^3}}\right)}{0.001313 \text{ m}^2} = 4.798 \text{ m/s}$$

The total equivalent length of the pipe and all fittings is

$$L = 10 \text{ m} + \frac{3.1 \text{ ft} + (2)(3.4 \text{ ft}) + (2)(1.2 \text{ ft})}{3.281 \frac{\text{ft}}{\text{m}}}$$
$$= 13.75 \text{ m}$$

The relative roughness is

$$\frac{\epsilon}{D} = \frac{6.0 \times 10^{-5} \text{ m}}{0.04089 \text{ m}} = 0.0015$$

Calculate the Reynolds number.

Reynolds Number

$$\text{Re} = \frac{\text{v}D}{\nu}$$
$$= \frac{\left(4.798 \frac{\text{m}}{\text{s}}\right)(0.04089 \text{ m})}{0.222 \times 10^{-6} \frac{\text{m}^2}{\text{s}}}$$
$$= 8.84 \times 10^5$$

From a Moody diagram, the friction factor corresponding to this Reynolds number and relative roughness is approximately $f = 0.0215$.

Use the Darcy-Weisbach equation to find the friction head.

Darcy-Weisbach Equation

$$h_f = f\left(\frac{L}{D}\right)\left(\frac{\text{v}^2}{2g}\right)$$
$$= (0.0215)\left(\frac{13.75 \text{ m}}{0.04089 \text{ m}}\right)$$
$$\times \left(\frac{\left(4.798 \frac{\text{m}}{\text{s}}\right)^2}{(2)\left(9.807 \frac{\text{m}}{\text{s}^2}\right)}\right)$$
$$= 8.49 \text{ m}$$

From steam tables, the specific volume of water at 140°C is 0.0011 m³/kg. [Properties of Saturated Water and Steam (Temperature) - I-P Units]

The density of 140°C water is

$$\rho = \frac{1}{v_f} = \frac{1}{0.0011 \frac{\text{m}^3}{\text{kg}}} = 909.1 \text{ kg/m}^3$$

The net positive suction head available is

Centrifugal Pump Characteristics

Calculate the specific weight of water using the density value [Density, Specific Weight, and Specific Gravity].

$$\gamma = \rho g = \left(909.1 \frac{\text{kg}}{\text{m}^3}\right)\left(9.81 \frac{\text{m}}{\text{sec}^2}\right)\left(\frac{1 \text{ kN}}{1000 \text{ N}}\right)$$
$$= 8.92 \text{ kN/m}^3$$

$$NPSH_A = h_p + h_z - h_{vpa} - h_f$$

Calculate the equivalent pressure head at the water surface in the supply reservoir.

$$h_p = \frac{P_{surface}}{\gamma} = \frac{550 \text{ kPa}}{8.92 \frac{\text{kN}}{\text{m}^3}} = 61.66 \text{ m}$$

The elevation difference between the water surface in the supply reservoir and the center line of the pump is

$$h = 6 \text{ m}$$

Calculate the equivalent vapor pressure head of water at pumping temperature.

$$h_{vpa} = \frac{P_{vpa}}{\gamma} = \frac{(3.431 \text{ bar})\left(\frac{100 \text{ kPa}}{1 \text{ bar}}\right)}{8.92 \frac{\text{kN}}{\text{m}^3}} = 38.46 \text{ m}$$

The calculated value of the total friction head loss from the supply reservoir to the pump inlet is

$$h_f = 8.49 \text{ m}$$

Substitute the known values into Eqn.1.

$$\begin{aligned} NPSH_A &= h_p + h_z - h_{vpa} - h_f \\ &= 61.66 \text{ m} + 6 \text{ m} - 38.46 \text{ m} - 8.49 \text{ ft} \\ &= 20.71 \text{ m} \quad (21 \text{ m}) \end{aligned}$$

As $NPSH_A$ (21 m) is greater than the net positive suction head required (3 m), the pump will not cavitate.

The answer is (D).

8. *Customary U.S. Solution*

The solvent is the freshwater, and the solution is the seawater. The seawater contains 2.5% salt by weight, so 100 lbm of seawater contains 2.5 lbm of salt and 97.5 lbm of water. The molecular weight of salt (NaCl) is 22.990 lbm/lbmol + 35.453 lbm/lbmol = 58.44 lbm/lbmol. [Periodic Table of the Elements]

The number of moles of salt in 100 lbm of seawater is

$$n_{\text{salt}} = \frac{m}{M} = \frac{2.5 \text{ lbm}}{58.44 \frac{\text{lbm}}{\text{lbmol}}} = 0.04278 \text{ lbmol}$$

The molecular weight of water is 18 lbm/lbmol. [Thermal and Physical Properties of Ideal Gases (at Room Temperature)]

The number of moles of water is

$$n_{\text{water}} = \frac{97.5 \text{ lbm}}{18 \frac{\text{lbm}}{\text{lbmol}}} = 5.417 \text{ lbmol}$$

The mole fraction of water is

Ideal Gas Mixtures

$$x_i = \frac{n_i}{\sum n_i}$$

$$\begin{aligned} x_{\text{water}} &= \frac{n_{\text{water}}}{n_{\text{water}} + n_{\text{salt}}} \\ &= \frac{5.417 \text{ lbmol}}{5.417 \text{ lbmol} + 0.04278 \text{ lbmol}} \\ &= 0.9922 \end{aligned}$$

Cavitation will occur when the net available head is less than the net required head.

$$h_{\text{atm}} + h_{\text{depth}} - h_v < h_{\text{vapor}}$$

The density of seawater is given as 64.0 lbm/ft³. The atmospheric head is

$$\begin{aligned} h_{\text{atm}} &= \frac{pg_c}{\rho g} \\ &= \frac{\left(14.7 \frac{\text{lbf}}{\text{in}^2}\right)\left(12 \frac{\text{in}}{\text{ft}}\right)^2 \left(32.17 \frac{\text{lbm-ft}}{\text{lbf-sec}^2}\right)}{\left(64.0 \frac{\text{lbm}}{\text{ft}^3}\right)\left(32.17 \frac{\text{ft}}{\text{sec}^2}\right)} \\ &= 33.08 \text{ ft} \quad [\text{of seawater}] \end{aligned}$$

The velocity head is

$$h_v = \frac{v_{\text{propeller}}^2}{2g} = \frac{(4.2v_{\text{boat}})^2}{(2)\left(32.17 \frac{\text{ft}}{\text{sec}^2}\right)}$$

$$= (0.2739 \text{ sec}^2/\text{ft})v_{\text{boat}}^2$$

From a table of water properties, the vapor pressure of freshwater at 70°F is 0.36 psi, and the water's density is 62.30 lbm/ft³. [Properties of Water (I-P Units)]

Henry's law predicts the actual vapor pressure of the solution.

Henry's Law at Constant Temperature

$$p_i = py_i$$

$$\begin{aligned} p_{\text{vapor,seawater}} &= p_{\text{vapor,freshwater}} x_{\text{water}} \\ &= \left(0.36 \frac{\text{lbf}}{\text{in}^2}\right)(0.992) \\ &= 0.3571 \text{ lbf/in}^2 \end{aligned}$$

The vapor pressure head is

$$h_{vapor} = \frac{p_{vapor,seawater} g_c}{\rho g}$$

$$= \frac{\left(0.3571 \frac{lbf}{in^2}\right)\left(12 \frac{in}{ft}\right)^2 \times \left(32.17 \frac{lbm\text{-}ft}{lbf\text{-}sec^2}\right)}{\left(62.30 \frac{lbm}{ft^3}\right)\left(32.17 \frac{ft}{sec^2}\right)}$$

$$= 0.8254 \text{ ft}$$

Use the equation for cavitation and solve for the maximum boat velocity.

$$h_{vapor} = h_{atm} + h_{depth} - h_v$$

$$= h_{atm} + h_{depth} - \left(0.2739 \frac{sec^2}{ft}\right) v_{boat}^2$$

$$v_{boat} = \sqrt{\frac{h_{atm} + h_{depth} - h_{vapor}}{0.2739 \frac{sec^2}{ft}}}$$

$$= \sqrt{\frac{33.08 \text{ ft} + 8 \text{ ft} - 0.8254 \text{ ft}}{0.2739 \frac{sec^2}{ft}}}$$

$$= 12.12 \text{ ft/sec} \quad (12 \text{ ft/sec})$$

SI Solution

The solvent is the freshwater, and the solution is the seawater. The seawater contains 2.5% salt by weight, so 100 kg of seawater contains 2.5 kg of salt and 97.5 kg of water. The molecular weight of salt (NaCl) is 22.990 kg/kmol + 35.453 kg/kmol = 58.44 kg/kmol. [Periodic Table of the Elements]

The number of moles of salt in 100 kg of seawater is

$$n_{salt} = \frac{m}{M} = \frac{2.5 \text{ kg}}{58.44 \frac{kg}{kmol}}$$

$$= 0.04278 \text{ kmol}$$

The molecular weight of water is 18 kg/kmol. [Thermal and Physical Properties of Ideal Gases (at Room Temperature)]

The number of moles of water is

$$n_{water} = \frac{97.5 \text{ kg}}{18 \frac{kg}{kmol}} = 5.417 \text{ kmol}$$

The mole fraction of water is

Ideal Gas Mixtures

$$x_i = \frac{n_i}{\sum n_i}$$

$$x_{water} = \frac{n_{water}}{n_{water} + n_{salt}}$$

$$= \frac{5.417 \text{ kmol}}{5.417 \text{ kmol} + 0.04278 \text{ kmol}}$$

$$= 0.9922$$

Cavitation will occur when the net available head is less than the net required head.

$$h_{atm} + h_{depth} - h_v < h_{vapor}$$

The density of seawater is given as 1024 kg/m³. The atmospheric head is

$$h_{atm} = \frac{p}{\rho g}$$

$$= \frac{(101.3 \text{ kPa})\left(1000 \frac{Pa}{kPa}\right)}{\left(1024 \frac{kg}{m^3}\right)\left(9.807 \frac{m}{s^2}\right)}$$

$$= 10.09 \text{ m} \quad [\text{of seawater}]$$

The velocity head is

$$h_v = \frac{v_{propeller}^2}{2g} = \frac{(4.2 v_{boat})^2}{(2)\left(9.807 \frac{m}{s^2}\right)}$$

$$= (0.8994 \text{ s}^2/\text{m}) v_{boat}^2$$

From a table of water properties, the vapor pressure of freshwater at 20°C is 2.34 kPa, and the water's density is 998.2 kg/m³. [Properties of Water (SI Units)]

Henry's law predicts the actual vapor pressure of the solution.

Henry's Law at Constant Temperature

$$p_i = p y_i$$

$$p_{vapor,seawater} = p_{vapor,freshwater} x_{water}$$

$$= (2.34 \text{ kPa})(0.9922)$$

$$= 2.321 \text{ kPa}$$

The vapor pressure head is

$$h_{\text{vapor}} = \frac{p_{\text{vapor,seawater}}}{\rho g}$$

$$= \frac{(2.321 \text{ kPa})\left(1000 \frac{\text{Pa}}{\text{kPa}}\right)}{\left(1024 \frac{\text{kg}}{\text{m}^3}\right)\left(9.807 \frac{\text{m}}{\text{s}^2}\right)}$$

$$= 0.2311 \text{ m}$$

Use the equation for cavitation and solve for the maximum boat velocity.

$$h_{\text{vapor}} = h_{\text{atm}} + h_{\text{depth}} - h_{\text{v}}$$

$$= h_{\text{atm}} + h_{\text{depth}} - \left(0.8994 \frac{\text{s}^2}{\text{m}}\right) \text{v}_{\text{boat}}^2$$

$$\text{v}_{\text{boat}} = \sqrt{\frac{h_{\text{atm}} + h_{\text{depth}} - h_{\text{vapor}}}{0.8994 \frac{\text{s}^2}{\text{m}}}}$$

$$= \sqrt{\frac{10.09 \text{ m} + 3 \text{ m} - 0.2311 \text{ m}}{0.8994 \frac{\text{s}^2}{\text{m}}}}$$

$$= 3.781 \text{ m/s} \quad (3.8 \text{ m/s})$$

The answer is (B).

9. Use the pump affinity law relating power and impeller diameter.

Pump Affinity Laws

$$\text{bhp}_2 = \text{bhp}_1 \left(\frac{N_2}{N_1}\right)^3$$

$$= (0.5 \text{ hp}) \left(\frac{2000 \frac{\text{rev}}{\text{min}}}{1750 \frac{\text{rev}}{\text{min}}}\right)^3$$

$$= 0.7464 \text{ hp} \quad (0.75 \text{ hp})$$

SI Solution

Use the pump affinity law relating power and impeller diameter.

Pump Affinity Laws

$$\text{bhp}_2 = \text{bhp}_1 \left(\frac{N_2}{N_1}\right)^3$$

$$= (0.37 \text{ kW}) \left(\frac{2000 \frac{\text{rev}}{\text{min}}}{1750 \frac{\text{rev}}{\text{min}}}\right)^3$$

$$= 0.5523 \text{ kW} \quad (0.55 \text{ kW})$$

The answer is (D).

10. Because a turbine is essentially a pump running backward, the pump power equations can be used. Convert the flow rate to gallons per minute.

$$Q = \left(100 \frac{\text{ft}^3}{\text{sec}}\right)\left(7.481 \frac{\text{gal}}{\text{ft}^3}\right)\left(60 \frac{\text{sec}}{\text{min}}\right)$$

$$= 44{,}886 \text{ gal/min}$$

Use the equation for the water horsepower of a pump.

Pump Power Equation

$$\text{whp} = \frac{Q \Delta p}{1714}$$

$$= \frac{\left(44{,}886 \frac{\text{gal}}{\text{min}}\right)\left(\left(30 \frac{\text{lbf}}{\text{in}^2} - 5 \frac{\text{lbf}}{\text{in}^2}\right)\right)}{1714 \frac{\text{gal-lbf}}{\text{hp-in}^2\text{-min}}}$$

$$= 654.7 \text{ hp} \quad (650 \text{ hp})$$

The answer is (C).

11. Use the continuity equation to find the mass flow rate.

Continuity Equation

$$\dot{m} = \rho Q$$

$$= \left(62.4 \frac{\text{lbm}}{\text{ft}^3}\right)\left(1000 \frac{\text{ft}^3}{\text{sec}}\right)$$

$$= 62{,}400 \text{ lbm/sec}$$

The head available for work is

$$\Delta h = 625 \text{ ft} - 58 \text{ ft} = 567 \text{ ft}$$

Use the equation for calculating the water horsepower from mass flow rate in lbm/min.

Pump Power Equation

$$\text{whp} = \frac{\dot{m}\Delta h}{33{,}000}$$

$$= \frac{\left(62{,}400\ \dfrac{\text{lbm}}{\text{sec}}\right)\left(60\ \dfrac{\text{sec}}{\text{min}}\right)(567\ \text{ft})}{33{,}000\ \dfrac{\text{ft-lbm}}{\text{hp-min}}}$$

$$= 64{,}329\ \text{hp}$$

The turbine efficiency is 89%, so the actual power developed by the turbines is

$$P_{\text{turbines}} = (\text{whp})\eta_{\text{turbine}}$$
$$= (64{,}329\ \text{hp})(0.89)$$
$$= 57{,}253\ \text{hp}$$

Convert from horsepower to kilowatts. [Measurement Relationships]

$$P_{\text{turbines}} = \frac{57{,}253\ \text{hp}}{1.341\ \dfrac{\text{hp}}{\text{kW}}}$$
$$= 42{,}694\ \text{kW} \quad (43\ \text{MW})$$

The answer is (C).

12. *Customary U.S. Solution*

The density of water at 60°F is 62.4 lbm/ft³. [Properties of Water at Standard Conditions]

Use the equation for fluid pressure, and solve for the head extracted by the turbine.

Units

$$p = \frac{\rho g h}{g_c}$$

$$h_{\text{turbine}} = \frac{\Delta p\, g_c}{\rho g}$$

$$= \frac{\left(500\ \dfrac{\text{lbf}}{\text{in}^2} - 30\ \dfrac{\text{lbf}}{\text{in}^2}\right)\left(12\ \dfrac{\text{in}}{\text{ft}}\right)^2 \times \left(32.17\ \dfrac{\text{lbm-ft}}{\text{lbf-sec}^2}\right)}{\left(62.4\ \dfrac{\text{lbm}}{\text{ft}^3}\right)\left(32.17\ \dfrac{\text{ft}}{\text{sec}^2}\right)}$$

$$= 1085\ \text{ft}$$

The specific speed of the turbine is

$$N_{\text{specific}} = \frac{N_{\text{turbine}}\sqrt{P}}{h_{\text{turbine}}^{1.25}}$$

$$= \frac{\left(1750\ \dfrac{\text{rev}}{\text{min}}\right)\sqrt{250\ \text{hp}}}{(1085\ \text{ft})^{1.25}}$$

$$= 4.443 \quad (4.4)$$

SI Solution

From a table of water properties, the density of water at 15°C is 999.1 kg/m³. [Properties of Water (SI Units)]

Use the equation for fluid pressure, and solve for the head extracted by the turbine.

Units

$$p = \rho g h$$

$$h_{\text{turbine}} = \frac{\Delta p}{\rho g}$$

$$= \frac{(3.5\ \text{MPa})\left(10^6\ \dfrac{\text{Pa}}{\text{MPa}}\right) - (210\ \text{kPa})\left(1000\ \dfrac{\text{Pa}}{\text{kPa}}\right)}{\left(999.1\ \dfrac{\text{kg}}{\text{m}^3}\right)\left(9.807\ \dfrac{\text{m}}{\text{s}^2}\right)}$$

$$= 335.8\ \text{m}$$

The specific speed of the turbine is

$$N_{\text{specific}} = \frac{N_{\text{turbine}}\sqrt{P}}{h_{\text{turbine}}^{1.25}}$$

$$= \frac{\left(1750\ \dfrac{\text{rev}}{\text{min}}\right)\sqrt{185\ \text{kW}}}{(335.8\ \text{m})^{1.25}}$$

$$= 16.56 \quad (17)$$

The answer is (A).

13. *Customary U.S. Solution*

From a table of water properties, the density of water at 60°F is 62.37 lbm/ft³. [Properties of Water (I-P Units)]

The cross-sectional area of the nozzle is

$$A = \frac{\pi D^2}{4} = \frac{\pi\left(\dfrac{4\ \text{in}}{12\ \dfrac{\text{in}}{\text{ft}}}\right)^2}{4} = 0.08727\ \text{ft}^2$$

The analysis is for a single blade, not the entire turbine, so only a portion of the water will catch up with the blade. Calculate the effective velocity.

$$v_{\text{eff}} = v_{\text{jet}} - v_{\text{blade}} = 35 \; \frac{\text{ft}}{\text{sec}} - 10 \; \frac{\text{ft}}{\text{sec}}$$
$$= 25 \; \text{ft/sec}$$

Use the continuity equation to find the volumetric flow rate.

Continuity Equation

$$Q = Av$$
$$= A v_{\text{eff}}$$
$$= (0.08727 \; \text{ft}^2)\left(25 \; \frac{\text{ft}}{\text{sec}}\right)$$
$$= 2.182 \; \text{ft}^3/\text{sec}$$

Find the forces on a moving blade in the x- and y-directions.

Moving Blade

$$-F_x = \frac{Q\rho(v_{\text{jet}} - v_{\text{blade}})(1 - \cos\alpha)}{g_c}$$
$$= \frac{\left(2.182 \; \frac{\text{ft}^3}{\text{sec}}\right)\left(62.37 \; \frac{\text{lbm}}{\text{ft}^3}\right) \times \left(35 \; \frac{\text{ft}}{\text{sec}} - 10 \; \frac{\text{ft}}{\text{sec}}\right)(1 - \cos 80°)}{32.17 \; \frac{\text{lbm-ft}}{\text{lbf-sec}^2}}$$
$$= 87.39 \; \text{lbf}$$
$$F_x = -87.39 \; \text{lbf}$$

$$F_y = \frac{Q\rho(v_{\text{jet}} - v_{\text{blade}})\sin\alpha}{g_c}$$
$$= \frac{\left(2.182 \; \frac{\text{ft}^3}{\text{sec}}\right)\left(62.37 \; \frac{\text{lbm}}{\text{ft}^3}\right) \times \left(35 \; \frac{\text{ft}}{\text{sec}} - 10 \; \frac{\text{ft}}{\text{sec}}\right)\sin 80°}{32.17 \; \frac{\text{lbm-ft}}{\text{lbf-sec}^2}}$$
$$= 104.2 \; \text{lbf}$$

Find the total force acting on the blade.

Impulse-Momentum Principle

$$F = \sqrt{F_x^2 + F_y^2}$$
$$= \sqrt{(-87.39 \; \text{lbf})^2 + (104.2 \; \text{lbf})^2}$$
$$= 136.0 \; \text{lbf} \quad (140 \; \text{lbf})$$

SI Solution

The cross-sectional area of the nozzle is

$$A = \frac{\pi D^2}{4} = \frac{\pi \left(\dfrac{100 \; \text{mm}}{1000 \; \frac{\text{mm}}{\text{m}}}\right)^2}{4} = 0.007854 \; \text{m}^2$$

The analysis is for a single blade, not the entire turbine, so only a portion of the water will catch up with the blade. Calculate the effective velocity.

$$v_{\text{eff}} = v_{\text{jet}} - v_{\text{blade}} = 10.5 \; \frac{\text{m}}{\text{s}} - 3 \; \frac{\text{m}}{\text{s}}$$
$$= 7.5 \; \text{m/s}$$

Use the continuity equation to find the volumetric flow rate.

Continuity Equation

$$Q = Av$$
$$= A v_{\text{eff}}$$
$$= (0.007854 \; \text{m}^2)\left(7.5 \; \frac{\text{m}}{\text{s}}\right)$$
$$= 0.05891 \; \text{m}^3/\text{s}$$

Find the forces on a moving blade in the x- and y-directions. From a table of water properties, the density of water at 15°C is 999.1 kg/m³. [Properties of Water (SI Units)]

Moving Blade

$$-F_x = Q\rho(v_{\text{jet}} - v_{\text{blade}})(1 - \cos\alpha)$$
$$= \left(0.05891 \; \frac{\text{m}^3}{\text{s}}\right)\left(999.1 \; \frac{\text{kg}}{\text{m}^3}\right) \times \left(10.5 \; \frac{\text{m}}{\text{s}} - 3 \; \frac{\text{m}}{\text{s}}\right)(1 - \cos 80°)$$
$$= 364.8 \; \text{N}$$
$$F_x = -364.8 \; \text{N}$$

$$F_y = Q\rho(\text{v}_{\text{jet}} - \text{v}_{\text{blade}})\sin\alpha$$
$$= \left(0.05891\ \frac{\text{m}^3}{\text{s}}\right)\left(999.1\ \frac{\text{kg}}{\text{m}^3}\right)$$
$$\times \left(10.5\ \frac{\text{m}}{\text{s}} - 3\ \frac{\text{m}}{\text{s}}\right)\sin 80°$$
$$= 434.7\ \text{N}$$

Find the total force acting on the blade.

Impulse-Momentum Principle

$$F = \sqrt{F_x^2 + F_y^2}$$
$$= \sqrt{(-364.8\ \text{N})^2 + (434.7\ \text{N})^2}$$
$$= 567.5\ \text{N}\quad (570\ \text{N})$$

The answer is (B).

19 Hydraulic and Pneumatic Systems

Content in blue refers to the NCEES Handbook.

PRACTICE PROBLEMS

1. A hydraulic cylinder consists of a circular tube, piston, rod, and gland assembly; these are retained by a metal retaining ring that fits into a groove machined around the inside diameter of the cylinder tube. The cylinder must be disassembled for maintenance, but after removing the retaining ring, the gland assembly does not slide out easily. Which of these methods should be used to remove the stubborn gland assembly?

(A) Pressurize the cylinder with inert gas from a tank, allowing the piston to push the gland assembly out.

(B) Pressurize the cylinder with compressed air from an air compressor, allowing the piston to push the gland assembly out.

(C) Pressurize the cylinder with hydraulic fluid, allowing the piston to push the gland assembly out.

(D) Use a rosebud end with an acetylene torch to thermally expand the gland-end of the tube.

2. A hydraulic motor displaces 10 cm³ of hydraulic fluid per radian of rotation. Fluid pressure is increased by 8 MPa. The torque developed is most nearly

(A) 28 N·m
(B) 80 N·m
(C) 640 N·m
(D) 5120 N·m

3. A hydraulic fluid at 25°C has a bulk modulus of 1.72 GPa and a density of 870 kg/m³. The speed of sound in the fluid is most nearly

(A) 790 m/s
(B) 950 m/s
(C) 1100 m/s
(D) 1400 m/s

4. A water tower feeds an irrigation system. To prevent soil erosion, a gate valve with a K-value of 0.19 is used to limit the water discharge to 60 gpm. When fully open, the maximum pressure drop across the valve is limited to 5 lbf/in². The equation for flow through the value is

$$Q_{gpm} = \frac{29.9 D_{in}^2}{\sqrt{K}} \sqrt{\frac{\Delta p_{psi}}{SG}}$$

The minimum size gate valve that can operate fully open is most nearly

(A) 0.50 in
(B) 0.75 in
(C) 1.0 in
(D) 1.3 in

5. Water with a flow coefficient of 300 and rangeability of 9.5:1 flows through a 5 in globe control valve. The full flow (wide-open valve) differential pressure across the valve is 5 psig. The nearest uncontrollable flow rate of water is most nearly

(A) 71 gpm
(B) 86 gpm
(C) 100 gpm
(D) 120 gpm

6. Water flows through a 3 in globe valve that operates at 102 psia and 80°F. The outlet pressure at which the flow can be expected to become choked and result in cavitation inside the valve is most nearly?

(A) 11 psi
(B) 21 psi
(C) 48 psi
(D) 51 psi

SOLUTIONS

1. Filling the cylinder with hydraulic fluid and using a small hand pump to gradually increase the pressure (up to the maximum pressure rating of the cylinder, if necessary) is a common practice. Hydraulic fluid is incompressible, so a small movement of the gland assembly will eliminate the pressure in the cylinder. Use of compressed gas can result in explosive projectile motion of the gland, piston, and rod. Combining flame with flammable fluid and materials creates a fire and/or explosion hazard.

The answer is (C).

2. The torque is the product of the pressure and the displacement per radian during one complete rotation.

$$T = \Delta p \left(\frac{\text{displacement}}{2\pi} \right)(1 \text{ rotation})$$

$$= \left((8 \text{ MPa}) \left(10^6 \frac{\text{Pa}}{\text{MPa}} \right) \right)$$

$$\times \left(\frac{\left(10 \frac{\text{cm}^3}{\text{rad}} \right)}{2\pi \left(100 \frac{\text{cm}}{\text{m}} \right)^3} \right) (2\pi)$$

$$= 80 \text{ N·m}$$

The answer is (B).

3. The speed of sound is

Mach Number

$$c = \sqrt{\frac{B}{\rho}} = \sqrt{\frac{(1.72 \text{ GPa})\left(10^9 \frac{\text{Pa}}{\text{GPa}}\right)}{870 \frac{\text{kg}}{\text{m}^3}}}$$

$$= 1406 \text{ m/s} \quad (1400 \text{ m/s})$$

The answer is (D).

4. Rearranging the flow rate equation and solving for the diameter yields

$$D_{\text{in}} = \sqrt{\frac{Q_{\text{gpm}}\sqrt{K}}{29.9\sqrt{\frac{\Delta p}{\text{SG}}}}} = \sqrt{\frac{\left(60 \frac{\text{gal}}{\text{min}}\right)\sqrt{0.19}}{29.9\sqrt{\frac{5 \frac{\text{lbf}}{\text{in}^2}}{1}}}}$$

$$= 0.625 \text{ in}$$

The valve size must be at least 0.625 in. Of the options, 0.75 in is the minimum size gate valve that can operate fully open.

The answer is (B).

5. The valve flow coefficient formula is

Valve Flow Coefficient

$$C_v = \frac{Q}{\sqrt{\Delta p}}$$

The rangeability formula is [Valve Rangeability]

$$\text{rangeability} = \frac{Q_{\text{max}}}{Q_{\text{min}}}$$

Using both equations, calculate the minimum flow rate.

$$C_v = \frac{Q}{\sqrt{\Delta p}}$$

$$Q = C_v \sqrt{\Delta p}$$

$$Q_{\text{min}} = \frac{Q_{\text{max}}}{\text{rangeability}}$$

$$Q_{\text{min}} = \frac{C_v \sqrt{\Delta p}}{\text{rangeability}}$$

$$= \frac{300 \sqrt{5 \frac{\text{lbf}}{\text{in}^2}}}{9.5}$$

$$= 70.61 \text{ gpm} \quad (71 \text{ gpm})$$

The answer is (A).

6. From a table of water properties, the vapor pressure of water at 80°F is 0.51 psi. [Properties of Water (I-P Units)]

The valve recovery coefficient for a 3 in valve is 0.5. [Valve Cavitation]

Calculate the maximum allowable pressure drop to avoid cavitation.

Valve Cavitation

$$\Delta p_{\text{allowable}} = K_M(p_i - p_v)$$

$$= (0.5)\left(102 \frac{\text{lbf}}{\text{in}^2} - 0.51 \frac{\text{lbf}}{\text{in}^2} \right)$$

$$= 50.7 \text{ lbf/in}^2$$

Calculate the minimum outlet pressure to avoid cavitation.

$$\Delta p_{\text{allowable}} = p_i - p_o$$
$$\begin{aligned} p_o &= p_i - \Delta p_{\text{allowable}} \\ &= 102\ \frac{\text{lbf}}{\text{in}^2} - 50.7\ \frac{\text{lbf}}{\text{in}^2} \\ &= 51.3\ \text{lbf/in}^2 \quad (51\ \text{psi}) \end{aligned}$$

The answer is (D).

Inorganic Chemistry

Content in blue refers to the NCEES Handbook.

PRACTICE PROBLEMS

1. The gravimetric analysis of a compound is 40% carbon, 6.7% hydrogen, and 53.3% oxygen. The simplest formula for the compound is most nearly

(A) HCO
(B) HCO_2
(C) CH_2O
(D) CHO_2

2. A sacrificial galvanic protection system is proposed for a water storage tank made of low-carbon steel. Which of the following materials is most suitable as the sacrificial anode?

(A) brass
(B) nickel
(C) stainless steel
(D) zinc

3. The ionic concentrations of a municipal water supply are shown.

formula	concentration	$CaCO_3$ factor
Al^{+3}	0.5 mg/L	5.56
Ca^{+2}	80.2 mg/L	2.50
Cl^-	85.9 mg/L	2.66
CO_2	19 mg/L	2.27
CO_3^{-2}	(none)	N/A
Fe^{+2}	1.0 mg/L	1.79
Fl^-	(none)	N/A
HCO_3^-	185 mg/L	0.82
Mg^{+2}	24.3 mg/L	4.10
NO_3^-	(none)	N/A
SO_4^{-2}	125 mg/L	1.04

The total hardness is most nearly

(A) 160 mg/L as $CaCO_3$
(B) 200 mg/L as $CaCO_3$
(C) 260 mg/L as $CaCO_3$
(D) 300 mg/L as $CaCO_3$

4. A municipal water supply has the following ionic concentrations.

formula	concentration	$CaCO_3$ factor
Al^{+3}	0.5 mg/L	5.56
Ca^{+2}	80.2 mg/L	2.50
Cl^-	85.9 mg/L	2.66
CO_2	19 mg/L	2.27
Fe^{+2}	1.0 mg/L	2.66
HCO_3^-	185 mg/L	0.82
Mg^{+2}	24.3 mg/L	4.10

The $CaCO_3$ factor for $Ca(OH)_2$ is 1.35. Approximately how much slaked lime is required to combine with the carbonate hardness?

(A) 45 mg/L as substance
(B) 90 mg/L as substance
(C) 130 mg/L as substance
(D) 150 mg/L as substance

5. A municipal water supply contains the following ionic concentrations.

Ca(HCO$_3$)$_2$	137 mg/L as CaCO$_3$
MgSO$_4$	72 mg/L as CaCO$_3$
CO$_2$	(none)

The CaCO$_3$ factor for Ca(OH)$_2$ is 1.35. Approximately how much slaked lime is required to soften 1,000,000 gal of this water to a hardness of 100 mg/L if 30 mg/L of excess lime is used?

(A) 930 lbm
(B) 1200 lbm
(C) 1300 lbm
(D) 1700 lbm

SOLUTIONS

1. Calculate the relative mole ratios of the atoms by assuming there are 100 g of sample.

For 100 g of sample,

substance	mass	$n = \dfrac{m}{M}$	relative mole ratio
C	40 g	$\dfrac{40}{12} = 3.33$	1
H	6.7 g	$\dfrac{6.7}{1} = 6.7$	2
O	53.3 g	$\dfrac{53.3}{16} = 3.33$	1

The empirical formula is CH$_2$O.

The answer is (C).

2. From a galvanic series table of commercial metals and alloys, brass, nickel, and stainless steel have lower potentials than low-carbon steel; therefore, they will be cathodic, not anodic. Zinc has a higher potential than low-carbon steel, so it can be used as the sacrificial anode. [Galvanic Series of Some Commercial Metals and Alloys in Seawater]

The answer is (D).

3. Hardness is the sum of the concentrations of all doubly and triply charged positive ions, expressed as CaCO$_3$. Find the total hardness by multiplying the mg/L as substance by the CaCO$_3$ factor.

	mg/L as substance		factor CaCO$_3$		
Ca^{+2}:	80.2	×	2.50	=	200.5 mg/L
Mg^{+2}:	24.3	×	4.10	=	99.63 mg/L
Fe^{+2}:	1.0	×	1.79	=	1.79 mg/L
Al^{+3}:	0.5	×	5.56	=	2.78 mg/L
					304.7 mg/L
			hardness	=	(300 mg/L) as CaCO$_3$

The answer is (D).

4. Add lime to remove the carbonate hardness. It does not matter whether the HCO_3^- comes from Mg^{+2}, Ca^{+2}, or Fe^{+2}; adding lime will remove it.

There may be Mg^{+2}, Ca^{+2}, or Fe^{+2} ions left over in the form of noncarbonate hardness, but the problem asked for carbonate hardness. Converting from mg/L of substance to mg/L as $CaCO_3$,

$$CO_2: \quad \left(19 \ \frac{mg}{L}\right)(2.27) = 43.13 \ mg/L \ as \ CaCO_3$$

$$HCO_3^-: \quad \left(185 \ \frac{mg}{L}\right)(0.82) = 151.7 \ mg/L \ as \ CaCO_3$$

The total equivalents to be neutralized are

$$43.13 \ \frac{mg}{L} + 151.7 \ \frac{mg}{L} = 194.83 \ mg/L \ as \ CaCO_3$$

Convert $Ca(OH)_2$ to substance using the $CaCO_3$ factor.

$$\frac{mg}{L} \ of \ Ca(OH)_2 = \frac{194.83 \ \frac{mg}{L}}{1.35}$$
$$= 144.3 \ mg/L \ as \ substance$$
$$(150 \ mg/L \ as \ substance)$$

The answer is (D).

5. $Ca(HCO_3)_2$ and $MgSO_4$ both contribute to hardness. Since 100 mg/L of hardness is the goal, leave all $MgSO_4$ in the water. Take out 137 mg/L + 72 mg/L − 100 mg/L = 109 mg/L of $Ca(HCO_3)_2$. From the $CaCO_3$ (including the excess even though the reaction is not complete),

$$pure \ Ca(OH)_2 = 30 \ \frac{mg}{L} + \frac{109 \ \frac{mg}{L}}{1.35}$$
$$= 110.74 \ mg/L$$

$$\left(110.74 \ \frac{mg}{L}\right)\left(8.345 \ \frac{lbm\text{-}L}{mg\text{-}MG}\right) = 924 \ lbm/MG$$
$$(930 \ lbm/MG)$$

The answer is (A).

21 Fuels and Combustion

Content in blue refers to the *NCEES Handbook*.

PRACTICE PROBLEMS

1. 7 ft³ (200 L) of methane is at 60°F (30°C) and atmospheric pressure. The methane, which has a heating value of 24,000 Btu/lbm (56 MJ/kg), is burned with 50% efficiency to heat water from 60°F to 200°F (15°C to 95°C). The mass of the water that is heated is most nearly

(A) 25 lbm (11 kg)
(B) 35 lbm (16 kg)
(C) 50 lbm (23 kg)
(D) 95 lbm (43 kg)

2. 15 lbm/hr (6.8 kg/h) of propane is burned stoichiometrically in air. The products are cooled to 70°F (21°C) at atmospheric pressure. The volumetric rate of dry carbon dioxide formed is most nearly

(A) 180 ft³/hr (5.0 m³/hr)
(B) 270 ft³/hr (7.6 m³/hr)
(C) 390 ft³/hr (11 m³/hr)
(D) 450 ft³/hr (13 m³/hr)

3. In a particular installation, 30% excess air at 15 psia (103 kPa) and 100°F (40°C) is needed for the combustion of methane. The methane is burned at the rate of 4000 ft³/hr (31 L/s). The mass rate of nitrogen passing through the furnace is most nearly

(A) 270 lbm/hr (0.033 kg/s)
(B) 930 lbm/hr (0.11 kg/s)
(C) 1800 lbm/hr (0.22 kg/s)
(D) 2700 lbm/hr (0.34 kg/s)

4. Propane (C_3H_8) is burned with 20% excess air. The percentage by mass of carbon dioxide in the flue gas is most nearly

(A) 7.8%
(B) 11%
(C) 15%
(D) 22%

5. The ultimate analysis of a coal is 75% carbon, 5% hydrogen, 4% oxygen, 2% nitrogen, and the rest ash. The oxygen in coal is associated with moisture. Atmospheric air is 60°F (16°C) and at standard pressure. Neglect dissociation. 40% excess air is used to combust the coal, and 75% of the heat is transferred to the boiler. The lower heating value of the coal is most nearly

(A) 12,600 Btu/lbm
(B) 12,900 Btu/lbm
(C) 13,200 Btu/lbm
(D) 13,500 Btu/lbm

6. A coal is 65% carbon by weight. During combustion, 3% of the coal is lost in the ash pit. Combustion uses 9.87 lbm of air per pound of fuel. The flue gas analysis (dry basis) is 81.5% nitrogen, 9.5% carbon dioxide, and 9% oxygen. The percentage of excess air is most nearly

(A) 10%
(B) 30%
(C) 70%
(D) 140%

7. A coal has an ultimate analysis of 67.34% carbon, 4.91% oxygen, 4.43% hydrogen, 4.28% sulfur, 1.08% nitrogen, and the rest ash. 3% of the carbon is lost during combustion. The flue gases are 81.9% nitrogen, 15.5% carbon dioxide, 1.6% carbon monoxide, and 1% oxygen by volume. The heat loss due to the formation of carbon monoxide is most nearly

(A) 620 Btu/lbm

(B) 780 Btu/lbm

(C) 970 Btu/lbm

(D) 1200 Btu/lbm

8. An electrical power-generating plant burns refuse-derived fuel (RDF). After sorting, incoming refuse is shredded and compressed before being fed into the combustor. The raw refuse averages 7% incombustible solids by weight. 5000 lbm/hr of processed RDF produces 20,070 lbm/hr of saturated steam at 200 lbf/in². The combustion products are used to heat incoming feedwater to a saturated temperature of 160°F before entering the combustor. 2000 lbm/hr of water vapor condenses in the feedwater heater at a partial pressure of 4 lbf/in² and is removed. Disregard all thermal losses. The higher heating value of the RDF is most nearly

(A) 4300 Btu/lbm

(B) 4700 Btu/lbm

(C) 4900 Btu/lbm

(D) 5100 Btu/lbm

9. The heats of formation for NO, O_2, and NO_2 gases are shown.

$$\Delta H_f^\circ(NO_{gas}) = 85.68 \text{ Btu}$$

$$\Delta H_f^\circ(O_{2\ gas}) = 0 \text{ Btu}$$

$$\Delta H_f^\circ(NO_{2\ gas}) = 32.08 \text{ Btu}$$

Using the data shown, what is the standard heat of reaction for the reaction of nitrogen monoxide gas (NO) with oxygen (O_2) to form nitrogen dioxide (NO_2) gas?

(A) 85 Btu

(B) −85 Btu

(C) −107 Btu

(D) 107 Btu

SOLUTIONS

1. *Customary U.S. Solution*

Find the absolute temperature of the methane. [Temperature Conversions]

$$T = 60°F + 460° = 520°R$$

Atmospheric pressure is 14.7 psia. [Standard Dry Air Conditions at Sea Level]

The specific gas constant of methane is 96.32 ft-lbf/lbm-°R. [Thermal and Physical Properties of Ideal Gases (at Room Temperature)]

Calculate the mass of 7 ft³ of methane by using the ideal gas law.

Ideal Gas

$$pV = mRT$$

$$m = \frac{pV}{RT}$$

$$= \frac{\left(\left(14.7 \frac{\text{lbf}}{\text{in}^2}\right)\left(12 \frac{\text{in}}{\text{ft}}\right)^2\right)(7 \text{ ft}^3)}{\left(96.32 \frac{\text{ft-lbf}}{\text{lbm-°R}}\right)(520°R)}$$

$$= 0.2958 \text{ lbm}$$

Use the equation for thermal efficiency to find the combustion energy available from the methane. Q_h is the higher heating value of the fuel gas per unit mass, and $Q_h - Q_{fl}$ is the useful heat per unit mass.

Stoichiometric Combustion of Fuels

$$\eta = 100\% \times \frac{Q_h - Q_{fl}}{Q_h}$$

$$Q_h - Q_{fl} = \frac{\eta Q_h}{100\%}$$

$$= \frac{(50\%)\left(24{,}000 \frac{\text{Btu}}{\text{lbm}}\right)}{100\%}$$

$$= 12{,}000 \text{ Btu/lbm}$$

Multiply by the mass of methane to get the total combustion energy available.

$$Q_{\text{avail}} = (Q_h - Q_{fl})m$$

$$= \left(12{,}000 \frac{\text{Btu}}{\text{lbm}}\right)(0.2958 \text{ lbm})$$

$$= 3550 \text{ Btu}$$

Use the equation for heat transfer and solve for the mass of water that can be heated from 60°F to 200°F with this energy. The specific heat of water at standard conditions is 1 Btu/lbm-°F. [Properties of Water at Standard Conditions]

$$Q = mc_p \Delta T$$

$$m = \frac{Q_{avail}}{c_p \Delta T} = \frac{3550 \text{ Btu}}{\left(1 \dfrac{\text{Btu}}{\text{lbm-°F}}\right)(200°F - 60°F)}$$

$$= 25.36 \text{ lbm} \quad (25 \text{ lbm})$$

SI Solution

Find the absolute temperature of the methane. [Temperature Conversions]

$$T = 30°C + 273° = 303K$$

Atmospheric pressure is 101,300 Pa. [Measurement Relationships]

The specific gas constant of methane is 0.5182 kJ/kg·K. [Thermal and Physical Properties of Ideal Gases (at Room Temperature)]

Calculate the mass of 200 L of methane by using the ideal gas law.

Ideal Gas

$$pV = mRT$$

$$m = \frac{pV}{RT}$$

$$= \frac{(101,300 \text{ Pa})\left(\dfrac{200 \text{ L}}{1000 \dfrac{\text{L}}{\text{m}^3}}\right)}{\left(\left(0.5182 \dfrac{\text{kJ}}{\text{kg·K}}\right)\left(1000 \dfrac{\text{J}}{\text{kJ}}\right)\right)(303K)}$$

$$= 0.1290 \text{ kg}$$

Use the equation for thermal efficiency to find the combustion energy available from the methane. Q_h is the higher heating value of the fuel gas per unit mass, and $Q_h - Q_{fl}$ is the useful heat per unit mass.

Stoichiometric Combustion of Fuels

$$\eta = 100\% \times \frac{Q_h - Q_{fl}}{Q_h}$$

$$Q_h - Q_{fl} = \frac{\eta Q_h}{100\%}$$

$$= \frac{(50\%)\left(56 \dfrac{\text{MJ}}{\text{kg}}\right)\left(1000 \dfrac{\text{kJ}}{\text{MJ}}\right)}{100\%}$$

$$= 28,000 \text{ kJ/kg}$$

Multiply by the mass of methane to get the total combustion energy available.

$$Q_{avail} = (Q_h - Q_{fl})m$$

$$= \left(28,000 \dfrac{\text{kJ}}{\text{kg}}\right)(0.1290 \text{ kg})$$

$$= 3612 \text{ kJ}$$

Use the equation for heat transfer and solve for the mass of water that can be heated from 15°C to 95°C with this energy. The specific heat of water at standard conditions is 4.180 kJ/kg·K. A difference of 1°C is equivalent to a difference of 1K. [Properties of Water at Standard Conditions]

$$Q = mc_p \Delta T$$

$$m = \frac{Q_{avail}}{c_p \Delta T} = \frac{3612 \text{ kJ}}{\left(4.180 \dfrac{\text{kJ}}{\text{kg·K}}\right)(95°C - 15°C)}$$

$$= 10.80 \text{ kg} \quad (11 \text{ kg})$$

The answer is (A).

2. *Customary U.S. Solution*

From a table of combustion reactions, the stoichiometric combustion of propane with air produces carbon dioxide (CO_2) at a rate of 2.994 lbm of CO_2 per 1 lbm of propane. [Combustion Reactions of Common Fuel Constituents]

Multiply this rate by the mass rate of propane to get the mass rate of CO_2 produced.

$$\dot{m}_{CO_2} = \left(15 \dfrac{\text{lbm}}{\text{hr}}\right)\left(2.994 \dfrac{\text{lbm}}{\text{lbm}}\right)$$

$$= 44.91 \text{ lbm/hr}$$

The specific gas constant of carbon dioxide is 35.11 ft-lbf/lbm-°R. [Thermal and Physical Properties of Ideal Gases (at Room Temperature)]

Use the ideal gas law to convert the mass rate to volumetric rate. The absolute temperature of the gas is 70°F + 460° = 530°R. [Temperature Conversions]

$$pV = mRT$$

$$Q = \frac{\dot{m}RT}{p}$$

$$= \frac{\left(44.91 \; \frac{\text{lbm}}{\text{hr}}\right)\left(35.11 \; \frac{\text{ft-lbf}}{\text{lbm-°R}}\right)(530°\text{R})}{\left(14.7 \; \frac{\text{lbf}}{\text{in}^2}\right)\left(12 \; \frac{\text{in}}{\text{ft}}\right)^2}$$

$$= 394.8 \; \text{ft}^3/\text{hr} \quad (390 \; \text{ft}^3/\text{hr})$$

SI Solution

From a table of combustion reactions, the stoichiometric combustion of propane with air produces carbon dioxide (CO_2) at a rate of 2.994 lbm of CO_2 per 1 lbm of propane, which equivalent to 2.994 kg of CO_2 per 1 kg of propane. [Combustion Reactions of Common Fuel Constituents]

Multiply this rate by the mass rate of propane to get the mass rate of CO_2 produced.

$$\dot{m}_{CO_2} = \left(6.8 \; \frac{\text{kg}}{\text{h}}\right)\left(2.994 \; \frac{\text{kg}}{\text{kg}}\right)$$

$$= 20.36 \; \text{kg/h}$$

The specific gas constant of carbon dioxide is 0.1889 kJ/kg·K. [Thermal and Physical Properties of Ideal Gases (at Room Temperature)]

Atmospheric pressure is 101,300 Pa. [Measurement Relationships]

Use the ideal gas law to convert the mass rate to volumetric rate. The absolute temperature of the gas is 21°C + 273° = 294K. [Temperature Conversions]

Ideal Gas

$$pV = mRT$$

$$Q = \frac{\dot{m}RT}{p}$$

$$= \frac{\left(20.36 \; \frac{\text{kg}}{\text{h}}\right) \times \left(\left(0.1889 \; \frac{\text{kJ}}{\text{kg·K}}\right)\left(1000 \; \frac{\text{J}}{\text{kJ}}\right)\right) \times (294\text{K})}{101{,}300 \; \text{Pa}}$$

$$= 11.16 \; \text{m}^3/\text{h} \quad (11 \; \text{m}^3/\text{h})$$

The answer is (C).

3. *Customary U.S. Solution*

From a table of combustion reactions, the volumetric ratio of air to fuel used in the stoichiometric combustion of methane is 9.57 ft³ of air per 1 ft³ of methane. [Combustion Reactions of Common Fuel Constituents]

From the problem statement, 30% excess air is used, so the actual volumetric ratio of air to fuel is

$$\left(9.57 \; \frac{\text{ft}^3 \; \text{air}}{\text{ft}^3 \; \text{fuel}}\right)(1.30) = 12.44 \; \frac{\text{ft}^3 \; \text{air}}{\text{ft}^3 \; \text{fuel}}$$

Multiply this by the volumetric rate of the methane to find the volumetric rate of the air.

$$Q_{\text{air}} = \left(12.44 \; \frac{\text{ft}^3 \; \text{air}}{\text{ft}^3 \; \text{fuel}}\right)\left(4000 \; \frac{\text{ft}^3 \; \text{fuel}}{\text{hr}}\right)$$

$$= 49{,}760 \; \text{ft}^3/\text{hr}$$

Air is 79% nitrogen by volume. Multiply this by the volumetric rate of the air to find the volumetric rate of the nitrogen. [Combustion and Fuels: General Information]

$$Q_{\text{nitrogen}} = (0.79)\left(49{,}760 \; \frac{\text{ft}^3}{\text{hr}}\right)$$

$$= 39{,}310 \; \text{ft}^3/\text{hr}$$

The specific gas constant for nitrogen is 55.16 ft-lbf/lbm-°R. [Thermal and Physical Properties of Ideal Gases (at Room Temperature)]

Use the ideal gas law to convert the volumetric rate to the mass rate. The absolute temperature of the air is 100°F + 460° = 560°R. [Temperature Conversions]

Ideal Gas

$$pV = mRT$$

$$\dot{m}_{\text{nitrogen}} = \frac{pQ_{\text{nitrogen}}}{RT}$$

$$= \frac{\left(\left(15 \; \frac{\text{lbf}}{\text{in}^2}\right)\left(12 \; \frac{\text{in}}{\text{ft}}\right)^2\right)\left(39{,}310 \; \frac{\text{ft}^3}{\text{hr}}\right)}{\left(55.16 \; \frac{\text{ft-lbf}}{\text{lbm-°R}}\right)(560°\text{R})}$$

$$= 2749 \; \text{lbm/hr} \quad (2700 \; \text{lbm/hr})$$

SI Solution

Convert the rate of methane to cubic meters per second. [Measurement Relationships]

$$\frac{31 \frac{\text{L}}{\text{s}}}{1000 \frac{\text{L}}{\text{m}^3}} = 0.031 \text{ m}^3/\text{s}$$

From a table of combustion reactions, the volumetric ratio of air to fuel used in the stoichiometric combustion of methane is 9.57 ft^3 of air per 1 ft^3 of methane, which is equivalent to 9.57 m^3 of air per 1 m^3 of methane. [Combustion Reactions of Common Fuel Constituents]

From the problem statement, 30% excess air is used, so the actual volumetric ratio of air to fuel is

$$\left(9.57 \frac{\text{m}^3 \text{ air}}{\text{m}^3 \text{ fuel}}\right)(1.30) = 12.44 \frac{\text{m}^3 \text{ air}}{\text{m}^3 \text{ fuel}}$$

Multiply this by the volumetric rate of the methane to find the volumetric rate of the air.

$$Q_{\text{air}} = \left(12.44 \frac{\text{m}^3 \text{ air}}{\text{m}^3 \text{ fuel}}\right)\left(\frac{31 \frac{\text{L fuel}}{\text{s}}}{1000 \frac{\text{L}}{\text{m}^3}}\right)$$
$$= 0.3856 \text{ m}^3/\text{s}$$

Air is 79% nitrogen by volume. Multiply this by the volumetric rate of the air to find the volumetric rate of the nitrogen. [Combustion and Fuels: General Information]

$$Q_{\text{nitrogen}} = (0.79)\left(0.3856 \frac{\text{m}^3}{\text{s}}\right)$$
$$= 0.3046 \text{ m}^3/\text{s}$$

The specific gas constant for nitrogen is 0.2968 kJ/kg·K. [Thermal and Physical Properties of Ideal Gases (at Room Temperature)]

Use the ideal gas law to convert the volumetric rate to the mass rate. The absolute temperature of the air is 40°C + 273° = 313K. [Temperature Conversions]

Ideal Gas

$$pV = mRT$$

$$\dot{m}_{\text{nitrogen}} = \frac{pQ_{\text{nitrogen}}}{RT}$$

$$= \frac{\left((103 \text{ kPa})\left(1000 \frac{\text{Pa}}{\text{kPa}}\right)\right) \times \left(0.3046 \frac{\text{m}^3}{\text{s}}\right)}{\left(\left(0.2968 \frac{\text{kJ}}{\text{kg·K}}\right)\left(1000 \frac{\text{J}}{\text{kJ}}\right)\right) \times (313\text{K})}$$

$$= 0.3377 \text{ kg/s} \quad (0.34 \text{ kg/s})$$

The answer is (D).

4. The propane is combusting with air, not just pure oxygen. The balanced stoichiometric reaction of propane is [Combustion Reactions of Common Fuel Constituents] [Combustion in Air]

$$\text{C}_3\text{H}_8 + 5(\text{O}_2 + (3.76)\text{N}_2) \rightarrow (3)\text{CO}_2 + (4)\text{H}_2\text{O} + (5)(3.76)\text{N}_2$$

Solve for 20% excess air.

$$\text{C}_3\text{H}_8 + (5)((1.2)\text{O}_2 + (1.2)(3.76)\text{N}_2) \rightarrow \begin{array}{l}(3)\text{CO}_2 + (4)\text{H}_2\text{O} \\ + (5)((1.2)(3.76)\text{N}_2 + (0.2)\text{O}_2)\end{array}$$

$$\text{C}_3\text{H}_8 + (6)\text{O}_2 + (22.56)\text{N}_2 \rightarrow \begin{array}{l}(3)\text{CO}_2 + (4)\text{H}_2\text{O} \\ + (22.56)\text{N}_2 + (1)\text{O}_2\end{array}$$

Find the molar mass of each flue gas constituent. [Thermal and Physical Properties of Ideal Gases (at Room Temperature)]

Calculate the mass percent of CO_2.

$$\%\text{mass}_{\text{CO}_2} = \frac{(3)m_{\text{CO}_2}}{(3)m_{\text{CO}_2} + (4)m_{\text{H}_2\text{O}} + (1)m_{\text{O}_2} + (22.56)m_{\text{N}_2}}(100\%)$$
$$= \frac{(3)(44)}{(3)(44) + (4)(18) + (1)(32) + (22.56)(28)}(100\%)$$
$$= 15.21\% \quad (15\%)$$

The answer is (C).

5. The combustible elements in the coal are the carbon and that portion of the hydrogen that is not associated with moisture (water); the hydrogen and oxygen that make up the water are not available for combustion, and the nitrogen and the ash are noncombustible.

To calculate the mass of the free hydrogen in the coal, start with the mass of oxygen, which is 0.04 lbm per 1 lbm of coal. In 1 lbmol of water, the mass of hydrogen is 2 lbm, and the mass of oxygen is 16 lbm. From this proportion, the mass of hydrogen that is combined with 0.04 lbm of oxygen to form water is

$$m_{H_2,\text{water}} = \left(\frac{2 \text{ lbm}}{16 \text{ lbm}}\right) m_{O_2,\text{water}}$$
$$= \left(\frac{2}{16}\right)(0.04 \text{ lbm})$$
$$= 0.005 \text{ lbm}$$

The mass of hydrogen per 1 lbm of coal that is available for combustion is

$$m_{H_2,\text{free}} = m_{H_2,\text{total}} - m_{H_2,\text{water}}$$
$$= 0.05 \text{ lbm} - 0.005 \text{ lbm}$$
$$= 0.045 \text{ lbm}$$

From a table of heating values, find the lower heating values of carbon and hydrogen. [Heating Values of Substances Occurring in Common Fuels]

$$\text{LHV}_C = 14{,}093 \text{ Btu/lbm}$$
$$\text{LHV}_{H_2} = 51{,}623 \text{ Btu/lbm}$$

Because the coal is 75% carbon, the mass of carbon in 1 lbm of coal is 0.75 lbm. The lower heating value in 1 lbm of coal is

$$\text{LHV}_{\text{coal}} = m_C(\text{LHV}_C) + m_{H_2,\text{free}}(\text{LHV}_{H_2})$$
$$= (0.75 \text{ lbm})\left(14{,}093 \frac{\text{Btu}}{\text{lbm}}\right)$$
$$+ (0.045 \text{ lbm})\left(51{,}623 \frac{\text{Btu}}{\text{lbm}}\right)$$
$$= 12{,}893 \text{ Btu} \quad (12{,}900 \text{ Btu/lbm})$$

The answer is (B).

6. 9.87 lbm of air is used per 1 lbm of coal. Divide by the molecular weight of air, 29 lbm/lbmol, to find the molar mass of the air per 1 lbm of coal. [Thermal and Physical Properties of Ideal Gases (at Room Temperature)]

$$\text{molar mass of air} = \frac{9.87 \text{ lbm air}}{29 \frac{\text{lbm}}{\text{lbmol}}}$$
$$= 0.3403 \text{ lbmol}$$

Air is 21% oxygen (O_2) and 79% nitrogen (N_2) by volume and also by mole. Calculate the molar masses of the O_2 and N_2 per 1 lbm of coal. [Combustion and Fuels: General Information]

$$\text{molar mass of } O_2 = (0.21)(0.3403 \text{ lbmol})$$
$$= 0.0715 \text{ lbmol}$$
$$\text{molar mass of } N_2 = (0.79)(0.3403 \text{ lbmol})$$
$$= 0.2688 \text{ lbmol}$$

Write out the combustion equation, assigning variables to unknown quantities.

$$C_xH_y + 0.0715 O_2 + 0.2688 N_2$$
$$\rightarrow a CO_2 + b H_2O + c N_2 + d O_2$$

From the balance of N_2, c equals 0.2688. From the flue gas analysis,

$$\frac{c}{d} = \frac{81.5}{9} = 9.056$$
$$d = \frac{c}{9.056} = \frac{0.2688}{9.056} = 0.0297$$

There is 0.0297 lbmol of O_2 in the flue gas, so this is the amount of excess O_2. The stoichiometric O_2 needed is

$$O_{2,\text{stoichiometric}} = O_{2,\text{total}} - O_{2,\text{excess}}$$
$$= 0.0715 \text{ lbmol} - 0.0297 \text{ lbmol}$$
$$= 0.0418 \text{ lbmol}$$

The percentage of excess air is the same as the percentage of excess O_2, which is

$$\%_{O_2,\text{excess}} = \frac{O_{2,\text{excess}}}{O_{2,\text{stoichiometric}}} \times 100\%$$
$$= \frac{0.0297 \text{ lbmol}}{0.0418 \text{ lbmol}} \times 100\%$$
$$= 71.05\% \quad (70\%)$$

The answer is (C).

7. *Customary U.S. Solution*

1 lbm of coal contains 67.34% or 0.6734 lbm carbon. 3% of the carbon is lost during combustion, so the carbon available for combustion in 1 lbm of coal is

$$m_{C,\text{avail}} = (0.97)(0.6734 \text{ lbm})$$
$$= 0.6532 \text{ lbm}$$

Write out the actual combustion reaction based on the flue gas analysis, using variables for unknown numbers.

$$C_xH_y + aO_2 + 3.76aN_2$$
$$\rightarrow bCO_2 + cCO + dN_2 + eO_2 + fH_2O$$

From the flue gas analysis,

$$\frac{b}{c} = \frac{15.5}{1.6} = 9.69$$
$$b = 9.69c$$

The carbon in the flue gas consists of b lbmol found in the carbon dioxide (CO_2) and c lbmol found in the carbon monoxide (CO).

$$\text{molar mass of C} = (b+c) \text{ lbmol}$$
$$= (9.69c + c) \text{ lbmol}$$
$$= 10.69c \text{ lbmol}$$

To get the mass of carbon in the flue gas, multiply the molar mass by the molecular weight of carbon, which is 12 lbm/lbmol.

$$m_{C,\text{flue gas}} = M_C(\text{molar mass of C})$$
$$= \left(12 \frac{\text{lbm}}{\text{lbmol}}\right)(10.69c \text{ lbmol})$$
$$= 128.28c \text{ lbm}$$

From a carbon balance, per 1 lbm of coal,

$$m_{C,\text{avail}} = m_{C,\text{flue gas}}$$
$$0.6532 \text{ lbm} = 128.28c \text{ lbm}$$
$$c = \frac{0.6532 \text{ lbm}}{128.28 \text{ lbm}} = 0.0051$$
$$b = 9.69c = (9.69)(0.0051)$$
$$= 0.0494$$

The mass of carbon per 1 lbm of coal that is converted to CO_2 is

$$m_{C \text{ to } CO_2} = M_C(b \text{ lbmol})$$
$$= \left(12 \frac{\text{lbm}}{\text{lbmol}}\right)(0.0494 \text{ lbmol})$$
$$= 0.5928 \text{ lbm}$$

The mass of carbon per 1 lbm of coal that is converted to CO is

$$m_{C \text{ to } CO} = M_C(c \text{ lbmol})$$
$$= \left(12 \frac{\text{lbm}}{\text{lbmol}}\right)(0.0051 \text{ lbmol})$$
$$= 0.0612 \text{ lbm}$$

From a table of heating values, find the higher heating values of carbon converted to CO_2 and carbon converted to CO. [Heating Values of Substances Occurring in Common Fuels]

$$\text{HHV}_{C \text{ to } CO_2} = 14{,}093 \text{ Btu/lbm}$$
$$\text{HHV}_{C \text{ to } CO} = 3950 \text{ Btu/lbm}$$

The rate of heat loss due to the formation of CO is

$$h_{\text{lost}} = \text{HHV}_{C \text{ to } CO_2} - \text{HHV}_{C \text{ to } CO}$$
$$= 14{,}093 \frac{\text{Btu}}{\text{lbm}} - 3950 \frac{\text{Btu}}{\text{lbm}}$$
$$= 10{,}143 \text{ Btu/lbm}$$

The amount of heat lost per 1 lbm of coal is

$$H_{\text{lost per lbm coal}} = m_{C \text{ to } CO} h_{\text{lost}}$$
$$= (0.0612 \text{ lbm})\left(10{,}143 \frac{\text{Btu}}{\text{lbm}}\right)$$
$$= 620.8 \text{ Btu} \quad (620 \text{ Btu/lbm})$$

The answer is (A).

8. The feedwater is saturated liquid at 160°F, so its specific enthalpy, h_f, is 128.00 Btu/lbm. [Properties of Saturated Water and Steam (Temperature) - I-P Units]

The saturated steam that is produced is at a pressure of 200 lbf/in^2, so its specific enthalpy, h_g, is 1198.77 Btu/lbm. [Properties of Saturated Water and Steam (Pressure) - I-P Units]

Calculate the heat rate needed to turn the feedwater into steam.

$$q = \dot{m}(h_g - h_f)$$
$$= \left(20{,}070 \frac{\text{lbm}}{\text{hr}}\right)$$
$$\times \left(1198.77 \frac{\text{Btu}}{\text{lbm}} - 128.00 \frac{\text{Btu}}{\text{lbm}}\right)$$
$$= 21.5 \times 10^6 \text{ Btu/hr}$$

This heat rate is produced from burning 5000 lbm/hr of fuel, so the lower heating value of the fuel is

$$\text{LHV} = \frac{q}{\dot{m}_{\text{fuel}}} = \frac{21.5 \times 10^6 \frac{\text{Btu}}{\text{hr}}}{5000 \frac{\text{lbm}}{\text{hr}}}$$
$$= 4300 \text{ Btu/lbm}$$

The difference between lower heating value, LHV, and higher heating value, HHV, is the heat of vaporization of water formed due to combustion. Water vapor is generated by the combustion of hydrogen in the fuel. The mass of water in the combustion products per 1 lbm of fuel is

$$m_{\text{water}} = \frac{\dot{m}_{\text{water}}}{\dot{m}_{\text{fuel}}} = \frac{2000 \ \frac{\text{lbm water}}{\text{hr}}}{5000 \ \frac{\text{lbm fuel}}{\text{hr}}}$$

$$= 0.4 \ \text{lbm water/lbm fuel}$$

From steam tables, at a partial pressure of 4 lbf/in^2, the heat of vaporization, h_{fg}, is 1006.04 Btu/lbm. [Properties of Saturated Water and Steam (Pressure) - I-P Units]

The higher heating value of the fuel is

$$\text{HHV} = \text{LHV} + m_{\text{water}} h_{fg}$$

$$= 4300 \ \frac{\text{Btu}}{\text{lbm}} + \left(0.4 \ \frac{\text{lbm}}{\text{lbm}}\right)$$

$$\times \left(1006.04 \ \frac{\text{Btu}}{\text{lbm}}\right)$$

$$= 4702 \ \text{Btu/lbm} \quad (4700 \ \text{Btu/lbm})$$

The answer is (B).

9. Write the balanced equation for the reaction.

$$2\text{NO} + \text{O}_2 \rightarrow 2\text{NO}_2$$

Find the standard heat of reaction for the standard heats of formation.

Heats of Reaction

$$\Delta H_r^\circ = \sum_{\text{products}} \nu_i (\Delta H_f^\circ)_i - \sum_{\text{reactants}} \nu_i (\Delta H_f^\circ)_i$$

$$= (2 \ \text{mol NO}_2)(32.08 \ \text{Btu})$$
$$- (2 \ \text{mol NO})(85.68 \ \text{Btu})$$
$$- (1 \ \text{mol O}_2)(0)$$
$$= -107 \ \text{Btu}$$

The standard heat of reaction is -107 Btu.

The answer is (C).

22 Energy, Work, and Power

Content in blue refers to the *NCEES Handbook*.

PRACTICE PROBLEMS

1. A solid, cast-iron sphere with a 5 in (12.5 cm) radius travels without friction at 30 ft/sec (9 m/s) horizontally. The cast iron used to make the sphere has a density of 0.256 lbm/in³ (7090 kg/m³). Its kinetic energy is most nearly

(A) 900 ft-lbf (1.2 kJ)
(B) 1200 ft-lbf (1.6 kJ)
(C) 1600 ft-lbf (2.0 kJ)
(D) 1900 ft-lbf (2.3 kJ)

2. A 100 lbm (50 kg) weight is dropped from 8 ft (2 m) onto a spring with a stiffness of 33.33 lbf/in (5.837 × 10³ N/m). The deflection is most nearly

(A) 27 in (0.67 m)
(B) 34 in (0.85 m)
(C) 39 in (0.90 m)
(D) 45 in (1.1 m)

3. A force of 550 lbf (2500 N) making a 40° angle (upward) from the horizontal pushes a box 20 ft (6 m) across the floor. The work done is most nearly

(A) 2200 ft-lbf (3.0 kJ)
(B) 3700 ft-lbf (5.2 kJ)
(C) 4200 ft-lbf (6.0 kJ)
(D) 8400 ft-lbf (12 kJ)

4. A frictionless pulley is suspended by two truss members as shown in the following illustration.

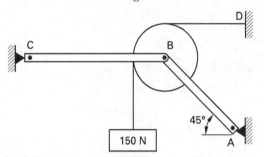

The pulley holds a 150 N weight, and the pulley itself weighs 50 N. What is most nearly the force in member AB?

(A) 50 N (tension)
(B) 50 N (compression)
(C) 100 N (tension)
(D) 300 N (compression)

5. The power in horsepower (kW) that is required to lift a 3300 lbm (1500 kg) mass 250 ft (80 m) vertically in 14 sec is most nearly

(A) 40 hp (30 kW)
(B) 70 hp (53 kW)
(C) 90 hp (68 kW)
(D) 110 hp (84 kW)

6. A car is traveling at 65 mi/hr (105 km/h) through air at a temperature of 70°F (20°C). The total resisting force is 157.3 lbf (708.2 N). The power manifested by the car is most nearly

(A) 15 Btu/sec (16 000 W)
(B) 16 Btu/sec (18 000 W)
(C) 19 Btu/sec (21 000 W)
(D) 24 Btu/sec (26 000 W)

7. A car traveling through 70°F (20°C) air has the characteristics shown.

engine thermal efficiency	28%
fuel heating value	115,000 Btu/gal (32 MJ/L)
power	19.27 Btu/sec (20 679 W)

The fuel consumption of the car at a velocity of 65 mi/hr (105 km/h) is most nearly

(A) 0.03 gal/mi (0.08 L/km)

(B) 0.07 gal/mi (0.16 L/km)

(C) 0.10 gal/mi (0.23 L/km)

(D) 0.11 gal/mi (0.25 L/km)

8. A car traveling through 70°F (20°C) air has the following characteristics.

engine thermal efficiency	28%
fuel heating value	115,000 Btu/gal (32 MJ/L)

The power required is 19.27 Btu/sec (20 679 W) at 65 mi/hr (105 km/h) and 12.65 Btu/sec (13 960 W) at 55 mi/hr (90 km/h). The percentage increase in fuel consumption at 65 mi/hr (105 km/h) compared to 55 mi/hr (90 km/h) is most nearly

(A) 10%

(B) 20%

(C) 30%

(D) 40%

SOLUTIONS

1. *Customary U.S. Solution*

Since there is no friction, there is no rotation. The sphere slides. Combine the equation for kinetic energy with the equation for mass and the equation for volume of a sphere and solve,

Kinetic Energy

$$KE = \frac{1}{2}mv^2$$

Circular Segment

$$V = 4\frac{\pi r^3}{3}$$

$$m = V\rho$$

The kinetic energy is

$$KE = \frac{1}{2}\left(\frac{m}{g_c}\right)v^2 = \frac{1}{2}\left(\frac{V\rho}{g_c}\right)v^2$$

$$= \left(\frac{1}{2}\right)\left(\frac{4}{3}\pi r^3\right)\left(\frac{\rho}{g_c}\right)v^2$$

$$= \frac{2}{3}\pi r^3\left(\frac{\rho}{g_c}\right)v^2$$

$$= \frac{2}{3}\pi(5 \text{ in})^3\left(\frac{0.256 \frac{\text{lbm}}{\text{in}^3}}{32.174 \frac{\text{lbm-ft}}{\text{lbf-sec}^2}}\right)\left(30 \frac{\text{ft}}{\text{sec}}\right)^2$$

$$= 1873 \text{ ft-lbf} \quad (1900 \text{ ft-lbf})$$

SI Solution

Since there is no friction, there is no rotation. The sphere slides. Combine the equation for kinetic energy with the equation for mass and the equation for volume of a sphere and solve.

Kinetic Energy

$$KE = \frac{1}{2}mv^2$$

Circular Segment

$$V = 4\frac{\pi r^3}{3}$$

$$m = V\rho$$

The kinetic energy is

$$KE = \frac{1}{2}mv^2 = \frac{1}{2}(V\rho)v^2$$
$$= \left(\frac{1}{2}\right)\left(\frac{4}{3}\pi r^3\right)\rho v^2$$
$$= \frac{2}{3}\pi r^3 \rho v^2$$
$$= \frac{2}{3}\pi (0.125 \text{ m})^3 \left(7090 \frac{\text{kg}}{\text{m}^3}\right)\left(9 \frac{\text{m}}{\text{s}}\right)^2$$
$$= 2349 \text{ J} \quad (2.3 \text{ kJ})$$

The answer is (D).

2. *Customary U.S. Solution*

Combine the law of conservation of energy and the equation for work done by a spring.

Conservation of Energy Law
$$KE_1 + PE_1 = KE_2 + PE_2 + W$$

Work
$$W_s = \frac{1}{2}k(s_1^2 - s_2^2)$$

The weight does not initially compress the spring, so the initial displacement is 0. Simplifying and combining,

$$PE = W_s$$
$$m\frac{g}{g_c}(\Delta h + \Delta x) = \frac{1}{2}k(s_1^2 - s_2^2)$$

Rearranging and using $\Delta x^2 = (s_1^2 - s_2^2)$,

$$\frac{1}{2}k(\Delta x)^2 - m\frac{g}{g_c}\Delta x - m\frac{g}{g_c}\Delta h = 0$$

$$\left(\frac{1}{2}\right)\left(33.33 \frac{\text{lbf}}{\text{in}}\right)(\Delta x)^2$$
$$- (100 \text{ lbm})\left(\frac{32.174 \frac{\text{ft}}{\text{sec}^2}}{32.174 \frac{\text{lbm-ft}}{\text{lbf-sec}^2}}\right)\Delta x$$
$$- (100 \text{ lbm})\left(\frac{32.174 \frac{\text{ft}}{\text{sec}^2}}{32.174 \frac{\text{lbm-ft}}{\text{lbf-sec}^2}}\right)(8 \text{ ft})\left(12 \frac{\text{in}}{\text{ft}}\right) = 0$$

$$16.665(\Delta x)^2 - 100\Delta x = 9600$$

Complete the square.

$$(\Delta x)^2 - 6\Delta x = 576$$
$$(\Delta x - 3)^2 = 576 + 9$$
$$\Delta x - 3 = \pm\sqrt{585} = \pm 24.2$$
$$\Delta x = 27.2 \text{ in} \quad (27 \text{ in})$$

SI Solution

Combine the law of conservation of energy and the equation for work done by a spring.

Conservation of Energy Law
$$KE_1 + PE_1 = KE_2 + PE_2 + W$$

Work
$$W_s = \frac{1}{2}k(s_1^2 - s_2^2)$$

$$PE = W_s$$
$$mg(\Delta h + \Delta x) = \frac{1}{2}k(s_1^2 - s_2^2)$$

Rearranging and using $\Delta x^2 = (s_1^2 - s_2^2)$,

$$\frac{1}{2}k(\Delta x)^2 - mg\Delta x - mg\Delta h = 0$$

$$\left(\frac{1}{2}\right)\left(5.837 \times 10^3 \frac{\text{N}}{\text{m}}\right)(\Delta x)^2$$
$$- (50 \text{ kg})\left(9.81 \frac{\text{m}}{\text{s}^2}\right)\Delta x$$
$$- (50 \text{ kg})\left(9.81 \frac{\text{m}}{\text{s}^2}\right)(2 \text{ m}) = 0$$

$$2918.5(\Delta x)^2 - 490.5\Delta x - 981.0 = 0$$
$$(\Delta x)^2 - 0.1681\Delta x = 0.3361$$

Complete the square.

$$(\Delta x - 0.08403)^2 = 0.3361 + (0.08403)^2$$
$$= 0.3432$$
$$\Delta x - 0.08403 = \pm\sqrt{0.3432} = \pm 0.5858$$
$$\Delta x = 0.6699 \text{ m} \quad (0.67 \text{ m})$$

The answer is (A).

3. *Customary U.S. Solution*
From the equation for work done by a constant force, the work done is

Work
$$W_{\text{done on box}} = F_c (\cos \theta) \Delta s$$
$$= (550 \text{ lbf})(\cos 40°)(20 \text{ ft})$$
$$= 8426 \text{ ft-lbf} \quad (8400 \text{ ft-lbf})$$

SI Solution
From the equation for work done by a constant force, the work done is

Work
$$W_{\text{done on box}} = F_c (\cos \theta) \Delta s$$
$$= \frac{(2500 \text{ N})(\cos 40°)(6 \text{ m})}{1000 \dfrac{\text{J}}{\text{kJ}}}$$
$$= 11.49 \text{ kJ} \quad (12 \text{ kJ})$$

The answer is (D).

4. Draw the free-body diagram for the whole system at point B.

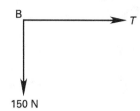

Member BC cannot carry any load in the y direction.

$$\sum F_y = F_{AB} \sin \theta - F_{\text{pulley}} - F_{\text{block}} = 0$$
$$F_{AB} = \frac{F_{\text{pulley}} + F_{\text{block}}}{\sin \theta} = \frac{50 \text{ N} + 150 \text{ N}}{\sin 45°}$$
$$= 280 \text{ N} \quad (300 \text{ N}) \quad [\text{compression}]$$

The answer is (D).

5. *Customary U.S. Solution*
Combining the potential energy equation and the equations for power and efficiency, the work required is

Potential Energy
$$PE = mgh$$

Power and Efficiency
$$P = \frac{dW}{dt} = \mathbf{F} \cdot \mathbf{v}$$
$$\epsilon = \frac{P_{\text{out}}}{P_{\text{in}}} = \frac{W_{\text{out}}}{W_{\text{in}}}$$

$$W = P \Delta t = \frac{mg \Delta h}{g_c}$$
$$P = \frac{mg \Delta h}{g_c \Delta t}$$
$$= \frac{(3300 \text{ lbm})\left(32.2 \dfrac{\text{ft}}{\sec^2}\right)(250 \text{ ft})}{\left(32.2 \dfrac{\text{lbm-ft}}{\sec^2\text{-lbf}}\right)(14 \sec)\left(550 \dfrac{\text{ft-lbf}}{\text{hp-sec}}\right)}$$
$$= 107 \text{ hp} \quad (110 \text{ hp})$$

SI Solution
Combining the potential energy equation and the equations for power and efficiency, the work required is

Potential Energy
$$PE = mgh$$

Power and Efficiency
$$P = \frac{dW}{dt} = \mathbf{F} \cdot \mathbf{v}$$
$$\epsilon = \frac{P_{\text{out}}}{P_{\text{in}}} = \frac{W_{\text{out}}}{W_{\text{in}}}$$

$$W = P \Delta t = mg \Delta h$$
$$P = \frac{mg \Delta h}{\Delta t} = \frac{(1500 \text{ kg})\left(9.81 \dfrac{\text{m}}{\text{s}^2}\right)(80 \text{ m})}{(14 \text{ s})\left(1000 \dfrac{\text{W}}{\text{kW}}\right)}$$
$$= 84.1 \text{ kW} \quad (84 \text{ kW})$$

The answer is (D).

6. *Customary U.S. Solution*
Convert the units for the velocity to ft/sec.

$$\text{v} = \frac{\left(65 \dfrac{\text{mi}}{\text{hr}}\right)\left(5280 \dfrac{\text{ft}}{\text{hr}}\right)}{3600 \dfrac{\sec}{\text{hr}}}$$
$$= 95.33 \text{ ft/sec}$$

The power manifested by virtue of the car's velocity is

ENERGY, WORK, AND POWER   22-5

Power and Efficiency

$$P = Fv = \frac{(157.3 \text{ lbf})\left(95.33 \dfrac{\text{ft}}{\text{sec}}\right)}{778 \dfrac{\text{ft-lbf}}{\text{Btu}}}$$

$$= 19.27 \text{ Btu/sec} \quad (19 \text{ Btu/sec})$$

SI Solution

Convert the units for the velocity to m/s.

$$v = \frac{\left(105 \dfrac{\text{km}}{\text{h}}\right)\left(1000 \dfrac{\text{m}}{\text{km}}\right)}{3600 \dfrac{\text{s}}{\text{h}}}$$

$$= 29.1 \text{ m/s}$$

The power manifested by virtue of the car's velocity is

Power and Efficiency

$$P = Fv = (708.2 \text{ N})\left(29.1 \dfrac{\text{m}}{\text{s}}\right)$$

$$= 20\,608 \text{ W} \quad (21\,000 \text{ W})$$

The answer is (C).

7. *Customary U.S. Solution*

The energy available from the fuel is

$$E_A = (\text{energy thermal efficiency})(\text{fuel heating value})$$

$$= (0.28)\left(115{,}000 \dfrac{\text{Btu}}{\text{gal}}\right)$$

$$= 32{,}200 \text{ Btu/gal}$$

The fuel consumption at 65 mi/hr is

$$\frac{P}{E_A v} = \frac{\left(19.27 \dfrac{\text{Btu}}{\text{sec}}\right)\left(3600 \dfrac{\text{sec}}{\text{hr}}\right)}{\left(32{,}200 \dfrac{\text{Btu}}{\text{gal}}\right)\left(65 \dfrac{\text{mi}}{\text{hr}}\right)}$$

$$= 0.0331 \text{ gal/mi} \quad (0.03 \text{ gal/mi})$$

SI Solution

The energy available from the fuel is

$$E_A = (\text{energy thermal efficiency})(\text{fuel heating value})$$

$$= (0.28)\left(32 \times 10^6 \dfrac{\text{J}}{\text{L}}\right)$$

$$= 8.96 \times 10^6 \text{ J/L}$$

The fuel consumption at 105 km/h is [Measurement Relationships]

$$\frac{P}{E_A v} = \frac{(20\,679 \text{ J/s})\left(3600 \dfrac{\text{s}}{\text{h}}\right)}{\left(8.96 \times 10^6 \dfrac{\text{J}}{\text{L}}\right)\left(105 \dfrac{\text{km}}{\text{h}}\right)}$$

$$= 0.0791 \text{ L/km} \quad (0.08 \text{ L/km})$$

The answer is (A).

8. *Customary U.S. Solution*

The energy available from the fuel is

$$E_A = (\text{energy thermal efficiency})(\text{fuel heating value})$$

$$= (0.28)\left(115{,}000 \dfrac{\text{Btu}}{\text{gal}}\right)$$

$$= 32{,}200 \text{ Btu/gal}$$

The fuel consumption at 65 mi/hr is

$$\frac{P}{E_A v} = \frac{\left(19.27 \dfrac{\text{Btu}}{\text{sec}}\right)\left(3600 \dfrac{\text{sec}}{\text{hr}}\right)}{\left(32{,}200 \dfrac{\text{Btu}}{\text{gal}}\right)\left(65 \dfrac{\text{mi}}{\text{hr}}\right)}$$

$$= 0.0331 \text{ gal/mi}$$

The fuel consumption at 55 mi/hr is

$$\frac{P}{E_A v} = \frac{\left(12.65 \dfrac{\text{Btu}}{\text{sec}}\right)\left(3600 \dfrac{\text{sec}}{\text{hr}}\right)}{\left(32{,}200 \dfrac{\text{Btu}}{\text{gal}}\right)\left(55 \dfrac{\text{mi}}{\text{hr}}\right)}$$

$$= 0.0257 \text{ gal/mi}$$

The relative difference between the fuel consumptions at 55 mi/hr and 65 mi/hr is

$$\text{relative difference} = \frac{0.0331 \dfrac{\text{gal}}{\text{mi}} - 0.0257 \dfrac{\text{gal}}{\text{mi}}}{0.0257 \dfrac{\text{gal}}{\text{mi}}}$$

$$= 0.288 \quad (30\%)$$

SI Solution

The energy available from the fuel is

$$E_A = \text{(energy thermal efficiency)(fuel heating value)}$$
$$= (0.28)\left(32 \times 10^6 \ \frac{\text{J}}{\text{L}}\right)$$
$$= 8.96 \times 10^6 \ \text{J/L}$$

The fuel consumption at 105 km/h is [Measurement Relationships]

$$\frac{P}{E_A \text{v}} = \frac{(20\,679 \ \text{J/s})\left(3600 \ \frac{\text{s}}{\text{h}}\right)}{\left(8.96 \times 10^6 \ \frac{\text{J}}{\text{L}}\right)\left(105 \ \frac{\text{km}}{\text{h}}\right)}$$
$$= 0.0791 \ \text{L/km}$$

The fuel consumption at 90 km/h is [Measurement Relationships]

$$\frac{P}{E_A \text{v}} = \frac{(13\,960 \ \text{J/s})\left(3600 \ \frac{\text{s}}{\text{h}}\right)}{\left(8.96 \times 10^6 \ \frac{\text{J}}{\text{L}}\right)\left(90 \ \frac{\text{km}}{\text{h}}\right)}$$
$$= 0.0623 \ \text{L/km}$$

The relative difference between the fuel consumptions at 90 km/h and 105 km/h is

$$\text{relative difference} = \frac{0.0791 \ \frac{\text{L}}{\text{km}} - 0.0623 \ \frac{\text{L}}{\text{km}}}{0.0623 \ \frac{\text{L}}{\text{km}}}$$
$$= 0.272 \ (30\%)$$

The answer is (C).

23 Thermodynamic Properties of Substances

Content in blue refers to the *NCEES Handbook*.

PRACTICE PROBLEMS

1. The molar enthalpy of 250°F (120°C) steam with a quality of 92% is most nearly

(A) 16,000 Btu/lbmol (37 MJ/kmol)
(B) 18,000 Btu/lbmol (41 MJ/kmol)
(C) 20,000 Btu/lbmol (46 MJ/kmol)
(D) 22,000 Btu/lbmol (51 MJ/kmol)

2. What is the thermodynamic state of water at 600°F and 300 psia?

(A) subcooled liquid
(B) saturated vapor
(C) superheated vapor
(D) real or ideal gas

3. 106°F water flows in a closed feedwater heater. At the pressure in the feedwater heater, water has a saturation temperature of 293°F and a saturation enthalpy of 262 Btu/lbm. The specific enthalpy of the water is most nearly

(A) 74 Btu/lbm
(B) 89 Btu/lbm
(C) 110 Btu/lbm
(D) 130 Btu/lbm

4. An Otto cycle has a compression ratio of 7. The pressure and temperature at the beginning of the isentropic compression process are 14.7 psia (101.3 kPa) and 77°F (25°C). The pressure and temperature after compression are most nearly

(A) 110 psia (0.76 MPa) and 620°F (330°C)
(B) 220 psia (1.5 MPa) and 710°F (380°C)
(C) 220 psia (1.5 MPa) and 1170°F (630°C)
(D) 340 psia (2.3 MPa) and 710°F (380°C)

5. 1 lbmol (1 kmol) of carbon dioxide is stored in a cylinder at 1000 psia (7 MPa) and 100°F (40°C). Using the generalized compressibility factor chart for nonideal behavior, the volume of the cylinder is most nearly

(A) 2.6 ft^3 (0.16 m^3)
(B) 3.5 ft^3 (0.21 m^3)
(C) 4.8 ft^3 (0.30 m^3)
(D) 6.1 ft^3 (0.36 m^3)

6. The density of a saturated steam mixture that is 87% liquid at 50 psia (350 kPa) is most nearly

(A) 0.75 lbm/ft^3 (12 kg/m^3)
(B) 0.89 lbm/ft^3 (14 kg/m^3)
(C) 0.94 lbm/ft^3 (15 kg/m^3)
(D) 1.07 lbm/ft^3 (17 kg/m^3)

SOLUTIONS

1. *Customary U.S. Solution*

From steam tables, for 250°F steam, the specific enthalpy of saturated liquid is, $h_f = 218.63$ Btu/lbm, and the heat of vaporization is $h_{fg} = 945.41$ Btu/lbm. [Properties of Saturated Water and Steam (Temperature) - I-P Units]

Calculate the specific enthalpy of a liquid-vapor mixture with a quality of 92%.

Properties for Two-Phase (Vapor-Liquid) Systems

$$h = h_f + x h_{fg}$$
$$= 218.63 \frac{\text{Btu}}{\text{lbm}} + (0.92)\left(945.41 \frac{\text{Btu}}{\text{lbm}}\right)$$
$$= 1088 \text{ Btu/lbm}$$

Calculate the molar enthalpy, H_m. The molecular weight, M, of steam is 18. [Thermal and Physical Properties of Ideal Gases (at Room Temperature)]

$$H_m = hM = \left(1088 \frac{\text{Btu}}{\text{lbm}}\right)\left(18 \frac{\text{lbm}}{\text{lbmol}}\right)$$
$$= 19{,}592 \text{ Btu/lbmol}$$
$$(20{,}000 \text{ Btu/lbmol})$$

SI Solution

From steam tables, for 120°C steam, the specific enthalpy of saturated liquid is $h_f = 503.81$ kJ/kg, and the heat of vaporization is $h_{fg} = 2202.1$ kJ/kg. [Properties of Saturated Water and Steam (Temperature) - SI Units]

Calculate the specific enthalpy of a liquid-vapor mixture with a quality of 92%.

$$h = h_f + x h_{fg}$$
$$= 503.81 \frac{\text{kJ}}{\text{kg}} + (0.92)\left(2202.1 \frac{\text{kJ}}{\text{kg}}\right)$$
$$= 2530 \text{ kJ/kg}$$

Calculate the molar enthalpy, H_m. The molecular weight, M, of steam is 18. [Thermal and Physical Properties of Ideal Gases (at Room Temperature)]

$$H_m = hM = \left(2530 \frac{\text{kJ}}{\text{kg}}\right)\left(18 \frac{\text{kg}}{\text{kmol}}\right)$$
$$= 45{,}540 \text{ kJ/kmol} \quad (46 \text{ MJ/kmol})$$

The answer is (C).

2. From steam tables, the saturation temperature for 300 psia steam is 417.33°F. The water's temperature is higher than this, so the water is either a superheated vapor or a gas. [Properties of Saturated Water and Steam (Temperature) - I-P Units]

The critical temperature for water is 1165°R (705°F). The water's temperature is lower than this, so the water can't be considered a gas. Therefore, it is a superheated vapor. [Critical Properties]

The answer is (C).

3. The temperature of the water, 106°F, is less than the saturation temperature of 293°F, so the state of water is subcooled liquid. For a subcooled liquid, enthalpy is governed by the temperature, and the enthalpy of subcooled liquid is very close to the enthalpy of saturated liquid at the same temperature. [Properties of Saturated Water and Steam (Temperature) - I-P Units]

$$h_{106°F} \approx h_{f,106°F}$$
$$= 74.02 \text{ Btu/lbm} \quad (74 \text{ Btu/lbm})$$

The answer is (A).

4. *Customary U.S. Solution*

The compression ratio, r, is given as 7.

Otto Cycle (Gasoline Engine)

$$r = \frac{v_1}{v_2}$$
$$= 7$$

The pressure and temperature after the compression process can be calculated with the p-V-T relationships for an isentropic process. The fluid undergoing the compression process in Otto cycle is primarily air. For air, the ratio of specific heats is $k = 1.40$. [Thermal and Physical Properties of Ideal Gases (at Room Temperature)]

The final pressure is

Ideal Gas

$$p_2 = p_1 \left(\frac{v_1}{v_2}\right)^k$$
$$= \left(14.7 \frac{\text{lbf}}{\text{in}^2}\right)(7)^{1.40}$$
$$= 224.11 \text{ lbf/in}^2 \quad (220 \text{ psia})$$

Calculate the absolute temperature at the beginning of the process. [Temperature Conversions]

$$T_1 = 77°\text{F} + 460° = 537°\text{R}$$

The final temperature is

Ideal Gas
$$T_2 = T_1\left(\frac{v_1}{v_2}\right)^{k-1}$$
$$= (537°R)(7)^{1.40-1}$$
$$= 1170°R$$

Convert to degrees Fahrenheit. [Temperature Conversions]

$$T_2 = 1170°R - 460° = 710°F$$

SI Solution

The compression ratio, r, is given as 7.

Otto Cycle (Gasoline Engine)
$$r = \frac{v_1}{v_2}$$
$$= 7$$

The pressure and temperature after the compression process can be calculated with the p-V-T relationships for an isentropic process. The fluid undergoing the compression process in an Otto cycle is primarily air. For air, the ratio of specific heats is $k = 1.40$. [Thermal and Physical Properties of Ideal Gases (at Room Temperature)]

The final pressure is

Ideal Gas
$$p_2 = p_1\left(\frac{v_1}{v_2}\right)^k$$
$$= \frac{(101.3 \text{ kPa})(7)^{1.40}}{1000 \dfrac{\text{kPa}}{\text{MPa}}}$$
$$= 1.544 \text{ MPa} \quad (1.5 \text{ MPa})$$

Calculate the absolute temperature at the beginning of the process. [Temperature Conversions]

$$T_1 = 25°C + 273° = 298K$$

The final temperature is

Ideal Gas
$$T_2 = T_1\left(\frac{v_1}{v_2}\right)^{k-1}$$
$$= (298K)(7)^{1.40-1}$$
$$= 649K$$

Convert to degrees Celsius. [Temperature Conversions]

$$T_2 = 649K - 273° = 376°C \quad (380°C)$$

The answer is (B).

5. *Customary U.S. Solution*

From a table of critical properties, find the critical pressure and temperature of carbon dioxide. [Critical Properties]

$$p_C = (72.9 \text{ atm})\left(14.7 \dfrac{\dfrac{\text{lbf}}{\text{in}^2}}{\text{atm}}\right) = 1072 \text{ psia}$$

$$T_C = 548°R$$

Calculate the absolute temperature of the cylinder. [Temperature Conversions]

$$T = 100°F + 460° = 560°R$$

Calculate the reduced temperature and reduced pressure.

Compressibility Factor and Charts
$$T_R = \frac{T}{T_C}$$
$$= \frac{560°R}{548°R}$$
$$= 1.02$$

$$p_R = \frac{p}{p_C}$$
$$= \frac{1000 \text{ psia}}{1072 \text{ psia}}$$
$$= 0.933$$

From the generalized compressibility chart, for these values of T_R and p_R, the compressibility factor is $Z = 0.55$. [Compressibility Factor and Charts]

Use the compressibility equation of state, and solve for the volume of the cylinder. The mass, m, divided by the molecular weight of carbon dioxide, M, gives the number of moles, which is given as 1 lbmol. The specific gas constant of carbon dioxide, R, multiplied by the molecular weight of carbon dioxide gives the universal gas constant, $\bar{R}$.

Equations of State (EOS)

$$p = \left(\frac{RT}{v}\right)Z$$

$$= \left(\frac{RT}{\frac{V}{m}}\right)Z$$

$$V = \frac{mZRT}{p} = \frac{\left(\frac{m}{M}\right)Z(RM)T}{p} = \frac{nZ\overline{R}T}{P}$$

$$= \frac{(1 \text{ lbmol})(0.55)\left(1545 \frac{\text{ft-lbf}}{\text{lbmol-}°\text{R}}\right)(560°\text{R})}{\left(\left(1000 \frac{\text{lbf}}{\text{in}^2}\right)\left(12 \frac{\text{in}}{\text{ft}}\right)^2\right)}$$

$$= 3.305 \text{ ft}^3 \quad (3.5 \text{ ft}^3)$$

SI Solution

From a table of critical properties, find the critical pressure and temperature of carbon dioxide. [Critical Properties]

$$p_C = (72.9 \text{ atm})\left(\frac{1.013 \times 10^5 \frac{\text{Pa}}{\text{atm}}}{10^6 \frac{\text{Pa}}{\text{MPa}}}\right)$$

$$= 7.385 \text{ MPa}$$
$$T_C = 304.3 \text{K}$$

Calculate the absolute temperature of the cylinder. [Temperature Conversions]

$$T = 40°\text{C} + 273° = 313\text{K}$$

Calculate the reduced temperature and reduced pressure.

Compressibility Factor and Charts

$$T_R = \frac{T}{T_C}$$
$$= \frac{313\text{K}}{304.3\text{K}}$$
$$= 1.03$$

$$p_R = \frac{p}{p_C}$$
$$= \frac{7 \text{ MPa}}{7.385 \text{ MPa}}$$
$$= 0.948$$

From the generalized compressibility chart, for these values of T_R and p_R, the compressibility factor is $Z = 0.56$. [Compressibility Factor and Charts]

Use the compressibility equation of state, and solve for the volume of the cylinder. The mass, m, divided by the molecular weight of carbon dioxide, M, gives the number of moles, which is given as 1 kmol. The specific gas constant of carbon dioxide, R, multiplied by the molecular weight of carbon dioxide gives the universal gas constant, $\overline{R}$.

Equations of State (EOS)

$$p = \left(\frac{RT}{v}\right)Z$$

$$= \left(\frac{RT}{\frac{V}{m}}\right)Z$$

$$V = \frac{mZRT}{p} = \frac{\left(\frac{m}{M}\right)Z(RM)T}{p} = \frac{nZ\overline{R}T}{P}$$

$$= \frac{(1 \text{ kmol})(0.56)\left(8.314 \frac{\text{kPa·m}^3}{\text{kmol·K}}\right)(313\text{K})}{\left((7 \text{ MPa})\left(1000 \frac{\text{kPa}}{\text{MPa}}\right)\right)}$$

$$= 0.2081 \text{ m}^3 \quad (0.21 \text{ m}^3)$$

The answer is (B).

6. *Customary U.S. Solution*

The steam is 87% liquid, so the remaining 13% is vapor and the quality is $x = 0.13$.

Find the specific volume at 50 psia from the steam tables. [Properties of Saturated Water and Steam (Pressure) - I-P Units]

$$v_f = 0.0173 \text{ ft}^3/\text{lbm}$$
$$v_{fg} = 8.50 \text{ ft}^3/\text{lbm}$$

The specific volume is

Properties for Two-Phase (Vapor-Liquid) Systems

$$v = v_f + xv_{fg}$$

$$= 0.0173 \frac{\text{ft}^3}{\text{lbm}} + (0.13)\left(8.50 \frac{\text{ft}^3}{\text{lbm}}\right)$$

$$= 1.122 \text{ ft}^3/\text{lbm}$$

Therefore, the density is

$$\rho = \frac{1}{v} = \frac{1}{1.122 \ \frac{\text{ft}^3}{\text{lbm}}}$$

$$= 0.891 \ \text{lbm/ft}^3 \quad (0.89 \ \text{lbm/ft}^3)$$

SI Solution

The steam is 87% liquid, so the remaining 13% is vapor and the quality is $x = 0.13$.

Interpolate to find the specific volume at 350 kPa (0.35 MPa) from the steam tables. [Properties of Saturated Water and Steam (Pressure) - SI Units]

$$v_f = 0.0011 \ \text{m}^3/\text{kg}$$
$$v_{fg} = 0.533 \ \text{m}^3/\text{kg}$$

Therefore, the specific volume is

Properties for Two-Phase (Vapor-Liquid) Systems

$$v = v_f + x v_{fg}$$

$$= 0.0011 \ \frac{\text{m}^3}{\text{kg}} + (0.13)\left(0.533 \ \frac{\text{m}^3}{\text{kg}}\right)$$

$$= 0.0703 \ \text{m}^3/\text{kg}$$

The density is

$$\rho = \frac{1}{v} = \frac{1}{0.0703 \ \frac{\text{m}^3}{\text{kg}}}$$

$$= 14.22 \ \text{kg/m}^3 \quad (14 \ \text{kg/m}^3)$$

The answer is (B).

24 Changes in Thermodynamic Properties

Content in blue refers to the NCEES Handbook.

PRACTICE PROBLEMS

1. Which statement about thermodynamics is FALSE?

(A) The availability of a system depends on the surrounding atmospheric conditions.

(B) In the absence of friction and other irreversibilities, a heat engine cycle can have a thermal efficiency of 100%.

(C) The temperature of a gas always decreases when the gas expands isentropically.

(D) Water cannot exist in a liquid state at any temperature if the pressure is less than the triple point pressure.

2. Which statement is true?

(A) Entropy does not change in an adiabatic process.

(B) The entropy of a closed system cannot decrease.

(C) Entropy increases when a refrigerant passes through a throttling valve.

(D) The entropy of air inside a closed room with a running, electrically driven fan will always increase over time.

3. Which of these processes CANNOT be modeled as an isenthalpic (throttling) process?

(A) viscous drag on an object moving through air

(B) an ideal gas accelerating to supersonic speed through a converging-diverging nozzle

(C) refrigerant passing through a pressure-reducing throttling valve

(D) high-pressure steam escaping through a spring-loaded pressure-relief (safety) valve

4. Air expands isentropically in a steady-flow process from 700°F and 400 psia to 50 psia.

From an air table, what is most nearly the change in specific enthalpy?

(A) -680 Btu/lbm

(B) -130 Btu/lbm

(C) -90 Btu/lbm

(D) -14 Btu/lbm

5. 8.0 ft^3 (0.25 m^3) of air at 180°F and 14.7 psia (82°C and 101.3 kPa) is cooled to 100°F (38°C) in a constant-pressure process. The amount of work done is most nearly

(A) -2100 ft-lbf (-3.1 kJ)

(B) -1500 ft-lbf (-2.3 kJ)

(C) -1100 ft-lbf (-1.5 kJ)

(D) -900 ft-lbf (-1.3 kJ)

6. An isentropic process uses steam with an initial quality of 95% at 300 psia. The steam at the process outlet is 50 psia with a specific enthalpy of 1031 Btu/lbm. Most nearly, the change in enthalpy of the process is

(A) -100 Btu/lbm

(B) -130 Btu/lbm

(C) -210 Btu/lbm

(D) -340 Btu/lbm

7. A closed air heater receives 540°F, 100 psia (280°C, 700 kPa) air and heats it to 1540°F (840°C). The outside temperature is 100°F (40°C). The pressure of the air drops 20 psi (150 kPa) as it passes through the heater.

The absolute temperatures at the inlet, T_1, and outlet, T_2, of the air heater are most nearly

(A) $T_1 = 460°R$, $T_2 = 1500°R$
 ($T_1 = 260K$, $T_2 = 860K$)

(B) $T_1 = 540°R$, $T_2 = 1000°R$
 ($T_1 = 300K$, $T_2 = 550K$)

(C) $T_1 = 1000°R$, $T_2 = 2000°R$
 ($T_1 = 550K$, $T_2 = 1100K$)

(D) $T_1 = 1500°R$, $T_2 = 1000°R$
 ($T_1 = 860K$, $T_2 = 550K$)

8. A closed air heater's maximum work is found from the equation shown.

$$W_{\max,p\text{ loss}} = h_1 - h_2 + T_{L,\mathcal{R}}\left(\phi_2 - \phi_1 - \frac{R}{J}\ln\left(\frac{p_2}{p_1}\right)\right)$$

A given closed air heater receives 540°F, 100 psia air and heats it to 1540°F. The outside temperature is 100°F. The pressure of the air drops 20 psi as it passes through the heater. The maximum work done by the heater is most nearly

(A) −240 Btu/lbm

(B) −200 Btu/lbm

(C) −170 Btu/lbm

(D) −150 Btu/lbm

9. A closed air heater receives 540°F, 100 psia (280°C, 700 kPa) air and heats it to 1540°F (840°C). The outside temperature is 100°F (40°C). The pressure of the air drops 20 psi (150 kPa) as it passes through the heater. The available per unit mass, before the pressure drop is −162.02 Btu/lbm (−378.17 kJ/kg), and the available per unit mass after the pressure drop is −153.46 Btu/lbm (−356.50 kJ/kg). The percentage loss in available energy due to the pressure drop is most nearly

(A) 5.0%

(B) 12%

(C) 18%

(D) 34%

10. Xenon gas at 20 psia and 70°F (150 kPa and 21°C) is compressed to 3800 psia and 70°F (25 MPa and 21°C) by a compressor/heat exchanger combination. The compressed gas is stored at 70°F (21°C) in a 100 ft³ (3 m³) rigid tank initially charged with xenon gas at 20 psia (150 kPa). The gas constant of xenon gas is 11.77 ft-lbf/lbm-°R (63.32 J/kg·K). The critical temperature and pressure of xenon are 521.9°R and 58.2 atm, respectively. The compressor fills the tank in exactly one hour. The average mass flow rate of xenon gas into the tank is most nearly

(A) 6300 lbm/hr (0.88 kg/s)

(B) 9700 lbm/hr (1.3 kg/s)

(C) 12,000 lbm/hr (1.6 kg/s)

(D) 14,000 lbm/hr (1.9 kg/s)

11. Xenon gas ($MW_{Xenon} = 131$ g/mol) at 20 psia and 70°F (150 kPa and 21°C) is compressed to 3800 psia and 70°F (25 MPa and 21°C) by a compressor/heat exchanger combination. The compressed gas is stored at 70°F (21°C) in a 100 ft³ (3 m³) rigid tank initially charged with xenon gas at 20 psia (150 kPa). The gas constant for xenon is 11.77 ft-lbf/lbm-°R (63.32 J/kg·K). The average mass flow rate of xenon is 14,341 lbm/hr (1.88 kg/s). Filling the tank takes exactly one hour, and electricity costs $0.045 per kW-hr. The cost of filling the tank is most nearly

(A) $8.00

(B) $14

(C) $27

(D) $35

12. A ball is inflated to a pressure of 14.7 psi (101.3 kPa). The initial volume of the a ball is 12.58 ft³. The temperature increases and the volume of the air is increased to 13 ft³. The pressure does not change. The work performed on the system boundary is most nearly

(A) 0.5 Btu

(B) 1 Btu

(C) 2 Btu

(D) 4 Btu

13. Which of the following statements regarding the Clausius-Clapeyron equation is FALSE?

(A) It is the slope of the vapor pressure curve.
(B) The vapor phase follows the ideal gas law.
(C) The equation holds true for high pressures.
(D) The integrated equation is
$$\ln\frac{p_2}{p_1} = \left(\frac{h_{fg}}{\overline{R}}\right)\left(\frac{1}{T_1} - \frac{1}{T_2}\right)$$

SOLUTIONS

1. Availability depends on the temperature of the environment, T_L. Option A is true.

A heat engine's maximum efficiency is that of a Carnot engine cycle, which is always less than 100%. Option B is false.

An expansion includes a drop in enthalpy, and since $h = u + pv$, both u (manifested as temperature) and p decrease. Option C is true.

Liquid water cannot exist below the triple point pressure. Option D is true.

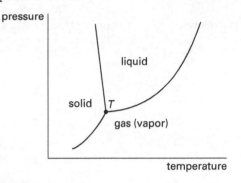

The answer is (B).

2. Entropy does not change in an isentropic (reversible adiabatic) process. However, not all adiabatic processes are reversible. Option A is false.

The entropy of a closed system can be decreased by decreasing the temperature. Option B is false.

When a refrigerant is throttled, the enthalpy remains constant, the pressure drops, and the entropy increases. Option C is true.

The fan and motor will certainly increase the entropy of the air inside the room. However, if the air temperature is decreased by heat loss to the outside, the entropy will decrease. Option D is false.

The answer is (C).

3. A throttling process (including flow through a control valve, safety relief valve, throttling valve, nozzle, or orifice) is modeled as an isenthalpic process. Consider steam expanding through a safety relief valve. To squeeze through the narrow restriction between the disk and the valve seat, steam has to accelerate to a high speed. It does so by converting enthalpy into kinetic energy. The process is not frictionless, but passage past the valve seat occurs so quickly as to be essentially so, and the process is considered to be isentropic. Once past the narrow restriction, the steam expands into the lower pressure region in the valve outlet. The steam decelerates as the flow area increases from the valve passageway to the downstream pipe. This decrease in velocity (kinetic energy) is manifested as an increase in temperature and enthalpy. The enthalpy drop associated with the initial increase in kinetic energy is reclaimed (except for a small portion lost due to the effects of friction). Therefore, the final process is essentially isenthalpic.

Viscous drag of an object in any fluid is an adiabatic process without work being performed on or by the fluid. The duration of the contact between fluid and object is short. Therefore, viscous drag is isenthalpic.

The increase in velocity in a supersonic nozzle comes entirely at the expense of enthalpy.

The answer is (B).

4. From a table of air properties, for 700°F air, the relative pressure is 21.18. [Properties of Air at Low Pressure, per Pound]

Find the relative pressure after isentropic expansion.

$$\frac{p_{r,1}}{p_{r,2}} = \frac{p_1}{p_2}$$

$$p_{r,2} = \frac{p_{r,1} p_2}{p_1} = \frac{(21.18)\left(50 \; \frac{\text{lbf}}{\text{in}^2}\right)}{400 \; \frac{\text{lbf}}{\text{in}^2}} = 2.648$$

From the table of air properties, interpolating as needed, this relative pressure corresponds to a temperature of 189.5°F, and the specific enthalpies of air at 700°F and 189.5°F are 281.14 Btu/lbm and 155.4 Btu/lbm, respectively. The change in specific enthalpy is

$$h_2 - h_1 = 155.4 \; \frac{\text{Btu}}{\text{lbm}} - 281.14 \; \frac{\text{Btu}}{\text{lbm}}$$
$$= -125.7 \; \text{Btu/lbm} \quad (-130 \; \text{Btu/lbm})$$

The answer is (B).

5. *Customary U.S. Solution*

Calculate the absolute temperature of the air before and after the cooling process. [Temperature Conversions]

$$T_1 = 180°F + 460° = 640°R$$
$$T_2 = 100°F + 460° = 560°R$$

Use the ideal gas law, and solve for the mass of air.

Ideal Gas

$$pV = mRT$$

$$m = \frac{p_1 V_1}{R T_1} = \frac{\left(14.7 \; \frac{\text{lbf}}{\text{in}^2}\right)\left(12 \; \frac{\text{in}}{\text{ft}}\right)^2 (8.0 \; \text{ft}^3)}{\left(53.35 \; \frac{\text{ft-lbf}}{\text{lbm-°R}}\right)(640°R)}$$
$$= 0.4960 \; \text{lbm}$$

Use the equation for work per unit mass in a constant-pressure process, and solve for the total work.

Special Cases of Closed Systems (With No Change in Kinetic or Potential Energy)

$$w = p\Delta v$$
$$= R\Delta T$$
$$W = mR\Delta T$$
$$= (0.4960 \; \text{lbm})\left(53.35 \; \frac{\text{ft-lbf}}{\text{lbm-°R}}\right)$$
$$\times (560°R - 640°R)$$
$$= -2117 \; \text{ft-lbf} \quad (-2100 \; \text{ft-lbf})$$

This is negative because work is done on the system.

SI Solution

Calculate the absolute temperature of the air before and after the cooling process. [Temperature Conversions]

$$T_1 = 82°C + 273° = 355K$$
$$T_2 = 38°C + 273° = 311K$$

Use the ideal gas law, and solve for the mass of air. The specific gas constant for air is 0.2870 kJ/kg·K. [Thermal and Physical Properties of Ideal Gases (at Room Temperature)]

Ideal Gas

$$pV = mRT$$

$$m = \frac{p_1 V_1}{R T_1}$$

$$= \frac{(101.3 \; \text{kPa}) \times \left(1000 \; \frac{\text{Pa}}{\text{kPa}}\right)(0.25 \; \text{m}^3)}{\left(0.2870 \; \frac{\text{kJ}}{\text{kg·K}}\right)\left(1000 \; \frac{\text{J}}{\text{kJ}}\right)(355K)}$$

$$= 0.2486 \; \text{kg}$$

Use the equation for work per unit mass in a constant-pressure process, and solve for the total work.

Special Cases of Closed Systems (With No Change in Kinetic or Potential Energy)

$$w = p\Delta v$$
$$= R\Delta T$$
$$W = mR\Delta T$$
$$= (0.2486 \; \text{kg})\left(0.2870 \; \frac{\text{kJ}}{\text{kg·K}}\right)$$
$$\times (311K - 355K)$$
$$= -3.139 \; \text{kJ} \quad (-3.1 \; \text{kJ})$$

This is negative because work is done on the system.

The answer is (A).

6. From steam tables, for a pressure of 300 psia, the enthalpy of a saturated liquid, h_f, is 393.93 Btu/lbm and the heat of vaporization, h_{fg}, is 809.42 Btu/lbm. [Properties of Saturated Water and Steam (Pressure) - I-P Units]

The total enthalpy is

Properties for Two-Phase (Vapor-Liquid) Systems

$$h_1 = h_f + xh_{fg} = 393.93 \; \frac{\text{Btu}}{\text{lbm}} + (0.95)\left(809.42 \; \frac{\text{Btu}}{\text{lbm}}\right)$$
$$= 1162.9 \; \text{Btu/lbm}$$

Calculate the change in enthalpy.

$$\Delta h = h_2 - h_1 = 1031 \; \frac{\text{Btu}}{\text{lbm}} - 1162.9 \; \frac{\text{Btu}}{\text{lbm}}$$
$$= -131.9 \; \text{Btu/lbm} \quad (-130 \; \text{Btu/lbm})$$

The answer is (B).

7. *Customary U.S. Solution*

The absolute temperature at the inlet of the air heater is

Temperature Conversions

$$°R = °F + 459.69° = T_1$$
$$T_1 = 540°F + 459.69°$$
$$= 999.69°R \quad (1000°R)$$

The absolute temperature at the outlet of the air heater is

Temperature Conversions

$$°R = °F + 459.69° = T_2$$
$$T_2 = 1540°F + 459.69°$$
$$= 1999.69°R \quad (2000°R)$$

SI Solution

The absolute temperature at the inlet of the air heater is

Temperature Conversions

$$K = °C + 273.15° = T_1$$
$$T_1 = 280°C + 273.15°$$
$$= 553.15 K \quad (550 K)$$

The absolute temperature at the outlet of the air heater is

Temperature Conversions

$$K = °C + 273.15° = T_2$$
$$T_2 = 840°C + 273.15°$$
$$= 1113.15 K \quad (1100 K)$$

The answer is (C).

8. *Customary U.S. Solution*

Since pressure is low and temperature is high, h and ϕ can be found from a table of air properties at low pressure. At 1000°R, $h_1 = 240.98$ Btu/lbm and $\phi_1 = 0.75042$ Btu/lbm-°R, and at 2000°R, $h_2 = 504.71$ Btu/lbm and $\phi_2 = 0.93205$ Btu/lbm-°R. [Properties of Air at Low Pressure, per Pound]

The availability per unit mass is $T_L = 100°F + 459.69° = 559.69°R$. [Temperature Conversions]

With a pressure drop from 100 psia to 80 psia, the maximum work is

$$W_{\text{max},p \text{ loss}} = h_1 - h_2 + T_{L,°R}\left(\phi_2 - \phi_1 - \frac{R}{J}\ln\left(\frac{p_2}{p_1}\right)\right)$$

$$= 240.98 \; \frac{\text{Btu}}{\text{lbm}} - 504.71 \; \frac{\text{Btu}}{\text{lbm}}$$
$$+ (559.69°R)$$

$$\times \left(\begin{array}{c} 0.93205 \; \dfrac{\text{Btu}}{\text{lbm-°R}} - 0.75042 \; \dfrac{\text{Btu}}{\text{lbm-°R}} \\ - \left(\dfrac{53.3 \; \dfrac{\text{ft-lbf}}{\text{lbm-°R}}}{778 \; \dfrac{\text{ft-lbf}}{\text{Btu}}}\right) \ln\left(\dfrac{80 \; \text{psia}}{100 \; \text{psia}}\right) \end{array}\right)$$

$$= -153.51 \; \text{Btu/lbm} \quad (-150 \; \text{Btu/lbm})$$

The answer is (D).

9. *Customary U.S. Solution*

The percentage loss in available energy is

$$\frac{W_{\text{max}} - W_{\text{max},p \text{ loss}}}{W_{\text{max}}} \times 100\%$$

$$= \frac{-162.02 \; \frac{\text{Btu}}{\text{lbm}} - \left(-153.46 \; \frac{\text{Btu}}{\text{lbm}}\right)}{-162.02 \; \frac{\text{Btu}}{\text{lbm}}} \times 100\%$$

$$= 5.28\% \quad (5.0\%)$$

SI Solution

The percentage loss in available energy is

$$\frac{W_{max} - W_{max,p\ loss}}{W_{max}} \times 100\%$$

$$= \frac{-378.17\ \frac{kJ}{kg} - \left(-356.50\ \frac{kJ}{kg}\right)}{-378.17\ \frac{kJ}{kg}} \times 100\%$$

$$= 5.73\%\quad(5.0\%)$$

The answer is (A).

10. *Customary U.S. Solution*

The absolute temperature is

Temperature Conversions

$$T = 70°F + 459.69 = 529.69°R$$

At 20 psia,

Ideal Gas

$$pv = mRT$$

$$m = \frac{pV}{RT}$$

$$= \frac{\left(20\ \frac{lbf}{in^2}\right)\left(12\ \frac{in}{ft}\right)^2 (100\ ft^3)}{\left(11.77\ \frac{ft\text{-}lbf}{lbm\text{-}°R}\right)(529.69°R)}$$

$$= 46.19\ lbm$$

The reduced variables are

Compressibility Factor and Charts

$$T_r = \frac{T}{T_c} = \frac{529.69°R}{521.9°R} = 1.01$$

$$p_r = \frac{p}{p_c} = \frac{3800\ \frac{lbf}{in^2}}{(58.2\ atm)\left(14.7\ \frac{lbf}{in^2\text{-}atm}\right)} = 4.44$$

Interpolating from the generalized compressibility chart, Z is 0.61. [Compressibility Factor and Charts]

At 3800 psia,

Equations of State (EOS)

$$p = \left(\frac{RT}{v}\right)Z$$

The volume of the gas is known. Divide specific volume by mass to convert to volume and then solve for m.

$$m = \frac{pV}{ZRT}$$

$$= \frac{\left(3800\ \frac{lbf}{in^2}\right)\left(12\ \frac{in}{ft}\right)^2(100\ ft^3)}{(0.61)\left(11.77\ \frac{ft\text{-}lbf}{lbm\text{-}°R}\right)(529.69°R)}$$

$$= 14{,}388\ lbm$$

The average mass flow rate of xenon is

$$\dot{m} = \frac{14{,}388\ lbm - 46.19\ lbm}{1\ hr}$$

$$= 14{,}341\ lbm/hr\quad(14{,}000\ lbm/hr)$$

SI Solution

The absolute temperature is

Temperature Conversions

$$T = 21°C + 273.15° = 294.15°K$$

At 150 kPa,

Ideal Gas

$$pv = mRT$$

$$m = \frac{pV}{RT}$$

$$= \frac{(150\ kPa)\left(1000\ \frac{Pa}{kPa}\right)(3\ m^3)}{\left(63.32\ \frac{J}{kg\cdot K}\right)(294.15K)}$$

$$= 24.16\ kg$$

The reduced variables are

Compressibility Factor and Charts

$$T_r = \frac{T}{T_c} = \frac{294.15K}{289.9K} = 1.01$$

$$p_r = \frac{p}{p_c} = \frac{25\ MPa}{(58.2\ atm)\left(0.1013\ \frac{MPa}{atm}\right)} = 4.24$$

Interpolating from the generalized compressibility chart, Z is 0.59. [Compressibility Factor and Charts]

At 25 MPa,

Equations of State (EOS)

$$p = \left(\frac{RT}{v}\right)Z$$

The volume of the gas is known. Divide specific volume by mass to convert to volume and then solve for m.

$$m = \frac{pV}{ZRT}$$

$$= \frac{(25 \text{ MPa})\left(10^6 \frac{\text{Pa}}{\text{MPa}}\right)(3 \text{ m}^3)}{(0.59)\left(63.32 \frac{\text{J}}{\text{kg·K}}\right)(294.15 \text{K})}$$

$$= 6824 \text{ kg}$$

The average mass flow rate of xenon is

$$\dot{m} = \frac{6824 \text{ kg} - 24.16 \text{ kg}}{(1 \text{ h})\left(3600 \frac{\text{s}}{\text{h}}\right)}$$

$$= 1.88 \text{ kg/s} \quad (1.9 \text{ kg/s})$$

The answer is (D).

11. *Customary U.S. Solution*

The absolute temperature is

Temperature Conversions

$$T = 70°\text{F} + 459.69 = 529.69°\text{R}$$

For isothermal compression, the work required to fill the tank is calculated from Boyle's law. 131g/mol is equivalent to 131 lbm/lbmol.

$$R_{\text{Xenon}} = \frac{\left(1545 \frac{\text{ft-lbf}}{\text{lbmol-R}}\right)}{\left(131.3 \frac{\text{lbm}}{\text{lbmol}}\right)}$$

$$= 11.77 \frac{\text{ft-lbf}}{\text{lbm-R}}$$

Special Cases of Closed Systems (With No Change in Kinetic or Potential Energy)

$$W = mRT \ln\left(\frac{p_1}{p_2}\right)$$

Calculate the power required using the following equation

$$\dot{W} = \dot{m}RT \ln\left(\frac{P1}{P2}\right)$$

$$= \frac{\left(14{,}341 \frac{\text{lbm}}{\text{hr}}\right)\left(11.77 \frac{\text{ft-lbf}}{\text{lbm-°R}}\right)}{\left(778 \frac{\text{ft-lbf}}{\text{Btu}}\right)\left(3413 \frac{\frac{\text{Btu}}{\text{hr}}}{\text{kW}}\right)}$$
$$\times (529.69°\text{R})\ln\left(\frac{20 \text{ psia}}{3800 \text{ psia}}\right)$$

$$= -176.67 \text{ kW}$$

The cost of the electricity needed is

$$\left(0.045 \frac{\$}{\text{kW-hr}}\right)(176.67 \text{ kW})(1 \text{ hr}) = \$7.95 \quad (\$8.00)$$

SI Solution

The absolute temperature is

Temperature Conversions

$$T = 21°\text{C} + 273.15° = 294.15°\text{K}$$

For isothermal compression, the work required to fill the tank is

$$R_{\text{Xenon}} = \frac{\left(8314 \frac{\text{kJ}}{\text{kmol-K}}\right)}{\left(131.3 \frac{\text{kg}}{\text{kmol}}\right)}$$

$$= 63.32 \frac{\text{J}}{\text{kg-K}}$$

Special Cases of Closed Systems (With No Change in Kinetic or Potential Energy)

$$W = mRT \ln\left(\frac{p_1}{p_2}\right)$$

Calculate the power required using the following equation.

$$\dot{W} = \dot{m}RT\ln\left(\frac{P1}{P2}\right)$$

$$= \frac{\left(1.88\ \dfrac{\text{kg}}{\text{s}}\right)\left(63.32\ \dfrac{\text{J}}{\text{kg}\cdot\text{K}}\right)(294.15\,\text{K})}{\left(1000\ \dfrac{\text{W}}{\text{kW}}\right)}$$

$$\times \ln\left[\frac{150\ \text{kPa}}{(25\ \text{MPa})\left(1000\ \dfrac{\text{kPa}}{\text{MPa}}\right)}\right]$$

$$= -179.14\ \text{kW}$$

The cost of electricity is

$$\left(0.045\ \dfrac{\$}{\text{kW-h}}\right)(179.14\ \text{kW})(1\ \text{hr}) = \$8.06 \quad (\$8.00)$$

The answer is (A).

12. *Customary U.S. Solution*

For a constant pressure closed system, the total work is found from Charles' law.

Special Cases of Closed Systems (With No Change in Kinetic or Potential Energy)

$$W_b = P\Delta V$$

$$= \frac{\left(14.7\ \dfrac{\text{lbf}}{\text{in}^2}\right)\left(12\ \dfrac{\text{in}}{\text{ft}}\right)^2(13\ \text{ft}^3 - 12.58\ \text{ft}^3)}{778\ \dfrac{\text{ft-lbf}}{\text{Btu}}}$$

$$= 1.14\ \text{Btu} \quad (1\ \text{Btu})$$

The answer is (B).

13. The Clausius-Clapeyron equation assumes that h_{fg} is constant, which is not accurate at high pressures, where h_{fg} decreases as temperature increases from the triple point of the critical point.

The answer is (C).

25 Compressible Fluid Dynamics

Content in blue refers to the *NCEES Handbook*.

PRACTICE PROBLEMS

1. A round-nosed bullet travels at 2000 ft/sec through 32°F, 14.7 psia air. Most nearly, what is the Mach number, and is the bullet's speed subsonic or supersonic?

(A) subsonic (Ma = 0.90)
(B) supersonic (Ma = 1.10)
(C) supersonic (Ma = 1.40)
(D) supersonic (Ma = 1.80)

2. A round-nosed bullet travels through 32°F (0°C) air. Assume that the bullet is stationary, and the air is moving at Ma = 1.84. The total temperature at the tip of the bullet is most nearly

(A) 770°R
(B) 800°R
(C) 830°R
(D) 860°R

3. A round-nosed bullet travels through air. The stagnation temperature of the bullet is 900°R. The total enthalpy at the bullet face is most nearly

(A) 200 Btu/lbm
(B) 250 Btu/lbm
(C) 300 Btu/lbm
(D) 350 Btu/lbm

4. A well-hit golf ball with a diameter of 1.69 in and a weight of 0.0992 lbf travels at 200 ft/sec as it leaves the tee. The drag coefficient is 0.25. The drag on the golf ball is most nearly

(A) 0.11 slug-ft/sec^2
(B) 0.19 slug-ft/sec^2
(C) 0.33 slug-ft/sec^2
(D) 0.520 slug-ft/sec^2

5. A well-hit table tennis ball with a diameter of 1.50 in and weight of 0.00551 lbf travels at 60 ft/sec as it leaves the paddle. The drag coefficient is 0.50. The drag force on the table tennis ball is most nearly

(A) 0.01 lbf
(B) 0.02 lbf
(C) 0.03 lbf
(D) 0.04 lbf

6. Air with a total pressure of 100 psia and a total temperature of 70°F flows through a converging-diverging nozzle. Supersonic flow is achieved in the diverging section. At a particular point in the diverging section, the ratio of the flow area to the critical throat area is 1.5553. The Mach number at that point is most nearly

(A) 1.10
(B) 1.30
(C) 1.90
(D) 2.40

7. Air with a total pressure of 100 psia (0.7 MPa) and a total temperature of 70°F (21°C) flows through a converging-diverging nozzle. Supersonic flow is achieved in the diverging section. At a particular point in the diverging section, the ratio of flow area to the critical throat area is 1.5553. The temperature at that point is most nearly

(A) 290°R (160K)
(B) 310°R (170K)
(C) 340°R (190K)
(D) 360°R (200K)

8. Air with a total pressure of 100 psia (0.7 MPa) and a total temperature of 70°F (21°C) flows through a converging-diverging nozzle. Supersonic flow is achieved in the diverging section. At a particular point in the diverging section, the ratio of flow area to the critical throat area is 1.5553. The pressure at that point is most nearly

(A) 15 psia (0.10 MPa)

(B) 20 psia (0.14 MPa)

(C) 25 psia (0.18 MPa)

(D) 30 psia (0.21 MPa)

9. The initial temperature and pressure of air inside a high-pressure reservoir are 68°F and 470 lbf/in², respectively. When a valve is suddenly opened, the air discharges through a simple square-edged orifice with an initial effective velocity of 875 ft/sec. The valve is closed when the reservoir pressure is reduced to 100 lbf/in². The orifice's area is 0.05 in². The backpressure is 14.7 lbf/in² and is constant. The reservoir has a total volume of 100 ft³, and the discharged air is not replenished. The discharge is adiabatic. The air's ratio of specific heats is 1.40 and remains constant. What is most nearly the initial static temperature of the air flowing through the orifice?

(A) 390°R

(B) 440°R

(C) 480°R

(D) 530°R

10. The initial temperature and pressure of air inside a high-pressure reservoir are 68°F and 470 lbf/in², respectively. When a valve is suddenly opened, the air discharges through a simple square-edged orifice with an initial effective velocity of 875 ft/sec. The valve is closed when the reservoir pressure is reduced to 100 lbf/in². The orifice's area is 0.05 in². The backpressure is 14.7 lbf/in² and is constant. The reservoir has a total volume of 100 ft³, and the discharged air is not replenished. The discharge is adiabatic. The air's ratio of specific heats is 1.40 and remains constant. The static to total temperature ratio is 0.8333. What is most nearly the initial sonic velocity in the throat?

(A) 950 ft/sec

(B) 990 ft/sec

(C) 1000 ft/sec

(D) 1100 ft/sec

11. The initial temperature, pressure and velocity of air inside a high-pressure reservoir are 68°F, 470 lbf/in², and 875 ft/sec, respectively. The reservoir has a total volume of 100 ft³, and the discharged air is not replenished. The discharge is adiabatic. The throat velocity is 1028 ft/sec. What is most nearly the orifice coefficient?

(A) 0.53

(B) 0.62

(C) 0.69

(D) 0.85

12. The initial temperature and pressure of air inside a high-pressure reservoir are 68°F and 470 lbf/in², respectively. When a valve is suddenly opened, the air discharges through a simple square-edged orifice with an initial effective velocity of 875 ft/sec. The valve is closed when the reservoir pressure is reduced to 100 lbf/in². The orifice's area is 0.05 in². Flow through the valve travels at approximately the speed of sound. The reservoir has a total volume of 100 ft³, and the discharged air is not replenished. The discharge is adiabatic. The air's ratio of specific heats is 1.40 and remains constant. The coefficient of discharge is 0.851. What is most nearly the initial discharge rate?

(A) 0.46 lbm/sec

(B) 0.69 lbm/sec

(C) 0.85 lbm/sec

(D) 1.2 lbm/sec

13. The initial temperature and pressure of air inside a high-pressure reservoir are 68°F and 470 lbf/in², respectively. When a valve is suddenly opened, the air discharges through a simple square-edged orifice with an initial effective velocity of 875 ft/sec. The valve is closed when the reservoir pressure is reduced to 100 lbf/in². The orifice's area is 0.05 in². The reservoir has a total volume of 100 ft³, and the discharged air is not replenished. The discharge is adiabatic. The total mass of air discharged from the tank is most nearly

(A) 50 lbm

(B) 120 lbm

(C) 190 lbm

(D) 240 lbm

SOLUTIONS

1. *Customary U.S. Solution*

The shock wave is normal to the direction of flight, so the Mach number equations are applicable. The points to consider are shown.

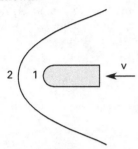

The temperature and pressure at point 1 are

Temperature Conversions
$$T_1 = 32°F + 460° = 492°R$$
$$p_1 = 14.7 \text{ lbf/in}^2$$

Combining the equation for the Mach number, the speed of sound, and the specific gas constant of air, and the molecular weight of air, the Mach number at point 1 is [Thermal and Physical Properties of Ideal Gases (at Room Temperature)]

Mach Number
$$\text{Ma} = \frac{\text{v}}{c}$$
$$c = \sqrt{kRT}$$
$$R = \frac{\overline{R}}{\text{molecular weight}}$$
$$= \frac{\text{v}}{\sqrt{\dfrac{kg_c \overline{R} T}{\text{molecular weight}}}}$$

$$= \frac{2000 \, \dfrac{\text{ft}}{\text{sec}}}{\sqrt{\dfrac{(1.40)\left(32.2 \, \dfrac{\text{lbm-ft}}{\text{lbf-sec}^2}\right)\left(1545 \, \dfrac{\text{ft-lbf}}{\text{lbmol-°R}}\right)(492°R)}{29.0 \, \dfrac{\text{lbm}}{\text{lbmol}}}}}$$

$$= 1.84 \quad (1.80)$$

The bullet speed is supersonic.

The answer is (D).

2. The ratio of static temperature before the shock to the total temperature after the shock is read from the normal shock table, interpolating as needed. [One-Dimensional Isentropic Compressible-Flow Functions $k = 1.4$]

For $\text{Ma}_1 = 1.84$, the temperature ratio is

$$\frac{T_1}{T_0} = 0.5963$$

The absolute temperature of the air is

Temperature Conversions
$$T_1 = 32°F + 460° = 492°R$$

Therefore, the stagnation total temperature is

$$T_0 = \frac{T_1}{0.5963} = \frac{492°R}{0.5963}$$
$$= 825.1°R \quad (830°R)$$

Since the shock wave is adiabatic, T_0 remains constant.

The answer is (C).

3. From the air tables, air at a temperature of 900°R has an enthalpy of 216.26 Btu/lbm (200 Btu/lbm). [Properties of Air at Low Pressure, per Pound]

The answer is (A).

4. The cross-sectional area of a ball is used to calculate the drag force. Using the equation of the area of a circle, the area of the golf ball is

$$A = \pi r^2 = \frac{\pi D^2}{4}$$
$$= \frac{\pi (1.69 \text{ in})^2}{(4)\left(12 \, \dfrac{\text{in}}{\text{ft}}\right)^2}$$
$$= 0.0156 \text{ ft}^2$$

Find the density of air at sea level in slug/ft^3. [Temperature and Altitude Corrections for Air]

$$\rho = \frac{0.075 \, \dfrac{\text{lbm}}{\text{ft}^3}}{32.2 \, \dfrac{\text{lbm}}{\text{slug}}}$$
$$= 2.33 \times 10^{-3} \text{ slug/ft}^3$$

The force due to drag can be found using the drag force equation.

$$F_D = \frac{C_D \rho v^2 A}{2}$$

$$= \frac{(0.25)\left(2.33 \times 10^{-3} \dfrac{\text{slug}}{\text{ft}^3}\right) \times \left(200 \dfrac{\text{ft}}{\text{sec}}\right)^2 (0.0156 \text{ ft}^2)}{2}$$

$$= 0.182 \text{ slug-ft/sec}^2 \quad (0.19 \text{ slug-ft/sec}^2)$$

The answer is (B).

5. The projected area of the spherical ball is the area of a circle, with the same diameter as the ball.

$$A = \pi r^2 = \frac{\pi D^2}{4}$$

$$= \frac{\pi (1.50 \text{ in})^2}{(4)\left(12 \dfrac{\text{in}}{\text{ft}}\right)^2}$$

$$= 0.0123 \text{ ft}^2$$

Calculate the density of air at sea level in slug/ft³. Convert to slug/ft³ to find the density in terms of mass, not weight. [Temperature and Altitude Corrections for Air]

$$\rho = \frac{0.075 \dfrac{\text{lbf}}{\text{ft}^3}}{32.2 \dfrac{\text{lbf}}{\text{slug}}}$$

$$= 2.33 \times 10^{-3} \text{ slug/ft}^3$$

The force due to drag can be solved with the drag force equation.

Drag Force

$$F_D = \frac{C_D \rho v^2 A}{2}$$

$$= \frac{(0.50)\left(2.33 \times 10^{-3} \dfrac{\text{slug}}{\text{ft}^3}\right)\left(\dfrac{1 \text{ lbf}}{\dfrac{\text{slug-ft}}{\text{sec}^2}}\right) \times \left(60 \dfrac{\text{ft}}{\text{sec}}\right)^2 (0.0123 \text{ ft}^2)}{2}$$

$$= 0.0258 \text{ lbf} \quad (0.03 \text{ lbf})$$

Note: 1 lbf = 1 slug × 1 ft/sec²

The answer is (C).

6. The ratio of flow area to the critical throat area is given as

$$\frac{A}{A^*} = 1.5553$$

From the isentropic flow tables, at $A/A^* = 1.5553$, the Mach number is 1.90. [One-Dimensional Isentropic Compressible-Flow Functions $k = 1.4$]

The answer is (C).

7. *U.S. Customary Solution*

The absolute total temperature is

Temperature Conversions

$$T_0 = 70°\text{F} + 460° = 530°\text{R}$$

From the isentropic flow tables, the ratio of static to total temperature at $A/A^* = 1.5553$ is [One-Dimensional Isentropic Compressible-Flow Functions $k = 1.4$]

$$\frac{T}{T_0} = 0.5807$$

The temperature at the diverging point is

$$T = \left(\frac{T}{T_0}\right)T_0 = (0.5807)(530°\text{R})$$

$$= 307.8°\text{R} \quad (310°\text{R})$$

SI Solution

The absolute total temperature is

Temperature Conversions

$$T_0 = 21°\text{C} + 273° = 294\text{K}$$

From the isentropic flow tables, the ratio of static to total temperature at $A/A^* = 1.5553$ is [One-Dimensional Isentropic Compressible-Flow Functions $k = 1.4$]

$$\frac{T}{T_0} = 0.5807$$

The temperature at the diverging point is

$$T = \left(\frac{T}{T_0}\right)T_0 = (0.5807)(294\text{K})$$

$$= 170.7\text{K} \quad (170\text{K})$$

The answer is (B).

8. *Customary U.S. Solution*
From the isentropic flow tables, the ratio of static to total pressure at $A/A^* = 1.5553$ is [One-Dimensional Isentropic Compressible-Flow Functions $k=1.4$]

$$\frac{p}{p_0} = 0.1492$$

The pressure at the diverging point is

$$p = \left(\frac{p}{p_0}\right)p_0 = (0.1492)\left(100\ \frac{\text{lbf}}{\text{in}^2}\right)$$
$$= 14.92\ \text{lbf/in}^2 \quad (15\ \text{psia})$$

SI Solution
From the isentropic flow tables, the ratio of static to total pressure at $A/A^* = 1.5553$ is [One-Dimensional Isentropic Compressible-Flow Functions $k=1.4$]

$$\frac{p}{p_0} = 0.1492$$

The pressure at the diverging point is

$$p = \left(\frac{p}{p_0}\right)p_0 = (0.1492)(0.7\ \text{MPa})$$
$$= 0.104\ \text{MPa} \quad (0.10\ \text{MPa})$$

The answer is (A).

9. The backpressure ratio is

$$\frac{p_{\text{back}}}{p_{\text{reservoir}}} = \frac{14.7\ \frac{\text{lbf}}{\text{in}^2}}{470\ \frac{\text{lbf}}{\text{in}^2}}$$
$$= 0.031$$

Since the backpressure ratio is less than 0.5283 (the critical backpressure ratio for gases with $k = 1.40$), flow in the orifice will be choked until the reservoir pressure reaches a ratio of 0.5283. The initial pressure when the ratio is 0.5283 is

$$\frac{p}{p_0} = 0.5283$$
$$p_0 = \frac{p}{0.5283}$$
$$= \frac{14.7\ \frac{\text{lbf}}{\text{in}^2}}{0.5283}$$
$$= 27.8\ \text{lbf/in}^2$$

This is lower than the pressure at which the valve closes. Therefore, the flow will be choked for the entire duration of discharge from the initial pressure of 470 lbf/in² down to the final pressure of 100 lbf/in². The discharge will be sonic throughout the discharge.

From the isentropic flow tables, the static to total temperature ratio for a pressure ratio of 0.5283 is [One-Dimensional Isentropic Compressible-Flow Functions $k = 1.4$]

$$\frac{T}{T_0} = 0.8333$$

The absolute temperature for the reservoir is

Temperature Conversions
$$T_{\text{reservoir}} = 68°\text{F} + 460° = 528°\text{R}$$

The static temperature where sonic flow is achieved is

$$T_{\text{throat}} = \left(\frac{T}{T_0}\right)T_{\text{reservoir}} = (0.8333)(528°\text{R})$$
$$= 440°\text{R}$$

The answer is (B).

10. The absolute temperature of the reservoir is

Temperature Conversions
$$T_{\text{reservoir}} = 68°\text{F} + 460° = 528°\text{R}$$

The throat temperature is

$$T_{\text{throat}} = \left(\frac{T}{T_0}\right)T_{\text{reservoir}} = (0.8333)(528°\text{R})$$
$$= 440°\text{R}$$

The discharge velocity will be limited by the speed of sound.

Mach Number
$$c = \sqrt{kRT} = \sqrt{kg_c R T_{\text{throat}}}$$
$$= \sqrt{(1.40)\left(32.2\ \frac{\text{lbm-ft}}{\text{lbf-sec}^2}\right)\left(53.3\ \frac{\text{ft-lbf}}{\text{lbm-°R}}\right)(440°\text{R})}$$
$$= 1028\ \text{ft/sec} \quad (1000\ \text{ft/sec})$$

The answer is (C).

11. The orifice coefficient is

$$C_d = C_c C_v \approx C_v = \frac{V_{actual}}{V_{ideal}}$$

$$= \frac{875 \ \frac{ft}{sec}}{1028 \ \frac{ft}{sec}}$$

$$= 0.851 \quad (0.85)$$

The answer is (D).

12. From the isentropic flow tables, the ratio of static pressure to total pressure for a Mach number of 1 is [One-Dimensional Isentropic Compressible-Flow Functions $k = 1.4$]

$$\frac{p}{p_0} = 0.5283$$

The static pressure at the throat for a Mach number of 1 is

$$p_{throat} = \left(\frac{p}{p_0}\right) p_{reservoir} = (0.5283)\left(470 \ \frac{lbf}{in^2}\right)$$

$$= 248.3 \ lbf/in^2$$

The ratio of static temperature to total temperature for a Mach number of 1 is [One-Dimensional Isentropic Compressible-Flow Functions $k = 1.4$]

$$\frac{T}{T_0} = 0.8333$$

The absolute temperature of the reservoir is

Temperature Conversions

$$T_{reservoir} = 68°F + 460° = 528°R$$

The throat temperature is

$$T_{throat} = \left(\frac{T}{T_0}\right) T_{reservoir} = (0.8333)(528°R)$$

$$= 440°R$$

Use the equation for the speed of sound to find the ideal velocity.

Mach Number

$$c = \sqrt{kRT} = \sqrt{kg_c R T_{throat}} = v_{ideal}$$

$$= \sqrt{(1.40)\left(32.2 \ \frac{lbm\text{-}ft}{lbf\text{-}sec^2}\right)\left(53.3 \ \frac{ft\text{-}lbf}{lbm\text{-}°R}\right)(440°R)}$$

$$= 1028 \ ft/sec$$

The initial mass flow rate of the air discharging through the orifice is

$$\dot{m}_{orifice} = c_d \dot{m}_{ideal} = c_d p_{throat} A v$$

$$= \frac{c_d p_{throat} A v_{ideal}}{RT_{throat}}$$

$$= \frac{(0.851)\left(248.3 \ \frac{lbf}{in^2}\right)(0.05 \ in^2)\left(1028 \ \frac{ft}{sec}\right)}{\left(53.3 \ \frac{ft\text{-}lbf}{lbm\text{-}°R}\right)(440°R)}$$

$$= 0.463 \ lbm/sec \quad (0.46 \ lbm/sec)$$

The answer is (A).

13. From the ideal gas law, the mass of air discharged from the reservoir is

Ideal Gas

$$pv = mRT$$

$$m_{discharged} = \frac{(p_1 - p_2)v}{RT_1}$$

$$= \frac{\left(470 \ \frac{lbf}{in^2} - 100 \ \frac{lbf}{in^2}\right)\left(12 \ \frac{in}{ft}\right)^2 (100 \ ft^3)}{\left(53.3 \ \frac{ft\text{-}lbf}{lbm\text{-}°R}\right)(68°F + 460°)}$$

$$= 189.3 \ lbm \quad (190 \ lbm)$$

The answer is (C).

26 Vapor Power Equipment

Content in blue refers to the *NCEES Handbook*.

PRACTICE PROBLEMS

1. A steam generator receives liquid water at 80°F and produces saturated steam at 300 psia. Most nearly, what is the ratio of boiler sensible to boiler latent heat supplied to the water and steam?

(A) 0.06
(B) 0.12
(C) 0.27
(D) 0.43

2. A 10 000 kW steam turbine operates on 400 psia, 750°F dry steam, expanding to 2 in Hg absolute. The maximum adiabatic heat drop available for power production is most nearly

(A) 460 Btu/lbm
(B) 610 Btu/lbm
(C) 740 Btu/lbm
(D) 930 Btu/lbm

3. A two-pass surface condenser receives 82,000 lbm/hr (10.3 kg/s) of steam from a turbine. Steam enters the condenser with an enthalpy of 980 Btu/lbm (2.280 MJ/kg). The condenser operates at a pressure of 1 in Hg (3.4 kPa) absolute. Water is circulated at 8 ft/sec (2.4 m/s) through an equivalent length of 120 ft (36 m) of extra strong 30 in (76.2 cm) steel pipe with a specific roughness of 0.0002 ft (6.0 × 10⁻⁵ m) and an inner diameter of 2.4167 ft (0.7366 m). The flow is fully turbulent. An additional head loss of 6 in wg (1.5 kPa) is incurred in the intake screens, which draw in the cooling water from a stationary reservoir.

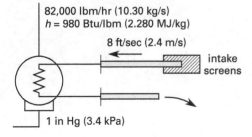

The head added by the circulating water pump is most nearly

(A) 2.1 ft (0.62 m)
(B) 3.1 ft (0.93 m)
(C) 6.2 ft (1.9 m)
(D) 8.5 ft (2.6 m)

4. A two-pass surface condenser receives 100,000 lbm/hr (12.56 kg/s) of steam from a turbine. Steam enters the condenser with an enthalpy of 980 Btu/lbm (2.280 MJ/kg). The condenser operates at a pressure of 1 in Hg (3.4 kPa) absolute.

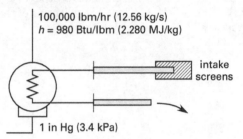

If the water temperature increases 10°F (5.6 K) across the condenser, what is most nearly the circulation rate of cooling water?

(A) 9000 gal/min (34 kL/min)
(B) 11,000 gal/min (42 kL/min)
(C) 13,000 gal/min (49 kL/min)
(D) 19,000 gal/min (70 kL/min)

5. 100 lbm/hr (0.013 kg/s) of 60°F (16°C) water is turned into 14.7 psia (101.3 kPa) saturated steam in an electric boiler. Radiation losses are 35% of the supplied energy. If electricity is $0.04 per kW-hr, the cost is most nearly

(A) $2/hr
(B) $3/hr
(C) $4/hr
(D) $5/hr

6. A gas burner produces 250 lbm/hr (0.032 kg/s) of 98% dry steam at 40 psia (300 kPa) from 60°F (16°C) feedwater. The fuel gas enters at 80°F (26°C) and a gage pressure 4 in Hg (13.6 kPa) and has a heating value of 550 Btu/ft^3 (20.5 MJ/m^3) at standard industrial conditions. The barometric pressure is 30.2 in Hg (102.4 kPa). 13.5 ft^3/min (6.4 L/s) of fuel gas is consumed. The efficiency of the boiler is most nearly

(A) 37%

(B) 43%

(C) 57%

(D) 66%

7. A boiler evaporates 8.23 lbm (8.23 kg) of 120°F (50°C) water per pound (per kilogram) of coal fired, producing 100 psia (700 kPa) saturated steam. The coal is 2% moisture by weight as fired, and dry coal is 5% ash. 1% of the coal is removed from the ash pit. (The ash pit loss has the same composition as unfired, dry coal.) Coal is initially at 60°F (16°C), and combustion occurs at 14.7 psia (101.3 kPa). The combustion products leave at 600°F (315°C). The heating value of the dry coal is 12,800 Btu/lbm (29.80 MJ/kg). The efficiency of the boiler is most nearly

(A) 53%

(B) 68%

(C) 73%

(D) 82%

8. 500 psia (3.5 MPa) steam is superheated to 1000°F (500°C) before expanding through a 75% efficient turbine to 5 psia (30 kPa). No subcooling occurs. The pump work is negligible compared to the 200 MW generated. The quantity of steam required is most nearly

(A) 8.6×10^5 lbm/hr

(B) 1.4×10^6 lbm/hr

(C) 2.0×10^6 lbm/hr

(D) 4.2×10^6 lbm/hr

9. A saturated water vapor at 210°F and a two-phase liquid-vapor water mixture at 300°F are combined in an open feedwater heater. The saturated vapor enters the feedwater heater at a rate of 28 lbm/sec, and the liquid-vapor mixture enters the feedwater heater at a rate of 3 lbm/sec. The liquid-vapor mixture has a quality of 0.904. If the mixture leaves the heater as a saturated vapor, the temperature of the exiting mixture is most nearly

(A) 196°F

(B) 210°F

(C) 270°F

(D) 290°F

10. The energy efficiency ratio of an air conditioner that draws 65 kW and is rated at 35 tons is most nearly

(A) 0.5

(B) 1.0

(C) 2.0

(D) 6.5

SOLUTIONS

1. Sensible heat changes temperature, while latent heat changes phase. The pressure throughout the steam generator is 300 psia. Water enters subcooled at 80°F and 300 psia. The water remains in liquid form as it is heated (sensible heat) to the saturation temperature corresponding to 300 psia, which (from steam tables) is 417.33°F. Additional energy (latent heat) converts the saturated water to saturated steam at 300 psia. [Properties of Saturated Water and Steam (Pressure) - I-P Units]

The enthalpy of 80°F liquid water is essentially a function of temperature only. From steam tables, $h_{80°F} = 48.06$ Btu/lbm. From steam tables, $h_{f,300\,psia} = 393.3$ Btu/lbm, and $h_{fg,300\,psia} = 809.42$ Btu/lbm. [Properties of Saturated Water and Steam (Pressure) - I-P Units] [Properties of Saturated Water and Steam (Temperature) - I-P Units]

The ratio of sensible heat to latent heat is

$$\frac{q_{\text{sensible}}}{q_{\text{latent}}} = \frac{h_{f,300\,psia} - h_{80°F}}{h_{fg,300\,psia}} = \frac{393.3\,\frac{\text{Btu}}{\text{lbm}} - 48.06\,\frac{\text{Btu}}{\text{lbm}}}{809.42\,\frac{\text{Btu}}{\text{lbm}}}$$

$$= 0.4265 \quad (0.43)$$

The answer is (D).

2. Interpolating from superheated steam tables, the approximate enthalpy, h_1, of dry steam at 400 psia and 750°F is 1390 Btu/lbm. [Properties of Superheated Steam - I-P Units]

Interpolating from the steam tables, s_1 is 1.6635 Btu/lbm-°R. The maximum possible adiabatic heat drop happens when the expansion to 2 in Hg (state 2) is isentropic.

For isentropic expansion from state 1 to state 2, $s_2 = s_1 = 1.6635$ Btu/lbm-R

Convert 2 in Hg absolute to psia [Measurement Relationships]

$$1\text{ psia} = 2.036\text{ in Hg} \simeq 2\text{ in Hg (state 2)}$$

From the steam tables, the value of s_2(1.6635 Btu/lbm-R) is between the values of s_f and s_g at 1 psia. Therefore, the steam will be a liquid – vapor mixture at state 2. Determine the quality of the steam at state 2 by using the relationships for liquid – vapor systems.

Properties for Two-Phase (Vapor-Liquid) Systems

$$s = s_f + xs_{fg}$$

From steam tables, at 1 psia,

$s_f = 0.1326$ Btu/lbm-R and $s_{fg} = 1.8451$ Btu/lbm-R

[Properties of Saturated Water and Steam (Pressure) - I-P Units]

Therefore,

$$x = \frac{s_2 - s_f}{s_{fg}} = \frac{1.6635\,\frac{\text{Btu}}{\text{lbm-R}} - 0.1326\,\frac{\text{Btu}}{\text{lbm-R}}}{1.8451\,\frac{\text{Btu}}{\text{lbm-R}}}$$

$$= 0.83$$

Use the quality and two-phase relationship to determine the enthalpy at state 2. From steam tables at 1 psia,
$h_f = 69.71$ Btu/lbm and $h_{fg} = 1035.71$ Btu/lbm

Therefore,

$$h_2 = h_f + xh_{fg} = 69.71\,\frac{\text{Btu}}{\text{lbm}} + 0.83 \times 1035.71\,\frac{\text{Btu}}{\text{lbm}}$$

$$= 929\text{ Btu/lbm}$$

Therefore, the maximum adiabatic heat drop is,

$$h_1 - h_2 = 1390\,\frac{\text{Btu}}{\text{lbm}} - 929\,\frac{\text{Btu}}{\text{lbm}} = 461\text{ Btu/lbm} \quad (460\text{ Btu/lbm})$$

The answer is (A).

3. *Customary U.S. Solution*

Find the relative roughness of the pipe.

$$\frac{\epsilon}{D} = \frac{0.0002\text{ ft}}{2.4167\text{ ft}} = 0.000083$$

From the Moody friction factor chart for fully turbulent flow, $f \approx 0.012$. [Moody Diagram (Stanton Diagram)]

The head loss from friction is

Darcy-Weisbach Equation

$$h_f = f \frac{L v^2}{2 D g}$$

$$= \frac{(0.012)(120 \text{ ft})\left(8 \frac{\text{ft}}{\text{sec}}\right)^2}{(2)(2.4167 \text{ ft})\left(32.2 \frac{\text{ft}}{\text{sec}^2}\right)}$$

$$= 0.59 \text{ ft}$$

The screen loss is

$$6 \text{ in wg} = \left(\frac{6 \text{ in}}{12 \frac{\text{in}}{\text{ft}}}\right) = 0.5 \text{ ft of water head}$$

$$\text{velocity head} = \frac{v^2}{2g} = \frac{\left(8 \frac{\text{ft}}{\text{sec}}\right)^2}{(2)\left(32.2 \frac{\text{ft}}{\text{sec}^2}\right)} = 0.99 \text{ ft}$$

There is inadequate information to evaluate pressure and elevation heads. The total head added by a coolant pump (not shown) not including losses inside the condenser is

$$0.59 \text{ ft} + 0.5 \text{ ft} + 0.99 \text{ ft} = 2.08 \text{ ft} \quad (2.1 \text{ ft})$$

SI Solution

Find the relative roughness of the pipe.

$$\frac{\epsilon}{D} = \frac{6.0 \times 10^{-5} \text{ m}}{0.7366 \text{ m}} = 0.0000815$$

From the Moody friction factor chart for fully turbulent flow, $f \approx 0.012$. [Moody Diagram (Stanton Diagram)]

The head loss from friction is

Darcy-Weisbach Equation

$$h_f = f \frac{L v^2}{2 D g}$$

$$= \frac{(0.012)(36 \text{ m})\left(2.4 \frac{\text{m}}{\text{s}}\right)^2}{(2)(0.7366 \text{ m})\left(9.81 \frac{\text{m}}{\text{s}^2}\right)}$$

$$= 0.1722 \text{ m}$$

The screen loss is 1.5 kPa.

$$h_{\text{screen}} = \frac{\Delta p}{\rho g} = \frac{(1.5 \text{ kPa})\left(1000 \frac{\text{Pa}}{\text{kPa}}\right)}{\left(1000 \frac{\text{kg}}{\text{m}^3}\right)\left(9.81 \frac{\text{m}}{\text{s}^2}\right)} = 0.1529 \text{ m}$$

The velocity head is

$$\frac{v^2}{2g} = \frac{\left(2.4 \frac{\text{m}}{\text{s}}\right)^2}{(2)\left(9.81 \frac{\text{m}}{\text{s}^2}\right)} = 0.2936 \text{ m}$$

There is inadequate information to evaluate pressure and elevation heads. The total head added by a coolant pump (not shown) not including losses inside the condenser is

$$0.1722 \text{ m} + 0.1529 \text{ m} + 0.2936 \text{ m} = 0.6187 \text{ m} \quad (0.62 \text{ m})$$

The answer is (A).

4. *Customary U.S. Solution*

The condenser pressure is

$$(1 \text{ in Hg})\left(0.491 \frac{\frac{\text{lbf}}{\text{in}^2}}{\text{in Hg}}\right) = 0.491 \text{ psia}$$

Interpolating from steam tables, the enthalpy of the saturated liquid is $h_f = 47.08$ Btu/lbm. [Properties of Saturated Water and Steam (Pressure) - I-P Units]

The specific heat of water is $c_{p,\text{water}} = 1$ Btu/lbm-°F. [Properties of Water at Standard Conditions]

The heat lost by the steam is equal to the heat gained by the water.

$$\dot{m}_{water}c_{p,water}\Delta T = \dot{m}_{steam}(h_2 - h_f)$$

$$\dot{m}_{water} = \frac{\left(100{,}000\ \dfrac{\text{lbm}}{\text{hr}}\right)\left(980\ \dfrac{\text{Btu}}{\text{lbm}} - 47.08\ \dfrac{\text{Btu}}{\text{lbm}}\right)}{\left(1\ \dfrac{\text{Btu}}{\text{lbm-}°\text{F}}\right)(10°\text{F})}$$

$$= 9.3292 \times 10^6\ \text{lbm/hr}$$

The density of water, ρ, is 62.4 lbm/ft^3. The circulation rate is [Properties of Water at Standard Conditions]

$$\dot{m}_{water} = Q\rho$$

$$Q = \frac{\dot{m}_{water}}{\rho} = \frac{\left(9.3292 \times 10^6\ \dfrac{\text{lbm}}{\text{hr}}\right)\left(7.48\ \dfrac{\text{gal}}{\text{ft}^3}\right)}{\left(62.4\ \dfrac{\text{lbm}}{\text{ft}^3}\right)\left(60\ \dfrac{\text{min}}{\text{hr}}\right)}$$

$$= 18{,}638.47\ \text{gal/min} \quad (19{,}000\ \text{gal/min})$$

(The flow rate can also be determined from the velocity and pipe area. However, this method does not use the 10°F data to perform an energy balance.)

SI Solution

The condenser pressure is 3.4 kPa.

Interpolating from steam tables, the enthalpy of saturated liquid is $h_1 = 109.8$ kJ/kg. [Properties of Saturated Water and Steam (Pressure) - SI Units]

The specific heat of water is $c_{p,water} = 4.180$ kJ/kg·K. [Properties of Water at Standard Conditions]

The heat lost by the steam is equal to the heat gained by the water.

$$\dot{m}_{water}c_{p,water}\Delta T = \dot{m}_{steam}(h_2 - h_f)$$

$$\dot{m}_{water}\left(4.180\ \dfrac{\text{kJ}}{\text{kg}\cdot\text{K}}\right)(5.6\ \text{K}) = \left(12.56\ \dfrac{\text{kg}}{\text{s}}\right)\left(\left(2.280\ \dfrac{\text{MJ}}{\text{kg}}\right)\right.$$

$$\left.\times\left(1000\ \dfrac{\text{kJ}}{\text{MJ}}\right) - 109.8\ \dfrac{\text{kJ}}{\text{kg}}\right)$$

$$\dot{m}_{water} = 1162.5\ \text{kg/s}$$

The density of water, ρ, is 1000 kg/m^3. The circulation rate is

$$Q = \frac{\dot{m}_{water}}{\rho} = \frac{\left(\dfrac{1162.5\ \dfrac{\text{kg}}{\text{s}}}{1000\ \dfrac{\text{kg}}{\text{m}^3}}\right)\left(1000\ \dfrac{\text{L}}{\text{m}^3}\right)\left(60\ \dfrac{\text{s}}{\text{min}}\right)}{1000\ \dfrac{\text{L}}{\text{kL}}}$$

$$= 69.7\ \text{kL/min} \quad (70\ \text{kL/min})$$

The answer is (D).

5. *Customary U.S. Solution*

From steam tables, the enthalpy of saturated liquid at 60°F is $h_1 = 28.08$ Btu/lbm. [Properties of Saturated Water and Steam (Temperature) - I-P Units]

From steam tables, the enthalpy of saturated steam at 14.7 psia is $h_2 = 1150.25$ Btu/lbm. [Properties of Saturated Water and Steam (Pressure) - I-P Units]

The heat transfer rate to the water is

$$q = \dot{m}(h_2 - h_1)$$

$$= \frac{\left(100\ \dfrac{\text{lbm}}{\text{hr}}\right)\left(1150.25\ \dfrac{\text{Btu}}{\text{lbm}} - 28.08\ \dfrac{\text{Btu}}{\text{lbm}}\right)}{\left(1000\ \dfrac{\text{W}}{\text{kW}}\right)\left(3.412\ \dfrac{\text{Btu}}{\text{W-hr}}\right)}$$

$$= 32.89\ \text{kW}$$

The cost is

$$C = \frac{(32.89\ \text{kW})\left(\dfrac{\$0.04}{\text{kW-hr}}\right)}{1 - 0.35}$$

$$= \$2.02/\text{hr} \quad (\$2/\text{hr})$$

SI Solution

From steam tables, for saturated liquid at 16°C, $h_1 = 67.17$ kJ/kg. [Properties of Saturated Water and Steam (Temperature) - SI Units]

From steam tables, the enthalpy of saturated steam at 101.3 kPa is $h_2 = 2675.6$ kJ/kg. [Properties of Saturated Water and Steam (Temperature) - I-P Units]

The heat transfer rate to the water is

$$q = \dot{m}(h_2 - h_1)$$
$$= \left(0.013 \; \frac{\text{kg}}{\text{s}}\right)\left(2675.6 \; \frac{\text{kJ}}{\text{kg}} - 67.17 \; \frac{\text{kJ}}{\text{kg}}\right)$$
$$= 33.907 \text{ kW}$$

The cost is

$$C = \frac{(33.907 \text{ kW})\left(\frac{\$0.04}{\text{kW}\cdot\text{h}}\right)}{1 - 0.35}$$
$$= \$2.09/\text{h} \quad (\$2/\text{h})$$

The answer is (A).

6. *Customary U.S. Solution*

At 30.2 in Hg (14.84 psia) and 60 F, water is a compressed liquid. However, the enthalpy of compressed liquid is approximately the same as the enthalpy of the saturated liquid **at the given temperature.**

From steam tables, for 40 psia steam, the enthalpy of saturated liquid, h_f, is 236.14 Btu/lbm. The heat of vaporization, h_{fg}, is 933.68 Btu/lbm. [Properties of Saturated Water and Steam (Pressure) - I-P Units]

From steam tables, the enthalpy of saturated liquid at 60°F is $h_1 = 28.08$ Btu/lbm. [Properties of Saturated Water and Steam (Temperature) - I-P Units]

The enthalpy is

Properties for Two-Phase (Vapor-Liquid) Systems

$$h_2 = h_f + x h_{fg}$$
$$= 236.14 \; \frac{\text{Btu}}{\text{lbm}} + (0.98)\left(933.68 \; \frac{\text{Btu}}{\text{lbm}}\right)$$
$$= 1151.15 \text{ Btu/lbm}$$

The heat transfer rate is

$$q = \dot{m}(h_2 - h_1)$$
$$= \frac{\left(250 \; \frac{\text{lbm}}{\text{hr}}\right)\left(1151.15 \; \frac{\text{Btu}}{\text{lbm}} - 28.08 \; \frac{\text{Btu}}{\text{lbm}}\right)}{60 \; \frac{\text{min}}{\text{hr}}}$$
$$= 4679.46 \text{ Btu/min}$$

Find the volume of gas used at standard conditions for a heating gas (60°F) by using the ideal gas law. The volume of an ideal gas is directly proportional to the absolute temperature and inversely proportional to the absolute pressure.

$$Q_{\text{std}} = Q\left(\frac{T_0}{T}\right)\left(\frac{p}{p_0}\right)$$
$$= \left(13.5 \; \frac{\text{ft}^3}{\text{min}}\right)\left(\frac{60°\text{F} + 460°}{80°\text{F} + 460°}\right)$$
$$\times \left(\frac{(4 \text{ in Hg} + 30.2 \text{ in Hg})\left(0.491 \; \frac{\text{lbf}}{\text{in}^2 \text{ in Hg}}\right)}{14.7 \; \frac{\text{lbf}}{\text{in}^2}}\right)$$
$$= 14.85 \text{ ft}^3/\text{min}$$

The efficiency of the boiler is

$$\eta = \frac{q}{\text{heat input}} = \frac{4679.46 \; \frac{\text{Btu}}{\text{min}}}{\left(14.85 \; \frac{\text{ft}^3}{\text{min}}\right)\left(550 \; \frac{\text{Btu}}{\text{ft}^3}\right)}$$
$$= 0.573 \quad (57\%)$$

SI Solution

At 104.2 kPa and 16 C, water is a compressed liquid. However, the enthalpy of compressed liquid is approximately the same as the enthalpy of the saturated liquid at the given temperature.

Interpolating from steam tables, the enthalpy of saturated liquid at 16°C is $h_1 = 67.17$ kJ/kg. [Properties of Saturated Water and Steam (Temperature) - I-P Units]

From steam tables, for 300 kPa steam, the enthalpy of saturated liquid, h_f, is 561.4 kJ/kg. The enthalpy of vaporization, h_{fg}, is 2163.5 kJ/kg. [Properties of Saturated Water and Steam (Pressure) - I-P Units]

The enthalpy is

Properties for Two-Phase (Vapor-Liquid) Systems

$$h_2 = h_f + x h_{fg}$$
$$= 561.4 \; \frac{\text{kJ}}{\text{kg}} + (0.98)\left(2163.5 \; \frac{\text{kJ}}{\text{kg}}\right)$$
$$= 2681.63 \text{ kJ/kg}$$

The heat transfer rate is

$$q = \dot{m}(h_2 - h_1)$$
$$= \left(0.032 \ \frac{\text{kg}}{\text{s}}\right)\left(2681.63 \ \frac{\text{kJ}}{\text{kg}} - 67.17 \ \frac{\text{kJ}}{\text{kg}}\right)$$
$$= 83.66 \ \text{kJ/s}$$

Find the volume of the gas used at standard conditions for a heating gas (16°C) by using ideal gas law. The volume of an ideal gas is directly proportional to the absolute temperature and inversely proportional to the absolute pressure.

$$Q_{\text{std}} = Q\left(\frac{T_0}{T}\right)\left(\frac{p}{p_0}\right)$$
$$= \frac{\left(6.4 \ \dfrac{\text{L}}{\text{s}}\right)\left(\dfrac{16°\text{C} + 273°}{26°\text{C} + 273°}\right)\left(\dfrac{13.6 \ \text{kPa} + 102.4 \ \text{kPa}}{101.3 \ \text{kPa}}\right)}{1000 \ \dfrac{\text{L}}{\text{m}^3}}$$
$$= 0.00708 \ \text{m}^3/\text{s}$$

The efficiency of the boiler is

$$\eta = \frac{q}{\text{heat input}}$$
$$= \frac{83.66 \ \dfrac{\text{kJ}}{\text{s}}}{\left(0.00708 \ \dfrac{\text{m}^3}{\text{s}}\right)\left(20.5 \ \dfrac{\text{MJ}}{\text{m}^3}\right)\left(1000 \ \dfrac{\text{kJ}}{\text{MJ}}\right)}$$
$$= 0.576 \quad (57\%)$$

The answer is (C).

7. *Customary U.S. Solution*

Determine the actual gravimetric analysis of the coal as fired. 1 lbm of coal contains 0.02 lbm moisture, leaving 0.98 lbm dry coal. 1% is lost to the ash pit. The remainder is 0.98 lbm − 0.01 lbm = 0.97 lbm.

Calculate the heating value of the remaining coal.

$$\text{HV} = (0.97 \ \text{lbm})\left(12{,}800 \ \frac{\text{Btu}}{\text{lbm}}\right) = 12{,}416 \ \text{Btu}$$

From steam tables, the enthalpy of water at 120°F is $h_1 = 88.00$ Btu/lbm. [Properties of Saturated Water and Steam (Temperature) - SI Units]

From steam tables, the enthalpy of saturated steam at 100 psia is $h_2 = 1187.50$ Btu/lbm. [Properties of Superheated Steam - SI Units]

Find the energy, Q, needed to produce steam.

$$Q = m(h_2 - h_1)$$
$$= (8.23 \ \text{lbm})\left(1187.50 \ \frac{\text{Btu}}{\text{lbm}} - 88.00 \ \frac{\text{Btu}}{\text{lbm}}\right)$$
$$= 9048.9 \ \text{Btu}$$

The combustion efficiency is

$$\eta = \frac{Q}{\text{HV}} = \frac{9048.9 \ \text{Btu}}{12{,}416 \ \text{Btu}} = 0.729 \quad (73\%)$$

SI Solution

Determine the actual gravimetric analysis of the coal as fired. 1 lbm of coal contains 0.02 lbm moisture, leaving 0.98 lbm dry coal. 1% is lost to the ash pit. The remainder is 0.98 kg − 0.01 kg = 0.97 kg.

Calculate the heating value of the remaining coal.

$$\text{HV} = (0.97 \ \text{kg})\left(29.80 \ \frac{\text{MJ}}{\text{kg}}\right)\left(1000 \ \frac{\text{kJ}}{\text{MJ}}\right) = 28\,906 \ \text{kJ}$$

From steam tables, the enthalpy of water at 50°C is $h_1 = 209.34$ kJ/kg. [Properties of Saturated Water and Steam (Temperature) - SI Units]

From steam tables, the enthalpy of saturated steam at 700 kPa is $h_2 = 2762.8$ kJ/kg. [Properties of Superheated Steam - SI Units]

Find the energy, Q, needed to produce steam.

$$Q = m(h_2 - h_1)$$
$$= (8.23 \ \text{kg})\left(2762.8 \ \frac{\text{kJ}}{\text{kg}} - 209.34 \ \frac{\text{kJ}}{\text{kg}}\right)$$
$$= 21\,015 \ \text{kJ}$$

The combustion efficiency is

$$\eta = \frac{Q}{\text{HV}} = \frac{21\,015 \ \text{kJ}}{28\,906 \ \text{kJ}} = 0.727 \quad (73\%)$$

The answer is (C).

8. *Customary U.S. Solution*

Refer to the diagram of the Rankine cycle.

At point 3 (leaving the boiler and entering the turbine), the enthalpy, h_3, and entropy, s_3, can be obtained from superheated steam tables. [Properties of Superheated Steam - I-P Units]

$$h_3 = 1521.0 \text{ Btu/lbm}$$
$$s_3 = 1.738 \text{ Btu/lbm-}°\text{F}$$

For isentropic expansion through the turbine, $s_3 = s_4$. At this entropy, use the steam tables to find the enthalpy of saturated liquid, h_f, the enthalpy of evaporation, h_{fg}, the entropy of saturated liquid, s_f, and the entropy of evaporation, s_{fg}. [Properties of Saturated Water and Steam (Pressure) - I-P Units]

$$h_f = 130.13 \text{ Btu/lbm}$$
$$h_{fg} = 1000.57 \text{ Btu/lbm}$$
$$s_f = 0.2348 \text{ Btu/lbm-}°\text{F}$$
$$s_{fg} = 1.6092 \text{ Btu/lbm-}°\text{F}$$

The quality of the mixture for an isentropic process is given by (at point 4)

Properties for Two-Phase (Vapor-Liquid) Systems
$$s_4 = s_f + x_4 s_{fg}$$
$$x_4 = \frac{s_4 - s_f}{s_{fg}}$$
$$= \frac{1.738 \frac{\text{Btu}}{\text{lbm-}°\text{F}} - 0.2348 \frac{\text{Btu}}{\text{lbm-}°\text{F}}}{1.6092 \frac{\text{Btu}}{\text{lbm-}°\text{F}}}$$
$$= 0.9341$$

The isentropic enthalpy is given by (at point 4)

Properties for Two-Phase (Vapor-Liquid) Systems
$$h_{4s} = h_f + x_4 h_{fg}$$
$$= 130.13 \frac{\text{Btu}}{\text{lbm}} + (0.9341)\left(1000.57 \frac{\text{Btu}}{\text{lbm}}\right)$$
$$= 1064.76 \text{ Btu/lbm}$$

The actual enthalpy of steam at point 4 is

$$h_4 = h_3 - \eta_T (h_3 - h_{4s})$$
$$= 1521.0 \frac{\text{Btu}}{\text{lbm}} - (0.75)\left(1521.0 \frac{\text{Btu}}{\text{lbm}} - 1064.76 \frac{\text{Btu}}{\text{lbm}}\right)$$
$$= 1178.82 \text{ Btu/lbm}$$

Since the pump work is negligible, the mass flow rate of steam is

$$\dot{m} = \frac{P}{W_{\text{turbine}}} = \frac{P}{h_3 - h_4}$$
$$= \frac{(200 \text{ MW})\left(10^6 \frac{\text{W}}{\text{MW}}\right)\left(3.412 \frac{\text{Btu}}{\text{hr-W}}\right)}{1521.0 \frac{\text{Btu}}{\text{lbm}} - 1178.82 \frac{\text{Btu}}{\text{lbm}}}$$
$$= 1.994 \times 10^6 \text{ lbm/hr} \quad (2.0 \times 10^6 \text{ lbm/hr})$$

The answer is (C).

9. *U.S. Customary Solution*

From steam tables, the enthalpy of the vapor water at 300°F is 1179.95 Btu/lbm, and the enthalpy of the fluid at at 300°F is 269.73 Btu/lbm. [Properties of Saturated Water and Steam (Temperature) - I-P Units]

The enthalpy of the liquid-vapor mixture at the inlet is

Properties for Two-Phase (Vapor-Liquid) Systems
$$h_2 = x h_g + (1 - x) h_f$$
$$= (0.904)\left(1179.95 \frac{\text{Btu}}{\text{lbm}}\right) + (1 - 0.904)\left(269.73 \frac{\text{Btu}}{\text{lbm}}\right)$$
$$= 1092.57 \text{ Btu/lbm}$$

From saturated steam tables, the value for the enthalpy of the saturated vapor at the inlet, at 210°F h_1, is 1149.54 Btu/lbm. [Properties of Saturated Water and Steam (Temperature) - I-P Units]

Find the enthalpy of the saturated vapor leaving the feedwater heater.

Open (Mixing) Feedwater Heater
$$\dot{m}_1 h_1 + \dot{m}_2 h_2 = h_3 (\dot{m}_1 + \dot{m}_2)$$
$$h_3 = \frac{\dot{m}_1 h_1 + \dot{m}_2 h_2}{\dot{m}_1 + \dot{m}_2}$$
$$= \frac{\left(28 \frac{\text{lbm}}{\text{sec}}\right)\left(1149.54 \frac{\text{Btu}}{\text{lbm}}\right) + \left(3 \frac{\text{lbm}}{\text{sec}}\right)\left(1092.57 \frac{\text{Btu}}{\text{lbm}}\right)}{28 \frac{\text{lbm}}{\text{sec}} + 3 \frac{\text{lbm}}{\text{sec}}}$$
$$= 1144.03 \text{ Btu/lbm}$$

From saturated steam tables, for the properties nearest to an enthalpy value of 1144.03 Btu/lbm, the temperature of the saturated vapor is 196°F at 1144.19 Btu/lbm. [Properties of Saturated Water and Steam (Temperature) - I-P Units]

The answer is (A).

10. Convert the rating from tons to Btu/hr.

Measurement Relationships

$$1 \text{ ton (refrigeration)} = 12{,}000 \text{ Btu/hr}$$

$$(35 \text{ tons})\left(12{,}000 \ \frac{\text{Btu}}{\text{hr-ton}}\right) = 420{,}000 \text{ Btu/hr}$$

The energy efficiency ratio is

Efficiency

$$\text{EER} = \frac{\text{output cooling energy (Btu/hr)}}{\text{input electrical energy (W)}}$$

$$= \frac{420{,}000 \ \dfrac{\text{Btu}}{\text{hr}}}{(65 \text{ kW})\left(1000 \ \dfrac{\text{W}}{\text{kW}}\right)}$$

$$= 6.46 \quad (6.5)$$

The answer is (D).

27 Vapor Power Cycles

Content in blue refers to the *NCEES Handbook*.

PRACTICE PROBLEMS

1. A steam Rankine cycle operates between 580°F and 100°F. The engineer designing this cycle attempts to use the Carnot cycle as the basis of operation by designing the compressor outlet as a saturated liquid and the turbine inlet as a saturated vapor. The turbine and compressor (pump) isentropic efficiencies are 90% and 80%, respectively. The thermal efficiency is most nearly

(A) 35%

(B) 37%

(C) 42%

(D) 48%

2. A steam turbine cycle produces 600 MWe (megawatts of electrical power). The condenser load is 3.07×10^9 Btu/hr (900 MW). The thermal efficiency is most nearly

(A) 32%

(B) 37%

(C) 40%

(D) 44%

3. A turbine and the boiler feed pumps in a reheat cycle have isentropic efficiencies of 88% and 96%, respectively. The cycle starts with water at 60°F at the entrance to the boiler feed pump and produces steam at 600°F and 600 psia. The steam is reheated when its pressure drops during the first expansion to 20 psia. The thermal efficiency of the cycle is most nearly

(A) 28%

(B) 34%

(C) 39%

(D) 42%

4. A reheat steam cycle operates as shown. The pressure and temperature at point B are 200 psia and 500 F respectively.

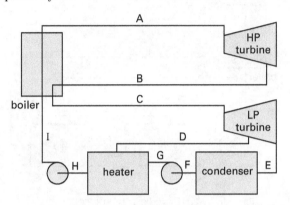

At point A, the pressure is 900 psia and the temperature is 800°F. The isentropic efficiency of the high-pressure turbine is most nearly

(A) 54%

(B) 61%

(C) 72%

(D) 76%

5. A reheat steam cycle operates as shown. The pump work between points F and G is 0.15 Btu/lbm (0.3 kJ/kg).

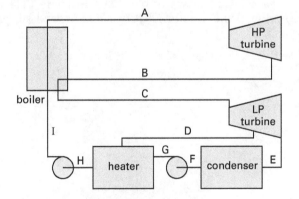

At F: 69.73 Btu/lbm (162.5 kJ/kg)

At H: 250.2 Btu/lbm (583.0 kJ/kg), 0.0173 ft³/lbm (0.0011 m³/kg)

At I: 253.1 Btu/lbm (589.7 kJ/kg)

The enthalpy at point G (after the pump) is most nearly

(A) 70 Btu/lbm (160 kJ/kg)

(B) 250 Btu/lbm (580 kJ/kg)

(C) 1100 Btu/lbm (2500 kJ/kg)

(D) 1300 Btu/lbm (3000 kJ/kg)

6. A reheat steam cycle operates as shown. The pump work between points F and G is 0.15 Btu/lbm (0.3 kJ/kg).

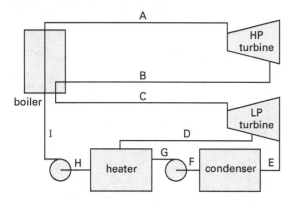

at A: 900 psia (6.2 MPa), 800°F (420°C)

at B: 200 psia (1.5 MPa), 1270 Btu/lbm (2960 kJ/kg)

at C: 190 psia (1.4 MPa), 800°F (420°C)

at D: 50 psia (350 kPa), 1280 Btu/lbm (2980 kJ/kg)

at E: 2 in Hg absolute (6.8 kPa), 1075 Btu/lbm (2500 kJ/kg)

at F: 69.73 Btu/lbm (162.5 kJ/kg)

at H: 250.2 Btu/lbm (583.0 kJ/kg), 0.0173 ft³/lbm (0.0011 m³/kg)

at I: 253.1 Btu/lbm (589.7 kJ/kg)

The enthalpy is known to be 1393.9 Btu/lbm (3235.0 kJ/kg) at point A and 1426.0 Btu/lbm (3301.3 kJ/kg) at point C.

The thermal efficiency of the cycle is most nearly

(A) 22%

(B) 29%

(C) 34%

(D) 41%

SOLUTIONS

1. Refer to the Carnot cycle. [Common Thermodynamic Cycles]

Find the enthalpy and entropy at point 2 using steam tables for a saturated liquid at 580°F. [Properties of Saturated Water and Steam (Temperature) - I-P Units]

$$h_2 = 589.05 \text{ Btu/lbm}$$
$$s_2 = 0.7876 \text{ Btu/lbm-°R}$$

Find the enthalpy and entropy at point 3 using steam tables for saturated vapor at 580°F. [Properties of Saturated Water and Steam (Temperature) - I-P Units]

$$h_3 = 1178.26 \text{ Btu/lbm}$$
$$s_3 = 1.3543 \text{ Btu/lbm-°R}$$

At point 4,

$$T_4 = 100°F$$
$$s_4 = s_3$$
$$= 1.3543 \text{ Btu/lbm-°F}$$

Find the enthalpies and entropies of fluid and vapor at 100°F. [Properties of Saturated Water and Steam (Temperature) - I-P Units]

$$s_f = 0.1296 \text{ Btu/lbm-°R}$$
$$s_g = 1.9819 \text{ Btu/lbm-°R}$$
$$h_f = 68.03 \text{ Btu/lbm}$$
$$h_{fg} = 1036.7 \text{ Btu/lbm}$$

The quality of the mixture at point 4 can be solved using the entropies of the fluid and vapor.

Properties for Two-Phase (Vapor-Liquid) Systems

$$x_4 = \frac{s_4 - s_f}{s_g - s_f}$$

$$= \frac{1.3543 \frac{\text{Btu}}{\text{lbm-°R}} - 0.1296 \frac{\text{Btu}}{\text{lbm-°R}}}{1.9819 \frac{\text{Btu}}{\text{lbm-°R}} - 0.1296 \frac{\text{Btu}}{\text{lbm-°R}}}$$

$$= 0.661$$

Find the enthalpy at point 4.

Properties for Two-Phase (Vapor-Liquid) Systems

$$h_4 = h_f + x_4 h_{fg}$$
$$= 68.03 \frac{\text{Btu}}{\text{lbm}} + (0.661)\left(1036.7 \frac{\text{Btu}}{\text{lbm}}\right)$$
$$= 753.28 \text{ Btu/lbm}$$

Find the mass fraction of the vapor phase at point 1.

$$T_1 = 100°F$$
$$s_1 = s_2 = 0.7876 \text{ Btu/lbm-°R}$$

Find the mass fraction of the vapor phase at point 1.

$$x_1 = \frac{s_1 - s_f}{s_g - s_f}$$

$$= \frac{0.7876 \frac{\text{Btu}}{\text{lbm-°R}} - 0.1296 \frac{\text{Btu}}{\text{lbm-°R}}}{1.9819 \frac{\text{Btu}}{\text{lbm-°R}} - 0.1296 \frac{\text{Btu}}{\text{lbm-°R}}}$$

$$= 0.355$$

Find the enthalpy at point 1.

Properties for Two-Phase (Vapor-Liquid) Systems

$$h_1 = h_f + x_1 h_{fg}$$

$$= 68.03 \frac{\text{Btu}}{\text{lbm}} + (0.355)\left(1036.7 \frac{\text{Btu}}{\text{lbm}}\right)$$

$$= 436.1 \text{ Btu/lbm}$$

The actual enthalpy at point 4 can be solved using the efficiency of the turbine.

Turbines

$$\eta_{\text{turbine}} = \frac{w_a}{w_s} = \frac{h_3 - h_4'}{h_3 - h_4}$$

$$h_4' = h_3 - \eta_{\text{turbine}}(h_3 - h_4)$$

$$= 1178.26 \frac{\text{Btu}}{\text{lbm}} - (0.9)$$

$$\times \left(1178.26 \frac{\text{Btu}}{\text{lbm}} - 753.28 \frac{\text{Btu}}{\text{lbm}}\right)$$

$$= 795.78 \text{ Btu/lbm}$$

Work is the difference in enthalpy due to the inefficiency of the pump. Find the actual enthalpy at point 2.

$$h_2' = h_1 + \frac{h_2 - h_1}{\eta_{s,\text{pump}}}$$

$$= 436.1 \frac{\text{Btu}}{\text{lbm}} + \frac{589.05 \frac{\text{Btu}}{\text{lbm}} - 436.1 \frac{\text{Btu}}{\text{lbm}}}{0.8}$$

$$= 627.3 \text{ Btu/lbm}$$

The thermal efficiency of the entire cycle is defined as the net work produced divided by the heat input.

$$\eta_{\text{th}} = \frac{(h_3 - h_4') - (h_2' - h_1)}{h_3 - h_2'}$$

$$= \frac{\left(1178.26 \frac{\text{Btu}}{\text{lbm}} - 795.78 \frac{\text{Btu}}{\text{lbm}}\right) - \left(627.3 \frac{\text{Btu}}{\text{lbm}} - 436.1 \frac{\text{Btu}}{\text{lbm}}\right)}{1178.26 \frac{\text{Btu}}{\text{lbm}} - 627.3 \frac{\text{Btu}}{\text{lbm}}}$$

$$= 0.347 \quad (35\%)$$

The answer is (A).

2. *Customary U.S. Solution*

The condenser load is $Q_L = 3.07 \times 10^9$ Btu/hr. The net amount of work is

Basic Cycles

$$\frac{W}{Q_H} = \frac{Q_H - Q_L}{Q_H}$$
$$W = Q_H - Q_L$$

The boiler load is

$$Q_H = W + Q_L$$
$$= (600 \text{ MW})\left(1000 \frac{\text{kW}}{\text{MW}}\right)\left(3412 \frac{\text{Btu}}{\text{kW-hr}}\right)$$
$$+ 3.07 \times 10^9 \frac{\text{Btu}}{\text{hr}}$$
$$= 5.12 \times 10^9 \text{ Btu/hr}$$

The thermal efficiency is

Basic Cycles

$$\eta = \frac{Q_H - Q_L}{Q_H}$$

$$= \frac{5.12 \times 10^9 \frac{\text{Btu}}{\text{hr}} - 3.07 \times 10^9 \frac{\text{Btu}}{\text{hr}}}{5.12 \times 10^9 \frac{\text{Btu}}{\text{hr}}}$$

$$= 0.400 \quad (40\%)$$

SI Solution

The condenser load is $Q_L = 900$ MW. The boiler load is

Basic Cycles

$$\frac{W}{Q_H} = \frac{Q_H - Q_L}{Q_H}$$
$$W = Q_H - Q_L$$

The boiler load is

$$\begin{aligned}Q_H &= W + Q_L \\ &= 600 \text{ MW} + 900 \text{ MW} \\ &= 1500 \text{ MW}\end{aligned}$$

The thermal efficiency is

Basic Cycles

$$\begin{aligned}\eta &= \frac{Q_H - Q_L}{Q_H} \\ &= \frac{1500 \text{ MW} - 900 \text{ MW}}{1500 \text{ MW}} \\ &= 0.400 \quad (40\%)\end{aligned}$$

The answer is (C).

3. Recall the reheat cycle.

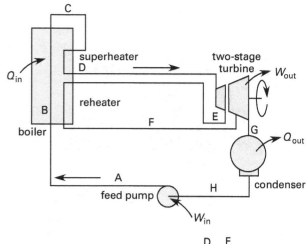

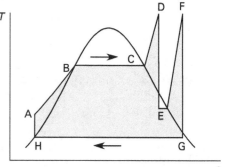

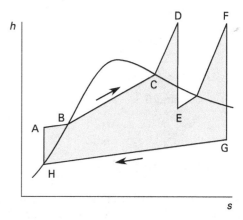

At point B (entrance into the boiler), the pressure is 600 psia. The enthalpy of the saturated liquid is $h_B = 471.75$ Btu/lbm. [Properties of Saturated Water and Steam (Pressure) - I-P Units]

At point C (exit from the boiler), the pressure is 600 psia. The enthalpy of the saturated vapor is $h_C = 1203.83$ Btu/lbm. [Properties of Saturated Water and Steam (Pressure) - I-P Units]

At point D (after the superheater), the temperature is 600°F and the pressure is 600 psia. [Properties of Superheated Steam - I-P Units]

The enthalpy and entropy of superheated vapor are

$$h_D = 1289.8 \text{ Btu/lbm}$$
$$s_D = 1.533 \text{ Btu/lbm-°R}$$

At point E (after first turbine stage),

$$p_E = 20 \text{ psia}$$
$$s_E = s_D$$
$$= 1.533 \text{ Btu/lbm-°R}$$

The various saturation properties at 20 psia are [Properties of Saturated Water and Steam (Pressure) - I-P Units]

$$s_f = 0.3358 \text{ Btu/lbm-°R}$$
$$s_{fg} = 1.3962 \text{ Btu/lbm-°R}$$
$$h_f = 196.25 \text{ Btu/lbm}$$
$$h_{fg} = 959.94 \text{ Btu/lbm}$$

Find the mass fraction of the vapor phase at point E.

$$s_{fg} = s_g - s_f$$

$$x_E = \frac{s_E - s_f}{s_g - s_f} = \frac{1.533 \frac{\text{Btu}}{\text{lbm-°R}} - 0.3358 \frac{\text{Btu}}{\text{lbm-°R}}}{1.3962 \frac{\text{Btu}}{\text{lbm-°R}}}$$
$$= 0.857$$

Find the enthalpy at point E.

Properties for Two-Phase (Vapor-Liquid) Systems
$$h_E = h_f + x_E h_{fg}$$
$$= 196.25 \frac{\text{Btu}}{\text{lbm}} + (0.857)\left(959.94 \frac{\text{Btu}}{\text{lbm}}\right)$$
$$= 1018.9 \text{ Btu/lbm}$$

Due to the inefficiency of the turbine, the enthalpy is

Turbines
$$\eta = \frac{w_a}{w_s} = \frac{h_3 - h_4'}{h_3 - h_4}$$
$$h_E' = h_D - \eta_{s,\text{turbine}}(h_D - h_E)$$
$$= 1289.8 \frac{\text{Btu}}{\text{lbm}} - (0.88)$$
$$\times \left(1289.8 \frac{\text{Btu}}{\text{lbm}} - 1018.9 \frac{\text{Btu}}{\text{lbm}}\right)$$
$$= 1051.4 \text{ Btu/lbm}$$

At point F (reheating stage), the temperature has returned to 600°F, but the pressure stays at the expansion pressure, p_E.

$$p_F = 20 \text{ psia}$$
$$T_F = 600°\text{F}$$

Find the enthalpy and entropy of the superheated vapor. [Properties of Superheated Steam - I-P Units]

$$h_F = 1334.9 \text{ Btu/lbm}$$
$$s_F = 1.940 \text{ Btu/lbm-°R}$$

At point G (after second turbine stage),

$$T_G = 60°\text{F}$$
$$s_G = s_F$$
$$= 1.940 \text{ Btu/lbm-°R}$$

Using steam tables, find the various saturation properties. [Properties of Saturated Water and Steam (Temperature) - I-P Units]

$$s_f = 0.0555 \text{ Btu/lbm-°R}$$
$$s_g = 2.0940 \text{ Btu/lbm-°R}$$
$$h_f = 28.08 \text{ Btu/lbm}$$
$$h_{fg} = 1059.35 \text{ Btu/lbm}$$

Find the mass fraction at point G.

$$x_G = \frac{s_G - s_f}{s_g - s_f}$$
$$= \frac{1.940 \frac{\text{Btu}}{\text{lbm-°R}} - 0.0555 \frac{\text{Btu}}{\text{lbm-°R}}}{2.0940 \frac{\text{Btu}}{\text{lbm-°R}} - 0.0555 \frac{\text{Btu}}{\text{lbm-°R}}}$$
$$= 0.924$$

Find the enthalpy at point G.

Properties for Two-Phase (Vapor-Liquid) Systems
$$h_G = h_f + x_G h_{fg}$$
$$= 28.08 \frac{\text{Btu}}{\text{lbm}} + (0.924)(1059.35)$$
$$= 1007.4 \text{ Btu/lbm}$$

Due to the inefficiency of the turbine, the enthalpy is

$$h'_G = h_F - \eta_{s,\text{turbine}}(h_F - h_G)$$
$$= 1334.9 \, \frac{\text{Btu}}{\text{lbm}} - (0.88)\left(1334.9 \, \frac{\text{Btu}}{\text{lbm}} - 1007.4 \, \frac{\text{Btu}}{\text{lbm}}\right)$$
$$= 1046.7 \, \text{Btu/lbm}$$

At point H, the temperature is 60°F. The saturation pressure, enthalpy, and specific volume of the saturated liquid are [Properties of Saturated Water and Steam (Temperature) - I-P Units]

$$p_H = 0.26 \, \text{psia}$$
$$h_H = 28.08 \, \text{Btu/lbm}$$
$$v_H = 0.0160 \, \text{ft}^3/\text{lbm}$$

At point A (after the feed pump), the pressure is 600 psia. Use the relationship between pressure and enthalpy to find the enthalpy at point A.

$$h_A \approx h_H + v_H(p_A - p_H)$$
$$= 28.08 \, \frac{\text{Btu}}{\text{lbm}}$$
$$+ \frac{\left(0.0160 \, \frac{\text{ft}^3}{\text{lbm}}\right) \times \left(600 \, \frac{\text{lbf}}{\text{in}^2} - 0.26 \, \frac{\text{lbf}}{\text{in}^2}\right)}{\left(778 \, \frac{\text{ft-lbf}}{\text{Btu}}\right)\left(12 \, \frac{\text{in}}{\text{ft}}\right)^2}$$
$$= 29.86 \, \text{Btu/lbm}$$

Due to the inefficiency of the pump, the actual enthalpy at point A is

$$h'_A = h_H + \frac{h_A - h_H}{\eta_{s,\text{pump}}}$$
$$= 28.08 \, \frac{\text{Btu}}{\text{lbm}} + \frac{29.86 \, \frac{\text{Btu}}{\text{lbm}} - 28.08 \, \frac{\text{Btu}}{\text{lbm}}}{0.96}$$
$$= 29.93 \, \text{Btu/lbm}$$

The thermal efficiency of the cycle for a nonisentropic process for the turbine and the pump is

$$\eta_{th} = \frac{Q_H - Q_L}{Q_H}$$
$$= \frac{(h_D - h'_A) + (h_F - h'_E) - (h'_G - h_H)}{(h_D - h'_A) + (h_F - h'_E)}$$

$$= \frac{\left(1289.8 \, \frac{\text{Btu}}{\text{lbm}} - 29.93 \, \frac{\text{Btu}}{\text{lbm}}\right) + \left(1334.9 \, \frac{\text{Btu}}{\text{lbm}} - 1051.4 \, \frac{\text{Btu}}{\text{lbm}}\right) - \left(1046.7 \, \frac{\text{Btu}}{\text{lbm}} - 28.08 \, \frac{\text{Btu}}{\text{lbm}}\right)}{\left(1289.8 \, \frac{\text{Btu}}{\text{lbm}} - 29.93 \, \frac{\text{Btu}}{\text{lbm}}\right) + \left(1334.9 \, \frac{\text{Btu}}{\text{lbm}} - 1051.4 \, \frac{\text{Btu}}{\text{lbm}}\right)}$$

$$= 0.340 \quad (34\%)$$

The answer is (B).

4. *Customary U.S. Solution*

Refer to the diagram in the problem statement.

At point A,

$$p_A = 900 \, \text{psia}$$
$$T_A = 800°\text{F}$$

Find the enthalpy and specific entropy at point A. [Properties of Superheated Steam - I-P Units]

$$h_A = 1394 \, \text{Btu/lbm}$$
$$s_A = 1.582 \, \text{Btu/lbm-°R}$$

Assuming isentropic expansion to 200 psia, determine the enthalpy of steam at point B by using $P_B = 200$ psia and $s_B = s_A = 1.582$ Btu/lbm-R. [Properties of Superheated Steam - I-P Units]

$h_{B,s}$ represents the enthalpy at point B after isentropic expansion.

$$h_{B,s} = 1{,}229 \, \text{Btu/lbm}$$

Determine the actual enthalpy of steam after expansion at 200 psia and 500 F [Properties of Superheated Steam - I-P Units]

$$h_B = 1{,}269 \, \text{Btu/lbm}$$

Isentropic efficiency (turbine)

$$\text{Isentropic efficiency (turbine)} = \left(\frac{h_i - h_e}{h_i - h_{e,s}}\right)100 = \left(\frac{h_A - h_B}{h_A - h_{B,s}}\right)100$$

$$= \left(\frac{1394\frac{\text{Btu}}{\text{lbm}} - 1269\frac{\text{Btu}}{\text{lbm}}}{1394\frac{\text{Btu}}{\text{lbm}} - 1229\frac{\text{Btu}}{\text{lbm}}}\right)100$$

$$= 75.8\% \quad (76\%)$$

The answer is (D).

5. *Customary U.S. Solution*

Refer to the following illustration and to the diagram in the problem statement.

At point F,

$$h_F = 69.73 \text{ Btu/lbm}$$

At point G,

$$W_{\text{pump}} = 0.15 \text{ Btu/lbm}$$
$$W_{\text{pump}} = h'_G - h_F$$
$$h'_G = W_{\text{pump}} + h_F$$
$$= 0.15\frac{\text{Btu}}{\text{lbm}} + 69.73\frac{\text{Btu}}{\text{lbm}}$$
$$= 69.88 \text{ Btu/lbm} \quad (70 \text{ Btu/lbm})$$

SI Solution

Refer to the customary U.S. solution illustration and to the diagram in the problem statement.

At point F,

$$h_F = 162.5 \text{ kJ/kg}$$

At point G,

$$W_{\text{pump}} = 0.3 \text{ kJ/kg}$$
$$= h'_G - h_F$$
$$h'_G = W_{\text{pump}} + h_F$$
$$= 0.3\frac{\text{kJ}}{\text{kg}} + 162.5\frac{\text{kJ}}{\text{kg}}$$
$$= 162.8 \text{ kJ/kg} \quad (160 \text{ kJ/kg})$$

The answer is (A).

6. *Customary U.S. Solution*

Refer to the following illustration and to the diagram in the problem statement.

From an energy balance in the heater,

Properties for Two-Phase (Vapor-Liquid) Systems

$$xh'_D + (1-x)h'_G = h_H$$
$$x(h'_D - h'_G) = h_H - h'_G$$

$$x = \frac{h_H - h'_G}{h'_D - h'_G} = \frac{250.2\frac{\text{Btu}}{\text{lbm}} - 69.88\frac{\text{Btu}}{\text{lbm}}}{1280\frac{\text{Btu}}{\text{lbm}} - 69.88\frac{\text{Btu}}{\text{lbm}}}$$

$$= 0.149$$

The thermal efficiency of the cycle is

$$\eta_{\text{th}} = \frac{W_{\text{out}} - W_{\text{in}}}{Q_{\text{in}}}$$

$$= \frac{\begin{pmatrix}(h_A - h'_B) + (h_C - h'_D) + (1-x)(h'_D - h'_E)\\ -(h'_I - h_H) - (1-x)(h'_G - h_F)\end{pmatrix}}{(h_A - h'_I) + (h_C - h'_B)}$$

$$= \frac{\begin{pmatrix}1393.9\frac{\text{Btu}}{\text{lbm}} - 1270\frac{\text{Btu}}{\text{lbm}}\end{pmatrix} + \begin{pmatrix}1426.0\frac{\text{Btu}}{\text{lbm}} - 1280\frac{\text{Btu}}{\text{lbm}}\end{pmatrix} + (1-0.149)\begin{pmatrix}1280\frac{\text{Btu}}{\text{lbm}} - 1075\frac{\text{Btu}}{\text{lbm}}\end{pmatrix} - \begin{pmatrix}253.1\frac{\text{Btu}}{\text{lbm}} - 250.2\frac{\text{Btu}}{\text{lbm}}\end{pmatrix} - (1-0.149)\begin{pmatrix}69.88\frac{\text{Btu}}{\text{lbm}} - 69.73\frac{\text{Btu}}{\text{lbm}}\end{pmatrix}}{\begin{pmatrix}1393.9\frac{\text{Btu}}{\text{lbm}} - 253.1\frac{\text{Btu}}{\text{lbm}}\end{pmatrix} + \begin{pmatrix}1426.0\frac{\text{Btu}}{\text{lbm}} - 1270\frac{\text{Btu}}{\text{lbm}}\end{pmatrix}}$$

$$= 0.340 \quad (34\%)$$

SI Solution

Refer to the customary U.S. solution illustration and to the diagram in the problem statement.

At point G,

$$W_{\text{pump}} = 0.3 \text{ kJ/kg}$$
$$= h'_G - h_F$$
$$h'_G = W_{\text{pump}} + h_F$$
$$= 0.3 \frac{\text{kJ}}{\text{kg}} + 162.5 \frac{\text{kJ}}{\text{kg}}$$
$$= 162.8 \text{ kJ/kg}$$

From an energy balance in the heater,

$$h_H = xh'_D + (1-x)h'_G$$
$$x = \frac{h_H - h'_G}{h'_D - h'_G}$$
$$= \frac{583.0 \frac{\text{kJ}}{\text{kg}} - 162.8 \frac{\text{kJ}}{\text{kg}}}{2980 \frac{\text{kJ}}{\text{kg}} - 162.8 \frac{\text{kJ}}{\text{kg}}}$$
$$= 0.149$$

The thermal efficiency of the cycle is

$$\eta_{\text{th}} = \frac{W_{\text{out}} - W_{\text{in}}}{Q_{\text{in}}}$$

$$= \frac{\begin{array}{c}(h_A - h'_B) + (h_C - h'_D) + (1-x)(h'_D - h'_E)\\ -(h'_I - h_H) - (1-x)(h'_G - h_F)\end{array}}{(h_A - h'_I) + (h_C - h'_B)}$$

$$= \frac{\begin{array}{c}\left(3235.0 \frac{\text{kJ}}{\text{kg}} - 2960 \frac{\text{kJ}}{\text{kg}}\right)\\ + \left(3301.3 \frac{\text{kJ}}{\text{kg}} - 2980 \frac{\text{kJ}}{\text{kg}}\right)\\ + (1 - 0.149)\left(2980 \frac{\text{kJ}}{\text{kg}} - 2500 \frac{\text{kJ}}{\text{kg}}\right)\\ - \left(589.7 \frac{\text{kJ}}{\text{kg}} - 583.0 \frac{\text{kJ}}{\text{kg}}\right)\\ - (1 - 0.149)\left(162.8 \frac{\text{kJ}}{\text{kg}} - 162.5 \frac{\text{kJ}}{\text{kg}}\right)\end{array}}{\begin{array}{c}\left(3235.0 \frac{\text{kJ}}{\text{kg}} - 589.7 \frac{\text{kJ}}{\text{kg}}\right)\\ + \left(3301.3 \frac{\text{kJ}}{\text{kg}} - 2960 \frac{\text{kJ}}{\text{kg}}\right)\end{array}}$$

$$= 0.334 \quad (34\%)$$

The answer is (C).

28 Reciprocating Combustion Engine Cycles

Content in blue refers to the NCEES Handbook.

PRACTICE PROBLEMS

1. A 10 in (250 mm) bore, 18 in (460 mm) stroke, two-cylinder, four-stroke internal combustion engine operates with an indicated mean effective pressure of 95 psi (660 kPa) at 200 rpm. The actual torque developed is 600 ft-lbf (820 N·m). The engine completes 100 power strokes per minute. The friction horsepower is most nearly

(A) 17 hp (13 kW)
(B) 22 hp (17 kW)
(C) 45 hp (33 kW)
(D) 78 hp (60 kW)

2. A four-cycle internal combustion engine has a bore of 3.1 in (80 mm) and a stroke of 3.8 in (97 mm). When the engine is running at 4000 rpm, the air-fuel mixture enters the cylinders during the intake stroke with an average velocity of 100 ft/sec (30 m/s). The intake valve is open 9.167×10^{-3} sec/rev. The volumetric efficiency is 65%. The effective area of the intake valve is most nearly

(A) 0.012 ft^2 (0.0012 m^2)
(B) 0.023 ft^2 (0.0023 m^2)
(C) 0.034 ft^2 (0.0034 m^2)
(D) 0.058 ft^2 (0.0058 m^2)

3. An engine runs on the Otto cycle. The compression ratio is 8:1. The total intake volume of the cylinders is 10 ft^3. Air enters at 14.2 psia and 90°F. In two revolutions of the crank, after the compression strokes, 150 Btu of energy is added. The temperature after the isentropic expansion is most nearly

(A) 1120°R
(B) 1290°R
(C) 1350°R
(D) 1520°R

4. An engine runs on the Otto cycle. The compression ratio is 10:1. The total intake volume of the cylinders is 11 ft^3 (0.3 m^3). Air enters at 14.2 psia and 80°F (98 kPa and 27°C). In two revolutions of the crank, after the compression strokes, 160 Btu (179 kJ) of energy is added. The thermal efficiency of the cycle is most nearly

(A) 45%
(B) 52%
(C) 57%
(D) 64%

5. A fully loaded diesel engine operating at sea level runs at 2000 rpm and develops 1000 bhp (750 kW). The initial operating temperature of the engine is 518.7°R (288.15K). Metered fuel injection is used, and the specific fuel consumption of the brake is 0.45 lbm/bhp-hr (76 kg/GJ). The engine then operates at an altitude of 5000 ft (1500 m) with a pressure of 12.225 psia (0.8456 bar) and an operating temperature of 500.9°R (278.4K). The mechanical efficiency of the engine is 80%, independent of altitude. The brake horsepower at the higher altitude is most nearly

(A) 810 hp (630 kW)
(B) 860 hp (650 kW)
(C) 890 hp (690 kW)
(D) 950 hp (740 kW)

6. A gasoline-fueled internal combustion engine runs at 4600 rpm. The engine is four-stroke and V-8 in configuration with a displacement of 265 in³ (4.3 L). The indicated work required to compress the air-fuel mixture is 1200 ft-lbf (1.6 kJ) per cycle. The indicated work done by the exhaust gases in expansion is 1500 ft-lbf (2.0 kJ) per cycle. The input energy from fuel combustion is 1.27 Btu (1.33 kJ) per cycle. Atmospheric air is at 14.7 psia and 70°F (101.3 kPa and 21°C). The air-fuel ratio is 20:1. The heating value of gasoline is 18,900 Btu/lbm (44 MJ/kg). Neglect the effects of friction. The thermal efficiency is most nearly

(A) 30%

(B) 35%

(C) 39%

(D) 46%

7. A mixture of carbon dioxide and helium in an engine undergoes the following processes in a cycle.

A to B: compression and heat removal
B to C: constant volume heating
C to A: isentropic expansion

The temperature and pressure at each point are shown in the table.

point	temperature	pressure
A	520°R (290K)	14.7 psia (101.3 kPa)
B	1240°R (690K)	unknown
C	1600°R (890K)	568.6 psia (3.920 MPa)

Both gases are ideal gases. The gravimetric fractions of the helium and carbon dioxide in the mixture are most nearly

(A) He, 0.12; CO_2, 0.88

(B) He, 0.23; CO_2, 0.77

(C) He, 0.35; CO_2, 0.65

(D) He, 0.46; CO_2, 0.54

8. When the atmospheric conditions are 14.7 psia and 518.7°R (101.3 kPa and 288.2K), a diesel engine with metered fuel injection has the following operating characteristics.

brake horsepower:	200 bhp (150 kW)
brake specific fuel consumption:	0.48 lbm/hp-hr (81 kg/GJ)
air-fuel ratio:	22:1
mechanical efficiency:	86%

The engine is moved to an altitude where the atmospheric conditions are 12.2 psia and 519.7°R (84.1 kPa and 288.7K). The running speed is unchanged. The new net power is most nearly

(A) 140 hp (100 kW)

(B) 170 hp (130 kW)

(C) 260 hp (200 kW)

(D) 330 hp (240 kW)

SOLUTIONS

1. *Customary U.S. Solution*

The actual brake horsepower is

Torques
$$T_{\text{ft-lbf}} = 5250 \times \frac{\text{horsepower}}{\text{rpm}}$$

$$\text{bhp} = \frac{NT_{\text{lbf-ft}}}{5250}$$

$$= \frac{\left(200 \dfrac{\text{rev}}{\text{min}}\right)(600 \text{ ft-lbf})}{5250 \dfrac{\text{ft-lbf}}{\text{hp-min}}}$$

$$= 22.85 \text{ hp}$$

The stroke in feet is

$$L = \frac{18 \text{ in}}{12 \dfrac{\text{in}}{\text{ft}}} = 1.5 \text{ ft}$$

Because there are two cylinders, the bore area is multiplied by 2. This value represents the A and N terms in the PLAN formula.

$$\text{AN} = (2)\left(\frac{\pi}{4}\right)(10 \text{ in})^2 = 157.08 \text{ in}^2$$

The ideal (indicated) horsepower is

Internal Combustion Engines
$$\text{hp} = (\text{MEP})\left(\frac{LAN}{K}\right) = (\text{MEP})\left(\frac{LAN}{33{,}000}\right)$$

$$= \frac{\left(95 \dfrac{\text{lbf}}{\text{in}^2}\right)(1.5 \text{ ft})(157.08 \text{ in}^2)\left(100 \dfrac{\text{power strokes}}{\text{min}}\right)}{33{,}000 \dfrac{\text{ft-lbf}}{\text{hp-min}}}$$

$$= 67.83 \text{ hp}$$

The friction horsepower is

$$\begin{aligned}\text{fhp} &= \text{ihp} - \text{bhp} \\ &= \text{ideal hp} - \text{actual bhp} \\ &= 67.83 \text{ hp} - 22.85 \text{ hp} \\ &= 44.98 \text{ hp} \quad (45 \text{ hp})\end{aligned}$$

SI Solution

From Eq. 18.30 actual brake horsepower is

$$\text{bkW} = \frac{nT}{9549} = \frac{\left(200 \dfrac{\text{rev}}{\text{min}}\right)(820 \text{ N·m})}{9549 \dfrac{\text{N·m}}{\text{kW·min}}}$$

$$= 17.17 \text{ kW}$$

The stroke in meters is

$$L = \frac{460 \text{ mm}}{1000 \dfrac{\text{mm}}{\text{m}}} = 0.46 \text{ m}$$

Because there are two cylinders, the bore area is multiplied by 2. This value represents the A and N terms in the PLAN formula.

$$\text{AN} = (2)\left(\frac{\pi}{4}\right)\left(\frac{250 \text{ mm}}{1000 \dfrac{\text{mm}}{\text{m}}}\right)^2 = 9.818 \times 10^{-2} \text{ m}^2$$

The variation of the PLAN formula can be used to give the power in kilowatts. The constant K is 60. From Eq. 28.52, the ideal (indicated) horsepower is

Internal Combustion Engines
$$p_{\text{kW}} = (\text{MEP})\left(\frac{LAN}{K}\right) = (\text{MEP})\left(\frac{LAN}{60}\right)$$

$$= \frac{(660 \text{ kPa})(0.46 \text{ m}) \times (9.818 \times 10^{-2} \text{ m}^2)\left(100 \dfrac{\text{power strokes}}{\text{min}}\right)}{60}$$

$$= 49.68 \text{ kW}$$

The friction horse power is

$$\begin{aligned}\text{ideal kW} - \text{actual kW} &= 49.68 \text{ kW} - 17.17 \text{ kW} \\ &= 32.51 \text{ kW} \quad (33 \text{ kW})\end{aligned}$$

The answer is (C).

2. *Customary U.S. Solution*

The displacement is

$$\left(\frac{\pi}{4}\right)(\text{bore})^2(\text{stroke}) = \left(\frac{\left(\dfrac{\pi}{4}\right)(3.1 \text{ in})^2}{\left(12 \dfrac{\text{in}}{\text{ft}}\right)^2}\right)\left(\frac{3.8 \text{ in}}{12 \dfrac{\text{in}}{\text{ft}}}\right)$$

$$= 0.0166 \text{ ft}^3$$

The actual incoming volume per intake stroke is

$$V = \text{(volumetric efficiency)}\text{(displacement)}$$
$$= (0.65)(0.0166 \text{ ft}^3)$$
$$= 0.01079 \text{ ft}^3$$

The area is

$$A = \frac{V}{\text{v}t} = \frac{0.01079 \text{ ft}^3}{\left(100 \dfrac{\text{ft}}{\text{sec}}\right)(9.167 \times 10^{-3} \text{ sec})}$$
$$= 0.0118 \text{ ft}^2 \quad (0.012 \text{ ft}^2)$$

SI Solution

The displacement is

$$\left(\frac{\pi}{4}\right)(\text{bore})^2(\text{stroke}) = \left(\frac{\pi}{4}\right)(0.08 \text{ m})^2(0.097 \text{ m})$$
$$= 4.876 \times 10^{-4} \text{ m}^3$$

The actual incoming volume per intake stroke is

$$V = \text{(volumetric efficiency)}\text{(displacement)}$$
$$= (0.65)(4.876 \times 10^{-4} \text{ m}^3)$$
$$= 3.169 \times 10^{-4} \text{ m}^3$$

The area is

$$A = \frac{V}{\text{v}t} = \frac{3.169 \times 10^{-4} \text{ m}^3}{\left(30 \dfrac{\text{m}}{\text{s}}\right)(9.167 \times 10^{-3} \text{ s})}$$
$$= 1.152 \times 10^{-3} \text{ m}^2 \quad (0.0012 \text{ m}^2)$$

The answer is (A).

3. Refer to the air-standard Otto cycle diagram. [Otto Cycle (Gasoline Engine)]

At point 1, the volume is

$$V_1 = 10 \text{ ft}^3$$

The absolute temperature is

Temperature Conversions

$$T_1 = 90°\text{F} + 460° = 550°\text{R}$$

The mass of the air in the intake volume can be solved using the ideal gas law.

Ideal Gas

$$pV = mRT$$
$$m = \frac{pV}{RT}$$
$$= \frac{\left(14.2 \dfrac{\text{lbf}}{\text{in}^2}\right)\left(12 \dfrac{\text{in}}{\text{ft}}\right)^2(10 \text{ ft}^3)}{\left(53.3 \dfrac{\text{ft-lbf}}{\text{lbm-°R}}\right)(550°\text{R})}$$
$$= 0.698 \text{ lbm}$$

From air tables, find the relative volume and specific internal energy at 550°R. [Air at Low Pressure, per Pound]

$$V_{r,1} = 137.85$$
$$u_1 = 93.76 \text{ Btu/lbm}$$

At point 2:

The compression ratio is a ratio of volumes.

$$V_2 = \tfrac{1}{8}V_1 = \left(\frac{1}{8}\right)(10 \text{ ft}^3)$$
$$= 1.25 \text{ ft}^3$$

Since the compression from 1 to 2 is isentropic,

$$V_{r,2} = \frac{V_{r,1}}{8} = \frac{137.85}{8}$$
$$= 17.231$$

Interpolating from the air tables, find the specific internal energy for $V_r = 17.231$. [Air at Low Pressure, per Pound]

$$u_2 \approx 215.05 \text{ Btu/lbm}$$

At point 3:

$$u_3 = u_2 + \frac{Q_{\text{in},2\text{-}3}}{m}$$
$$= 215.05 \frac{\text{Btu}}{\text{lbm}} + \frac{150 \text{ Btu}}{0.698 \text{ lbm}}$$
$$= 429.95 \text{ Btu/lbm}$$

Interpolating from air tables, find the relative volume for $u = 429.95$ Btu/lbm. [Air at Low Pressure, per Pound]

$$V_{r,3} = 2.786$$

At point 4:

Since expansion is isentropic and the ratio of volumes is the same,

$$V_{r,4} = 8 V_{r,3} = (8)(2.786)$$
$$= 22.29$$

Interpolating from air tables, find the temperature at $V_r = 22.29$. [Properties of Air at Low Pressure, per Pound]

$$T_4 = 1120°R$$

The answer is (A).

4. Refer to the air-standard Otto cycle diagram. [Otto Cycle (Gasoline Engine)]

At point 1, the volume is

$$V_1 = 11 \text{ ft}^3$$

The absolute temperature is

Temperature Conversions

$$T_1 = 80°F + 460° = 540°R$$

The mass of the air in the intake volume can be solved using the ideal gas law.

Ideal Gas

$$pV = mRT$$
$$m = \frac{pV}{RT}$$
$$= \frac{\left(14.2 \frac{\text{lbf}}{\text{in}^2}\right)\left(12 \frac{\text{in}}{\text{ft}}\right)^2 (11 \text{ ft}^3)}{\left(53.3 \frac{\text{ft-lbf}}{\text{lbm-°R}}\right)(540°R)}$$
$$= 0.781 \text{ lbm}$$

From air tables, find the relative volume and specific internal energy at 540°R. [Properties of Air at Low Pressure, per Pound]

$$V_{r,1} = 144.32$$
$$u_1 = 92.04 \text{ Btu/lbm}$$

At point 2:

The compression ratio is a ratio of volumes.

$$V_2 = \tfrac{1}{10} V_1 = \left(\frac{1}{10}\right)(11 \text{ ft}^3)$$
$$= 1.1 \text{ ft}^3$$

Since the compression from 1 to 2 is isentropic,

$$V_{r,2} = \frac{V_{r,1}}{10} = \frac{144.32}{10}$$
$$= 14.432$$

Interpolating from the air tables, find the properties for $V_r = 14.432$. [Properties of Air at Low Pressure, per Pound]

$$T_2 \approx 1314°R$$
$$u_2 \approx 230.5 \text{ Btu/lbm}$$

At point 3:

$$u_3 = u_2 + \frac{Q_{\text{in},2-3}}{m}$$
$$= 230.5 \frac{\text{Btu}}{\text{lbm}} + \frac{160 \text{ Btu}}{0.781 \text{ lbm}}$$
$$= 435.4 \text{ Btu/lbm}$$

Interpolating from the air tables, find the properties for $u = 435.4$ Btu/lbm. [Properties of Air at Low Pressure, per Pound]

$$V_{r,3} = 2.694$$

At point 4:

Since expansion is isentropic and the ratio of volumes is the same,

$$V_{r,4} = 10 V_{r,3} = (10)(2.694)$$
$$= 26.94$$

Interpolating from the air tables, find the properties at $V_r = 26.94$. [Properties of Air at Low Pressure, per Pound]

$$u_4 = 180.38 \text{ Btu/lbm}$$

The heat input is

$$q_{\text{in}} = \frac{Q}{m} = \frac{160 \text{ Btu}}{0.781 \text{ lbm}}$$
$$= 204.9 \text{ Btu/lbm}$$

The heat rejected during a constant volume process is $q_{\text{out}} = \Delta u$.

Heat is rejected between point 4 and point 1. Therefore,

$$q_{\text{out}} = u_4 - u_1 = 180.38 \frac{\text{Btu}}{\text{lbm}} - 92.04 \frac{\text{Btu}}{\text{lbm}}$$
$$= 88.34 \text{ Btu/lbm}$$

The thermal efficiency is

$$\eta_{th} = \frac{q_{in} - q_{out}}{q_{in}} = \frac{204.9\ \frac{\text{Btu}}{\text{lbm}} - 88.34\ \frac{\text{Btu}}{\text{lbm}}}{204.9\ \frac{\text{Btu}}{\text{lbm}}}$$

$$= 0.569 \quad (57\%)$$

Basic Cycles

The answer is (C).

5. *Customary U.S. Solution*

In this problem, the efficiency is known to be constant with altitude. The efficiency can be used to calculate the brake horsepower from the indicated horsepower at higher altitude. The friction horsepower is not needed in this problem since it is only used to calculate the indicated horsepower.

Calculate the frictionless power at sea level.

$$\text{ihp}_1 = \frac{\text{bhp}_1}{\eta_{m1}} = \frac{1000\ \text{hp}}{0.80}$$

$$= 1250\ \text{hp}$$

From the ideal gas law,

Ideal Gas

$$pv = RT$$
$$\frac{1}{v} = \frac{p}{RT}$$
$$\rho_1 = \frac{p}{RT}$$

$$= \frac{\left(14.696\ \frac{\text{lbf}}{\text{in}^2}\right)\left(12\ \frac{\text{in}}{\text{ft}}\right)^2}{\left(53.35\ \frac{\text{lbf-ft}}{\text{lbm-°R}}\right)(518.7°\text{R})}$$

$$= 0.0765\ \text{lbm/ft}^3$$

Similarly,

$$\rho_2 = \frac{p}{RT} = \frac{\left(12.225\ \frac{\text{lbf}}{\text{in}^2}\right)\left(12\ \frac{\text{in}}{\text{ft}}\right)^2}{\left(53.35\ \frac{\text{ft-lbf}}{\text{lbm-°R}}\right)(500.9°\text{R})}$$

$$= 0.0659\ \text{lbm/ft}^3$$

Calculate the frictionless power at 5000 ft.

$$\text{ihp}_2 = \text{ihp}_1\left(\frac{\rho_2}{\rho_1}\right) = (1250\ \text{hp})\left(\frac{0.0659\ \frac{\text{lbm}}{\text{ft}^3}}{0.0765\ \frac{\text{lbm}}{\text{ft}^3}}\right)$$

$$= 1076.8\ \text{hp}$$

Calculate the net power at 5000 ft.

$$\text{bhp}_2 = \eta_{m2}(\text{ihp}_2) = (0.80)(1076.8\ \text{hp})$$
$$= 861.4\ \text{hp} \quad (860\ \text{hp})$$

SI Solution

In this problem, the efficiency is known to be constant with altitude. The efficiency can be used to calculate the brake horsepower from the indicated horsepower at higher altitude. The friction horsepower is not needed in this problem since it is only used to calculate the indicated horsepower.

Calculate the frictionless power at sea level.

$$\text{ihp}_1 = \frac{\text{bhp}_1}{\eta_{m1}} = \frac{750\ \text{kW}}{0.80} = 937.5\ \text{kW}$$

From the ideal gas law, the air densities are

Ideal Gas

$$pv = RT$$
$$\frac{1}{v} = \frac{p}{RT}$$
$$\rho_1 = \frac{p}{RT}$$

$$= \frac{(1.01325\ \text{bar})\left(10^5\ \frac{\text{Pa}}{\text{bar}}\right)}{\left(287.03\ \frac{\text{J}}{\text{kg·K}}\right)(288.15\text{K})}$$

$$= 1.225\ \text{kg/m}^3$$

Similarly,

$$\rho_2 = \frac{p}{RT} = \frac{(0.8456\ \text{bar})\left(10^5\ \frac{\text{Pa}}{\text{bar}}\right)}{\left(287.03\ \frac{\text{J}}{\text{kg·K}}\right)(278.4\text{K})}$$

$$= 1.058\ \text{kg/m}^3$$

Calculate the frictionless power at 1500 m.

$$\text{ihp}_2 = \text{ihp}_1\left(\frac{\rho_2}{\rho_1}\right) = (937.5 \text{ kW})\left(\frac{1.058 \frac{\text{kg}}{\text{m}^3}}{1.225 \frac{\text{kg}}{\text{m}^3}}\right)$$
$$= 809.7 \text{ kW}$$

Calculate the net power at 1500 m.

$$\text{bhp}_2 = \eta_{m2}(\text{ihp}_2) = (0.80)(809.7 \text{ kW})$$
$$= 647.8 \text{ kW} \quad (650 \text{ kW})$$

The answer is (B).

6. *Customary U.S. Solution*

The net work per cycle is

$$W_{\text{net}} = W_{\text{out}} - W_{\text{in}} = 1500 \text{ ft-lbf} - 1200 \text{ ft-lbf}$$
$$= 300 \text{ ft-lbf}$$

The thermal efficiency is

Basic Cycles

$$\eta_{\text{th}} = \frac{W_{\text{net}}}{Q_H} = \frac{W_{\text{out}} - W_{\text{in}}}{Q_{\text{in}}}$$
$$= \frac{1500 \frac{\text{ft-lbf}}{\text{cycle}} - 1200 \frac{\text{ft-lbf}}{\text{cycle}}}{\left(1.27 \frac{\text{Btu}}{\text{cycle}}\right)\left(778 \frac{\text{ft-lbf}}{\text{Btu}}\right)}$$
$$= 0.304 \quad (30\%)$$

SI Solution

The net work per cycle is

$$W_{\text{net}} = W_{\text{out}} - W_{\text{in}} = 2.0 \frac{\text{kJ}}{\text{cycle}} - 1.6 \frac{\text{kJ}}{\text{cycle}}$$
$$= 0.4 \text{ kJ/cycle}$$

The thermal efficiency is

Basic Cycles

$$\eta_{\text{th}} = \frac{W_{\text{net}}}{Q_H} = \frac{W_{\text{out}} - W_{\text{in}}}{Q_{\text{in}}}$$
$$= \frac{2.0 \frac{\text{kJ}}{\text{cycle}} - 1.6 \frac{\text{kJ}}{\text{cycle}}}{1.33 \frac{\text{kJ}}{\text{cycle}}}$$
$$= 0.304 \quad (30\%)$$

The answer is (A).

7. At A:

$$T_A = 520°\text{R}$$
$$p_A = 14.7 \text{ psia}$$

At C:

$$T_C = 1600°\text{R}$$
$$p_C = 568.6 \text{ psia}$$

For isentropic process C-A, the pressure at A can be found.

Isentropic Flow Relationships

$$\frac{p_0}{p} = \left(\frac{T_0}{T}\right)^{k/(k-1)}$$

$$p_A = p_C\left(\frac{T_A}{T_C}\right)^{k/(k-1)}$$

Therefore,

$$14.7 \frac{\text{lbf}}{\text{in}^2} = \left(568.6 \frac{\text{lbf}}{\text{in}^2}\right)\left(\frac{520°\text{R}}{1600°\text{R}}\right)^{k/(k-1)}$$

$$\log\left(\frac{14.7 \frac{\text{lbf}}{\text{in}^2}}{568.6 \frac{\text{lbf}}{\text{in}^2}}\right) = \left(\frac{k}{k-1}\right)\log\left(\frac{520°\text{R}}{1600°\text{R}}\right)$$

$$\log(0.02585) = \left(\frac{k}{k-1}\right)\log(0.325)$$

$$\frac{k}{k-1} = \frac{\log(0.02585)}{\log(0.325)}$$

$$k = 1.444$$

The molar heat capacity of the mixture is

$$C_{p,\text{mixture}} = \frac{R^* k}{k-1}$$

$$= \frac{\left(1545 \frac{\text{ft-lbf}}{\text{lbmol-°R}}\right)(1.444)}{\left(778 \frac{\text{ft-lbf}}{\text{Btu}}\right)(1.444 - 1)}$$

$$= 6.459 \text{ Btu/lbmol-°R}$$

The specific heats and molecular weights of helium and carbon dioxide are [Thermal and Physical Properties of Ideal Gases (at Room Temperature)]

$$(c_p)_{He} = 1.250 \text{ Btu/lbm-°R}$$
$$(M)_{He} = 4 \text{ lbm/lbmol}$$
$$(c_p)_{CO_2} = 0.203 \text{ Btu/lbm-°R}$$
$$(M)_{CO_2} = 44 \text{ lbm/lbmol}$$

The molar heat capacity for helium and carbon dioxide are

$$C_{m,He} = (M)_{He}(c_p)_{He}$$
$$= \left(4 \frac{\text{lbm}}{\text{lbmol}}\right)\left(1.250 \frac{\text{Btu}}{\text{lbm-°R}}\right)$$
$$= 5.0 \text{ Btu/lbmol-°R}$$
$$C_{m,CO_2} = (M)_{CO_2}(c_p)_{CO_2}$$
$$= \left(44 \frac{\text{lbm}}{\text{lbmol}}\right)\left(0.203 \frac{\text{Btu}}{\text{lbm-°R}}\right)$$
$$= 8.932 \text{ Btu/lbmol-°R}$$

Let x be the mole fraction of helium in the mixture.

$$x = \frac{n_{He}}{n_{He} + n_{CO_2}}$$

Find the total moles of the mixture.

$$C_{p,\text{mixture}} = x(C_m)_{He} + (1-x)(C_m)_{CO_2}$$
$$6.459 \frac{\text{Btu}}{\text{lbmol-°R}} = x\left(5.0 \frac{\text{Btu}}{\text{lbmol-°R}}\right)$$
$$+ (1-x)\left(8.932 \frac{\text{Btu}}{\text{lbmol-°R}}\right)$$
$$x = 0.606 \text{ lbmol}$$

On a per mole basis, the mass of helium is

$$m_{He} = x(M)_{He} = (0.606 \text{ lbmol})\left(4 \frac{\text{lbm}}{\text{lbmol}}\right)$$
$$= 2.424 \text{ lbm}$$

Similarly, the mass of carbon dioxide on a per mole basis is

$$m_{CO_2} = (1-x)(M)_{CO_2}$$
$$= (1 \text{ lbmol} - 0.606 \text{ lbmol})\left(44 \frac{\text{lbm}}{\text{lbmol}}\right)$$
$$= 17.336 \text{ lbm}$$

The gravimetric (mass) fraction of the gases is

$$G_{He} = \frac{m_{He}}{m_{He} + m_{CO_2}}$$
$$= \frac{2.424 \text{ lbm}}{2.424 \text{ lbm} + 17.336 \text{ lbm}}$$
$$= 0.123 \quad (0.12)$$
$$G_{CO_2} = 1 - G_{He} = 1 - 0.123$$
$$= 0.877 \quad (0.88)$$

The answer is (A).

8. *Customary U.S. Solution*

From the ideal gas law, the density at altitude 1 is

Ideal Gas

$$pv = RT$$
$$\frac{1}{v} = \frac{p}{RT}$$
$$\rho_1 = \frac{p}{RT}$$
$$= \frac{\left(14.696 \frac{\text{lbf}}{\text{in}^2}\right)\left(12 \frac{\text{in}}{\text{ft}}\right)^2}{\left(53.35 \frac{\text{lbf-ft}}{\text{lbm-°R}}\right)(518.7°R)}$$
$$= 0.0765 \text{ lbm/ft}^3$$

Similarly, the density at altitude 2 is

$$\rho_2 = \frac{p}{RT} = \frac{\left(12.2 \frac{\text{lbf}}{\text{in}^2}\right)\left(12 \frac{\text{in}}{\text{ft}}\right)^2}{\left(53.35 \frac{\text{ft-lbf}}{\text{lbm-°R}}\right)(519.7°R)}$$
$$= 0.0634 \text{ lbm/ft}^3$$

Calculate the new brake horsepower (net horsepower). Note that the engine's efficiency doesn't change with altitude.

$$\text{bhp}_2 = \text{bhp}_1\left(\frac{\rho_2}{\rho_1}\right) = (200 \text{ hp})\left(\frac{0.0634 \frac{\text{lbm}}{\text{ft}^3}}{0.0765 \frac{\text{lbm}}{\text{ft}^3}}\right)$$
$$= 165.8 \text{ hp} \quad (170 \text{ hp})$$

SI Solution

From the ideal gas law, the density at altitude 1 is

Ideal Gas

$$pv = RT$$
$$\frac{1}{v} = \frac{p}{RT}$$
$$\rho_1 = \frac{p}{RT}$$
$$= \frac{(101.3 \text{ kPa})\left(1000 \frac{\text{Pa}}{\text{kPa}}\right)}{\left(287.03 \frac{\text{J}}{\text{kg·K}}\right)(288.2 \text{K})}$$
$$= 1.22 \text{ kg/m}^3$$

Similarly, the density at altitude 2 is

$$\rho_2 = \frac{p}{RT} = \frac{(84.1 \text{ kPa})\left(1000 \frac{\text{Pa}}{\text{kPa}}\right)}{\left(287.03 \frac{\text{kJ}}{\text{kg·K}}\right)(288.7 \text{K})}$$
$$= 1.01 \text{ kg/m}^3$$

Calculate the new brake horsepower (net horsepower). Note that the engine's efficiency doesn't change with altitude.

$$\text{bhp}_2 = \text{bhp}_1\left(\frac{\rho_2}{\rho_1}\right) = (150 \text{ kW})\left(\frac{1.01 \frac{\text{kg}}{\text{m}^3}}{1.22 \frac{\text{kg}}{\text{m}^3}}\right)$$
$$= 124.2 \text{ kW} \quad (130 \text{ kW})$$

The answer is (B).

29 Combustion Turbine Cycles

Content in blue refers to the NCEES Handbook.

PRACTICE PROBLEMS

1. Air expands at the rate of 10 ft^3/sec from a pressure of 200 psia at 1500°F to a pressure of 50 psia. The air's final temperature is most nearly

(A) 1280°R
(B) 1330°R
(C) 1380°R
(D) 1430°R

2. Air expands isentropically at a rate of 10 ft^3/sec from a pressure of 200 psia at a temperature of 1500°F to a pressure of 50 psia. Using air tables, the air's final volumetric flow rate is most nearly

(A) 30 ft^3/sec
(B) 34 ft^3/sec
(C) 39 ft^3/sec
(D) 45 ft^3/sec

3. Air expands isentropically at the rate of 10 ft^3/sec from a pressure of 200 psia at 1500°F to a pressure of 50 psia. Using air tables, the change in the enthalpy of the air after the expansion is most nearly

(A) −350 Btu/lbm
(B) −230 Btu/lbm
(C) −160 Btu/lbm
(D) −110 Btu/lbm

4. In an air-standard gas turbine, air at 14.7 psia and 60°F enters a compressor and is compressed through a volumetric ratio of 5:1. The compressor efficiency is 83%. Air enters the turbine at 1500°F and expands to 14.7 psia. The turbine efficiency is 92%. The turbine follows the Brayton gas turbine cycle, as shown.

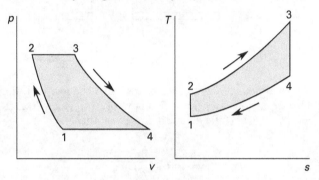

The thermal efficiency of the cycle is most nearly

(A) 33%
(B) 39%
(C) 44%
(D) 51%

5. In an air-standard gas turbine, air at 14.7 psia and 60°F enters a compressor and is compressed through a volume ratio of 5:1. The compressor efficiency is 83%. Air enters the turbine at 1500°F and expands to 14.7 psia. The turbine efficiency is 92%. Assume air to be ideal gas. (Solve using the ideal gas equation rather than air tables). The gas turbine follows the cycle shown.

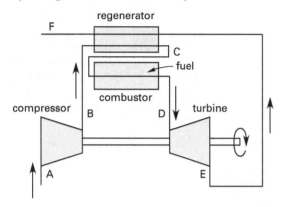

A 65% efficient regenerator is added to the gas turbine. The equation for the efficiency of a regenerator is shown. The specific heat remains constant.

$$\eta_{regenerator} = \frac{h_C - h_B'}{h_E' - h_B'}$$

The new thermal efficiency for the Brayton Cycle is most nearly

(A) 24%

(B) 28%

(C) 34%

(D) 41%

6. A precision air turbine is used to drive a small dentist's drill. 140°F air enters the turbine at the rate of 15 lbm/hr. The output of the turbine is 0.25 hp. The turbine exhausts to 15 psia. The flow is steady. The isentropic efficiency of the expansion process is 60%.

If the expansion is isentropic, the enthalpy of the air exiting the turbine is most nearly

(A) 70 Btu/lbm

(B) 100 Btu/lbm

(C) 140 Btu/lbm

(D) 200 Btu/lbm

7. A precision air turbine is used to drive a small dentist's drill. 140°F air enters the turbine at the rate of 15 lbm/hr. The output of the turbine is 0.25 hp. The turbine exhausts to 15 psia. The flow is steady. The isentropic efficiency of the expansion process is 60%.

The actual exhaust temperature is most nearly

(A) 370°R

(B) 420°R

(C) 450°R

(D) 610°R

8. A precision air turbine is used to drive a small dentist's drill. 140°F air enters the turbine at the rate of 15 lbm/hr. The output of the turbine is 0.25 hp. The turbine exhausts to 15 psia. The flow is steady. The isentropic efficiency of the expansion process is 60%. The equations for the power generated and the efficiency are shown.

$$P = \dot{m}(h_1 - h_2')$$

$$\eta_{s,turbine} = \frac{h_1 - h_2'}{h_1 - h_2}$$

The inlet air pressure is most nearly

(A) 160 psia

(B) 210 psia

(C) 240 psia

(D) 290 psia

9. A precision air turbine is used to drive a small dentist's drill. 140°F air enters the turbine at the rate of 15 lbm/hr. The output of the turbine is 0.25 hp. The turbine exhausts to 15 psia. The flow is steady. The isentropic efficiency of the expansion process is 60%. The equations for the power generated and the efficiency are shown.

$$P = \dot{m}(h_1 - h_2')$$

$$\eta_{s,turbine} = \frac{h_1 - h_2'}{h_1 - h_2}$$

The change in entropy through the turbine is most nearly

(A) 0.078 Btu/lbm-°R

(B) 0.10 Btu/lbm-°R

(C) 0.18 Btu/lbm-°R

(D) 0.32 Btu/lbm-°R

SOLUTIONS

1. The absolute initial temperature is

Temperature Conversions

$$T_1 = 1500°F + 460° = 1960°R$$

From air tables, air at 1960°R has a relative pressure of 160.37. [Properties of Air at Low Pressure, per Pound]

After expansion, the relative pressure is

$$p_{r,2} = p_{r,1}\left(\frac{p_2}{p_1}\right) = (160.37)\left(\frac{50 \,\frac{\text{lbf}}{\text{in}^2}}{200 \,\frac{\text{lbf}}{\text{in}^2}}\right)$$

$$= 40.09$$

Interpolating from air tables, a relative pressure of 40.09 corresponds to a temperature of 1375°R (1380°R). [Properties of Air at Low Pressure, per Pound]

The answer is (C).

2. The absolute temperature is

Temperature Conversions

$$T_1 = 1500°F + 460° = 1960°R$$

From air tables, air at 1960°R has an enthalpy of 493.64 Btu/lbm, a relative pressure of 160.37, and a relative volume of 4.527. [Properties of Air at Low Pressure, per Pound]

After expansion,

$$p_{r,2} = p_{r,1}\left(\frac{p_2}{p_1}\right) = (160.37)\left(\frac{50 \,\frac{\text{lbf}}{\text{in}^2}}{200 \,\frac{\text{lbf}}{\text{in}^2}}\right)$$

$$= 40.09$$

Interpolating from the air tables, a relative pressure of 40.09 corresponds to a temperature of 1375°R, and a relative volume of 12.71. [Properties of Air at Low Pressure, per Pound]

The air's final volumetric flow rate is

$$\dot{V}_2 = \dot{V}_1\left(\frac{V_{r,2}}{V_{r,1}}\right) = \left(10 \,\frac{\text{ft}^3}{\text{sec}}\right)\left(\frac{12.71}{4.527}\right)$$

$$= 28.1 \text{ ft}^3/\text{sec} \quad (30 \text{ ft}^3/\text{sec})$$

The answer is (A).

3. The absolute temperature is

Temperature Conversions

$$T_1 = 1500°F + 460° = 1960°R$$

From air tables, air at 1960°R has an enthalpy of 493.64 Btu/lbm, a relative pressure of 160.37. [Properties of Air at Low Pressure, per Pound]

After expansion,

$$p_{r,2} = p_{r,1}\left(\frac{p_2}{p_1}\right) = (160.37)\left(\frac{50 \,\frac{\text{lbm}}{\text{in}^2}}{200 \,\frac{\text{lbm}}{\text{in}^2}}\right)$$

$$= 40.09$$

Interpolating from the air tables, a relative pressure of 40.09 corresponds to a temperature of 1375°R, an enthalpy of 336.5 Btu/lbm. [Properties of Air at Low Pressure, per Pound]

The air's enthalpy change is

$$\Delta h = h_2 - h_1 = 336.5 \,\frac{\text{Btu}}{\text{lbm}} - 493.64 \,\frac{\text{Btu}}{\text{lbm}}$$

$$= -157.14 \text{ Btu/lbm} \quad (-160 \text{ Btu/lbm}) \quad [\text{decrease}]$$

The answer is (C).

4. At A,

Temperature Conversions

$$T_A = 60°F + 460° = 520°R \quad [\text{given}]$$

$$p_A = 14.7 \text{ psia} \quad [\text{given}]$$

From the air table, air at a temperature of 520°R has a relative volume of 158.58, a relative pressure of 1.2147, and an enthalpy of 124.27 Btu/lbm. [Air at Low Pressure, per Pound]

The process from A to B is isentropic, so the relative volume at B is

$$v_{r,B} = v_{r,A}\left(\frac{V_B}{V_A}\right) = (158.58)\left(\frac{1}{5}\right)$$

$$= 31.716$$

From the air table, a relative volume of 31.716 corresponds to a temperature of 980°R, an enthalpy of 236.02 Btu/lbm, and a relative pressure of 11.430. [Air at Low Pressure, per Pound]

Since process A-B is isentropic,

$$p_B = \left(\frac{p_{r,B}}{p_{r,A}}\right)p_A = \left(\frac{11.430}{1.2147}\right)\left(14.7 \,\frac{\text{lbf}}{\text{in}^2}\right)$$

$$= 138.3 \text{ lbf/in}^2$$

At C,

Temperature Conversions
$$T_C = 1500°F + 460° = 1960°R \quad \text{[given]}$$
$$p_C = p_B = 138.3 \text{ lbf/in}^2$$

Air at a temperature of 1960°R has an enthalpy of 493.64 Btu/lbm and a relative pressure of 160.37. [Air at Low Pressure, per Pound]

At D,
$$p_D = 14.7 \text{ psia}$$

Since process C-D is isentropic,

$$p_{r,D} = p_{r,C}\left(\frac{p_D}{p_C}\right) = (160.37)\left(\frac{14.7 \frac{\text{lbf}}{\text{in}^2}}{138.3 \frac{\text{lbf}}{\text{in}^2}}\right)$$
$$= 17.045$$

Interpolating from the air table, a relative pressure of 17.045 corresponds to a temperature of 1094°R and an enthalpy of 264.49 Btu/lbm. [Air at Low Pressure, per Pound]

Since the efficiency of compression is 83%, the enthalpy at point B is

$$h'_B = h_A + \frac{h_B - h_A}{\eta_{s,\text{compressor}}}$$
$$= 124.27 \frac{\text{Btu}}{\text{lbm}} + \frac{236.02 \frac{\text{Btu}}{\text{lbm}} - 124.27 \frac{\text{Btu}}{\text{lbm}}}{0.83}$$
$$= 258.9 \text{ Btu/lbm}$$

Since the efficiency of the expansion process is 92%, the enthalpy at point D is

$$h'_D = h_C - \eta_{s,\text{turbine}}(h_C - h_D)$$
$$= 493.64 \frac{\text{Btu}}{\text{lbm}}$$
$$\quad - (0.92)\left(493.64 \frac{\text{Btu}}{\text{lbm}} - 264.49 \frac{\text{Btu}}{\text{lbm}}\right)$$
$$= 282.8 \text{ Btu/lbm}$$

The thermal efficiency is

$$\eta_{\text{th}} = \frac{(h_C - h'_B) - (h'_D - h_A)}{h_C - h'_B}$$

$$= \frac{\left(493.64 \frac{\text{Btu}}{\text{lbm}} - 258.9 \frac{\text{Btu}}{\text{lbm}}\right) - \left(282.8 \frac{\text{Btu}}{\text{lbm}} - 124.27 \frac{\text{Btu}}{\text{lbm}}\right)}{493.64 \frac{\text{Btu}}{\text{lbm}} - 258.9 \frac{\text{Btu}}{\text{lbm}}}$$

$$= 0.325 \quad (33\%)$$

The answer is (A).

5. Since the specific heats are constant, use ideal gas equations rather than air tables.

At A,

Temperature Conversions
$$T_A = 60°F + 460° = 520°R \quad \text{[given]}$$
$$p_A = 14.7 \text{ lbf/in}^2 \quad \text{[given]}$$

At B,

$$T_B = T_A\left(\frac{V_A}{V_B}\right)^{k-1} = (520°R)(5)^{1.4-1}$$
$$= 989.9°R$$
$$p_B = p_A\left(\frac{V_A}{V_B}\right)^k = \left(14.7 \frac{\text{lbf}}{\text{in}^2}\right)(5)^{1.4}$$
$$= 139.9 \text{ lbf/in}^2$$

At D,

Temperature Conversions
$$T_D = 1500°F + 460° = 1960°R \quad \text{[given]}$$
$$p_D = p_B = 139.9 \text{ lbf/in}^2$$

At E,

$$p_E = 14.7 \text{ lbf/in}^2 \quad \text{[given]}$$

$$T_E = T_D\left(\frac{p_E}{p_D}\right)^{(k-1)/k} = (1960°R)\left(\frac{14.7 \frac{\text{lbf}}{\text{in}^2}}{139.9 \frac{\text{lbf}}{\text{in}^2}}\right)^{(1.4-1)/1.4}$$

$$= 1029.6°R$$

The actual temperature at point B is

$$T'_B = T_A + \frac{T_B - T_A}{\eta_{s,\text{compressor}}} = 520°R + \frac{989.9°R - 520°R}{0.83}$$
$$= 1086.1°R$$

The actual temperature at point E is

$$T'_E = T_D - \eta_{s,\text{turbine}}(T_D - T_E)$$
$$= 1960°R - (0.92)(1960°R - 1029.6°R)$$
$$= 1104.0°R$$

For constant specific heat, the temperature at point C can be found from the equation for the efficiency of a regenerator, substituting the temperatures at each point for the enthalpies.

$$\eta_{\text{regenerator}} = \frac{T_C - T'_B}{T'_E - T'_B}$$
$$0.65 = \frac{T_C - 1086.1°R}{1104.0°R - 1086.1°R}$$
$$T_C = 1097.7°R$$

Note that the simplified formula for Brayton Cycle thermal efficiency with regeneration shown in the *NCEES Handbook* only applies to a Brayton cycle with constant entropy across the turbine and compressor. The Brayton Cycle thermal efficiency with non-isentropic compressor and turbine is

$$\eta_{\text{th}} = \frac{(T_D - T'_E) - (T'_B - T_A)}{T_D - T_C}$$
$$= \frac{(1960°R - 1104.0°R) - (1086.1°R - 520°R)}{1960°R - 1097.7°R}$$
$$= 0.336 \quad (34\%)$$

The answer is (C).

6. The drill power in Btu/hr is

$$P = 0.25 \text{ hp} \quad [\text{given}]$$
$$= (0.25 \text{ hp})\left(2545 \frac{\text{Btu}}{\text{hp-hr}}\right)$$
$$= 636.25 \text{ Btu/hr}$$

The absolute inlet temperature is

Temperature Conversions
$$T_1 = 140°F + 460° = 600°R$$

From air tables, air at 600°R has an enthalpy of 143.47 Btu/lbm, a relative pressure of 2.005, and a specific heat of 0.62607 Btu/lbm-°R. [Air at Low Pressure, per Pound]

The power of an isentropic turbine is

Turbines
$$P = \dot{m}\eta_{s,\text{turb}}(h_i - h_e)$$
$$h_e = h_i - \frac{P}{\dot{m}\eta_{s,\text{turb}}}$$
$$= 143.47 \frac{\text{Btu}}{\text{lbm}} - \frac{636.25 \frac{\text{Btu}}{\text{hr}}}{\left(15 \frac{\text{lbm}}{\text{hr}}\right)(0.60)}$$
$$= 72.776 \text{ Btu/lbm} \quad (70 \text{ Btu/lbm})$$

The answer is (A).

7. The drill power in Btu/hr is

$$\dot{W}_{\text{turb}} = 0.25 \text{ hp} \quad [\text{given}]$$
$$= (0.25 \text{ hp})\left(2545 \frac{\text{Btu}}{\text{hp-hr}}\right)$$
$$= 636.25 \text{ Btu/hr}$$

The absolute inlet temperature is

Temperature Conversions
$$T_1 = 140°F + 460° = 600°R$$

From air tables, air at 600°R has an enthalpy of 143.47 Btu/lbm, a relative pressure of 2.005, and a specific heat of 0.62607 Btu/lbm-°R. [Air at Low Pressure, per Pound]

Rearrange the formula for the power of an isentropic turbine with the formula for isentropic efficiency to solve for the isentropic enthalpy at the exit. [Turbines]

$$\dot{W}_{\text{turb}} = \dot{m}\eta_T(h_i - h_{es})$$
$$h_{es} = h_i - \frac{\dot{W}_{\text{turb}}}{\dot{m}\eta_T}$$
$$= 143.47 \frac{\text{Btu}}{\text{lbm}} - \frac{636.25 \frac{\text{Btu}}{\text{hr}}}{\left(15 \frac{\text{lbm}}{\text{hr}}\right)(0.60)}$$
$$= 72.776 \text{ Btu/lbm}$$

Interpolating from the air tables, an enthalpy of 72.776 Btu/lbm corresponds to a temperature of 305°R and a relative pressure of 0.18851. [Air at Low Pressure, per Pound]

Rearrange the formula for turbine isentropic efficiency to solve for the enthalpy at the exit of the turbine. [Special Cases of the Steady-Flow Energy Equation]

$$\eta_T = \frac{h_i - h_e}{h_i - h_{es}}$$

$$h_e = h_i - \eta_T(h_i - h_{es})$$
$$= 143.47 \ \frac{\text{Btu}}{\text{lbm}} - (0.60)$$
$$\times \left(143.47 \ \frac{\text{Btu}}{\text{lbm}} - 72.776 \ \frac{\text{Btu}}{\text{lbm}}\right)$$
$$= 101.05 \ \text{Btu/lbm}$$

Interpolating from the air tables, an enthalpy of 101.05 Btu/lbm corresponds to a temperature of 423°R (420°R). [Air at Low Pressure, per Pound]

The answer is (B).

8. The drill power in Btu/hr is

$$P = 0.25 \ \text{hp} \quad [\text{given}]$$
$$= (0.25 \ \text{hp})\left(2545 \ \frac{\text{Btu}}{\text{hp-hr}}\right)$$
$$= 636.25 \ \text{Btu/hr}$$

The absolute inlet temperature is

Temperature Conversions

$$T_1 = 140°\text{F} + 460° = 600°\text{R}$$

From air tables, air at 600°R has an enthalpy of 143.47 Btu/lbm, a relative pressure of 2.005, and a specific heat of 0.62607 Btu/lbm-°R. [Air at Low Pressure, per Pound]

Combining the equations for efficiency and power generated, the enthalpy at the exit of the turbine is

$$P = \dot{m}\eta_{s,\text{turbine}}(h_1 - h_2)$$
$$h_2 = h_1 - \frac{P}{\dot{m}\eta_{s,\text{turbine}}}$$
$$= 143.47 \ \frac{\text{Btu}}{\text{lbm}} - \frac{636.25 \ \frac{\text{Btu}}{\text{hr}}}{\left(15 \ \frac{\text{lbm}}{\text{hr}}\right)(0.60)}$$
$$= 72.776 \ \text{Btu/lbm}$$

Interpolating from the air tables, an enthalpy of 72.776 Btu/lbm corresponds to a temperature of 305°R and a relative pressure of 0.18851. [Air at Low Pressure, per Pound]

The isentropic efficiency does not change the entrance and exit pressures, so the air tables can be used assuming $s_1 = s_2$. Since $p_1/p_2 = p_{r,1}/p_{r,2}$,

$$p_1 = p_2\left(\frac{p_{r,1}}{p_{r,2}}\right) = \left(15 \ \frac{\text{lbf}}{\text{in}^2}\right)\left(\frac{2.005}{0.18851}\right)$$
$$= 159.5 \ \text{lbf/in}^2 \quad (160 \ \text{psia})$$

The answer is (A).

9. The drill power in Btu/hr is

$$P = 0.25 \ \text{hp} \quad [\text{given}]$$
$$= (0.25 \ \text{hp})\left(2545 \ \frac{\text{Btu}}{\text{hp-hr}}\right)$$
$$= 636.25 \ \text{Btu/hr}$$

The absolute inlet temperature is

Temperature Conversions

$$T_1 = 140°\text{F} + 460° = 600°\text{R}$$

From air tables, air at 600°R has an enthalpy of 143.47 Btu/lbm, a relative pressure of 2.005, and a specific heat of 0.62607 Btu/lbm-°R. [Air at Low Pressure, per Pound]

Combining the equations for efficiency and power generated, the enthalpy at the exit of the turbine is

$$P = \dot{m}\eta_{s,\text{turbine}}(h_1 - h_2)$$
$$h_2 = h_1 - \frac{P}{\dot{m}\eta_{s,\text{turbine}}}$$
$$= 143.47 \ \frac{\text{Btu}}{\text{lbm}} - \frac{636.25 \ \frac{\text{Btu}}{\text{hr}}}{\left(15 \ \frac{\text{lbm}}{\text{hr}}\right)(0.60)}$$
$$= 72.776 \ \text{Btu/lbm}$$

Interpolating from the air tables, an enthalpy of 72.776 Btu/lbm corresponds to a temperature of 305°R and a relative pressure of 0.18851. [Air at Low Pressure, per Pound]

Due to the irreversibility of the expansion, the enthalpy at the exit of the turbine is

$$h_2' = h_1 - \eta_s(h_1 - h_2)$$
$$= 143.47 \ \frac{\text{Btu}}{\text{lbm}} - (0.60)\left(143.47 \ \frac{\text{Btu}}{\text{lbm}} - 72.776 \ \frac{\text{Btu}}{\text{lbm}}\right)$$
$$= 101.05 \ \text{Btu/lbm}$$

Interpolating from the air tables, an enthalpy of 101.05 Btu/lbm corresponds to a temperature of 423°R and a specific heat of 0.54225 Btu/lbm-°R. [Air at Low Pressure, per Pound]

The isentropic efficiency does not change the entrance and exit pressures, so the air tables can be used assuming $s_1 = s_2$. Since $p_1/p_2 = p_{r,1}/p_{r,2}$,

$$p_1 = p_2 \left(\frac{p_{r,1}}{p_{r,2}} \right) = \left(15 \ \frac{\text{lbf}}{\text{in}^2} \right) \left(\frac{2.005}{0.18851} \right)$$
$$= 159.5 \ \text{lbf/in}^2$$

The R-value for air at room temperature is 53.35 ft-lbf/lbm-°R. [Thermal and Physical Properties of Ideal Gases (at Room Temperature)]

The entropy change is

$$s_2 - s_1 = \phi_2 - \phi_1 - R \ln\left(\frac{p_2}{p_1}\right)$$
$$= 0.54225 \ \frac{\text{Btu}}{\text{lbm-°R}} - 0.62607 \ \frac{\text{Btu}}{\text{lbm-°R}}$$
$$- \left(\frac{53.35 \ \frac{\text{ft-lbf}}{\text{lbm-°R}}}{778 \ \frac{\text{ft-lbf}}{\text{Btu}}} \right) \ln\left(\frac{15 \ \frac{\text{lbf}}{\text{in}^2}}{159.5 \ \frac{\text{lbf}}{\text{in}^2}} \right)$$
$$= 0.07829 \ \text{Btu/lbm-°R} \quad (0.078 \ \text{Btu/lbm-°R})$$

The answer is (A).

30 Advanced and Alternative Power-Generating Systems

Content in blue refers to the NCEES Handbook.

PRACTICE PROBLEMS

1. An ocean thermal energy conversion plant draws 40°F (4.4°C) water from a depth of 1200 ft (360 m). The water temperature at the surface is 82°F (27.8°C). The maximum achievable thermal efficiency is most nearly

(A) 8%

(B) 11%

(C) 15%

(D) 21%

2. A wind turbine blade has a radius of 90 ft. The mass per unit length of the blade is 20 lbm/ft for the first 15 ft from the center of rotation. As the blade tapers out to the tip, the mass per unit length varies linearly from 20 lbm/ft to 2 lbm/ft. The blade turns at 250 rev/min. What is most nearly the tension in the blade root caused by the centrifugal force at the hub?

(A) 3.7×10^5 lbf

(B) 7.9×10^5 lbf

(C) 5.2×10^6 lbf

(D) 1.6×10^7 lbf

3. A wind turbine rotor with two tapered 20 ft long blades is locked in position for maintenance. Each blade is modeled as a flat projection 3 ft wide for the first 11 ft from the hub, followed by 9 ft that tapers uniformly from a flat projected width of 3 ft to a width of 1.5 ft. The blades are vertical in the locked position with the lower blade in the lee of the tower. Tower interference reduces the effective wind speed on the lower blade by 25%. The average drag coefficient is 0.4. The tower extends 40 ft vertically above its foundation. Air temperature and pressure are 70°F and 1 atm. Neglecting drag on the tower, what is most nearly the moment at the tower base when there is a 20 mi/hr wind in standard conditions?

(A) 370 ft-lbf

(B) 760 ft-lbf

(C) 1300 ft-lbf

(D) 1400 ft-lbf

4. A 1.2 V, 600 milliamp-hour battery is fully uncharged. A matched charger consists of a transformer-rectifier unit that has a transformation efficiency of 65%. The transformer primary uses 120 V, single-phase AC power. During the charge cycle, the primary draws 1 mA with a power factor of 0.8. The charger secondary delivers 1.5 V to the battery. The battery takes 18 hours to fully charge from the fully uncharged condition. The electrical transformation efficiency within the battery during the charge cycle is most nearly

(A) 42%

(B) 64%

(C) 88%

(D) 98%

5. A molten carbonate fuel cell system supplies electricity to a manufacturing plant. Waste heat from the system is used to generate steam for use in a manufacturing process. The fuel cell system consists of a fuel cell stack with internal reformer, an inverter to convert the DC power generated to AC power, and a heat exchanger to heat water to make steam. Methane gas fuels the system. The efficiency of the inverter is 90%, and the efficiency of the heat exchanger is 75%. The lower heating value (LHV) of methane is 913 Btu/ft^3.

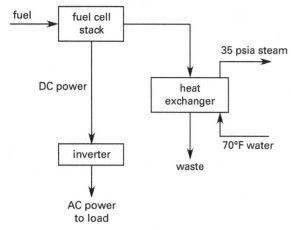

At full power, the system consumes 1200 ft^3/hr of methane gas at 60°F and atmospheric pressure, has an output of 150 kW, and heats 100 lbm/hr of 70°F water to generate 35 psia saturated steam. Feedwater enters the steam generator at 70°F. What is most nearly the fuel system's efficiency?

(A) 48%

(B) 55%

(C) 65%

(D) 87%

6. An experimental solid polymer fuel cell runs on hydrogen and powers an electric motor in a car. The 4000 lbm car has a maximum speed of 55 mi/hr. At that speed, the fuel cell consumes 0.2 ft^3/sec of hydrogen at 60°F and atmospheric pressure. The hydrogen has a lower heating value (LHV) of 275 Btu/ft^3. The rolling coefficient of friction is 0.01. The drag force on the car is 136 lbf. The motor and drive train are 80% efficient. The efficiency of the fuel cell is most nearly

(A) 23%

(B) 28%

(C) 33%

(D) 41%

7. A 20 mi/hr wind at standard temperature and pressure passes through a horizontal wind turbine. The wind is perpendicular to the plane of the blades. The turbine has a blade diameter of 30 ft. Air temperature and pressure are 70°F and 1 atm. What is most nearly the maximum power that the turbine can generate?

(A) 12 hp

(B) 38 hp

(C) 150 hp

(D) 1200 hp

SOLUTIONS

1. *Customary U.S. Solution*

The maximum achievable thermal efficiency is achieved with the Carnot cycle and is given by Eq. 27.9.

Basic Cycles

$$\eta_c = \frac{T_H - T_L}{T_H}$$

$$T_H = 82°F + 460° = 542°R$$
$$T_L = 40°F + 460° = 500°R$$
$$\eta_{th} = \frac{542°R - 500°R}{542°R} = 0.0775 \quad (8\%)$$

SI Solution

The maximum achievable thermal efficiency is achieved with the Carnot cycle and is given by Eq. 27.9.

Basic Cycles

$$\eta_c = \frac{T_H - T_L}{T_H}$$

$$T_H = 27.8°C + 273° = 300.8 K$$
$$T_L = 4.4°C + 273° = 277.4 K$$
$$\eta_{th} = \frac{300.8 K - 277.4 K}{300.8 K} = 0.0778 \quad (8\%)$$

The answer is (A).

2. The mass of the root of the blade is

$$m_{root} = (15 \text{ ft})\left(20 \frac{\text{lbm}}{\text{ft}}\right) = 300 \text{ lbm}$$

Measured from the center of rotation, the centroid of the root is located at

$$\bar{x}_{root} = \frac{15 \text{ ft}}{2} = 7.5 \text{ ft}$$

The mass of the tapered part of the blade is

$$m_{tapered} = (90 \text{ ft} - 15 \text{ ft})\left(\frac{20 \frac{\text{lbm}}{\text{ft}} + 2 \frac{\text{lbm}}{\text{ft}}}{2}\right)$$

$$= 825 \text{ lbm}$$

The mass per length of the tapered part of the blade is distributed as a trapezoid, which can be subdivided into triangular and rectangular shapes. Measured from the center of rotation, the centroid of the tapered part of the blade is located at

$$\bar{x}_{tapered} = 15 \text{ ft} + \frac{\left(\left(\frac{1}{3}\right)(75 \text{ ft})\right)\left(\left(\frac{1}{2}\right)(75 \text{ ft})\right)\left(20 \frac{\text{lbm}}{\text{ft}} - 2 \frac{\text{lbm}}{\text{ft}}\right) + \left(\left(\frac{1}{2}\right)(75 \text{ ft})\right)(75 \text{ ft})\left(2 \frac{\text{lbm}}{\text{ft}}\right)}{825 \text{ lbm}}$$

$$= 42.27 \text{ ft}$$

The total mass of the blade is

$$m_t = m_{root} + m_{tapered} = 300 \text{ lbm} + 825 \text{ lbm} = 1125 \text{ lbm}$$

The center of gravity, referenced to the root, of the blade is located at

$$\bar{x} = \frac{\sum \bar{x}_i m_i}{\sum m_i}$$
$$= \frac{(7.5 \text{ ft})(300 \text{ lbm}) + (42.27 \text{ ft})(825 \text{ lbm})}{1125 \text{ lbm}}$$
$$= 33.0 \text{ ft}$$

The centrifugal force is

$$F_c = \frac{m_t \bar{x} \omega^2}{g_c}$$

$$= \frac{(1125 \text{ lbm})(33.0 \text{ ft})\left(\frac{\left(250 \frac{\text{rev}}{\text{min}}\right)\left(2\pi \frac{\text{rad}}{\text{rev}}\right)}{60 \frac{\text{sec}}{\text{min}}}\right)^2}{32.2 \frac{\text{lbm-ft}}{\text{lbf-sec}^2}}$$

$$= 7.9 \times 10^5 \text{ lbf}$$

(Both blades experience this centrifugal force. However, the tensile force at the hub is not doubled.)

The answer is (B).

3. Use Eq. 17.182(b) to find the force on each blade. Air at 70°F and 1 atm has a density of approximately 0.075 lbm/ft².

The force on the untapered section of the top blade is

Drag Force
$$F_{top,untapered} = \frac{C_D A \rho v^2}{2g_c}$$

$$= \frac{(0.4)(11 \text{ ft})(3 \text{ ft})\left(0.075 \dfrac{\text{lbm}}{\text{ft}^3}\right) \times \left(20 \dfrac{\text{mi}}{\text{hr}}\right)^2 \left(5280 \dfrac{\text{ft}}{\text{mi}}\right)^2}{(2)\left(32.2 \dfrac{\text{lbm-ft}}{\text{lbf-sec}^2}\right)\left(3600 \dfrac{\text{sec}}{\text{hr}}\right)^2}$$

$$= 13.2 \text{ lbf}$$

The resultant of this force is located $40 \text{ ft} + (\frac{1}{2})(11 \text{ ft}) = 45.5$ ft above the tower foundation.

The area of the tapered top blade section is

$$A_{tapered} = \frac{(b_1 + b_2)h}{2} = \frac{(3 \text{ ft} + 1.5 \text{ ft})(9 \text{ ft})}{2}$$

$$= 20.25 \text{ ft}^2$$

The force on the tapered top blade section is

Drag Force
$$F_{top,tapered} = \frac{C_D A_{tapered} \rho v^2}{2g_c}$$

$$= \frac{(0.4)(20.25 \text{ ft}^2)\left(0.075 \dfrac{\text{lbm}}{\text{ft}^3}\right) \times \left(20 \dfrac{\text{mi}}{\text{hr}}\right)^2 \left(5280 \dfrac{\text{ft}}{\text{mi}}\right)^2}{(2)\left(32.2 \dfrac{\text{lbm-ft}}{\text{lbf-sec}^2}\right)\left(3600 \dfrac{\text{sec}}{\text{hr}}\right)^2}$$

$$= 8.1 \text{ lbf}$$

The centroid of the tapered section, measured from the 3 ft wide end, is located at

$$\frac{(9 \text{ ft})(1.5 \text{ ft})\left(\dfrac{9 \text{ ft}}{2}\right) + \left(\dfrac{1}{2}\right)(1.5 \text{ ft})(9 \text{ ft})\left(\dfrac{9 \text{ ft}}{3}\right)}{20.25 \text{ ft}^2} = 4 \text{ ft}$$

The resultant of this force is located $40 \text{ ft} + 11 \text{ ft} + 4 \text{ ft} = 55$ ft above the tower foundation.

The wind speed on the bottom blade sections is $100\% - 25\% = 75\%$ of the top blade. Since velocity is squared in the drag force equation, the force on the untapered part of the bottom blade is

$$F_{bottom,untapered} = (0.75)^2 (13.2 \text{ lbf}) = 7.4 \text{ lbf}$$

The resultant of this force is located $40 \text{ ft} + (\frac{1}{2})(11 \text{ ft}) = 34.5$ ft above the tower foundation.

The force on the tapered part of the bottom blade is

$$F_{bottom,tapered} = (0.75)^2 (8.1 \text{ lbf}) = 4.6 \text{ lbf}$$

The resultant of this force is located $40 \text{ ft} - 11 \text{ ft} - 4 \text{ ft} = 25$ ft above the tower foundation.

The moment at the base of the tower is

$$M = \sum Fr$$
$$= (13.2 \text{ lbf})(45.5 \text{ ft}) + (8.1 \text{ lbf})(55 \text{ ft})$$
$$\quad + (7.4 \text{ lbf})(34.5 \text{ ft}) + (4.6 \text{ lbf})(25 \text{ ft})$$
$$= 1416 \text{ ft-lbf} \quad (1400 \text{ ft-lbf})$$

The answer is (D).

4. The energy used by the primary side of the transformer is

$$E = Pt = VI(\text{pf})t = \frac{(120 \text{ V})(1 \text{ mA})(0.8)(18 \text{ hr})}{1000 \dfrac{\text{mA}}{\text{A}}}$$

$$= 1.728 \text{ W-hr}$$

The charger-rectifier efficiency is 65%, so the energy supplied to the battery is

$$E_{supplied} = \eta E = (0.65)(1.728 \text{ W-hr}) = 1.12 \text{ W-hr}$$

The energy storage of the battery is

$$E_{battery} = V_{battery} I_{battery} = (1.2 \text{ V})(0.600 \text{ A-hr})$$
$$= 0.72 \text{ W-hr}$$

The efficiency of the battery during the charge cycle is

$$\eta_{charge} = \frac{E_{battery}}{E_{supplied}} = \frac{0.72 \text{ W-hr}}{1.12 \text{ W-hr}} = 0.64 \quad (64\%)$$

The answer is (B).

5. The total power generated by the methane is

$$P_{methane} = \dot{V}_{methane}(\text{LHV}) = \left(1200 \dfrac{\text{ft}^3}{\text{hr}}\right)\left(913 \dfrac{\text{Btu}}{\text{ft}^3}\right)$$

$$= 1.1 \times 10^6 \text{ Btu/hr}$$

The AC electrical power output from the inverter is

$$P_{inverter,out} = (150 \text{ kW})\left(3412 \dfrac{\text{Btu}}{\text{kW-hr}}\right)$$

$$= 5.12 \times 10^5 \text{ Btu/hr}$$

The DC electrical power input to the inverter is

$$P_{\text{inverter,in}} = \frac{P_{\text{inverter,out}}}{\eta_{\text{inverter}}} = \frac{5.12 \times 10^5 \frac{\text{Btu}}{\text{hr}}}{0.9}$$
$$= 5.69 \times 10^5 \text{ Btu/hr}$$

From a saturated steam table, the enthalpy of the feedwater at 70°F is 38.08 Btu/lbm. [Properties of Saturated Water and Steam (Temperature) - I-P Units]

From a superheated steam table, the enthalpy of saturated steam at 35 psia is 1167.2 Btu/lbm. The power needed to generate steam is [Properties of Superheated Steam - I-P Units]

$$P_{\text{exchanger,out}} = \dot{m}(h_g - h_f)$$
$$= \left(100 \frac{\text{lbm}}{\text{hr}}\right)\left(1167.2 \frac{\text{Btu}}{\text{lbm}} - 38.08 \frac{\text{Btu}}{\text{lbm}}\right)$$
$$= 1.13 \times 10^5 \text{ Btu/hr}$$

The power delivered to the heat exchanger is

$$P_{\text{exhanger,in}} = \frac{P_{\text{exchanger,out}}}{\eta_{\text{exchanger}}} = \frac{1.13 \times 10^5 \frac{\text{Btu}}{\text{hr}}}{0.75}$$
$$= 1.51 \times 10^5 \text{ Btu/hr}$$

The overall efficiency of the fuel cell stack is

$$\eta_{\text{system}} = \frac{P_{\text{inverter,in}} + P_{\text{exchanger,in}}}{P_{\text{methane}}}$$
$$= \frac{5.69 \times 10^5 \frac{\text{Btu}}{\text{hr}} + 1.51 \times 10^5 \frac{\text{Btu}}{\text{hr}}}{1.1 \times 10^6 \frac{\text{Btu}}{\text{hr}}}$$
$$= 0.6545 \quad (65\%)$$

The answer is (C).

6. The force on the car due to rolling friction is

$$F_r = f_r m \times \frac{g}{g_c}$$
$$= (0.01)(4000 \text{ lbm}) \left(\frac{32.2 \frac{\text{ft}}{\text{sec}^2}}{32.2 \frac{\text{lbm-ft}}{\text{lbf-sec}^2}} \right)$$
$$= 40 \text{ lbf}$$

The total resisting force on the car is the drag force plus the force due to rolling friction.

$$F = F_D + F_r = 136 \text{ lbf} + 40 \text{ lbf} = 176 \text{ lbf}$$

The power required to maintain the car's velocity is

$$P_{\text{car}} = Fv = \frac{(176 \text{ lbf})\left(55 \frac{\text{mi}}{\text{hr}}\right)\left(5280 \frac{\text{ft}}{\text{mi}}\right)}{\left(3600 \frac{\text{sec}}{\text{hr}}\right)\left(778 \frac{\text{ft-lbf}}{\text{Btu}}\right)}$$
$$= 18.25 \text{ Btu/sec}$$

The power delivered to the drive train is

$$P_{\text{drive train}} = \frac{P_{\text{car}}}{\eta_{\text{drive train}}} = \frac{18.25 \frac{\text{Btu}}{\text{sec}}}{0.8} = 22.8 \text{ Btu/sec}$$

Energy is released by the hydrogen at a rate of

$$q = \dot{V}(\text{LHV}) = \left(0.2 \frac{\text{ft}^3}{\text{sec}}\right)\left(275 \frac{\text{Btu}}{\text{ft}^3}\right) = 55 \text{ Btu/sec}$$

The efficiency of the fuel cell is

$$\eta = \frac{P_{\text{drive train}}}{q} = \frac{22.8 \frac{\text{Btu}}{\text{sec}}}{55 \frac{\text{Btu}}{\text{sec}}} = 0.414 \quad (41\%)$$

The answer is (D).

7. Air at 70°F and 1 atm has a density, ρ, of approximately 0.075 lbm/ft³. From Eq. 30.6(b), the maximum power density, P_{max}, of the airstream is

$$P_{\text{max}} = \frac{\pi r_{\text{rotor}}^2 \rho v^3}{2g_c}$$
$$= \frac{\pi \left(\frac{30 \text{ ft}}{2}\right)^2 \left(0.075 \frac{\text{lbm}}{\text{ft}^3}\right)\left(20 \frac{\text{mi}}{\text{hr}}\right)^3 \left(5280 \frac{\text{ft}}{\text{mi}}\right)^3}{(2)\left(32.2 \frac{\text{lbm-ft}}{\text{lbf-sec}^2}\right)\left(3600 \frac{\text{sec}}{\text{hr}}\right)^3 \left(550 \frac{\text{ft-lbf}}{\text{hp-sec}}\right)}$$
$$= 37.8 \text{ hp} \quad (38 \text{ hp})$$

The answer is (B).

31 Gas Compression Cycles

Content in blue refers to the *NCEES Handbook*.

PRACTICE PROBLEMS

1. A reciprocating air compressor has a 7% clearance and discharges air at the rate of 48 lbm/min (0.36 kg/s). The polytropic exponent is 1.33. The compression ratio is 4.42. The equation for the volumetric efficiency is

$$\eta_v = 1 - \left(r_p^{1/n} - 1\right)\left(\frac{c}{100\%}\right)$$

The mass of air that is compressed each minute is most nearly

(A) 42 lbm/min (0.32 kg/s)
(B) 51 lbm/min (0.38 kg/s)
(C) 56 lbm/min (0.42 kg/s)
(D) 74 lbm/min (0.55 kg/s)

2. Air at 14.7 psia and 500°F is compressed in a centrifugal compressor to 6 atm. The isentropic efficiency of the compression process is 65%. The compression work is most nearly

(A) 140 Btu/lbm
(B) 190 Btu/lbm
(C) 230 Btu/lbm
(D) 350 Btu/lbm

3. Air at 14.7 psia and 500°F is compressed in a centrifugal compressor to 6 atm. The isentropic efficiency of the compression process is 65%. The final temperature is most nearly

(A) 1550°R
(B) 1650°R
(C) 1750°R
(D) 1850°R

4. Air at 14.7 psia and 500°F is compressed in a centrifugal compressor to 6 atm. The isentropic efficiency of the compression process is 65%. The increase in entropy is most nearly

(A) 0.048 Btu/lbm-°R
(B) 0.099 Btu/lbm-°R
(C) 0.23 Btu/lbm-°R
(D) 0.44 Btu/lbm-°R

5. Air at a pressure of 14.7 psia and 90°F enters a compressor at a flow rate of 300 ft³/min. The compressor discharges air into a water-cooled heat exchanger. The compressed air is stored at 300 psig and 90°F in a 1000 ft³ tank. The tank feeds three air-driven tools with the flow rates and properties shown. The pressures to the air-driven tools are regulated and remain constant at the minimum required operating pressure as the tank pressure changes. Air is an ideal gas, and compressibility effects are to be disregarded.

	tool 1	tool 2	tool 3
flow rate (cfm)	40	15	unknown
flow rate (lbm/min)	unknown	unknown	6
minimum pressure (psig)	90	50	80
temperature (°F)	90	85	80

The total mass flow leaving the system is most nearly

(A) 31 lbm/min
(B) 37 lbm/min
(C) 43 lbm/min
(D) 48 lbm/min

6. Air enters an adiabatic compressor at 250K and 1 atm pressure. The actual work input to the compressor is 400 kJ per kg of air flowing through the compressor. The compressor has an isentropic efficiency of 0.80. The exit temperature for isentropic compression is most nearly

(A) 400K

(B) 520K

(C) 567K

(D) 652K

7. Air enters a compressor at 250K and 1 atm pressure. The air is compressed to 10 atm. The efficiency of the compressor is 1. Assuming an isothermal and reversible process, the amount of work required is most nearly

(A) 40 kJ/kg

(B) 110 kJ/kg

(C) 135 kJ/kg

(D) 165 kJ/kg

8. Air enters a centrifugal compressor at rate of 12 kg/min. The velocity of the air is 12 m/s at the inlet and 90 m/s at the outlet, and the pressure of the air is 1 bar at the inlet and 8 bar at the outlet. The increase in enthalpy of the air when it passes through the compressor is 150 kJ/kg. Assume no heat loss to the surroundings. The motor power required to drive the compressor work is most nearly

(A) −31 kW

(B) 31 kW

(C) 45 kW

(D) 50 kW

SOLUTIONS

1.

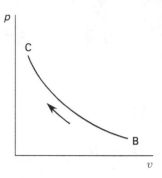

Customary U.S. Solution

The volumetric efficiency is

$$\eta_v = 1 - \left(r_p^{1/n} - 1\right)\left(\frac{c}{100\%}\right)$$
$$= 1 - (4.42^{1/1.33} - 1)\left(\frac{7\%}{100\%}\right)$$
$$= 0.856 \quad (85.6\%)$$

The mass of air compressed per minute is

$$\dot{m} = \frac{\dot{m}_{\text{actual}}}{\eta_v} = \frac{48 \, \frac{\text{lbm}}{\text{min}}}{0.856}$$
$$= 56.07 \text{ lbm/min} \quad (56 \text{ lbm/min})$$

SI Solution

The volumetric efficiency is

$$\eta_v = 1 - \left(r_p^{1/n} - 1\right)\left(\frac{c}{100\%}\right)$$
$$= 1 - (4.46^{1/1.33} - 1)\left(\frac{7\%}{100\%}\right)$$
$$= 0.855 \quad (85.5\%)$$

The mass of air compressed per minute is

$$\dot{m} = \frac{\dot{m}_{\text{actual}}}{\eta_v} = \frac{0.36 \, \frac{\text{kg}}{\text{s}}}{0.855}$$
$$= 0.4211 \text{ kg/s} \quad (0.42 \text{ kg/s})$$

The answer is (C).

2. Although the ideal gas laws can be used, it is more expedient to use air tables for the low pressures.

The absolute inlet temperature is [Temperature Conversions]

$$T_1 = 500°F + 460° = 960°R$$

From air tables, for an absolute temperature of 960°R, $h_1 = 231.06$ Btu/lbm, $p_{r,1} = 10.61$, and $\phi_1 = 0.74030$ Btu/lbm-°R. [Properties of Air at Low Pressure, per Pound]

Since $p_1/p_2 = p_{r,1}/p_{r,2}$ and the compression ratio is 6, for isentropic compression,

$$p_{r,2} = 6p_{r,1} = (6)(10.61) = 63.66$$

From air tables, for $p_{r,2} = 63.66$, $T_1 = 1552°R$ and $h_2 = 382.95$ Btu/lbm. [Properties of Air at Low Pressure, per Pound]

From the definition of isentropic efficiency, the actual enthalpy is

$$h_2' = h_1 + \frac{h_2 - h_1}{\eta_s}$$

$$= 231.06 \, \frac{\text{Btu}}{\text{lbm}} + \frac{382.95 \, \frac{\text{Btu}}{\text{lbm}} - 231.06 \, \frac{\text{Btu}}{\text{lbm}}}{0.65}$$

$$= 464.74 \, \text{Btu/lbm}$$

The compression work is

$$W = h_2' - h_1 = 464.74 \, \frac{\text{Btu}}{\text{lbm}} - 231.06 \, \frac{\text{Btu}}{\text{lbm}}$$

$$= 233.68 \, \text{Btu/lbm} \quad (230 \, \text{Btu/lbm})$$

The answer is (C).

3. Although the ideal gas laws can be used, it is more expedient to use air tables for the low pressures.

The absolute inlet temperature is [Temperature Conversions]

$$T_1 = 500°F + 460° = 960°R$$

From air tables, for an absolute temperature of 960°R, $h_1 = 231.06$ Btu/lbm, $p_{r,1} = 10.61$, and $\phi_1 = 0.74030$ Btu/lbm-°R. [Properties of Air at Low Pressure, per Pound]

Since $p_1/p_2 = p_{r,1}/p_{r,2}$ and the compression ratio is 6, for isentropic compression,

$$p_{r,2} = 6p_{r,1} = (6)(10.61) = 63.66$$

Interpolating from air tables, for $p_{r,2} = 63.66$, $T_1 = 1552°R$ and $h_2 = 382.95$ Btu/lbm. [Properties of Air at Low Pressure, per Pound]

From the definition of isentropic efficiency, the actual enthalpy is

$$h_2' = h_1 + \frac{h_2 - h_1}{\eta_s}$$

$$= 231.06 \, \frac{\text{Btu}}{\text{lbm}} + \frac{382.95 \, \frac{\text{Btu}}{\text{lbm}} - 231.06 \, \frac{\text{Btu}}{\text{lbm}}}{0.65}$$

$$= 464.74 \, \text{Btu/lbm}$$

Interpolating values from air tables, this corresponds to $T_2' = 1855°R$ (1850°R). [Properties of Air at Low Pressure, per Pound]

The answer is (D).

4. *Customary U.S. Solution*

Although the ideal gas laws can be used, it is more expedient to use air tables for the low pressures.

The absolute inlet temperature is [Temperature Conversions]

$$T_1 = 500°F + 460° = 960°R$$

From air tables, for an absolute temperature of 960°R, $h_1 = 231.06$ Btu/lbm, $p_{r,1} = 10.61$, and $\phi_1 = 0.74030$ Btu/lbm-°R. [Properties of Air at Low Pressure, per Pound]

Since $p_1/p_2 = p_{r,1}/p_{r,2}$ and the compression ratio is 6, for isentropic compression,

$$p_{r,2} = 6p_{r,1} = (6)(10.61) = 63.66$$

Interpolating from air tables, for $p_{r,2} = 63.66$, $T_2 = 1552°R$ and $h_2 = 382.95$ Btu/lbm. [Properties of Air at Low Pressure, per Pound]

From the definition of isentropic efficiency, the actual enthalpy is

$$h_2' = h_1 + \frac{h_2 - h_1}{\eta_s}$$

$$= 231.06 \, \frac{\text{Btu}}{\text{lbm}} + \frac{382.95 \, \frac{\text{Btu}}{\text{lbm}} - 231.06 \, \frac{\text{Btu}}{\text{lbm}}}{0.65}$$

$$= 464.74 \, \text{Btu/lbm}$$

Interpolating values from the air tables, this corresponds to $\phi_2' = 0.91129$ Btu/lbm-°R.

The increase in entropy is a function of the difference in entropy due to the temperature change and the difference in entropy due to the pressure change pressure.

For an ideal gas with constant specific heat,

$$\Delta s = (c_p)\ln\left(\frac{T_2}{T_1}\right) - R \ln\left(\frac{p_2}{p_1}\right)$$

Air tables are used for varying specific heat.

$$\Delta s = \phi_2' - \phi_1 - R \ln\left(\frac{p_2}{p_1}\right)$$

$$= 0.91129 \; \frac{\text{Btu}}{\text{lbm-°R}} - 0.74030 \; \frac{\text{Btu}}{\text{lbm-°R}}$$

$$- \left(\frac{53.3 \; \frac{\text{ft-lbf}}{\text{lbm-°R}}}{778 \; \frac{\text{ft-lbf}}{\text{Btu}}}\right) \ln\left(\frac{6 \text{ atm}}{1 \text{ atm}}\right)$$

$$= 0.04824 \text{ Btu/lbm-°R} \quad (0.048 \text{ Btu/lbm-°R})$$

The answer is (A).

5. Assuming that each tool operates at its minimum pressure, the mass leaving the system can be calculated as follows. First, find the mass flow rates for tool 1 and tool 2.

Tool 1:

The absolute pressure is

$$90 \; \frac{\text{lbf}}{\text{in}^2} + 14.7 \; \frac{\text{lbf}}{\text{in}^2} = 104.7 \text{ psia}$$

The absolute temperature is

$$90°\text{F} + 460° = 550°\text{R}$$

The mass flow rate is found from the ideal gas equation by substituting the specific volume with the volumetric flow rate and rearranging.

Ideal Gas

$$pv = mRT$$
$$p\dot{V} = \dot{m}RT$$

$$\dot{m}_{\text{tool 1}} = \frac{p\dot{V}}{RT}$$

$$= \frac{\left(104.7 \; \frac{\text{lbf}}{\text{in}^2}\right)\left(12 \; \frac{\text{in}}{\text{ft}}\right)^2 \left(40 \; \frac{\text{ft}^3}{\text{min}}\right)}{\left(53.3 \; \frac{\text{ft-lbf}}{\text{lbm-°R}}\right)(550°\text{R})}$$

$$= 20.57 \text{ lbm/min}$$

Tool 2:

The absolute pressure is

$$50 \; \frac{\text{lbf}}{\text{in}^2} + 14.7 \; \frac{\text{lbf}}{\text{in}^2} = 64.7 \text{ psia}$$

The absolute temperature is

$$85°\text{F} + 460° = 545°\text{R}$$

The mass flow rate is found from the ideal gas equation, the same as tool 1.

$$\dot{m}_{\text{tool 2}} = \frac{p\dot{V}}{RT}$$

$$= \frac{\left(64.7 \; \frac{\text{lbf}}{\text{in}^2}\right)\left(12 \; \frac{\text{in}}{\text{ft}}\right)^2 \left(15 \; \frac{\text{ft}^3}{\text{min}}\right)}{\left(53.3 \; \frac{\text{ft-lbf}}{\text{lbm-°R}}\right)(545°\text{R})}$$

$$= 4.81 \text{ lbm/min}$$

The total mass flow leaving the system is

$$\dot{m}_{\text{total}} = \dot{m}_{\text{tool 1}} + \dot{m}_{\text{tool 2}} + \dot{m}_{\text{tool 3}}$$

$$= 20.57 \; \frac{\text{lbm}}{\text{min}} + 4.81 \; \frac{\text{lbm}}{\text{min}} + 6 \; \frac{\text{lbm}}{\text{min}}$$

$$= 31.38 \text{ lbm/min} \quad (31 \text{ lbm/min})$$

The answer is (A).

6. From the equation for isentropic efficiency, calculate the work per unit mass for isentropic compression.

Compressors

$$\eta_C = \frac{w_s}{w_a}$$
$$w_s = \eta_C w_a$$
$$w_s = (0.80)\left(400 \; \frac{\text{kJ}}{\text{kg}}\right) = 320 \text{ kJ/kg}$$

Use the equation for compressor work per unit mass to find the exit temperature. Treat the work as positive. For air, $c_p = 1.004$ kJ/kg·K. [Thermal and Physical Properties of Ideal Gases (at Room Temperature)]

Compressors

$$w_{\text{comp}} = c_p(T_e - T_i)_s = w_s = 320 \ \frac{\text{kJ}}{\text{kg}}$$

$$T_e = \frac{w_{\text{comp}}}{c_p} + T_i$$

$$= \frac{320 \ \frac{\text{kJ}}{\text{kg}}}{1.004 \ \frac{\text{kJ}}{\text{kg} \cdot \text{K}}} + 250 \text{K}$$

$$= 568.7 \text{K} \quad (567 \text{K})$$

The answer is (C).

7. Use the equation for fluid or gas power in isothermal compression work and simplify to find the work required. The value of R for air is 0.287 kJ/kg·K. [Thermal and Physical Properties of Ideal Gases (at Room Temperature)]

Compressors

$$\dot{W}_{\text{comp}} = \frac{\overline{R} \, T_i}{M \eta_c} \ln \frac{p_e}{p_i} (\dot{m})$$

$$w_{\text{comp}} = RT \ln \frac{p_e}{p_i}$$

$$= \left(0.287 \ \frac{\text{kJ}}{\text{kg} \cdot \text{K}}\right)(250 \text{K})\left(\ln \frac{10}{1}\right)$$

$$= 165.21 \ \text{kJ/kg} \quad (165 \ \text{kJ/kg})$$

The answer is (D).

8. From the equation for compressor power,

Compressors

$$\dot{W}_{\text{comp}} = -\dot{m}\left(h_e - h_i + \frac{V_e^2 - V_i^2}{2}\right)$$

$$= -\left(\frac{12 \ \frac{\text{kg}}{\text{min}}}{60 \ \frac{\text{s}}{\text{min}}}\right)\left(150 \ \frac{\text{kJ}}{\text{kg}} + \frac{\left(90 \ \frac{\text{m}}{\text{s}}\right)^2 - \left(12 \ \frac{\text{m}}{\text{s}}\right)^2}{(2)\left(1000 \ \frac{\text{J}}{\text{kJ}}\right)}\right)$$

$$= -30.79 \ \text{kW} \quad (-31 \ \text{kW})$$

The answer is (A).

32 Refrigeration Cycles

Content in blue refers to the *NCEES Handbook*.

PRACTICE PROBLEMS

1. A refrigerator uses refrigerant R-12. The input power is 585 W. Heat absorbed from the cooled space is 450 Btu/hr (0.13 kW). The coefficient of performance is most nearly

(A) 0.2
(B) 0.4
(C) 0.7
(D) 0.9

2. An ammonia compressor is used in a heat pump cycle. The suction pressure is 30 psia, and the discharge pressure is 160 psia. The saturated liquid ammonia enters the throttle valve at 160 psia. The net refrigeration effect is 500 Btu/lbm. The coefficient of performance for the heat pump is most nearly

(A) 2
(B) 4
(C) 5
(D) 6

3. A refrigeration cycle uses refrigerant 22. The refrigerant leaves the evaporator as a saturated vapor at $-30°F$. The refrigerant leaves the condenser as a saturated liquid at $70°F$. Assume isentropic compression. The volume flow of refrigerant leaving the evaporator per ton of refrigeration is most nearly

(A) 5.6 ft³/min-ton
(B) 7.3 ft³/min-ton
(C) 8.5 ft³/min-ton
(D) 9.1 ft³/min-ton

4. Which of the following is NOT an advantage of top-feed versus bottom-feed configurations for refrigerant evaporators?

(A) smaller quantity of refrigerant needed
(B) higher defrost rates
(C) simpler design and layout
(D) refrigerant circulation does not need to overcome the static pressure

5. Which of the following refrigerants is likely to be the most energy efficient refrigerant at an evaporator temperature of 20°F and a condenser temperature of 86°F?

(A) ammonia
(B) carbon dioxide
(C) refrigerant 22
(D) refrigerant 134a

6. A refrigeration system uses refrigerant 134a and has a capacity of 30,000 Btu/hr. The evaporator operates at a pressure of 50 psia and consists of type L copper tubing. The appropriate line size of the suction tubing for the compressor is most nearly

(A) $\frac{5}{8}$ in OD
(B) $\frac{7}{8}$ in OD
(C) $1\frac{1}{8}$ in OD
(D) $1\frac{3}{8}$ in OD

7. Which of the following refrigerants has an occupational exposure limit of less than 400 ppm and exhibits flame propagation at 140°F and 14.7 psia?

(A) refrigerant 123
(B) refrigerant 134a
(C) refrigerant 22
(D) refrigerant 717

SOLUTIONS

1. *Customary U.S. Solution*

Use the equation for the coefficient of performance for a refrigerator. [Measurement Relationships]

Basic Cycles

$$\text{COP}_{\text{refrigerator}} = \frac{Q_L}{W}$$

$$= \frac{Q_{\text{in}}}{W_{\text{in}}} = \frac{450 \, \frac{\text{Btu}}{\text{hr}}}{(585 \, \text{W}) \left(3.413 \, \frac{\frac{\text{Btu}}{\text{hr}}}{\text{W}} \right)}$$

$$= 0.2254 \quad (0.2)$$

SI Solution

Use the equation for the coefficient of performance for a refrigerator.

Basic Cycles

$$\text{COP}_{\text{refrigerator}} = \frac{Q_L}{W}$$

$$= \frac{Q_{\text{in}}}{W_{\text{in}}} = \frac{(0.13 \, \text{kW}) \left(1000 \, \frac{\text{W}}{\text{kW}} \right)}{585 \, \text{W}}$$

$$= 0.2222 \quad (0.2)$$

The answer is (A).

2. The description of the state points in the refrigeration cycle are as follows:

- state 1: vapor entering the compressor at 30 psia
- state 2: superheated vapor leaving the compressor and entering the condenser at 160 psia
- state 3: saturated liquid leaving the condenser and entering the throttling valve at 160 psia
- state 4: liquid–vapor mixture leaving the throttling valve and entering the evaporator at 30 psia

Determine the enthalpies using a pressure versus enthalpy diagram for refrigerant 717 (ammonia). [Pressure Versus Enthalpy Curves for Refrigerant 717 (Ammonia)]

Since ammonia entering the throttling valve at state 3 (160 psia) is a saturated liquid,

$$h_3 = h_{\text{sat,liq}} = 135 \, \text{Btu/lbm}$$

Throttling is a constant enthalpy process. Therefore,

$$h_4 = h_3 = 135 \, \text{Btu/lbm}$$

Find the enthalpy at state 2 using the net refrigeration effect. [Refrigeration Cycle - Single Stage]

$$h_1 - h_4 = 500 \, \text{Btu/lbm}$$

$$h_1 = 500 \, \frac{\text{Btu}}{\text{lbm}} + h_4$$

$$= 500 \, \frac{\text{Btu}}{\text{lbm}} + 135 \, \frac{\text{Btu}}{\text{lbm}}$$

$$= 635 \, \text{Btu/lbm}$$

Determine the entropy at state 1 at the intersection of h_1 (635 Btu/lbm) and p_1 (30 psia).

$$s_1 = 1.38 \, \text{Btu/lbm-°R}$$

Since the compression from state 1 to state 2 is isentropic, move along the line of constant entropy to the discharge pressure (160 psia). The intersection point represents state 2.

$$h_2 = 750 \, \text{Btu/lbm}$$

Use the formula for the coefficient of performance (COP) for a heat pump. [Refrigeration Cycle - Single Stage]

$$\text{COP}_{\text{heat pump}} = \frac{h_2 - h_3}{h_2 - h_1} = \frac{750 \, \frac{\text{Btu}}{\text{lbm}} - 135 \, \frac{\text{Btu}}{\text{lbm}}}{750 \, \frac{\text{Btu}}{\text{lbm}} - 635 \, \frac{\text{Btu}}{\text{lbm}}}$$

$$= 5.35 \quad (5)$$

The answer is (C).

3. Determine the saturation pressures for refrigerant 22 at –30°F and 70°F from a table of properties for refrigerant 22. [Refrigerant 22 (Chlorodifluoromethane) Properties of Saturated Liquid and Saturated Vapor]

$$\text{evaporator pressure} = p_{\text{sat},-30°\text{F}} = 19.624 \, \text{psia}$$

$$\text{condenser pressure} = p_{\text{sat},70°\text{F}} = 136.130 \, \text{psia}$$

The description of the state points in the cycle are as follows:

- state 1: saturated vapor leaving the evaporator and entering the compressor at -30°F (19.624 psia)
- state 2: superheated vapor leaving the compressor and entering the condenser at 136.130 psia

- state 3: saturated liquid leaving the condenser and entering the throttling valve at 136.130 psia
- state 4: liquid–vapor mixture leaving the throttling valve and entering the evaporator at 19.624 psia

Determine the enthalpies from the pressure versus enthalpy diagram and the tables for refrigerant 22. [Refrigerant 22 (Chlorodifluoromethane) Properties of Saturated Liquid and Saturated Vapor] [Pressure Versus Enthalpy Curves for Refrigerant 22]

The refrigerant leaving the evaporator at state 1 (19.624 psia) is a saturated vapor.

$$h_1 = h_{\text{sat,vap}} = 101.44 \text{ Btu/lbm}$$

The refrigerant leaving the condenser at state 3 (136.130 psia) is a saturated liquid.

$$h_3 = h_{\text{sat,liq}} = 30.35 \text{ Btu/lbm}$$

Throttling is a constant enthalpy process. Therefore,

$$h_4 = h_3 = 30.35 \text{ Btu/lbm}$$

A refrigeration ton is equal to 12,000 BTU/hr.

Calculate the mass flow rate of the refrigerant per ton of refrigeration. [Refrigeration Cycle - Single Stage] [Measurement Relationships]

$$q_{4-1} = \dot{m}(h_1 - h_4)$$
$$\dot{m} = \frac{q_{4-1}}{(h_1 - h_4)}$$
$$= \frac{12{,}000 \dfrac{\text{Btu}}{\text{hr-ton}}}{\left(101.44 \dfrac{\text{Btu}}{\text{lbm}} - 30.35 \dfrac{\text{Btu}}{\text{lbm}}\right)}$$
$$= 168.80 \text{ lbm/hr-ton}$$

Find the specific volume at state 1 using a table of properties for refrigerant 22. [Refrigerant 22 (Chlorodifluoromethane) Properties of Saturated Liquid and Saturated Vapor]

At –30°F (19.624 psia), the specific volume of the saturated vapor is 2.5984 ft³/lbm.

Therefore, the volumetric flow rate of the refrigerant at the evaporator exit is

$$Q_1 = v_1 \dot{m} = \frac{\left(2.5984 \dfrac{\text{ft}^3}{\text{lbm}}\right)\left(168.80 \dfrac{\text{lbm}}{\text{hr-ton}}\right)}{60 \dfrac{\text{min}}{\text{hr}}}$$
$$= 7.310 \text{ ft}^3/\text{min–ton} \quad (7.3 \text{ ft}^3/\text{min-ton})$$

The answer is (B).

4. Typically, the charge of the refrigerant is small in top-feed evaporators. The rate of frost formation is also smaller in top-feed evaporators. Bottom-feed systems have to overcome the static pressure during circulation, which is not the case in a top-feed system. However, the layout and design of top-feed systems tend to be more complex. [Refrigeration Evaporator: Top-Feed Versus Bottom-Feed]

The answer is (C).

5. The refrigerant with the highest coefficient of performance (COP) will require the least amount of compressor power, W, for a fixed amount of refrigeration.

$$\text{COP}_{\text{ref}} = \frac{Q_L}{W} = \frac{Q_{\text{in}}}{W_c}$$

Use a table of comparative performances of different refrigerants at an evaporator temperature of 20°F and a condenser temperature of 86°F. [Comparative Refrigerant Performance per Ton of Refrigeration]

The COP of the four refrigerants in the answer options are shown.

refrigerant	COP
ammonia	6.254
carbon dioxide	3.514
refrigerant 22	5.799
refrigerant 134a	6.063

Since ammonia has the highest COP, it is the most energy efficient refrigerant.

The answer is (A).

6. Calculate the capacity in tons of refrigeration. [Measurement Relationships]

$$\frac{30{,}000 \dfrac{\text{Btu}}{\text{hr}}}{12{,}000 \dfrac{\text{Btu}}{\text{hr} \cdot \text{ton}}} = 2.5 \text{ tons}$$

The tubing from the evaporator is the suction line for the compressor. In the table of suction line capacities for refrigerant 134a, the capacities are based on the saturation temperature at suction. Determine the saturation temperature at 50 psia from the R-134a property tables. [Refrigerant 134a (1,1,1,2-Tetrafluoroethane) Properties of Saturated Liquid and Saturated Vapor]

$$T_{\text{sat}} = 40°F \quad [\text{at } 49.741 \text{ psia } (\approx 50 \text{ psia})]$$

From the table of suction line capacities for refrigerant 134a, for a saturated suction temperature of 40°F and a capacity 3.54 tons (which is greater than the required 2.5 tons), the appropriate size of the copper tubing from the evaporator is $1\frac{1}{8}$ in OD. [Suction Line Capacities in Tons for Refrigerant 134a (Single- or High-Stage Applications)]

The answer is (C).

7. Use a table of refrigerant data and safety classifications to determine the correct answer. [Refrigerant Data and Safety Classifications]

Among the answer options, refrigerant 717 (ammonia) is the only refrigerant with a safety classification of B2. The letter "B" denotes a toxicity class, and the numeral "2" denotes a flammability class. Class B refrigerants have an occupational exposure limit of less than 400 ppm. Class 2 refrigerants exhibit a flame propagation at 140°F and 14.7 psia.

The answer is (D).

33 Fundamental Heat Transfer

Content in blue refers to the NCEES Handbook.

PRACTICE PROBLEMS

1. Experiments have shown that the thermal conductivity, k, of a particular material varies with temperature, T, according to the following relationship.

$$k_T = (0.030)(1 + 0.0015\,T)$$

A material has a hot-side temperature of 350° (°F or °C) and a cold-side temperature of 150° (°F or °C). What is most nearly the value of k that should be used for a transfer of heat through the material?

(A) 0.04

(B) 0.06

(C) 0.10

(D) 0.20

2. The thermal gradient is 350°F (177°C). What is most nearly the heat flow through insulating brick 1.0 ft (30 cm) thick in an oven wall with a thermal conductivity of 0.038 Btu-ft/hr-ft²-°F (0.066 W/m·K)?

(A) 9 Btu/hr-ft² (27 W/m²)

(B) 13 Btu/hr-ft² (39 W/m²)

(C) 21 Btu/hr-ft² (53 W/m²)

(D) 28 Btu/hr-ft² (69 W/m²)

3. A composite wall is made up of 3.0 in (7.6 cm) of material A exposed to 1000°F (540°C), 5.0 in (13 cm) of material B, and 6.0 in (15 cm) of material C exposed to 200°F (90°C), as shown. Material A has a mean thermal conductivity of 0.06 Btu-ft/hr-ft²-°F (0.1 W/m·K), material B has a mean thermal conductivity of 0.5 Btu-ft/hr-ft²-°F (0.9 W/m·K), and material C has a mean thermal conductivity of 0.8 Btu-ft/hr-ft²-°F (1.4 W/m·K).

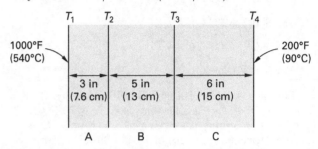

The equation for the heat flow through a composite wall is

$$q = \frac{A(T_1 - T_4)}{\sum_{i=1}^{n} \frac{L_i}{k_i}}$$

The temperature at the A-B material interface is most nearly

(A) 300°F (100°C)

(B) 350°F (150°C)

(C) 400°F (200°C)

(D) 500°F (250°C)

4. A composite wall is made up of 3.0 in (7.6 cm) of material A exposed to 1000°F (540°C), 5.0 in (13 cm) of material B, and 6.0 in (15 cm) of material C exposed to 200°F (90°C). Material A has a mean thermal conductivity of 0.06 Btu-ft/hr-ft²-°F (0.1 W/m·K), material B has a mean thermal conductivity of 0.5 Btu-ft/hr-ft²-°F (0.9 W/m·K), and material C has a mean thermal conductivity of 0.8 Btu-ft/hr-ft²-°F (1.4 W/m·K).

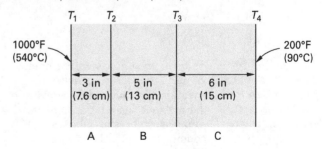

The equation for the heat flow through a composite wall is

$$q = \frac{A(T_1 - T_4)}{\sum_{i=1}^{n} \frac{L_i}{k_i}}$$

The temperature at the B-C material interface is most nearly

(A) 300°F (150°C)

(B) 350°F (175°C)

(C) 400°F (200°C)

(D) 500°F (250°C)

5. The heat supply of a large building is turned off at 5:00 p.m. when the interior temperature is 70°F (21°C). The outdoor temperature remains constant at 40°F (4°C). The thermal capacity of the building and its contents is 100,000 Btu/°F (60 MJ/K), and the conductance is 6500 Btu/hr-°F (1.1 kW/K). Assuming the lumped-capacitance system is valid, the interior temperature at 1:00 a.m. is most nearly

(A) 45°F (7°C)

(B) 50°F (10°C)

(C) 60°F (15°C)

(D) 65°F (20°C)

Hint: β is equal to the inverse of thermal resistance times thermal conductance.

6. Two long pieces of $1/16$ in (1.6 mm) copper wire are connected end-to-end with a hot soldering iron. The minimum melting temperature of the solder is 450°F (230°C). The surrounding air temperature is 80°F (27°C). The unit film coefficient is 3 Btu/hr-ft^2-°F (17 W/m^2·K). The value of k at 450°F is approximately 215 Btu-ft/hr-ft^2-°F (372.1 W/m·K). Radiation losses can be disregarded. The minimum rate of heat application to keep the solder molten is most nearly

(A) 4.3 Btu/hr (1.3 W)

(B) 11 Btu/hr (3.3 W)

(C) 20 Btu/hr (6.0 W)

(D) 47 Btu/hr (14 W)

7. A piping system is made up of pipes with a nominal size of 1 in and 2 in of insulation. Water is left standing in the pipes during winter. The initial temperature of the water is 42°F, the ambient air temperature in the basement where the piping system is installed is −18°F, and the thermal conductivity of the insulation is 0.30 Btu-in/(hr-ft^2-F). The amount of time it will take for the water in the piping system to freeze is most nearly

(A) 0.1 hr

(B) 0.4 hr

(C) 0.6 hr

(D) 1.3 hr

8. During preventative maintenance of a bare 3 in copper piping system, a 2 ft section of lagging is removed. The inside temperature of the pipe is 200°F, and the surrounding air temperature is 80°F. The approximate heat rate loss without the lagging is

(A) 71 Btu/hr

(B) 200 Btu/hr

(C) 400 Btu/hr

(D) 520 Btu/hr

9. A 2 ft long cylindrical fin with a diameter of 0.5 in is attached at one end to a heat source. The temperature of the heat source is 350°F, and the temperature of the ambient air is 75°F. The thermal conductivity of the fin is 128 Btu/hr-ft-°F, and the average film coefficient along the length of the fin is 1.3 Btu/hr-ft^2-°F. No heat is transferred from the exposed fin tip. Heat transfer from the heat source into the base of the fin is most nearly

(A) 22 Btu/hr

(B) 29 Btu/hr

(C) 36 Btu/hr

(D) 46 Btu/hr

10. A 2 ft long cylindrical fin with a diameter of 0.5 in is attached at one end to a heat source. The temperature of the heat source is 350°F, and the temperature of the ambient air is 75°F. The average film coefficient along the length of the fin is 1.3 Btu/hr-ft^2-°F. No heat is transferred from the exposed fin tip. The actual rate of heat transfer from the fin is 45.6 Btu/hr. Use Newton's law of cooling to determine the ideal rate of heat transfer. The efficiency of the fin is

(A) 20%

(B) 35%

(C) 50%

(D) 65%

SOLUTIONS

1. Use the value of k at an average temperature of $\frac{1}{2}(T_1 + T_2)$.

$$T = \left(\frac{1}{2}\right)(150° + 350°) = 250°$$
$$k = (0.030)(1 + 0.0015\,T)$$
$$= (0.030)\bigl(1 + (0.0015)(250°)\bigr)$$
$$= 0.04125 \quad (0.04)$$

The answer is (A).

2. *Customary U.S. Solution*

Use Fourier's law of heat conduction.

$$\text{Fourier's Law of Conduction}$$
$$q = -kA\frac{dT}{dx}$$

$$\frac{q_{1-2}}{A} = \frac{k\Delta T}{L}$$
$$= \frac{\left(0.038\,\dfrac{\text{Btu-ft}}{\text{hr-ft}^2\text{-°F}}\right)(350°\text{F})}{1.0\text{ ft}}$$
$$= 13.3\text{ Btu/hr-ft}^2 \quad (13\text{ Btu/hr-ft}^2)$$

SI Solution

Use Fourier's law of heat conduction.

$$\text{Fourier's Law of Conduction}$$
$$q = -kA\frac{dT}{dx}$$

The temperature difference across the wall will be the same in normal and absolute units. Since thermal difference is given, applying an addition or subtraction conversion factor (such as from Celsius to Kelvin) will yield the same answer.

$$\Delta T°\,\text{C} = \Delta T\,\text{K} = 177\text{ K}$$

$$\frac{q_{1-2}}{A} = \frac{k\Delta T}{x}$$
$$= \frac{\left(0.066\,\dfrac{\text{W}}{\text{m·K}}\right)(177\text{ K})\left(100\,\dfrac{\text{cm}}{\text{m}}\right)}{30\text{ cm}}$$
$$= 38.9\text{ W/m}^2 \quad (39\text{ W/m}^2)$$

The answer is (B).

3. *Customary U.S. Solution*

Since the wall temperatures are given, it is not necessary to consider films.

Rearranging the equation for the heat flow on a per unit area basis,

$$q = \frac{A(T_1 - T_4)}{\sum_{i=1}^{n}\dfrac{L_i}{k_i}}$$

$$\frac{q}{A} = \frac{T_1 - T_4}{\sum_{i=1}^{n}\dfrac{L_i}{k_i}} = \frac{T_1 - T_2}{\dfrac{L_A}{k_A}} = \frac{T_2 - T_3}{\dfrac{L_B}{k_B}}$$

$$= \frac{1000°\text{F} - 200°\text{F}}{\dfrac{3\text{ in}}{\left(0.06\,\dfrac{\text{Btu-ft}}{\text{hr-ft}^2\text{-°F}}\right)\left(12\,\dfrac{\text{in}}{\text{ft}}\right)}}$$
$$+ \dfrac{5\text{ in}}{\left(0.5\,\dfrac{\text{Btu-ft}}{\text{hr-ft}^2\text{-°F}}\right)\left(12\,\dfrac{\text{in}}{\text{ft}}\right)}$$
$$+ \dfrac{6\text{ in}}{\left(0.8\,\dfrac{\text{Btu-ft}}{\text{hr-ft}^2\text{-°F}}\right)\left(12\,\dfrac{\text{in}}{\text{ft}}\right)}$$
$$= 142.2\text{ Btu/hr-ft}^2$$

Rearrange the equation to find the temperature at the A-B interface, T_2.

$$\frac{q}{A} = \frac{T_1 - T_2}{\dfrac{L_A}{k_A}}$$

$$T_2 = T_1 - \left(\frac{q}{A}\right)\left(\frac{L_A}{k_A}\right)$$
$$= 1000°\text{F} - \left(142.2\,\dfrac{\text{Btu}}{\text{hr-ft}^2}\right)$$
$$\times \left(\dfrac{3\text{ in}}{\left(0.06\,\dfrac{\text{Btu-ft}}{\text{hr-ft}^2\text{-°F}}\right)\left(12\,\dfrac{\text{in}}{\text{ft}}\right)}\right)$$
$$= 407.5°\text{F} \quad (400°\text{F})$$

SI Solution

Since the wall temperatures are given, it is not necessary to consider films.

Rearranging the equation for the heat flow on a per unit area basis,

$$q = \frac{A(T_1 - T_4)}{\sum_{i=1}^{n} \frac{L_i}{k_i}}$$

$$\frac{q}{A} = \frac{T_1 - T_4}{\sum_{i=1}^{n} \frac{L_i}{k_i}} = \frac{T_1 - T_2}{\frac{L_A}{k_A}} = \frac{T_2 - T_3}{\frac{L_B}{k_B}}$$

$$= \frac{540°C - 90°C}{\frac{7.6 \text{ cm}}{\left(0.1 \frac{W}{m \cdot K}\right)\left(100 \frac{cm}{m}\right)} + \frac{13 \text{ cm}}{\left(0.9 \frac{W}{m \cdot K}\right)\left(100 \frac{cm}{m}\right)} + \frac{15 \text{ cm}}{\left(1.4 \frac{W}{m \cdot K}\right)\left(100 \frac{cm}{m}\right)}}$$

$$= 444.8 \text{ W/m}^2$$

Rearrange the equation to find the temperature at the A-B interface, T_2.

$$\frac{q}{A} = \frac{T_1 - T_2}{\frac{L_A}{k_A}}$$

$$T_2 = T_1 - \left(\frac{q}{A}\right)\left(\frac{L_A}{k_A}\right)$$

$$= 540°C - \left(444.8 \frac{W}{m^2}\right)\left(\frac{7.6 \text{ cm}}{\left(0.1 \frac{W}{m \cdot K}\right)\left(100 \frac{cm}{m}\right)}\right)$$

$$= 202.0°C \quad (200°C)$$

The answer is (C).

4. *Customary U.S. Solution*

Since the wall temperatures are given, it is not necessary to consider films.

Rearranging the equation for the heat flow on a per unit area basis,

$$q = \frac{A(T_1 - T_4)}{\sum_{i=1}^{n} \frac{L_i}{k_i}}$$

$$\frac{q}{A} = \frac{T_1 - T_4}{\sum_{i=1}^{n} \frac{L_i}{k_i}} = \frac{T_1 - T_2}{\frac{L_A}{k_A}} = \frac{T_2 - T_3}{\frac{L_B}{k_B}}$$

$$= \frac{1000°F - 200°F}{\frac{3 \text{ in}}{\left(0.06 \frac{\text{Btu-ft}}{\text{hr-ft}^2\text{-°F}}\right)\left(12 \frac{\text{in}}{\text{ft}}\right)} + \frac{5 \text{ in}}{\left(0.5 \frac{\text{Btu-ft}}{\text{hr-ft}^2\text{-°F}}\right)\left(12 \frac{\text{in}}{\text{ft}}\right)} + \frac{6 \text{ in}}{\left(0.8 \frac{\text{Btu-ft}}{\text{hr-ft}^2\text{-°F}}\right)\left(12 \frac{\text{in}}{\text{ft}}\right)}}$$

$$= 142.2 \text{ Btu/hr-ft}^2$$

Rearrange the equation to find the temperature at the A-B interface, T_2.

$$\frac{q}{A} = \frac{T_1 - T_2}{\frac{L_A}{k_A}}$$

$$T_2 = T_1 - \left(\frac{q}{A}\right)\left(\frac{L_A}{k_A}\right)$$

$$= 1000°F - \left(142.2 \frac{\text{Btu}}{\text{hr-ft}^2}\right)$$

$$\times \left(\frac{3 \text{ in}}{\left(0.06 \frac{\text{Btu-ft}}{\text{hr-ft}^2\text{-°F}}\right)\left(12 \frac{\text{in}}{\text{ft}}\right)}\right)$$

$$= 407.5°F$$

To find the temperature at the B-C interface, T_3, use

$$\frac{q}{A} = \frac{T_2 - T_3}{\frac{L_B}{k_B}}$$

$$T_3 = T_2 - \left(\frac{q}{A}\right)\left(\frac{L_B}{k_B}\right)$$

$$= 407.5°F - \left(142.2 \frac{\text{Btu}}{\text{hr-ft}^2}\right)$$

$$\times \left(\frac{5 \text{ in}}{\left(0.5 \frac{\text{Btu-ft}}{\text{hr-ft}^2\text{-°F}}\right)\left(12 \frac{\text{in}}{\text{ft}}\right)}\right)$$

$$= 289°F \quad (300°F)$$

SI Solution

Since the wall temperatures are given, it is not necessary to consider films.

Rearranging the equation for the heat flow on a per unit area basis,

$$q = \frac{A(T_1 - T_4)}{\sum_{i=1}^{n} \frac{L_i}{k_i}}$$

$$\frac{q}{A} = \frac{T_1 - T_4}{\sum_{i=1}^{n} \frac{L_i}{k_i}} = \frac{T_1 - T_2}{\frac{L_A}{k_A}} = \frac{T_2 - T_3}{\frac{L_B}{k_B}}$$

$$= \frac{540°C - 90°C}{\frac{7.6 \text{ cm}}{\left(0.1 \frac{\text{W}}{\text{m·K}}\right)\left(100 \frac{\text{cm}}{\text{m}}\right)} + \frac{13 \text{ cm}}{\left(0.9 \frac{\text{W}}{\text{m·K}}\right)\left(100 \frac{\text{cm}}{\text{m}}\right)}}$$

$$+ \frac{15 \text{ cm}}{\left(1.4 \frac{\text{W}}{\text{m·K}}\right)\left(100 \frac{\text{cm}}{\text{m}}\right)}$$

$$= 444.8 \text{ W/m}^2$$

Rearrange the equation to find the temperature at the A-B interface, T_2.

$$\frac{q}{A} = \frac{T_1 - T_2}{\frac{L_A}{k_A}}$$

$$T_2 = T_1 - \left(\frac{q}{A}\right)\left(\frac{L_A}{k_A}\right)$$

$$= 540°C - \left(444.8 \frac{\text{W}}{\text{m}^2}\right)\left(\frac{7.6 \text{ cm}}{\left(0.1 \frac{\text{W}}{\text{m·K}}\right)\left(100 \frac{\text{cm}}{\text{m}}\right)}\right)$$

$$= 202.0°C$$

To find the temperature at the B-C interface, T_3, use

$$\frac{q}{A} = \frac{T_2 - T_3}{\frac{L_B}{k_B}}$$

$$T_3 = T_2 - \left(\frac{q}{A}\right)\left(\frac{L_B}{k_B}\right)$$

$$= 202.0°C - \left(444.8 \frac{\text{W}}{\text{m}^2}\right)\left(\frac{13 \text{ cm}}{\left(0.9 \frac{\text{W}}{\text{m·K}}\right)\left(100 \frac{\text{cm}}{\text{m}}\right)}\right)$$

$$= 137.8°C \quad (150°C)$$

The answer is (A).

5. *Customary U.S. Solution*

This is a transient problem. The total time is from 5 p.m. to 1 a.m., which is eight hours.

Resistance and conductance are reciprocals. [Composite Plane Wall]

The thermal resistance is

$$R_e = \frac{1}{C} = \frac{1}{6500 \frac{\text{Btu}}{\text{hr-°F}}}$$

$$= 0.0001538 \text{ hr-°F/Btu}$$

From the equation for the temperature variation of the body with time, the interior temperature is

Constant Fluid Temperature
$$T - T_\infty = (T_i - T_\infty)e^{-\beta t}$$
$$T_t = T_\infty + (T_0 - T_\infty)e^{-t/R_e C_e}$$
$$T_{8\,\text{hr}} = 40°\text{F} + (70°\text{F} - 40°\text{F})$$
$$\times \exp\left(\frac{-8\,\text{hr}}{\left(0.0001538\,\frac{\text{hr-°F}}{\text{Btu}}\right)\left(100{,}000\,\frac{\text{Btu}}{°\text{F}}\right)}\right)$$
$$= 57.8°\text{F} \quad (60°\text{F})$$

SI Solution

The thermal capacitance is

$$C_e = \left(60\,\frac{\text{MJ}}{\text{K}}\right)\left(1000\,\frac{\text{kJ}}{\text{MJ}}\right) = 60\,000\,\text{kJ/K} \quad \text{[given]}$$

Resistance and conductance are reciprocals. The thermal resistance is

$$R_e = \frac{1}{\text{thermal conductance}} = \frac{1}{1.1\,\frac{\text{kW}}{\text{K}}} = 0.909\,\text{K/kW}$$

From the equation for the temperature variation of the body with time, the interior temperature is

Constant Fluid Temperature
$$T - T_\infty = (T_i - T_\infty)e^{-\beta t}$$
$$T_t = T_\infty + (T_0 - T_\infty)e^{-t/R_e C_e}$$
$$T_{8\,\text{hr}} = 4°\text{C} + (21°\text{C} - 4°\text{C})$$
$$\times \exp\left(\frac{(-8\,\text{h})\left(3600\,\frac{\text{s}}{\text{h}}\right)}{\left(0.909\,\frac{\text{K}}{\text{kW}}\right)\left(60\,000\,\frac{\text{kJ}}{\text{K}}\right)}\right)$$
$$= 14.0°\text{C} \quad (15°\text{C})$$

The answer is (C).

6. *Customary U.S. Solution*

Consider this to be two infinite cylindrical fins with

$$T_b = 450°\text{F}$$
$$T_\infty = 80°\text{F}$$
$$h = 3\,\text{Btu/hr-ft}^2\text{-°F}$$

The perimeter length of a pin fin is

Fins
$$P = \pi D = \frac{\pi\left(\frac{1}{16}\,\text{in}\right)}{12\,\frac{\text{in}}{\text{ft}}} = 0.01636\,\text{ft}$$

The cross-sectional area of the fin at its base is

Fins
$$A_c = \frac{\pi D^2}{4} = \frac{\pi\left(\frac{1}{16}\,\text{in}\right)^2}{(4)\left(12\,\frac{\text{in}}{\text{ft}}\right)^2}$$
$$= 2.131 \times 10^{-5}\,\text{ft}^2$$

The equation for the minimum rate of heat application to keep the solder molten for two fins joined at the middle is

Fins
$$q = \sqrt{hPkA_c}\,(T_b - T_\infty)\tanh mL_c$$

Because the fins are being treated as infinite, the tanh term can be disregarded. The minimum rate of heat application is

$$q = 2\sqrt{hPkA_c}\,(T_b - T_\infty)$$
$$= (2)\sqrt{\begin{array}{c}\left(3\,\frac{\text{Btu}}{\text{hr-ft}^2\text{-°F}}\right)(0.01636\,\text{ft})\\\times\left(215\,\frac{\text{Btu-ft}}{\text{hr-ft}^2\text{-°F}}\right)(2.131\times 10^{-5}\,\text{ft}^2)\end{array}}$$
$$\times (450°\text{F} - 80°\text{F})$$
$$= 11.1\,\text{Btu/hr} \quad (11\,\text{Btu/hr})$$

SI Solution

Consider this an infinite cylindrical fin with

$$T_b = 230°\text{C}$$
$$T_\infty = 27°\text{C}$$
$$h = 17\,\text{W/m}^2\cdot\text{K}$$

The perimeter length of a pin fin is

Fins
$$P = \pi D = \frac{\pi(1.6\,\text{mm})}{1000\,\frac{\text{mm}}{\text{m}}} = 5.027 \times 10^{-3}\,\text{m}$$

The cross-sectional area of the fin at its base is

$$A_c = \frac{\pi D^2}{4} = \frac{\left(\frac{\pi}{4}\right)(1.6 \text{ mm})^2}{\left(1000 \frac{\text{mm}}{\text{m}}\right)^2}$$

$$= 2.011 \times 10^{-6} \text{ m}^2$$

The equation for the minimum rate of heat application to keep the solder molten for two fins joined at the middle is

Fins

$$q = \sqrt{hPkA_c}\,(T_b - T_\infty)\tanh mL_c$$

Because the fins are being treated as infinite, the tanh term can be disregarded. The minimum rate of heat application is

$$q = 2\sqrt{hPkA_c}\,(T_b - T_\infty)$$

$$= (2)\sqrt{\begin{array}{c}\left(17\,\dfrac{\text{W}}{\text{m}^2\cdot\text{K}}\right)(5.027 \times 10^{-3} \text{ m}) \\ \times \left(372.1\,\dfrac{\text{W}}{\text{m}\cdot\text{K}}\right)(2.011 \times 10^{-6} \text{ m}^2)\end{array}}$$

$$\times (230°\text{C} - 27°\text{C})$$

$$= 3.25 \text{ W} \quad (3.3 \text{ W})$$

The answer is (B).

7. Use a table of freezing times for water to find the answer. For pipes with a nominal pipe size of 1 in and 2 in of insulation in the conditions described, it would take 0.6 hr for water to freeze in the piping. [Time Needed to Freeze Water, in Hours]

The answer is (C).

8. Interpolating from a table of heat loss in bare copper tubing, the approximate heat loss for a bare 3 in copper pipe with an inside temperature of 200°F is 198 Btu/hr-ft. [Heat Loss from Bare Copper Tubing in Still Air at 80°F]

For 2 ft of bare copper piping, the heat rate loss is

$$q = \left(198\,\frac{\text{Btu}}{\text{hr-ft}}\right)(2 \text{ ft}) = 396 \text{ Btu/hr} \quad (400 \text{ Btu/hr})$$

The answer is (C).

9. Use the equation for heat transfer from a fin with negligible heat transfer from the tip.

Fins

$$q = \sqrt{hPkA_c}\,(T_b - T_\infty)\tanh(mL_c)$$

Calculate the perimeter of the cylindrical fin.

$$P = \pi D = \frac{\pi(0.5 \text{ in})}{12\,\dfrac{\text{in}}{\text{ft}}}$$

$$= 0.13 \text{ ft}$$

Calculate the cross-sectional area of the fin.

Fins

$$A_c = \frac{\pi D^2}{4}$$

$$= \frac{\pi(0.5 \text{ in})^2}{(4)\left(12\,\dfrac{\text{in}}{\text{ft}}\right)^2}$$

$$= 0.0014 \text{ ft}^2$$

Evaluate the parameter, m.

Fins

$$m = \sqrt{\frac{hP}{kA_c}} = \sqrt{\frac{h(\pi D)}{k\left(\dfrac{\pi D^2}{4}\right)}}$$

$$= \sqrt{\frac{4h}{kD}}$$

$$= \sqrt{\frac{(4)\left(1.3\,\dfrac{\text{Btu}}{\text{hr-ft}^2\text{-}°\text{F}}\right)}{\left(128\,\dfrac{\text{Btu}}{\text{hr-ft-}°\text{F}}\right)(0.5 \text{ in})\left(\dfrac{1}{12}\,\dfrac{\text{ft}}{\text{in}}\right)}}$$

$$= 0.98 \text{ ft}^{-1}$$

Calculate the corrected length.

Fins

$$L_c = L + \frac{A_c}{P} = L + \frac{\left(\dfrac{\pi D^2}{4}\right)}{\pi D}$$

$$= L + \frac{D}{4}$$

$$= 2 \text{ ft} + \frac{(0.5 \text{ in})\left(\dfrac{1}{12}\,\dfrac{\text{ft}}{\text{in}}\right)}{4}$$

$$= 2.010 \text{ ft}$$

The heat transfer from the fin is

$$q = \sqrt{hPkA_c}\,(T_b - T_\infty)\tanh mL_c$$

Fins

$$= \left(\sqrt{\left(1.3\ \frac{\text{Btu}}{\text{hr-ft}^2\text{-}°\text{F}}\right)(0.13\ \text{ft}) \times \left(128\ \frac{\text{Btu}}{\text{hr-ft-}°\text{F}}\right)(0.0014\ \text{ft}^2)}\right.$$
$$\left. \times \left((350°\text{F} - 75°\text{F})\tanh\left(\left(0.98\ \frac{1}{\text{ft}}\right)(2.010\ \text{ft})\right)\right)\right)$$
$$= 45.6\ \text{Btu/hr} \quad (46\ \text{Btu/hr})$$

The answer is (D).

10. The ideal heat transfer from the fin is

Newton's Law of Cooling

$$q_{\text{ideal}} = hA_{s,\text{fin}}(T_b - T_\infty)$$

$$= \left(1.3\ \frac{\text{Btu}}{\text{hr-ft}^2\text{-}°\text{F}}\right)\pi\left(\frac{0.5\ \text{in}}{12\ \frac{\text{in}}{\text{ft}}}\right)(2\ \text{ft})$$
$$\times (350°\text{F} - 75°\text{F})$$
$$= 93.6\ \text{Btu/hr}$$

The fin efficiency is

$$\eta_f = \left(\frac{q_{\text{actual}}}{q_{\text{ideal}}}\right) \times 100\% = \left(\frac{45.6\ \frac{\text{Btu}}{\text{hr}}}{93.6\ \frac{\text{Btu}}{\text{hr}}}\right) \times 100\%$$
$$= 48.72\% \quad (50\%)$$

The answer is (C).

34 Natural Convection, Evaporation, and Condensation

Content in blue refers to the NCEES Handbook.

PRACTICE PROBLEMS

1. The viscosity of 100°F (38°C) water in units of lbm/hr-ft (kg/s·m) is most nearly

(A) 1.2 lbm/ft-hr (0.00052 kg/m·s)
(B) 1.4 lbm/ft-hr (0.00060 kg/m·s)
(C) 1.6 lbm/ft-hr (0.00068 kg/m·s)
(D) 1.8 lbm/ft-hr (0.00077 kg/m·s)

2. A fluid in a tank is maintained at 85°F (29°C) by a copper tube carrying hot water. The water decreases in temperature from 190°F (88°C) to 160°F (71°C) as it flows through the tube. The temperature at which the film coefficient of the fluid should be evaluated is most nearly

(A) 130°F (54°C)
(B) 160°F (71°C)
(C) 175°F (79°C)
(D) 190°F (88°C)

3. A pot of water sits on a 4 in diameter hot stove burner. The temperature at the bottom of the pot is the same as the water, except for a 4 in diameter circle that has a temperature of 320°F. After the water temperature has risen to 80°F, the convection heat transfer coefficient is 438 Btu/hr-ft²-°F. What is most nearly the heat transfer rate?

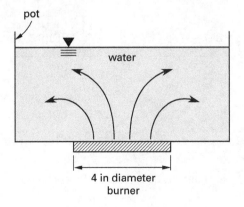

(A) 8500 Btu/hr
(B) 9200 Btu/hr
(C) 9400 Btu/hr
(D) 11,000 Btu/hr

4. A 320°F vertical electrical heating plate with a diameter of 4 in was placed in the center of a large pot of 70°F water. The thermal conductivity is 0.394 Btu-ft/hr-ft²-°F. At 200°F, the coefficient of thermal expansion is 4.0×10^{-4} 1/°F, and the Prandtl number is 1.88.

The water temperature in the pot rises to 80°F. The heat transfer rate is most nearly

(A) 5200 Btu/hr
(B) 6500 Btu/hr
(C) 6800 Btu/hr
(D) 9000 Btu/hr

5. A long, horizontal steel rod with an outside diameter of 1 in is heated to 300°F. The heated rod is placed in a large, stagnant pool of 60°F water. The water properties at the film temperature (180°F) are

thermal conductivity, k	0.388 Btu-ft/hr-ft²-°F
coefficient of thermal expansion, β	2.05×10^{-3} 1/°F
Prandtl number, Pr	2.04

The convective heat transfer coefficient is most nearly

(A) 480 Btu/hr-ft²-°F
(B) 540 Btu/hr-ft²-°F
(C) 620 Btu/hr-ft²-°F
(D) 710 Btu/hr-ft²-°F

SOLUTIONS

1. *Customary U.S. Solution*

Find the viscosity of water at 100°F using a table of properties of water. [Properties of Water (I-P Units)] Use the lbf to lbm conversion factor.

$$\mu = \left(1.424 \times 10^{-5} \frac{\text{lbf-sec}}{\text{ft}^2}\right)\left(32.17 \frac{\text{lbm-ft}}{\text{lbf-sec}^2}\right)\left(3600 \frac{\text{sec}}{\text{hr}}\right)$$
$$= 1.649 \text{ lbm/ft-hr} \quad (1.6 \text{ lbm/ft-hr})$$

SI Solution

Interpolate the value for the viscosity of water at 38°C using a table of properties for water. [Properties of Water (SI Units)]

$$\mu = 0.682 \times 10^{-3} \text{ kg/m·s} \quad (0.00068 \text{ kg/m·s})$$

The answer is (C).

2. *Customary U.S. Solution*

The midpoint tube temperature is

$$T_s = \left(\tfrac{1}{2}\right)(190°F + 160°F) = 175°F$$
$$T_\infty = 85°F \quad \text{[given]}$$

The film coefficient should be evaluated at a temperature of

Film Temperature of a Tube

$$T_{\text{film}} = \tfrac{1}{2}(T_s + T_\infty)$$
$$= \left(\tfrac{1}{2}\right)(175°F + 85°F)$$
$$= 130°F$$

SI Solution

The midpoint tube temperature is

$$T_s = \left(\tfrac{1}{2}\right)(88°C + 71°C) = 79.5°C$$
$$T_\infty = 29°C \quad \text{[given]}$$

The film coefficient should be evaluated at a temperature of

Film Temperature of a Tube

$$T_{\text{film}} = \tfrac{1}{2}(T_s + T_\infty)$$
$$= \left(\tfrac{1}{2}\right)(79.5°C + 29°C)$$
$$= 54.3°C \quad (54°C)$$

The answer is (A).

3. The heat transfer rate is [Newton's Law of Cooling]

$$\dot{Q} = hA(T_W - T_\infty)$$
$$= \frac{\left(438 \frac{\text{Btu}}{\text{hr-ft}^2\text{-°F}}\right)\left(\pi\left(\frac{4 \text{ in}}{2}\right)^2\right)(320°F - 80°F)}{\left(12 \frac{\text{in}}{\text{ft}}\right)^2}$$
$$= \boxed{9164 \text{ Btu/hr} \quad (9200 \text{ Btu/hr})}$$

The answer is (B).

4. The characteristic length is

$$L = \frac{4 \text{ in}}{12 \frac{\text{in}}{\text{ft}}}$$
$$= 0.33 \text{ ft}$$

The temperature gradient is

$$T_s - T_\infty = 320°F - 80°F$$
$$= 240°F$$

The film temperature is

Film Temperature of a Tube

$$T_{\text{film}} = \tfrac{1}{2}(T_s + T_\infty) = \left(\tfrac{1}{2}\right)(320°F + 80°F)$$
$$= 200°F$$

Using a table of properties of water, find the kinematic viscosity at 200°F. [Properties of Water (I-P Units)]

$$\nu = 0.341 \times 10^{-5} \text{ ft}^2/\text{sec}$$

The Rayleigh number is

Vertical Flat Plate in Large Body of Stationary Fluid

$$\text{Ra}_L = \frac{g\beta(T_s - T_\infty)L^3}{\nu^2}\text{Pr}$$
$$= \left(\frac{\left(32.17 \frac{\text{ft}}{\text{sec}^2}\right)\left(4.0 \times 10^{-4} \frac{1}{°F}\right)(240°F)(0.33 \text{ ft})^3}{\left(0.341 \times 10^{-5} \frac{\text{ft}^2}{\text{sec}}\right)^2}\right)(1.88)$$
$$= 1.79 \times 10^{10}$$

Since Ra_L is between $10^9 - 10^{13}$, $C = 0.10$. [Vertical Flat Plate in Large Body of Stationary Fluid]

The average convection heat transfer coefficient is

Vertical Flat Plate in Large Body of Stationary Fluid

$$\overline{h} = C\left(\frac{k}{L}\right)Ra_L^n$$

$$= (0.10)\left(\frac{0.394\ \frac{\text{Btu-ft}}{\text{hr-ft}^2\text{-°F}}}{0.33\ \text{ft}}\right)(1.79 \times 10^{10})^{1/3}$$

$$= 312.3\ \text{Btu/hr-ft}^2\text{-°F}$$

Find the heat transfer rate using Newton's law of cooling.

Newton's Law of Cooling

$$q = hA(T_s - T_\infty)$$

$$= \frac{\left(312.3\ \frac{\text{Btu}}{\text{hr-ft}^2\text{-°F}}\right)\pi\left(\frac{4\ \text{in}}{2}\right)^2(320°\text{F} - 80°\text{F})}{\left(12\ \frac{\text{in}}{\text{ft}}\right)^2}$$

$$= 6541\ \text{Btu/hr}\quad(6500\ \text{Btu/hr})$$

The answer is (B).

5. Using a table of properties for water, find the kinematic viscosity at 180°F. [Properties of Water (I-P Units)]

$$\nu = 0.385 \times 10^{-5}\ \text{ft}^2/\text{sec}$$

The Rayleigh number is

Long Horizontal Cylinder in Large Body of Stationary Fluid

$$Ra_D = \frac{g\beta(T_s - T_\infty)D^3}{\nu^2}Pr$$

$$= \frac{\left(32.17\ \frac{\text{ft}}{\text{sec}^2}\right)\left(2.05 \times 10^{-3}\ \frac{1}{°\text{F}}\right)}{\left(0.385 \times 10^{-5}\ \frac{\text{ft}^2}{\text{sec}}\right)^2}(2.04)$$

$$\times (300°\text{F} - 60°\text{F})(1\ \text{in})^3\left(\frac{1}{12}\ \frac{\text{ft}}{\text{in}}\right)^3$$

$$= 1.262 \times 10^9$$

Since Ra_D is between $10^7 - 10^{12}$, $C = 0.125$. [Long Horizontal Cylinder in Large Body of Stationary Fluid]

The convective heat transfer coefficient is

Long Horizontal Cylinder in Large Body of Stationary Fluid

$$\overline{h} = C\left(\frac{k}{D}\right)Ra_D^n$$

$$= (0.125)\left(\frac{0.388\ \frac{\text{Btu-ft}}{\text{hr-ft}^2\text{-°F}}}{(1\ \text{in})\left(\frac{1}{12}\ \frac{\text{ft}}{\text{in}}\right)}\right)(1.262 \times 10^9)^{0.333}$$

$$= 624.23\ \text{Btu/hr-ft}^2\text{-°F}\quad(620\ \text{Btu/hr-ft}^2\text{-°F})$$

The answer is (C).

35 Forced Convection and Heat Exchangers

Content in blue refers to the *NCEES Handbook*.

PRACTICE PROBLEMS

1. A U-tube surface feedwater heater with one shell pass and two tube passes is being designed to heat 500,000 lbm/hr (60 kg/s) of water from 200°F to 390°F (100°C to 200°C). The water flows at 5 ft/sec (1.5 m/s) through the tubes. The tubes in the heater are 7/8 in (2.2 cm) outside diameter with 1/16 in (1.6 mm) walls. The number of tubes needed is most nearly

(A) 80
(B) 110
(C) 140
(D) 160

2. A U-tube surface feedwater heater with one shell pass and two tube passes is being designed to heat 500,000 lbm/hr (60 kg/s) of water from 200°F to 390°F (100°C to 200°C). Dry, saturated steam at 400°F (205°C) is to be used as the heating medium. The heater is to operate straight condensing (i.e., the condensed steam will not be mixed with the heated water). Saturated water at 400°F (205°C) is removed from the heater. The overall heat transfer coefficient is estimated as 700 Btu/hr-ft²-°F (4 kW/m²·°C). The surface area of the tubes needed in the feedwater heater is most nearly

(A) 2200 ft² (200 m²)
(B) 2400 ft² (220 m²)
(C) 2600 ft² (240 m²)
(D) 2800 ft² (260 m²)

3. A single-pass heat exchanger is tested in a clean condition and is found to heat 100 gal/min (6.3 L/s) of water from 70°F (21°C) to 140°F (60°C). The hot side uses condensing steam at 230°F (110°C), which enters and leaves the exchanger at saturated conditions. The heat transfer surface area of the heat exchanger is 50 ft² (4.7 m²). After being used in the field for several months, the exit temperature of water drops to 122°F (50°C) due to fouling. The flow rate and the inlet temperature of water remains the same as before. The overall heat transfer coefficient under fouled conditions is most nearly

(A) 350 Btu/hr-ft²-°F (1900 W/m²·°C)
(B) 390 Btu/hr-ft²-°F (2200 W/m²·°C)
(C) 420 Btu/hr-ft²-°F (2400 W/m²·°C)
(D) 450 Btu/hr-ft²-°F (2600 W/m²·°C)

4. Cold water is used to cool hot water in a single-pass, tube-in-tube, counterflow heat exchanger. Compared to a plot of temperature versus length for the cold water, the plot for temperature versus length for the hot water is

(A) identical
(B) reversed end-to-end
(C) shifted (delayed)
(D) inverted top-to-bottom

5. Cold water is used to condense saturated steam in a single-pass, tube-in-tube, counterflow heat exchanger without subcooling. The temperature of the steam along the length of the heat exchanger from entrance to exit

(A) decreases linearly
(B) decreases parabolically
(C) decreases logarithmically
(D) remains constant

6. A single-pass, counter flow heat exchanger is designed to heat 60 gal/min (0.0038 m³/s) of city water from 60°F (16°C) to 130°F (55°C) using 300 gal/min (0.0189 m³/s) of hot water. The hot water enters the shell at 190°F (88°C) and exits at 176°F (80°C). The heat transfer surface area under design (fouled) conditions is 88.08 ft² (8.33 m²). The overall heat transfer coefficient under clean conditions is 636.54 Btu/hr-ft²-°F (3040 W/m²·°C). Assume that the overall heat transfer coefficient is not affected by temperature changes and that the inlet temperatures of the fluids and the heat transfer area remain the same whether under design (fouled) conditions or clean conditions. Under clean conditions, the outlet temperature of the water being heated is most nearly

(A) 139°F (59°C)

(B) 147°F (64°C)

(C) 164°F (73°C)

(D) 178°F (81°C)

SOLUTIONS

1. *Customary U.S. Solution*

The flow rate of water needs to be accommodated by the number of tubes needed. Use the continuity equation for calculating the mass flow rate through each tube.

Continuity Equation
$$\dot{m} = \rho Q = \rho A \text{v}$$

Find the inside diameter of the tubes.

$$D_i = D_o - 2t_w = \frac{7}{8} \text{ in} - (2)\left(\frac{1}{16} \text{ in}\right)$$
$$= 0.75 \text{ in}$$

Find the inside cross-sectional area of the tubes.

$$A_i = \frac{\pi D_i^2}{4} = \left(\frac{\pi}{4}\right)\left(\frac{0.75 \text{ in}}{12 \frac{\text{in}}{\text{ft}}}\right)^2$$
$$= 0.003068 \text{ ft}^2$$

Determine the specific volume of water at the average bulk temperature of 295°F (the highest temperature, which will require the most tubes) from steam tables. [Properties of Saturated Water and Steam (Temperature) - I-P Units]

$$v_{f,295°F} = 0.0174 \text{ ft}^3/\text{lbm}$$

The density of water at 295°F is

$$\rho = \frac{1}{v_{f,295°F}} = \frac{1}{0.0174 \frac{\text{ft}^3}{\text{lbm}}}$$
$$= 57.47 \text{ lbm/ft}^3$$

The mass flow rate of the water through the heater can be expressed as

$$\dot{m}_w = n\dot{m}_{\text{tube}} = n(\rho A_i \text{v})$$

Therefore, the number of tubes needed is

$$n = \frac{\dot{m}_w}{\rho A_i \mathrm{v}}$$

$$= \frac{500{,}000 \ \frac{\mathrm{lbm}}{\mathrm{hr}}}{\left(57.47 \ \frac{\mathrm{lbm}}{\mathrm{ft}^3}\right)(0.003068 \ \mathrm{ft}^2)\left(5 \ \frac{\mathrm{ft}}{\mathrm{sec}}\right)\left(3600 \ \frac{\mathrm{sec}}{\mathrm{hr}}\right)}$$

$$= 157.5 \quad (160)$$

SI Solution

The flow rate of water needs to be accommodated by the number of tubes needed. Use the continuity equation for calculating the mass flow rate through each tube.

Continuity Equation
$$\dot{m} = \rho Q = \rho A \mathrm{v}$$

Find the inside diameter of the tubes.

$$D_i = D_o - 2t_w = 22 \ \mathrm{mm} - (2)(1.6 \ \mathrm{mm})$$
$$= 18.80 \ \mathrm{mm}$$

Find the inside cross-sectional area of the tubes.

$$A_i = \frac{\pi D_i^2}{4} = \left(\frac{\pi}{4}\right)\left(\frac{18.80 \ \mathrm{mm}}{1000 \ \frac{\mathrm{mm}}{\mathrm{m}}}\right)^2$$

$$= 2.776 \times 10^{-4} \ \mathrm{m}^2$$

Determine the specific volume of water at the average bull temperature of 150°C from steam tables. [Properties of Saturated Water and Steam (Temperature) - SI Units]

$$v_{f,150°C} = 0.00109 \ \mathrm{m}^3/\mathrm{kg}$$

The density of water at 150°C is

$$\rho = \frac{1}{v_{f,150°C}} = \frac{1}{0.00109 \ \frac{\mathrm{m}^3}{\mathrm{kg}}}$$

$$= 917 \ \mathrm{kg/m}^3$$

The mass flow rate of the water through the heater can be expressed as

$$\dot{m}_w = n\dot{m}_{tube} = n(\rho A_i \mathrm{v})$$

Therefore, the number of tubes needed is

$$n = \frac{\dot{m}_w}{\rho A_i \mathrm{v}}$$

$$= \frac{60 \ \frac{\mathrm{kg}}{\mathrm{s}}}{\left(917 \ \frac{\mathrm{kg}}{\mathrm{m}^3}\right)(2.776 \times 10^{-4} \ \mathrm{m}^2)\left(1.5 \ \frac{\mathrm{m}}{\mathrm{s}}\right)}$$

$$= 157 \quad (160)$$

The answer is (D).

2. *Customary U.S. Solution*

Calculate the heat duty of the feedwater heater by using the heat balance equation.

Special Cases of the Steady-Flow Energy Equation
$$q = \dot{m}_1(h_{1,i} - h_{1,e}) = \dot{m}_2(h_{2,e} - h_{2,i})$$

In this case, the feedwater is the cold fluid. Therefore,

$$q = \dot{m}_w(h_{w,e} - h_{w,i})$$

Find the enthalpies of water from the steam tables. [Properties of Saturated Water and Steam (Temperature) - I-P Units]

$$h_{w,e} = h_{f,390°F} = 364.26 \ \mathrm{Btu/lbm}$$
$$h_{w,i} = h_{f,200°F} = 168.13 \ \mathrm{Btu/lbm}$$

From the heat balance equation,

$$q = \dot{m}_w(h_{w,e} - h_{w,i})$$
$$= \left(500{,}000 \ \frac{\mathrm{lbm}}{\mathrm{hr}}\right)\left(364.26 \ \frac{\mathrm{Btu}}{\mathrm{lbm}} - 168.13 \ \frac{\mathrm{Btu}}{\mathrm{lbm}}\right)$$
$$= 9.809 \times 10^7 \ \mathrm{Btu/hr}$$

Calculate the surface area of the tubes needed in the feedwater heater by using the heat exchanger design equation.

Rate of Heat Transfer
$$q = UA_s F \Delta T_{lm}$$

Therefore,

$$A_s = \frac{q}{UF\Delta T_{lm}}$$

For condensing steam, the temperature remains constant. Hence, the log mean temperature difference (LMTD) will be the same for both parallel and counter-flow. Therefore, the LMTD correction factor $F = 1.0$.

Calculate the log mean temperature difference for counterflow.

Log Mean Temperature Difference (LMTD)

$$\Delta T_{lm} = \frac{(T_{Ho} - T_{Ci}) - (T_{Hi} - T_{Co})}{2.3 \log_{10}\left(\dfrac{T_{Ho} - T_{Ci}}{T_{Hi} - T_{Co}}\right)}$$

$$= \frac{(400°F - 200°F) - (400°F - 390°F)}{2.3 \log_{10}\left(\dfrac{400°F - 200°F}{400°F - 390°F}\right)}$$

$$= 63.42°F$$

The surface area of the tubes needed in the feedwater heater is

$$A_s = \frac{q}{UF\Delta T_{lm}} = \frac{9.809 \times 10^7 \,\dfrac{\text{Btu}}{\text{hr}}}{\left(700 \,\dfrac{\text{Btu}}{\text{hr-ft}^2\text{-}°F}\right)(1)(63.42°F)}$$

$$= 2209.53 \text{ ft}^2 \quad (2200 \text{ ft}^2)$$

SI Solution

Calculate the heat duty of the feedwater heater by using the heat balance equation.

Special Cases of the Steady-Flow Energy Equation

$$q = \dot{m}_1(h_{1,i} - h_{1,e}) = \dot{m}_2(h_{2,e} - h_{2,i})$$

In this case, the feedwater is the cold fluid. Therefore,

$$q = \dot{m}_w(h_{w,e} - h_{w,i})$$

Find the enthalpies of water from the steam tables. [Properties of Saturated Water and Steam (Temperature) - SI Units]

$$h_{w,e} = h_{f,200°C} = 852.27 \text{ kJ/kg}$$
$$h_{w,i} = h_{f,100°C} = 419.17 \text{ kJ/kg}$$

From the heat balance equation,

$$q = \dot{m}_w(h_{w,e} - h_{w,i})$$

$$= \left(60 \,\frac{\text{kg}}{\text{s}}\right)\left(852.27 \,\frac{\text{kJ}}{\text{kg}} - 419.17 \,\frac{\text{kJ}}{\text{kg}}\right)$$

$$= 25\,986 \text{ kW}$$

Calculate the surface area of the tubes required for the feedwater heater by using the heat exchanger design equation.

Rate of Heat Transfer

$$q = UA_s F \Delta T_{lm}$$

Therefore,

$$A_s = \frac{q}{UF\Delta T_{lm}}$$

For condensing steam, the temperature remains constant. Hence, the log mean temperature difference (LMTD) will be the same for both parallel and counterflow. Therefore, the LMTD correction factor $F = 1.0$.

Calculate the log mean temperature difference for counterflow.

Log Mean Temperature Difference (LMTD)

$$\Delta T_{lm} = \frac{(T_{Ho} - T_{Ci}) - (T_{Hi} - T_{Co})}{2.3 \log_{10}\left(\dfrac{T_{Ho} - T_{Ci}}{T_{Hi} - T_{Co}}\right)}$$

$$= \frac{(205°C - 100°C) - (205°C - 200°C)}{2.3 \log_{10}\left(\dfrac{205°C - 100°C}{205°C - 200°C}\right)}$$

$$= 32.8°C$$

The surface area of the tubes needed in the feedwater heater is

$$A_s = \frac{q}{UF\Delta T_{lm}} = \frac{25\,986 \text{ kW}}{\left(4 \,\dfrac{\text{kW}}{\text{m}^2\text{·°C}}\right)(1)(32.8°C)}$$

$$= 198.06 \text{ m}^2 \quad (200 \text{ m}^2)$$

The answer is (A).

3. *Customary U.S. Solution*

Find the enthalpies of water under fouled conditions using steam tables. [Properties of Saturated Water and Steam (Temperature) - I-P Units]

$$h_{w,e} = h_{f,122°F} = 90.00 \text{ Btu/lbm}$$
$$h_{w,i} = h_{f,70°F} = 38.08 \text{ Btu/lbm}$$

Calculate the heat duty of the exchanger under fouled conditions after converting the volume flow rate of water to mass flow rate. [Measurement Relationships]

$$q_{fouled} = \dot{m}_w(h_{w,e} - h_{w,i})_{fouled} = Q_w \rho_w (h_{w,e} - h_{w,i})_{fouled}$$

$$= \left(100 \,\frac{\text{gal}}{\text{min}}\right)(60 \text{ min/hr})\left(8.34 \,\frac{\text{lbm}}{\text{gal}}\right)\left(90.0 \,\frac{\text{Btu}}{\text{lbm}} - 38.08 \,\frac{\text{Btu}}{\text{lbm}}\right)$$

$$= 2.598 \times 10^6 \text{ Btu / hr}$$

Calculate the log mean temperature difference (LMTD) under fouled conditions for a counterflow heat exchanger.

FORCED CONVECTION AND HEAT EXCHANGERS

Log Mean Temperature Difference (LMTD)

$$\Delta T_{\text{lm, fouled}} = \frac{(T_{\text{Ho}} - T_{\text{Ci}}) - (T_{\text{Hi}} - T_{\text{Co}})}{2.3\log_{10}\left(\dfrac{T_{\text{Ho}} - T_{\text{Ci}}}{T_{\text{Hi}} - T_{\text{Co}}}\right)}$$

$$= \frac{(230°F - 70°F) - (230°F - 122°F)}{2.3\log_{10}\left(\dfrac{230°F - 70°F}{230°F - 122°F}\right)}$$

$$= 132.3°F$$

For condensing steam, the temperature remains constant. Hence, the log mean temperature difference (LMTD) will be the same for both parallel and counterflow. Therefore, the LMTD correction factor is $F = 1.0$.

Calculate the overall heat transfer coefficient under fouled conditions by using the heat exchanger design equation.

Rate of Heat Transfer

$$q_{\text{fouled}} = UAF\Delta T_{\text{lm}}$$

$$U_{\text{fouled}} = \frac{q_{\text{fouled}}}{AF\Delta T_{\text{lm,fouled}}}$$

$$= \frac{2.598 \times 10^6 \,\dfrac{\text{Btu}}{\text{hr}}}{(50\text{ ft}^2)(1.0)(132.3°F)}$$

$$= 392.7\text{ Btu/hr-ft}^2\text{-}°F \quad (390\text{ Btu/hr-ft}^2\text{-}°F)$$

SI Solution

Find the enthalpies of water under fouled conditions using steam tables. [Properties of Saturated Water and Steam (Temperature) - SI Units]

$$h_{w,e} = h_{f,50°C} = 209.34\text{ kJ/kg}$$
$$h_{w,i} = h_{f,21°C} = 88.10\text{ kJ/kg}$$

The standard density of water is 1 kg / L. Therefore the mass flow rate of water is

$$\dot{m}_w = Q_w \rho_w = \left(6.30\,\frac{\text{L}}{\text{s}}\right)\left(1\,\frac{\text{kg}}{\text{L}}\right) = 6.30\text{ kg / s}$$

The heat duty of the exchanger under fouled conditions is

$$q_{\text{fouled}} = \dot{m}_w(h_{w,e} - h_{w,i})_{\text{fouled}}$$

$$= \left(6.30\,\frac{\text{kg}}{\text{s}}\right)\left(209.34\,\frac{\text{kJ}}{\text{kg}} - 88.10\,\frac{\text{kJ}}{\text{kg}}\right)\left(1000\,\frac{\text{J}}{\text{kJ}}\right)$$

$$= 7.638 \times 10^5\text{ W}$$

Calculate the log mean temperature difference (LMTD) under fouled conditions for a counterflow heat exchanger.

Log Mean Temperature Difference (LMTD)

$$\Delta T_{\text{lm, fouled}} = \frac{(T_{\text{Ho}} - T_{\text{Ci}}) - (T_{\text{Hi}} - T_{\text{Co}})}{2.3\log_{10}\left(\dfrac{T_{\text{Ho}} - T_{\text{Ci}}}{T_{\text{Hi}} - T_{\text{Co}}}\right)}$$

$$= \frac{(110°C - 21°C) - (110°C - 50°C)}{2.3\log_{10}\left(\dfrac{110°C - 21°C}{110°C - 50°C}\right)}$$

$$= 73.55°C$$

For condensing steam, the temperature remains constant. Hence, the log mean temperature difference (LMTD) will be the same for both parallel and counterflow. Therefore, the LMTD correction factor $F = 1.0$.

Calculate the overall heat transfer coefficient under fouled conditions by using the heat exchanger design equation.

Rate of Heat Transfer

$$q_{\text{fouled}} = UAF\Delta T_{\text{lm}}$$

$$U_{\text{fouled}} = \frac{q_{\text{fouled}}}{AF(\Delta T_{\text{lm}})_{\text{fouled}}}$$

$$= \frac{7.638 \times 10^5\text{ W}}{(4.7\text{ m}^2)(1)(73.55°C)}$$

$$= 2209.6\text{ W/m}^2\cdot°C \quad (2200\text{ W/m}^2\cdot°C)$$

The answer is (B).

4. As one moves along the length of the exchanger, the temperature of the hot water keeps decreasing due to heat transferred to the cold water. Since the fluid flow directions are counter to each other, moving along from the opposite end of the exchanger, the temperature of the cold water keeps increasing due to heat gained from the hot water. The heat transfer rate will be the same at any point in the heat exchanger. The plots for temperature versus length for the two fluids will be reversed end-to-end as shown in the figure.

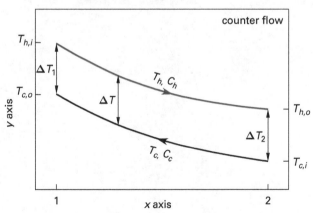

(It could be argued that from the standpoint of the thermodynamic sign convention, hot water has a negative heat transfer rate while cold water has a positive heat transfer rate. In that case, the temperature profiles would be reversed end-to-end and inverted top-to-bottom. However, that is not one of the options.)

The answer is (B).

5. Condensation of steam (or any vapor) is an isothermal (constant temperature) process involving phase change. There is no subcooling below the saturation temperature. Hence, the temperature of steam will be constant and will be equal to the saturation temperature along the length of the heat exchanger.

The answer is (D).

6. *Customary U.S. Solution*

The average temperature of the hot water is

$$T_{h,\text{avg}} = \frac{T_{\text{hi}} + T_{\text{ho}}}{2} = \frac{190°F + 176°F}{2} = 183°F$$

The average temperature of the cold water is

$$T_{c,\text{avg}} = \frac{T_{\text{co}} + T_{\text{ci}}}{2} = \frac{130°F + 60°F}{2} = 95°F$$

Using steam tables, find the specific volumes of the cold and hot water at their average temperatures. [Properties of Saturated Water and Steam (Temperature) - I-P Units]

$$v_h = v_{f,183°F} = 0.0165 \text{ ft}^3/\text{lbm}$$
$$v_c = v_{f,95°F} = 0.0161 \text{ ft}^3/\text{lbm}$$

Calculate the mass flow rates of the hot and cold water. [Measurement Relationships]

Continuity Equation

$$\dot{m}_h = \rho_h Q_h$$
$$= \frac{Q_h}{v_h}$$
$$= \frac{\left(300 \dfrac{\text{gal}}{\text{min}}\right)\left(0.134 \dfrac{\text{ft}^3}{\text{gal}}\right)\left(60 \dfrac{\text{min}}{\text{hr}}\right)}{0.0165 \dfrac{\text{ft}^3}{\text{lbm}}}$$
$$= 146{,}182 \text{ lbm/hr}$$

$$\dot{m}_c = \frac{Q_c}{v_c}$$
$$= \frac{\left(60 \dfrac{\text{gal}}{\text{min}}\right)\left(0.134 \dfrac{\text{ft}^3}{\text{gal}}\right)\left(60 \dfrac{\text{min}}{\text{hr}}\right)}{0.0161 \dfrac{\text{ft}^3}{\text{lbm}}}$$
$$= 29{,}963 \text{ lbm/hr}$$

The specific heat of water at standard conditions is 1 Btu/lbm-°F. [Properties of Water at Standard Conditions]

Find the heat capacity rate of the hot and cold water.

Heat Exchanger Effectiveness, ε

$$C = \dot{m}c_p$$

$$C_h = \dot{m}_h c_{p,h} = \left(146{,}182 \dfrac{\text{lbm}}{\text{hr}}\right)\left(1 \dfrac{\text{Btu}}{\text{lbm-°F}}\right) = 146{,}182 \text{ Btu/hr-°F}$$

$$C_c = \dot{m}_c c_{p,c} = \left(29{,}963 \dfrac{\text{lbm}}{\text{hr}}\right)\left(1 \dfrac{\text{Btu}}{\text{lbm-°F}}\right) = 29{,}963 \text{ Btu/hr-°F}$$

Therefore,

$$C_{\min} = C_c = 29{,}963 \text{ Btu/hr-°F}$$
$$C_{\max} = C_h = 146{,}182 \text{ Btu/hr-°F}$$

Calculate the heat capacity ratio.

Effectiveness-NTU Relations

$$C_r = \frac{C_{\min}}{C_{\max}} = \frac{29{,}963 \dfrac{\text{Btu}}{\text{hr-°F}}}{146{,}182 \dfrac{\text{Btu}}{\text{hr-°F}}} = 0.2050$$

Calculate the number of transfer units (NTU) under clean conditions.

Number of Exchanger Transfer Units (NTU)

$$\text{NTU} = \frac{A_s U_{\text{avg}}}{C_{\min}} = \frac{A_s U_{\text{clean}}}{C_{\min}}$$

$$= \frac{(88.08 \text{ ft}^2)\left(636.54 \dfrac{\text{Btu}}{\text{hr-ft}^2\text{-°F}}\right)}{29{,}963 \dfrac{\text{Btu}}{\text{hr-°F}}}$$

$$= 1.871$$

Calculate the heat exchanger effectiveness under clean conditions.

Effectiveness-NTU Relations

$$\epsilon = \frac{1 - \exp\bigl(-\text{NTU}(1 - C_r)\bigr)}{1 - C_r \exp\bigl(-\text{NTU}(1 - C_r)\bigr)}$$

$$= \frac{1 - e^{(-1.871(1-0.2050))}}{1 - 0.2050 \times e^{(-1.871(1-0.2050))}}$$

$$= 0.81164$$

Calculate the exit temperature of the cold water using the equation for heat exchanger effectiveness. In this case, $C_c = C_{\min}$. Therefore, the exit temperature of cold water under clean conditions is

Heat Exchanger Effectiveness, ε

$$\epsilon = \frac{C_c(T_{\text{co}} - T_{\text{ci}})}{C_{\min}(T_{\text{hi}} - T_{\text{ci}})}$$

$$T_{\text{co}} = T_{\text{ci}} + \epsilon(T_{\text{hi}} - T_{\text{ci}})$$

$$= 60°\text{F} + (0.81164)(190°\text{F} - 60°\text{F})$$

$$= 165.51°\text{F} \quad (164°\text{F})$$

SI Solution

Calculate the average temperature of the hot water.

$$T_{h,\text{avg}} = \frac{T_{\text{hi}} + T_{\text{ho}}}{2} = \frac{88°\text{C} + 80°\text{C}}{2}$$

$$= 84°\text{C}$$

The average temperature of the cold water is

$$T_{c,\text{avg}} = \frac{T_{\text{co}} + T_{\text{ci}}}{2} = \frac{16°\text{C} + 55°\text{C}}{2}$$

$$= 35.5°\text{C}$$

Using steam tables, find the specific volumes of the cold and hot water at their average temperatures. [Properties of Saturated Water and Steam (Temperature) - SI Units]

$$v_h = v_{f,84°\text{C}} = 0.001 \text{ m}^3/\text{kg}$$

$$v_c = v_{f,35.5°\text{C}} = 0.001 \text{ m}^3/\text{kg}$$

Calculate the mass flow rates of the hot and cold water. [Measurement Relationships]

Continuity Equation

$$\dot{m} = \rho Q$$

$$\dot{m}_h = \frac{Q_h}{v_h} = \frac{\left(0.0189 \dfrac{\text{m}^3}{\text{s}}\right)}{0.001 \dfrac{\text{m}^3}{\text{kg}}}$$

$$= 18.9 \text{ kg/s}$$

$$\dot{m}_c = \frac{Q_c}{v_c} = \frac{\left(0.0038 \dfrac{\text{m}^3}{\text{s}}\right)}{0.001 \dfrac{\text{m}^3}{\text{kg}}}$$

$$= 3.8 \text{ kg/s}$$

The specific heat of water at standard conditions is 4.180 kJ/kg·K (or °C). [Properties of Water at Standard Conditions]

Find the heat capacity rate of the hot and cold water.

Heat Exchanger Effectiveness, ε

$$C = \dot{m} c_p$$

$$C_h = \dot{m}_h c_{ph} = \left(18.9 \dfrac{\text{kg}}{\text{s}}\right)\left(4.180 \dfrac{\text{kJ}}{\text{kg}\cdot°\text{C}}\right)\left(1000 \dfrac{\text{J}}{\text{kJ}}\right)$$

$$= 79\,002 \text{ W/°C}$$

$$C_c = \dot{m}_c c_{pc} = \left(3.8 \dfrac{\text{kg}}{\text{s}}\right)\left(4.180 \dfrac{\text{kJ}}{\text{kg}\cdot°\text{C}}\right)\left(1000 \dfrac{\text{J}}{\text{kJ}}\right)$$

$$= 15\,884 \text{ W/°C}$$

Therefore,

$$C_{\min} = 15\,884 \text{ W/°C}$$

$$C_{\max} = 79\,002 \text{ W/°C}$$

Calculate the heat capacity ratio.

Effectiveness-NTU Relations

$$C_r = \frac{C_{\min}}{C_{\max}} = \frac{15\,884\ \frac{W}{°C}}{79\,002\ \frac{W}{°C}} = 0.2011$$

Calculate the number of transfer units (NTU) under clean conditions.

Number of Exchanger Transfer Units (NTU)

$$\text{NTU} = \frac{A_s U_{\text{avg}}}{C_{\min}} = \frac{A_s U_{\text{clean}}}{C_{\min}}$$

$$= \frac{(8.33\ \text{m}^2)\left(3040\ \frac{W}{\text{m}^2\cdot°C}\right)}{15\,884\ \frac{W}{°C}}$$

$$= 1.594$$

Calculate the heat exchanger effectiveness under clean conditions.

Effectiveness-NTU Relations

$$\epsilon = \frac{1 - \exp(-\text{NTU}(1 - C_r))}{1 - C_r \exp(-\text{NTU}(1 - C_r))}$$

$$= \frac{1 - e^{(-1.594(1-0.2011))}}{1 - 0.2011 \times e^{(-1.594(1-0.2011))}}$$

$$= 0.7630$$

Calculate the exit temperature of the cold water using the equation for heat exchanger effectiveness. In this case, $C_c = C_{\min}$. Therefore, the exit temperature of cold water under clean conditions is

Heat Exchanger Effectiveness, ε

$$\epsilon = \frac{C_c(T_{co} - T_{ci})}{C_{\min}(T_{hi} - T_{ci})}$$

$$T_{co} = T_{ci} + \epsilon(T_{hi} - T_{ci})$$

$$= 16°C + (0.7630)(88°C - 16°C)$$

$$= 70.94°C \quad (73°C)$$

The answer is (C).

36 Radiation and Combined Heat Transfer

Content in blue refers to the *NCEES Handbook*.

PRACTICE PROBLEMS

1. A 6 in (15 cm) thick furnace wall has a 3 in (8 cm) square peephole. The interior of the furnace is at 2200°F (1200°C). The surrounding air temperature is 70°F (20°C). The graph for black body shape factors is shown.

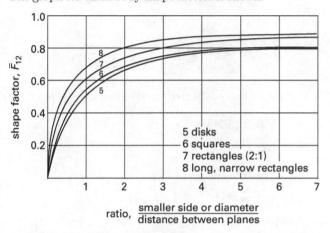

The heat loss due to radiation when the peephole is open is most nearly

(A) 450 Btu/hr (150 W)

(B) 1300 Btu/hr (440 W)

(C) 2000 Btu/hr (680 W)

(D) 7900 Btu/hr (2.7 kW)

2. A 9 in (23 cm) diameter duct is painted with white enamel. The surface of the duct is at 200°F (95°C). The duct carries hot air through a room whose walls are 70°F (20°C). The duct emissivity is 0.90. The convective film coefficient on the outside of the duct is approximated using the following equations:

h (Btu / hr-ft^2-R), D (ft), T (R)

$$h_{\text{convective}} = 0.27 \left(\frac{T_{\text{Duct}} - T_\infty}{D} \right)^{\frac{1}{4}} \quad [\text{US}]$$

h in Btu / hr-ft^2-°R, T in °R, and D in ft

$$h_{\text{convective}} = 1.32 \left(\frac{T_{\text{Duct}} - T_\infty}{D} \right)^{\frac{1}{4}} \quad [\text{SI}]$$

h in W / m^2 - K, T in K, and D in m

The air in the room is at 80°F (27°C). The heat transfer per unit length of the duct is most nearly

(A) 650 Btu/hr-ft (650 W/m)

(B) 900 Btu/hr-ft (900 W/m)

(C) 1100 Btu/hr-ft (1100 W/m)

(D) 1500 Btu/hr-ft (1500 W/m)

3. Dry air at 1 atm flows through 15 m of uninsulated duct. The flow rate of air through the duct is 0.25 m^3/s, and the diameter of the duct is 30 cm. The film coefficient for air flowing inside the duct is given by the following equation.

$$h_i \approx \frac{3.52 v_{\text{m/s}}^{0.8}}{D_{\text{m}}^{0.2}}$$

The film coefficient for air flowing inside the duct is most nearly

(A) 12 W/m^2·K

(B) 17 W/m^2·K

(C) 28 W/m^2·K

(D) 51 W/m^2·K

4. Dry air at 1 atmospheric pressure flows at 500 ft³/min (0.25 m³/s) through 50 ft (15 m) of 12 in (30 cm) diameter uninsulated duct. Air enters the duct at 45°F (7°C). The duct surface is at 68°F (20°C) and the room air is at 80°F (27°C). The air leaving the duct is at 50°F (10°C). The film coefficient for air flowing inside the duct is given by the following equations.

$$h_{i,\text{US}} \approx \left(0.00351 + (1.583 \times 10^{-6}) T_{°\text{F}}\right) \left(\frac{G^{0.8}_{\text{lbm/hr-ft}^2}}{D^{0.2}_{\text{ft}}}\right)$$

$$h_{i,\text{SI}} \approx \frac{3.52 v^{0.8}_{\text{m/s}}}{D^{0.2}_{\text{m}}}$$

The film coefficient for a horizontal cylinder is given by the following equations.

$$h_{o,\text{US}} = 0.27 \left(\frac{T_\infty - T_{\text{surface}}}{D}\right)^{1/4}$$

$$h_{o,\text{SI}} \approx 1.32 \left(\frac{T_\infty - T_{\text{surface}}}{D}\right)^{1/4}$$

The heat transfer due to convection is most nearly

(A) 640 Btu/hr (200 W)

(B) 1600 Btu/hr (490 W)

(C) 2000 Btu/hr (610 W)

(D) 2200 Btu/hr (670 W)

5. Dry air at 1 atmospheric pressure flows at 500 ft³/min (0.25 m³/s) through 50 ft (15 m) of 12 in (30 cm) diameter uninsulated duct. The emissivity of the duct surface is 0.28. Air enters the duct at 45°F (7°C). The walls, air, and contents of the room through which the duct passes are at 80°F (27°C). The air leaving the duct is at 50°F (10°C). The surface temperature is 68°F (20°C). The heat transfer due to convection is 2006.3 Btu/hr (614.6 W). The total heat transfer to the air is most nearly

(A) 2200 Btu/hr (690 W)

(B) 2600 Btu/hr (780 W)

(C) 2700 Btu/hr (840 W)

(D) 3600 Btu/hr (1120 W)

6. A semiconductor device is modeled as a circular cylinder 0.75 in (19 mm) in diameter and 1.5 in (38 mm) high standing on one end. The device emits 5.0 W and is cooled by a combination of natural convection and radiation. The surface emissivity is 0.65. The base is insulated and transmits no heat. The air and surroundings are at 14.7 psia (101 kPa) and 75°F (24°C). The film coefficient is 1.65 Btu/hr-ft²-°F (9.4 W/m²·K). The film coefficient for top and sides of the surface are

$$h_{\text{sides}} = 0.29 \left(\frac{T_s - T_\infty}{L}\right)^{1/4} \quad [\text{US}]$$

$$h_{\text{top}} = 0.27 \left(\frac{T_s - T_\infty}{D}\right)^{1/4} \quad [\text{US}]$$

$$h_{\text{top}} = 1.32 \left(\frac{T_s - T_\infty}{D}\right)^{1/4} \quad [\text{SI}]$$

$$h_{\text{sides}} = 1.37 \left(\frac{T_s - T_\infty}{L}\right)^{1/4} \quad [\text{SI}]$$

The surface temperature of the device is most nearly

(A) 650°R (360K)

(B) 700°R (390K)

(C) 740°R (410K)

(D) 750°R (415K)

7. The temperature of a gas in a duct with 600°F (315°C) walls is evaluated with a 0.5 in (13 mm) diameter thermocouple probe. The emissivity of the probe is 0.8. The gas flow rate is 3480 lbm/hr-ft² (4.7 kg/s·m²). The gas velocity is 400 ft/min (2 m/s). The film coefficient on the probe is given empirically as

$$h = \frac{0.024 G^{0.8}}{D^{0.4}} \quad [\text{US}]$$

$$h = \frac{17 G^{0.8}}{D^{0.4}} \quad [\text{SI}]$$

Where G is in units of lbm/hr-ft² or kg/s-m² and D is in ft or m, h will be in Btu/hr-ft²-°F or W/m²-°K. If the actual gas temperature is 300°F (150°C), the probe's reading is most nearly

(A) 700°R (390K)

(B) 740°R (410K)

(C) 760°R (420K)

(D) 780°R (430K)

8. The temperature of a gas in a duct with 600°F (315°C) walls is evaluated with a 0.5 in (13 mm) diameter thermocouple probe. The emissivity of the probe is 0.8. The gas flow rate is 3480 lbm/hr-ft^2 (4.7 kg/s·m^2), and the gas velocity is 400 ft/min (2 m/s). The film coefficient on the probe is given empirically as

$$h = \frac{0.024\,G^{0.8}}{D^{0.4}} \quad \text{[US]}$$

$$h = \frac{17\,G^{0.8}}{D^{0.4}} \quad \text{[SI]}$$

The probe reading indicates that the gas temperature is 300°F (150°C). The actual gas temperature is most nearly

(A) 650°R (360K)

(B) 740°R (410K)

(C) 770°R (430K)

(D) 810°R (450K)

SOLUTIONS

1. *Customary U.S. Solution*

Find the absolute temperatures. [Temperature Conversions]

$$T_{\text{furnace}} = 2200°F + 459.69 = 2659.69°R$$
$$T_{\infty} = 70°F + 459.69 = 529.69°R$$

Assuming that the walls are reradiating, nonconducting, and varying in temperature from 2200°F at the inside to 70°F at the outside, curve 6 for square shapes from the given graph in the problem statement can be used to find F_{12} using $x = 3\text{ in}/6\text{ in} = 0.5$; $F_{12} = 0.38$.

The Stefan-Boltzman constant is 0.1713×10^{-8} Btu/ft^2-hr-°R^4. [Fundamental Constants]

The radiation heat loss is

Net Energy Exchange by Radiation Between Two Black Bodies

$$q = A\sigma F_{12}(T_{\text{furnace}}^4 - T_{\infty}^4)$$

$$= \frac{(3\text{ in})^2 \left(0.1713 \times 10^{-8}\,\dfrac{\text{Btu}}{\text{hr-ft}^2\text{-°R}^4}\right) \times (0.38)\big((2659.69°R)^4 - (529.69°R)^4\big)}{\left(12\,\dfrac{\text{in}}{\text{ft}}\right)^2}$$

$$= 2032.6\text{ Btu/hr} \quad (2000\text{ Btu/hr})$$

SI Solution

Find the absolute temperatures. [Temperature Conversions]

$$T_{\text{furnace}} = 1200°C + 273.15° = 1473.15\text{K}$$
$$T_{\infty} = 20°C + 273.15° = 293.15\text{K}$$

Assuming that the walls are reradiating, nonconducting, and varying in temperature from 1200°C at the inside to 20°C at the outside, curve 6 for square shapes from the given graph in the problem statement can be used to find F_{12} using $x = 8\text{ cm}/15\text{ cm} = 0.533$; $F_{12} = 0.4$.

The Stefan-Boltzman constant is 5.67×10^{-8} W/m^2·K^4. [Fundamental Constants]

The radiation heat loss is

Net Energy Exchange by Radiation Between Two Black Bodies

$$q = A\sigma F_{12}(T_{\text{furnace}}^4 - T_\infty^4)$$

$$= \frac{(8 \text{ cm})^2 \left(5.67 \times 10^{-8} \dfrac{\text{W}}{\text{m}^2 \cdot \text{K}^4}\right) \times (0.4)\left((1473.15\text{K})^4 - (293.15\text{K})^4\right)}{\left(100 \dfrac{\text{cm}}{\text{m}}\right)^2}$$

$$= 682.5 \text{ W} \quad (680 \text{ W})$$

The answer is (C).

2. *Customary U.S. Solution*

Find the absolute temperatures. [Temperature Conversions]

$$T_\infty = 80°\text{F} + 459.69° = 539.69°\text{R}$$
$$T_{\text{duct}} = 200°\text{F} + 459.69° = 659.69°\text{R}$$
$$T_{\text{wall}} = 70°\text{F} + 459.69° = 529.69°\text{R}$$

The characteristic length of a cylinder is its diameter. Assume laminar flow. The convective film coefficient on the outside of the duct is approximately

$$h_{\text{convective}} = 0.27\left(\frac{T_{\text{duct}} - T_\infty}{D}\right)^{1/4}$$

$$= (0.27)\left(\frac{(659.69°\text{R} - 539.69°\text{R})\left(12 \dfrac{\text{in}}{\text{ft}}\right)}{9 \text{ in}}\right)^{1/4}$$

$$= 0.96 \text{ Btu/hr-ft}^2\text{-}°\text{F}$$

The duct surface area per unit length is

$$\frac{A}{L} = \pi D = \frac{\pi(9 \text{ in})}{12 \dfrac{\text{in}}{\text{ft}}}$$

$$= 2.356 \text{ ft}^2/\text{ft}$$

Find the convection losses per unit length. [Newton's Law of Cooling]

$$\frac{q_{\text{convection}}}{L} = h\left(\frac{A}{L}\right)\Delta T = h\left(\frac{A}{L}\right)(T_{\text{duct}} - T_\infty)$$

$$= \left(0.96 \dfrac{\text{Btu}}{\text{hr-ft}^2\text{-}°\text{F}}\right)\left(2.356 \dfrac{\text{ft}^2}{\text{ft}}\right) \times (659.69°\text{R} - 539.69°\text{R})$$

$$= 271.4 \text{ Btu/hr-ft}$$

The duct emissivity, ϵ_{duct_t}, is 0.90. The Stefan-Boltzman constant is 0.1713×10^{-8} Btu/hr-ft^2-°R^4. [Fundamental Constants]

Find the radiation losses per unit length. [Net Energy Exchange by Radiation Between Two Bodies]

$$\frac{q_{\text{radiation}}}{L} = \frac{E_{\text{net}}}{L}$$

$$= \left(\frac{A}{L}\right)\sigma\epsilon(T_{\text{duct}}^4 - T_{\text{wall}}^4)$$

$$= (2.356 \text{ ft}^2/\text{ft})\left(0.1713 \times 10^{-8} \dfrac{\text{Btu}}{\text{hr-ft}^2\text{-}°\text{R}^4}\right)$$
$$\times (0.90)\left((659.69°\text{R})^4 - (529.69°\text{R})^4\right)$$

$$= 401.9 \text{ Btu/hr-ft}$$

The total heat transfer per unit length is

$$\frac{q_{\text{total}}}{L} = \frac{q_{\text{convection}}}{L} + \frac{q_{\text{radiation}}}{L}$$

$$= 271.4 \dfrac{\text{Btu}}{\text{hr-ft}} + 401.9 \dfrac{\text{Btu}}{\text{hr-ft}}$$

$$= 673.3 \text{ Btu/hr-ft length} \quad (650 \text{ Btu/hr-ft length})$$

SI Solution

Find the absolute temperatures. [Temperature Conversions]

$$T_{\text{duct}} = 95°\text{C} + 273.15° = 368.15\text{K}$$
$$T_{\text{wall}} = 20°\text{C} + 273.15° = 293.15\text{K}$$
$$T_\infty = 27°\text{C} + 273.15° = 300.15\text{K}$$

The characteristic length of a cylinder is its diameter. Assume laminar flow. The convective film coefficient on the outside of the duct is approximately

$$h_{\text{convective}} = 1.32\left(\frac{T_{\text{duct}} - T_\infty}{D}\right)^{1/4}$$

$$= (1.32)\left(\frac{(368.15\text{K} - 300.15\text{K})\left(100 \dfrac{\text{cm}}{\text{m}}\right)}{23 \text{ cm}}\right)^{1/4}$$

$$= 5.47 \text{ W/m}^2\cdot\text{K}$$

The duct area per unit length is

$$\frac{A}{L} = \pi D = \frac{\pi(23 \text{ cm})}{100 \dfrac{\text{cm}}{\text{m}}}$$

$$= 0.723 \text{ m}^2/\text{m}$$

Find the convection losses per unit length. [Newton's Law of Cooling]

$$\frac{q_{convection}}{L} = h\left(\frac{A}{L}\right)(T_{duct} - T_\infty)$$

$$= \left(5.47 \ \frac{W}{m^2 \cdot K}\right)\left(0.723 \ \frac{m^2}{m}\right)(368.15K - 300.15K)$$

$$= 268.9 \ W/m$$

The duct emissivity, ϵ_{duct}, is 0.90. The Stefan-Boltzman constant is 5.67×10^{-8} W/m²·K⁴. [Fundamental Constants]

Find the radiation losses per unit length. [Net Energy Exchange by Radiation Between Two Bodies]

$$\frac{q_{radiation}}{L} = \frac{E_{net}}{L} = \left(\frac{A}{L}\right)\sigma\epsilon(T_{duct}^4 - T_{wall}^4)$$

$$= \left(0.723 \ \frac{m^2}{m}\right)\left(5.67 \times 10^{-8} \ \frac{W}{m^2 \cdot K^4}\right)$$

$$\times (0.90)\left((368.15K)^4 - (293.15K)^4\right)$$

$$= 405.2 \ W/m$$

The total heat transfer per unit length is

$$\frac{q_{total}}{L} = \frac{q_{convection}}{L} + \frac{q_{radiation}}{L}$$

$$= 268.9 \ \frac{W}{m} + 405.2 \ \frac{W}{m}$$

$$= 674.1 \ W/m \ \text{length} \quad (650 \ W/m \ \text{length})$$

The answer is (A).

3. The diameter of the duct in centimeters is

$$D = \frac{30 \ cm}{100 \ \frac{cm}{m}} = 0.30 \ m$$

The velocity of air entering the duct is

$$v = \frac{Q}{A_{flow}} = \frac{0.25 \ \frac{m^3}{s}}{\left(\frac{\pi}{4}\right)(0.30 \ m)^2}$$

$$= 3.537 \ m/s$$

The film coefficient for air flowing inside the duct is

$$h_i \approx \frac{3.52 v_{m/s}^{0.8}}{D_m^{0.2}}$$

$$= \frac{(3.52)\left(3.537 \ \frac{m}{s}\right)^{0.8}}{(0.30 \ m)^{0.2}}$$

$$= 12.3 \ W/m^2 \cdot K \quad (12 \ W/m^2 \cdot K)$$

The answer is (A).

4. *Customary U.S. Solution*

Find the absolute temperature of air entering the duct. [Temperature Conversions]

$$45°F + 459.69 = 504.69°R$$

At normal temperature and pressure, the pressure of air is 14.696 lbf/in². The gas constant, R, is 53.3 ft-lbf/lbm-°R. [Standard Dry Air Conditions at Sea Level] [Fundamental Constants]

From the ideal gas law, calculate the density of air entering the duct.

Ideal Gas

$$pV = mRT$$

$$\rho = \frac{m}{V} = \frac{p}{RT}$$

$$\rho = \frac{p}{RT} = \frac{\left(14.696 \ \frac{lbf}{in^2}\right)\left(12 \ \frac{in}{ft}\right)^2}{\left(53.3 \ \frac{ft\text{-}lbf}{lbm\text{-}°R}\right)(504.69°R)}$$

$$= 0.07867 \ lbm/ft^3$$

The mass flow rate of air entering the duct is

Continuity Equation

$$\dot{m} = \rho Q$$

$$= \left(0.07867 \ \frac{lbm}{ft^3}\right)\left(500 \ \frac{ft^3}{min}\right)\left(60 \ \frac{min}{hr}\right)$$

$$= 2360.1 \ lbm/hr$$

The mass flow rate of air per unit area entering the duct is

$$G = \frac{\dot{m}}{A_{flow}} = \frac{\left(2360.1 \ \frac{lbm}{hr}\right)\left(12 \ \frac{in}{ft}\right)^2}{\left(\frac{\pi}{4}\right)(12 \ in)^2}$$

$$= 3006.49 \ lbm/hr\text{-}ft^2$$

The characteristic length is the diameter of the duct.

$$L = \frac{12 \text{ in}}{12 \frac{\text{in}}{\text{ft}}} = 1 \text{ ft}$$

To calculate the initial film coefficients, estimate the bulk temperature.

$$T_{\text{bulk,air}} = \frac{1}{2}(T_{\text{air,in}} + T_{\text{air,out}}) = \left(\frac{1}{2}\right)(45°\text{F} + 50°\text{F})$$
$$= 47.5°\text{F}$$
$$T_{\text{surface}} = 68°\text{F}$$

The film coefficient for air flowing inside the duct is

$$h_i \approx \left(0.00351 + (1.583 \times 10^{-6})T_{°\text{F}}\right)\left(\frac{G_{\text{lbm/hr-ft}^2}^{0.8}}{D_{\text{ft}}^{0.2}}\right)$$

$$= \left(0.00351 + (1.583 \times 10^{-6})(47.5°\text{F})\right)$$

$$\times \left(\frac{\left(3006.49 \frac{\text{lbm}}{\text{hr-ft}^2}\right)^{0.8}}{\left(\frac{12 \text{ in}}{12 \frac{\text{in}}{\text{ft}}}\right)^{0.2}}\right)$$

$$= 2.17 \text{ Btu/hr-ft}^2\text{-°F}$$

The film coefficient for a horizontal cylinder is

$$h_o = 0.27\left(\frac{T_\infty - T_{\text{surface}}}{D}\right)^{1/4} = (0.27)\left(\frac{80°\text{F} - 68°\text{F}}{1 \text{ ft}}\right)^{1/4}$$
$$= 0.48 \text{ Btu/hr-ft}^2\text{-°F}$$

Neglecting the wall resistance, the overall film coefficient is

$$\frac{1}{U} = \frac{1}{h_i} + \frac{1}{h_o} = \frac{1}{2.17 \frac{\text{Btu}}{\text{hr-ft}^2\text{-°F}}} + \frac{1}{0.48 \frac{\text{Btu}}{\text{hr-ft}^2\text{-°F}}}$$
$$= 2.544 \text{ hr-ft}^2\text{-°F/Btu}$$
$$U = \frac{1}{2.544 \frac{\text{hr-ft}^2\text{-°F}}{\text{Btu}}} = 0.393 \text{ Btu/hr-ft}^2\text{-°F}$$

Find the heat transfer due to convection.

$$q_{\text{convection}} = UA_{\text{surface}}(T_\infty - T_{\text{bulk,air}})$$
$$= U(\pi DL)(T_\infty - T_{\text{bulk,air}})$$
$$= \left(0.393 \frac{\text{Btu}}{\text{hr-ft}^2\text{-°F}}\right)\pi(12 \text{ in})\left(\frac{1}{12}\frac{\text{ft}}{\text{in}}\right)$$
$$\times (50 \text{ ft})(80°\text{F} - 47.5°\text{F})$$
$$= 2006.3 \text{ Btu/hr} \quad (2000 \text{ Btu/hr})$$

SI Solution

The diameter of the duct is

$$D = \frac{30 \text{ cm}}{100 \frac{\text{cm}}{\text{m}}} = 0.30 \text{ m}$$

The velocity of air entering the duct is

$$\text{v} = \frac{Q}{A_{\text{flow}}} = \frac{0.25 \frac{\text{m}^3}{\text{s}}}{\left(\frac{\pi}{4}\right)(0.30 \text{ m})^2}$$
$$= 3.537 \text{ m/s}$$

To calculate the initial film coefficients, estimate the bulk temperature.

$$T_{\text{bulk,air}} = \frac{1}{2}(T_{\text{air,in}} + T_{\text{air,out}}) = \left(\frac{1}{2}\right)(7°\text{C} + 10°\text{C})$$
$$= 8.5°\text{C}$$
$$T_{\text{surface}} = 20°\text{C}$$

The film coefficient for air flowing inside the duct is

$$h_i \approx \frac{3.52\text{v}_{\text{m/s}}^{0.8}}{D_{\text{m}}^{0.2}}$$

$$= \frac{(3.52)\left(3.537 \frac{\text{m}}{\text{s}}\right)^{0.8}}{(0.30 \text{ m})^{0.2}} = 12.3 \text{ W/m}^2\cdot\text{K}$$

The film coefficient for a horizontal cylinder is

$$h_o \approx 1.32\left(\frac{T_\infty - T_{\text{surface}}}{D}\right)^{1/4}$$
$$= (1.32)\left(\frac{27°\text{C} - 20°\text{C}}{0.30 \text{ m}}\right)^{1/4}$$
$$= 2.90 \text{ W/m}^2\cdot\text{K}$$

Neglecting the wall resistance, the overall film coefficient is

$$\frac{1}{U} = \frac{1}{h_i} + \frac{1}{h_o} = \frac{1}{12.3 \ \frac{W}{m^2 \cdot K}} + \frac{1}{2.90 \ \frac{W}{m^2 \cdot K}}$$

$$= 0.426 \ m^2 \cdot K/W$$

$$U = \frac{1}{0.426 \ \frac{m^2 \cdot K}{W}} = 2.35 \ W/m^2 \cdot K$$

Find the heat transfer due to convection.

$$q_{\text{convection}} = UA_{\text{surface}}(T_\infty - T_{\text{bulk,air}})$$
$$= U\pi DL(T_\infty - T_{\text{bulk,air}})$$
$$= \left(2.35 \ \frac{W}{m^2 \cdot K}\right)\pi(0.30 \ m)(15 \ m)$$
$$\times (27°C - 8.5°C)$$
$$= 614.6 \ W \quad (610 \ W)$$

The answer is (C).

5. *Customary U.S. Solution*

The room and duct have an unobstructed view of each other. The duct emissivity, ϵ, is 0.28. The Stefan-Boltzman constant is 0.1713×10^{-8} Btu/ft²-hr-°R⁴. [Fundamental Constants]

Find the absolute temperatures. [Temperature Conversions]

$$T_\infty = 80°F + 459.69 = 539.69°R$$
$$T_{\text{surface}} = 68°F + 459.69 = 527.69°R$$

Find the heat transfer due to radiation.

Net Energy Exchange by Radiation Between Two Bodies

$$q_{\text{radiation}} = \epsilon\sigma A_{\text{surface}}(T_\infty^4 - T_{\text{surface}}^4)$$

$$= \frac{(0.28)\left(0.1713 \times 10^{-8} \ \frac{\text{Btu}}{\text{hr-ft}^2\text{-°R}^4}\right)}{12 \ \frac{\text{in}}{\text{ft}}}$$
$$\times \pi(12 \ \text{in})(50 \ \text{ft})\left((539.69°R)^4 - (527.69°R)^4\right)$$

$$= 549.8 \ \text{Btu/hr}$$

The total heat transfer to the air is

$$q_{\text{total}} = q_{\text{convection}} + q_{\text{radiation}}$$
$$= 2006.3 \ \frac{\text{Btu}}{\text{hr}} + 549.8 \ \frac{\text{Btu}}{\text{hr}}$$
$$= 2556.1 \ \text{Btu/hr} \quad (2600 \ \text{Btu/hr})$$

SI Solution

The diameter of the duct is

$$D = \frac{30 \ \text{cm}}{100 \ \frac{\text{cm}}{\text{m}}} = 0.30 \ \text{m}$$

The room and duct have an unobstructed view of each other. The duct emissivity, ϵ, is 0.28. The Stefan-Boltzman constant is 5.67×10^{-8} W/m²·K⁴. [Fundamental Constants]

Find the absolute temperatures. [Temperature Conversions]

$$T_\infty = 27°C + 273.15 = 300.15 \ \text{K}$$
$$T_{\text{surface}} = 20°C + 273.15 = 293.15 \ \text{K}$$

Find the heat transfer due to radiation.

Net Energy Exchange by Radiation Between Two Bodies

$$q_{\text{radiation}} = \epsilon\sigma A_{\text{surface}}(T_\infty^4 - T_{\text{surface}}^4)$$
$$= (0.28)\left(5.67 \times 10^{-8} \ \frac{W}{m^2 \cdot K^4}\right)\pi$$
$$\times (0.30 \ m)(15 \ m)\left((300.15 \ K)^4 - (293.15 \ K)^4\right)$$
$$= 163.8 \ W$$

The total heat transfer to the air is

$$q_{\text{total}} = q_{\text{convection}} + q_{\text{radiation}} = 614.6 \ W + 163.8 \ W$$
$$= 778.4 \ W \quad (780 \ W)$$

The answer is (B).

6. *Customary U.S. Solution*

Heat is lost from the top and sides by radiation and convection. Find the absolute temperature of the surroundings. [Temperature Conversions]

$$T_\infty = 75°F + 459.69 = 534.69°R$$

Find the area of the top and sides.

$$A_{\text{sides}} = \pi DL$$
$$= \frac{\pi(0.75 \ \text{in})(1.5 \ \text{in})}{\left(12 \ \frac{\text{in}}{\text{ft}}\right)^2}$$
$$= 0.0245 \ \text{ft}^2$$

$$A_{\text{top}} = \left(\frac{\pi}{4}\right)D^2$$

$$= \frac{\left(\frac{\pi}{4}\right)(0.75 \text{ in})^2}{\left(12 \frac{\text{in}}{\text{ft}}\right)^2}$$

$$= 0.003068 \text{ ft}^2$$

The total heat from convection and radiation can be expressed using the following equation.

Newton's Law of Cooling

$$\dot{Q} = hA(T_w - T_\infty)$$

Net Energy Exchange by Radiation Between Two Bodies

$$\dot{Q}_{12} = \epsilon \sigma A (T_1^4 - T_2^4)$$

$$q_{\text{total}} = q_{\text{convection}} + q_{\text{radiation}}$$
$$= h_{\text{sides}} A_{\text{sides}} (T_s - T_\infty) + h_{\text{top}} A_{\text{top}} (T_s - T_\infty)$$
$$+ \sigma \epsilon (A_{\text{sides}} + A_{\text{top}})(T_s^4 - T_\infty^4)$$

First, find the surface temperature, then adjust for the film coefficient to get the final answer. For the first approximation of T_s, $h_{\text{sides}} = h_{\text{top}} = 1.65$ Btu/hr-ft²-°R. The emissivity, ϵ, is 0.65. The Stefan-Boltzman constant is 0.1713×10^{-8} Btu/ft²-hr-°R⁴. [Fundamental Constants]

$$(5.0 \text{ W})\left(\frac{3.142 \frac{\text{Btu}}{\text{hr}}}{1 \text{ W}}\right) = \left(1.65 \frac{\text{Btu}}{\text{hr-ft}^2\text{-°F}}\right)(0.0245 \text{ ft}^2)$$
$$(T_s - 534.69\text{°R})$$
$$+ \left(1.65 \frac{\text{Btu}}{\text{hr-ft}^2\text{-°F}}\right)$$
$$(0.003068 \text{ ft}^2)(T_s - 534.69\text{°R})$$
$$+ \left(0.1713 \times 10^{-8} \frac{\text{Btu}}{\text{hr-ft}^2\text{-°R}^4}\right)(0.65)$$
$$\times (0.0245 \text{ ft}^2 + 0.003068 \text{ ft}^2)$$
$$(T_s^4 - (534.69\text{°R})^4)$$

By trial and error, $T_s \approx 750$°R.

$$h_{\text{sides}} = 0.29 \left[\frac{T_s - T_\infty}{L}\right]^{1/4}$$

$$= (0.29)\left[\frac{(750\text{°R} - 534.69\text{°R})\left(12 \frac{\text{in}}{\text{ft}}\right)}{1.5 \text{ in}}\right]^{1/4}$$

$$= 1.87 \text{ Btu/hr-ft}^2\text{-°F}$$

The film coefficient for a horizontal surface is

$$h_{\text{top}} = 0.27 \left[\frac{T_s - T_\infty}{D}\right]^{1/4}$$

$$= (0.27)\left[\frac{(750\text{°R} - 534.69\text{°R})\left(12 \frac{\text{in}}{\text{ft}}\right)}{(0.9)(0.75 \text{ in})}\right]^{1/4}$$

$$= 2.12 \text{ Btu/hr-ft}^2\text{-°F}$$

Substitute the calculated values of h_{top} and h_{sides} into the total heat transfer equation.

$$q_{\text{total}} = (5.0 \text{ W})\left(\frac{3.412 \frac{\text{Btu}}{\text{hr}}}{1 \text{ W}}\right)$$

$$= \left(1.87 \frac{\text{Btu}}{\text{hr-ft}^2\text{-°F}}\right)(0.0245 \text{ ft}^2)(T_s - 534.69\text{°R})$$
$$+ \left(2.12 \frac{\text{Btu}}{\text{hr-ft}^2\text{-°F}}\right)(0.003068 \text{ ft}^2)(T_s - 534.69\text{°R})$$
$$+ \left(0.1713 \times 10^{-8} \frac{\text{Btu}}{\text{hr-ft}^2\text{-°R}^4}\right)(0.65)$$
$$\times (0.0245 \text{ ft}^2 + 0.003068 \text{ ft}^2)(T_s^4 - (534.69\text{°R})^4)$$

By trial and error, $T_s \approx 736$°R   (740°R)

SI Solution

Heat is lost from the top and sides by radiation and convection. Find the absolute temperature of the surrounding. [Temperature Conversions]

$$T_\infty = 24\text{°C} + 273.15 = 297.15 \text{K}$$

Find the area of the top and sides.

$$A_{\text{sides}} = \pi D L = \frac{\pi (19 \text{ mm})(38 \text{ mm})}{\left(1000 \frac{\text{mm}}{\text{m}}\right)^2}$$

$$= 0.00227 \text{ m}^2$$

$$A_{\text{top}} = \left(\frac{\pi}{4}\right)D^2$$

$$= \frac{\left(\frac{\pi}{4}\right)(19 \text{ mm})^2}{\left(1000 \frac{\text{mm}}{\text{m}}\right)^2}$$

$$= 0.000284 \text{ m}^2$$

First, find the surface temperature, then adjust for the film coefficient to get the final answer. The total heat from convection and radiation can be expressed using the following equation.

Newton's Law of Cooling
$$\dot{Q} = hA(T_w - T_\infty)$$

Net Energy Exchange by Radiation Between Two Bodies
$$\dot{Q}_{12} = \epsilon \sigma A(T_1^4 - T_2^4)$$

$$\begin{aligned}q_{total} &= q_{convection} + q_{radiation} \\ &= h_{sides}A_{sides}(T_s - T_\infty) + h_{top}A_{top}(T_s - T_\infty) \\ &\quad + \sigma\epsilon(A_{sides} + A_{top})(T_s^4 - T_\infty^4)\end{aligned}$$

First, find the surface temperature, then adjust for the film coefficient to get the final answer. For a first approximation of T_s, $h_{sides} = h_{top} = 9.4$ W/m²·K. The emissivity, ϵ, is 0.65. The Stefan-Boltzman constant is 5.67×10^{-8} W/m²·K⁴. [Fundamental Constants]

$$\begin{aligned}5.0 \text{ W} &= \left(9.4 \frac{\text{W}}{\text{m}^2 \cdot \text{K}}\right)(0.00227 \text{ m}^2)(T_s - 297.15\text{K}) \\ &\quad + \left(9.4 \frac{\text{W}}{\text{m}^2 \cdot \text{K}}\right)(0.000284 \text{ m}^2)(T_s - 297.15\text{K}) \\ &\quad + \left(5.67 \times 10^{-8} \frac{\text{W}}{\text{m}^2 \cdot \text{K}^4}\right)(0.65) \\ &\quad \times (0.00227 \text{ m}^2 + 0.000284 \text{ m}^2) \\ &\quad \times (T_s^4 - (297.15\text{K})^4)\end{aligned}$$

By trial and error, $T_s = 416.75$K.

The film coefficient for a horizontal surface is

$$\begin{aligned}h_{top} &= 1.32\left(\frac{T_s - T_\infty}{D}\right)^{1/4} \\ &= (1.32)\left[\frac{(416.75\text{K} - 297.15\text{K})\left(1000 \frac{\text{mm}}{\text{m}}\right)}{(0.9)(19 \text{ mm})}\right]^{1/4} \\ &= 12.07 \text{ W/m}^2\cdot\text{K}\end{aligned}$$

The film coefficient for a vertical surface is

$$\begin{aligned}h_{sides} &= 1.37\left(\frac{T_s - T_\infty}{L}\right)^{1/4} \\ &= (1.37)\left[\frac{(416.75\text{K} - 297.15\text{K})\left(1000 \frac{\text{mm}}{\text{m}}\right)}{38 \text{ mm}}\right]^{1/4} \\ &= 10.26 \text{ W/m}^2\cdot\text{K}\end{aligned}$$

Substitute the calculated values of h_{top} and h_{sides} into the total heat transfer equation.

$$\begin{aligned}q_{total} &= 5.0 \text{ W} \\ &= \left(10.26 \frac{\text{W}}{\text{m}^2 \cdot \text{K}}\right)(0.00227 \text{ m}^2)(T_s - 297.15\text{K}) \\ &\quad + \left(12.07 \frac{\text{W}}{\text{m}^2 \cdot \text{K}}\right)(0.000284 \text{ m}^2)(T_s - 297.15\text{K}) \\ &\quad + \left(5.67 \times 10^{-8} \frac{\text{W}}{\text{m}^2 \cdot \text{K}^4}\right)(0.65) \\ &\quad \times (0.00227 \text{ m}^2 + 0.000284 \text{ m}^2) \\ &\quad \times (T_s^4 - (297.15\text{K})^4)\end{aligned}$$

By trial and error, $T_s = 411$K. (410K)

The answer is (C).

7. *Customary U.S. Solution*

The velocity is relatively low, so incompressible flow can be assumed.

The film coefficient on the probe is

$$\begin{aligned}h &= \frac{0.024\, G^{0.8}}{D^{0.4}} \\ &= \frac{(0.024)\left(3480 \frac{\text{lbm}}{\text{hr-ft}^2}\right)^{0.8}}{\left((0.5 \text{ in})\left(\frac{1}{12} \frac{\text{ft}}{\text{in}}\right)\right)^{0.4}} \\ &= 58.3 \text{ Btu/hr-ft}^2\text{-°F}\end{aligned}$$

Find the absolute temperature of the and the gas. [Temperature Conversions]

$$T_{walls} = 600°\text{F} + 459.69 = 1059.69°\text{R}$$

$$T_{gas} = 300°\text{F} + 459.69 = 759.69°\text{R}$$

Neglect conduction and the insignificant kinetic energy loss. The thermocouple gains heat through radiation from the walls and loses heat through convection to the gas. [Net Energy Exchange by Radiation Between Two Bodies]

$$q_{\text{convection}} = q_{\text{radiation}}$$
$$hA(T_{\text{probe}} - T_{\text{gas}}) = A\sigma\epsilon(T_{\text{walls}}^4 - T_{\text{probe}}^4)$$
$$h(T_{\text{probe}} - T_{\text{gas}}) = \sigma\epsilon(T_{\text{walls}}^4 - T_{\text{probe}}^4)$$

The Stefan-Boltzman constant is 0.1713×10^{-8} Btu/ft^2-hr-°R^4. [Fundamental Constants]

$$\left(58.3 \frac{\text{Btu}}{\text{hr-ft}^2\text{-°F}}\right)$$
$$\times (T_{\text{probe}} - 759.69\text{°R}) = \left(0.1713 \times 10^{-8} \frac{\text{Btu}}{\text{hr-ft}^2\text{-°R}^4}\right)$$
$$\times (0.8)\left((1059.69\text{°R})^4 - T_{\text{probe}}^4\right)$$

$$(1.37 \times 10^{-9})$$
$$\times T_{\text{probe}}^4 + (58.3) T_{\text{probe}} = 46{,}038$$

By trial and error, the temperature of the probe is

$$T_{\text{probe}} = 781\text{°R} \quad (780\text{°R})$$

SI Solution

The velocity is relatively low, so incompressible flow can be assumed.

The film coefficient on the probe is

$$h = \frac{17 G^{0.8}}{D^{0.4}} = \frac{(17)\left(4.7 \frac{\text{kg}}{\text{s·m}^2}\right)^{0.8}}{\left((13 \text{ mm})\left(\frac{1}{1000} \frac{\text{m}}{\text{mm}}\right)\right)^{0.4}}$$
$$= 333.1 \text{ W/m}^2\text{·K}$$

Find the absolute temperature of the walls. [Temperature Conversions]

$$T_{\text{walls}} = 315\text{°C} + 273.15 = 588.15\text{K}$$

Neglect conduction and the insignificant kinetic energy loss. The thermocouple gains heat through radiation from the walls and loses heat through convection to the gas. [Net Energy Exchange by Radiation Between Two Bodies]

$$q_{\text{convection}} = q_{\text{radiation}}$$
$$h(T_{\text{probe}} - T_{\text{gas}}) = \sigma\epsilon(T_{\text{walls}}^4 - T_{\text{probe}}^4)$$

The Stefan-Boltzman constant is 5.67×10^{-8} W/m^2·K^4. The actual gas temperature is $150\text{°C} + 273.15 = 423.15\text{K}$. [Fundamental Constants]

$$\left(333.1 \frac{\text{W}}{\text{m}^2\text{·K}}\right)$$
$$(T_{\text{probe}} - 423.15\text{K}) = \left(5.67 \times 10^{-8} \frac{\text{W}}{\text{m}^2\text{·K}^4}\right)(0.8)$$
$$\times \left((588.15\text{K})^4 - T_{\text{probe}}^4\right)$$

$$(4.536 \times 10^{-8}) T_{\text{probe}}^4$$
$$+ (333.1) T_{\text{probe}} = 146{,}323$$

By trial and error, the temperature of the probe is

$$T_{\text{probe}} = 435\text{K} \quad (430\text{K})$$

The answer is (D).

8. *Customary U.S. Solution*

The velocity is relatively low, so incompressible flow can be assumed. The film coefficient on the probe is

$$h = \frac{0.024 G^{0.8}}{D^{0.4}}$$
$$= \frac{(0.024)\left(3480 \frac{\text{lbm}}{\text{hr-ft}^2}\right)^{0.8}}{\left((0.5 \text{ in})\left(\frac{1}{12} \frac{\text{ft}}{\text{in}}\right)\right)^{0.4}}$$
$$= 58.3 \text{ Btu/hr-ft}^2\text{-°F}$$

Find the absolute temperature of the and the probe reading. [Temperature Conversions]

$$T_{\text{walls}} = 600\text{°F} + 459.69 = 1059.69\text{°R}$$
$$T_{\text{probe}} = 300\text{°F} + 459.69 = 759.69\text{°R}$$

Neglect conduction and the insignificant kinetic energy loss. The thermocouple gains heat through radiation from the walls and loses heat through convection to the gas. [Net Energy Exchange by Radiation Between Two Bodies]

$$q_{\text{convection}} = q_{\text{radiation}}$$
$$hA(T_{\text{probe}} - T_{\text{gas}}) = A\sigma\epsilon(T_{\text{walls}}^4 - T_{\text{probe}}^4)$$
$$h(T_{\text{probe}} - T_{\text{gas}}) = \sigma\epsilon(T_{\text{walls}}^4 - T_{\text{probe}}^4)$$

The Stefan-Boltzman constant is 0.1713×10^{-8} Btu/ft^2-hr-°R^4. [Fundamental Constants]

The actual gas temperature is

$$h(T_{probe} - T_{gas}) = \sigma\epsilon(T_{walls}^4 - T_{probe}^4)$$

$$T_{gas} = \frac{hT_{probe} - \sigma\epsilon(T_{walls}^4 - T_{probe}^4)}{h}$$

$$= \frac{\left(58.3 \; \dfrac{\text{Btu}}{\text{hr-ft}^2\text{-}°\text{F}}\right)(759.69°\text{R})}{58.3 \; \dfrac{\text{Btu}}{\text{hr-ft}^2\text{-}°\text{F}}}$$

$$\frac{- \left(0.1713 \times 10^{-8} \; \dfrac{\text{Btu}}{\text{hr-ft}^2\text{-}°\text{R}^4}\right)}{}$$

$$\frac{\times (0.8)\left((1059.69°\text{R})^4 - (759.69°\text{R})^4\right)}{58.3 \; \dfrac{\text{Btu}}{\text{hr-ft}^2\text{-}°\text{F}}}$$

$$= 737.8°\text{R} \quad (740°\text{R})$$

SI Solution

The velocity is relatively low, so incompressible flow can be assumed. The film coefficient on the probe is

$$h = \frac{17G^{0.8}}{D^{0.4}} = \frac{(17)\left(4.7 \; \dfrac{\text{kg}}{\text{s}\cdot\text{m}^2}\right)^{0.8}}{\left(\dfrac{13 \text{ mm}}{1000 \; \dfrac{\text{mm}}{\text{m}}}\right)^{0.4}}$$

$$= 333.1 \; \text{W/m}^2\text{·K}$$

Find the absolute temperature of the walls. [Temperature Conversions]

$$T_{walls} = 315°\text{C} + 273.15 = 588.15\text{K}$$

$$T_{probe} = 150°\text{C} + 273.15 = 423.15°\text{K}$$

Neglect conduction and the insignificant kinetic energy loss. The thermocouple gains heat through radiation from the walls and loses heat through convection to the gas. [Net Energy Exchange by Radiation Between Two Bodies]

$$q_{convection} = q_{radiation}$$

$$h(T_{probe} - T_{gas}) = \sigma\epsilon(T_{walls}^4 - T_{probe}^4)$$

The Stefan-Boltzman constant is 5.67×10^{-8} W/m²·K⁴. [Fundamental Constants]

The actual gas temperature is

$$\left(333.1 \; \dfrac{\text{W}}{\text{m}^2\text{·K}}\right)$$

$$(423.15\text{K} - T_{gas}) = \left(5.67 \times 10^{-8} \; \dfrac{\text{W}}{\text{m}^2\text{·K}^4}\right)(0.8)$$

$$\times \left((588.15\text{K})^4 - (423.15\text{K})^4\right)$$

$$T_{gas} = 411.1\text{K} \quad (410\text{K})$$

The answer is (B).

37 Psychrometrics

Content in blue refers to the *NCEES Handbook*.

PRACTICE PROBLEMS

1. A room contains air at 80°F dry-bulb temperature and 67°F wet-bulb temperature. The total pressure is 1 atm. The enthalpy is most nearly

(A) 30.2 Btu/lbm
(B) 30.8 Btu/lbm
(C) 31.6 Btu/lbm
(D) 32.2 Btu/lbm

2. A room contains air at 80°F dry-bulb temperature and 67°F wet-bulb temperature. The total pressure is 1 atm. The specific heat of the air in the room is most nearly

(A) 0.234 Btu/lbm-°R
(B) 0.237 Btu/lbm-°R
(C) 0.239 Btu/lbm-°R
(D) 0.242 Btu/lbm-°R

3. If one layer of cooling coils effectively bypasses one-third of the air passing through it, the theoretical bypass factor for four layers of identical cooling coils in series is most nearly

(A) 0.01
(B) 0.09
(C) 0.33
(D) 0.67

4. Air at 60°F dry-bulb temperature and 45°F wet-bulb temperature passes through an air washer with a humidifying efficiency of 70%. The effective bypass factor of the system is most nearly

(A) 0.30
(B) 0.50
(C) 0.67
(D) 0.70

5. Air at 60°F (16°C) dry-bulb temperature and 45°F (7°C) wet-bulb temperature passes through an air washer with a humidifying efficiency of 70%. The dry-bulb temperature of the air leaving the washer is most nearly

(A) 45°F (7°C)
(B) 50°F (10°C)
(C) 54°F (12°C)
(D) 57°F (14°C)

6. During performances, a theater experiences a sensible heat load of 500,000 Btu/hr and a moisture load of 175 lbm/hr. Air enters the theater at 65°F and 55% relative humidity and is removed when it reaches 75°F or 60% relative humidity, whichever comes first. The condition of the air leaving the theater is most nearly

(A) $T_{db} = 65°F$, RH = 44%
(B) $T_{db} = 65°F$, RH = 55%
(C) $T_{db} = 75°F$, RH = 44%
(D) $T_{db} = 75°F$, RH = 55%

7. An HVAC system removes 500 ft³/min of air from a room. Some of the removed air enters an air conditioner at 80°F dry-bulb temperature and 70% relative humidity and leaves as saturated air at 50°F. The flow rate of the air passing through the air conditioner is 150 ft³/min. The remainder of the removed air bypasses the air conditioner and mixes with the conditioned air at 1 atm. The relative humidity of the mixed air is most nearly

(A) 45%
(B) 57%
(C) 73%
(D) 81%

8. An HVAC system removes 500 ft³/min of air from a room. Some of the removed air enters an air conditioner at 80°F dry-bulb temperature and 70% relative humidity and leaves as saturated air at 50°F. The flow rate of the air passing through the air conditioner is 150 ft³/min. The remainder of the removed air bypasses the air conditioner and mixes with the conditioned air at 1 atm. The heat load of the air conditioner is most nearly

(A) 0.90 tons

(B) 1.3 tons

(C) 2.4 tons

(D) 2.9 tons

9. A dehumidifier takes 5000 ft³/min of air at 95°F dry-bulb temperature and 70% relative humidity and discharges it at 60°F dry-bulb and 95% relative humidity. The dehumidifier uses a wet R-12 refrigeration cycle operating between 100°F (saturated) and 50°F. The rate of heat removed from the air is most nearly

(A) 4000 Btu/min

(B) 5500 Btu/min

(C) 6500 Btu/min

(D) 8550 Btu/min

10. A dehumidifier takes 5000 ft³/min of air at 95°F dry-bulb temperature and 70% relative humidity and discharges it at 60°F dry-bulb and 95% relative humidity. The dehumidifier uses refrigerant-12 (R-12), and the refrigeration cycle operates between 100°F (saturated) and 50°F. Draw the temperature-entropy and enthalpy-entropy diagrams for the refrigeration cycle.

11. Combustion products leaving a gas turbine combustor are released at a temperature of 180°F into the atmosphere. The combustion products have a relative humidity of 20%. What is most nearly the dew-point temperature of the combustion products?

(A) 110°F

(B) 115°F

(C) 120°F

(D) 125°F

12. Each hour, 100 lbm of methane are burned with excess air. The combustion products pass through a heat exchanger used to heat water. The combustion products enter the heat exchanger at 340°F, and they leave the heat exchanger at 110°F. The total pressure of the combustion products in the heat exchanger is 19 psia. The humidity ratio of the combustion products is 560 gr/lbm (7000 grain = 1 lbm). The dew-point temperature of the combustion products is most nearly

(A) 90°F

(B) 100°F

(C) 120°F

(D) 130°F

13. An HVAC unit maintains air at 75°F dry-bulb temperature and 50% relative humidity. Outside air is supplied to the system at 3000 ft³/min at 91°F dry-bulb and 77°F wet-bulb. The flow of the return air through the system is 7000 ft³/min. The supply air leaves the air conditioner with an enthalpy of 22 Btu/lbm. The system utilizes an energy recovery ventilator with an efficiency of 75% for the outside airstream exhaust to condition the fresh 3000 ft³/min incoming airstream before it mixes with the 7000 ft³/min of return air. Assume the supply and exhaust specific heats are approximately equal, and assume the specific volume of the air is approximately constant at 13.33 ft³/lbm across the states. The resulting cooling load is most nearly

(A) 310,500 Btu/hr

(B) 390,000 Btu/hr

(C) 400,000 Btu/hr

(D) 454,000 Btu/hr

14. An HVAC system is used to precool 9000 ft³/min of supply outdoor air at a temperature of 95°F dry-bulb and 81°F wet-bulb using an energy-recovery ventilator (ERV) with a total effectiveness of 55%. The return air entering the ERV is at 75°F dry-bulb and 63°F wet-bulb. Assume the density of air is 0.069 lbm/ft³. The energy recovered by the ERV is most nearly

(A) 210,000 Btu/hr

(B) 290,000 Btu/hr

(C) 330,000 Btu/hr

(D) 410,000 Btu/hr

SOLUTIONS

1.

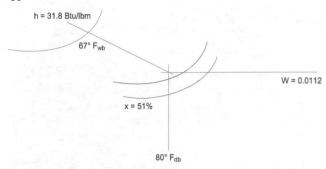

The problem indicates a single thermodynamic state at the given 80° F_{db} and 67° F_{wb}. These two characteristics are sufficient to identify the thermodynamic state.

[ASHRAE Psychrometric Chart No. 1 - Normal Temperature at Sea Level]

Locate the dry bulb temperature on the bottom (80° F_{db}) and wet bulb temperature (67° F_{wb}) on the top-left curve of saturation, and draw a line along each of these temperature lines. The intersection of these two lines provides us with the quality $x = 51\%$ for this problem. It also points to the humidity ratio of $w = 0.0112$.

Then, following the line traced along the wet-bulb temperature of 67° F_{wb}, we cross the enthalpy (h) line above the saturation line, and find h, the enthalpy of the cited state.

$h = 31.8$ Btu/lbm (31.6 Btu/lbm)

Alternate Solution

[Thermodynamic Properties of Moist Air at Standard Atmospheric Pressure, 14.696 Psia]

Locate the dry bulb temperature of 80°F on the left column, and see that the corresponding $h_f = 19.222$ Btu/lbm and $h_{fg} = 24.475$ Btu/lbm. Then, calculate h for the cited thermodynamic state.

$$h = h_f + (x)h_{fg}$$
$$= 19.222 \text{ Btu/lbm} + (0.51)\, 24.479 \text{ Btu/lbm}$$
$$= 31.71 \text{ Btu/lbm} \quad (31.6 \text{ Btu/lbm})$$

The answer is (C).

2. Locate the intersection of 80°F dry-bulb temperature and 67°F wet-bulb temperature on the psychrometric chart. The humidity ratio is 0.0112 lbm of moisture/lbm of dry air. [ASHRAE Psychrometric Chart No. 1 - Normal Temperature at Sea Level]

Specific heat, c_p, is gravimetrically weighted. Find the gravimetric fraction, G, of the dry air and the steam.

$$G_{\text{air}} = \frac{1 \text{ lbm}_{\text{dry air}}}{1 \text{ lbm}_{\text{dry air}} + 0.0112 \text{ lbm}_{\text{moist air}}} = 0.989$$

$$G_{\text{steam}} = \frac{0.0112 \text{ lbm}_{\text{moist air}}}{1 \text{ lbm}_{\text{dry air}} + 0.0112 \text{ lbm}_{\text{moist air}}} = 0.011$$

From a table of thermal properties of ideal gases, the specific heat for air is 0.240 Btu/lbm-°R and the specific heat for steam is 0.445 Btu/lbm-°R. [Thermal and Physical Properties of Ideal Gases (at Room Temperature)]

Find the specific heat of the mixed air.

$$c_{p,\text{mixture}} = G_{\text{air}} c_{p,\text{air}} + G_{\text{steam}} c_{p,\text{steam}}$$
$$= (0.989)\left(0.240\ \frac{\text{Btu}}{\text{lbm-°R}}\right)$$
$$\quad + (0.011)\left(0.445\ \frac{\text{Btu}}{\text{lbm-°R}}\right)$$
$$= 0.242 \text{ Btu/lbm-°R}$$

The answer is (D).

3. The bypass factor is the percentage of air that does not pass through a cooling coil. Find the bypass factor by

$$\text{BF}_{n\text{ layers}} = (\text{BF}_{1\text{ layer}})^n = \left(\frac{1}{3}\right)^4 = 0.0123 \quad (0.01)$$

Only 1% of the air will be untreated.

The answer is (A).

4. The bypass factor is the percentage of air that doesn't pass through a cooling/heating coil. The efficiency is the complement of the bypass factor. Hence, 1 – efficiency will give the bypass factor value.

$$\text{BF} = 1 - \eta_{\text{sat}}$$
$$= 1 - 0.70$$
$$= 0.30$$

The answer is (A).

5. *Customary U.S. Solution*

Use the equation for determining the direct saturation efficiency to find the dry-bulb temperature of air leaving the washer.

Direct Evaporative Air Coolers

$$\varepsilon_e = 100\left(\frac{t_1 - t_2}{t_1 - t'_s}\right)$$

$$t_2 = t_1 - \left(\frac{\varepsilon_e}{100}\right)(t_1 - t'_s)$$

$$= 60°F - \left(\frac{70}{100}\right)(60°F - 45°F)$$

$$= 49.5°F \quad (50°F)$$

SI Solution

Use the equation for determining the direct saturation efficiency to find the dry-bulb temperature of air leaving the washer.

Direct Evaporative Air Coolers

$$\varepsilon_e = 100\left(\frac{t_1 - t_2}{t_1 - t'_s}\right)$$

$$t_2 = t_1 - \left(\frac{\varepsilon_e}{100}\right)(t_1 - t'_s)$$

$$= 16°C - \left(\frac{70}{100}\right)(16°C - 7°C)$$

$$= 9.7°C \quad (10°C)$$

The answer is (B).

6. From the psychrometric chart, for incoming air at 65°F and 55% relative humidity, $W_1 = 0.0072$ lbm moisture/lbm air. [ASHRAE Psychrometric Chart No. 1 - Normal Temperature at Sea Level]

From a table of thermal properties of ideal gases, the specific heat for air is 0.240 Btu/lbm-°R and 0.445 Btu/lbm-°R for water. [Thermal and Physical Properties of Ideal Gases (at Room Temperature)]

With sensible heating as a limiting factor, calculate the mass flow rate of the air entering the theater. [Space Heat Absorption and Moist-Air Moisture Gains]

$$q = \dot{m}_{da}(c_{p,\text{air}} + Wc_{p,\text{moisture}})(T_2 - T_1)$$

$$\dot{m}_{da} = \frac{q}{(c_{p,\text{air}} + Wc_{p,\text{moisture}})(T_2 - T_1)}$$

$$= \frac{500{,}000 \frac{\text{Btu}}{\text{hr}}}{\left(\left(0.240 \frac{\text{Btu}}{\text{lbm-°F}} + (0.0072 \frac{\text{lbm moisture}}{\text{lbm air}})\right.\right.}$$
$$\left.\left.\times \left(0.445 \frac{\text{Btu}}{\text{lbm-°F}}\right)\right)(75°F - 65°F)\right)$$

$$= 2.056 \times 10^5 \text{ lbm air/hr}$$

Assume that this air absorbs all the moisture. Find the final humidity ratio.

Moist-Air Cooling and Dehumidification

$$\dot{m}_w = \dot{m}_{da}(W_2 - W_1)$$

$$W_2 = \left(\frac{\dot{m}_w}{\dot{m}_{da}}\right) + W_1$$

$$= \frac{175 \frac{\text{lbm moisture}}{\text{hr}}}{2.056 \times 10^5 \frac{\text{lbm air}}{\text{hr}}} + 0.0072 \frac{\text{lbm moisture}}{\text{lbm air}}$$

$$= 0.00805 \text{ lbm moisture/lbm air}$$

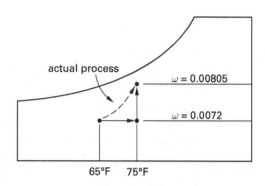

Find the final condition of the air leaving the theater.

$$T_{db} = 75°F \quad [\text{given}]$$
$$W_2 = 0.00805 \text{ lbm moisture/lbm air}$$

From a psychrometric chart, the relative humidity is 44%. This is below 60%. [ASHRAE Psychrometric Chart No. 1 - Normal Temperature at Sea Level]

The answer is (C).

7. Use the illustration shown.

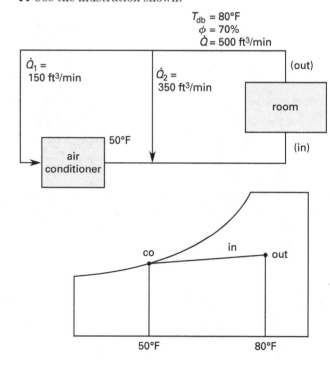

Locate point "out" at 80°F dry-bulb temperature and relative humidity is 70%, and point "co" (saturated at 50°F) on the psychrometric chart. [ASHRAE Psychrometric Chart No. 1 - Normal Temperature at Sea Level]

$$v_{\text{out}} = 13.95 \text{ ft}^3/\text{lbm air}$$
$$h_{\text{out}} = 36.2 \text{ Btu/lbm air}$$

At point "co,"

$$h_{\text{co}} = 20.3 \text{ Btu/lbm air}$$

The mass flow rate of air through the air conditioner is

$$\dot{m}_1 = \frac{Q_1}{v_1} = \frac{Q_1}{v_{\text{out}}} = \frac{150 \dfrac{\text{ft}^3}{\text{min}}}{13.95 \dfrac{\text{ft}^3}{\text{lbm air}}} = 10.75 \text{ lbm air/min}$$

The mass flow rate of the bypass air is

$$\dot{m}_2 = \frac{Q_2}{v} = \frac{\left(500 \dfrac{\text{ft}^3}{\text{min}} - 150 \dfrac{\text{ft}^3}{\text{min}}\right)}{13.95 \dfrac{\text{ft}^3}{\text{lbm air}}}$$
$$= 25.09 \text{ lbm air/min}$$

The percentage of bypass air is

$$x = \frac{25.09 \dfrac{\text{lbm air}}{\text{min}}}{10.75 \dfrac{\text{lbm air}}{\text{min}} + 25.09 \dfrac{\text{lbm air}}{\text{min}}} = 0.70$$

Using the lever rule and the fact that all of the temperature scales are linear,

$$T_{\text{db,in}} = T_{\text{co}} + (0.70)(T_{\text{out}} - T_{\text{co}})$$
$$= 50°F + (0.70)(80°F - 50°F)$$
$$= 71°F$$

At that point (placed on the line between two points), the dry-bulb temperature is 71°F and the relative humidity is 81%.

The answer is (D).

8. Use the illustration shown.

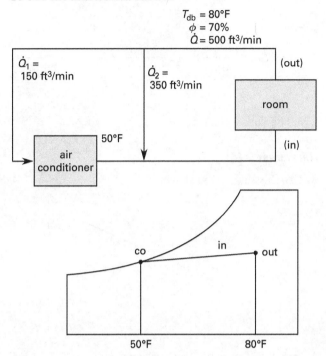

Locate point "out" at 80°F dry-bulb temperature and at 70% relative humidity, and locate point "co" (saturated at 50°F) on the psychrometric chart. [ASHRAE Psychrometric Chart No. 1 - Normal Temperature at Sea Level]

At point "out,"

$$v_{\text{out}} = 13.95 \text{ ft}^3/\text{lbm air}$$
$$h_{\text{out}} = 36.2 \text{ Btu/lbm air}$$

At point "co,"

$$h_{\text{co}} = 20.3 \text{ Btu/lbm air}$$

The mass flow rate of air through the air conditioner is

$$\dot{m}_1 = \frac{Q_1}{v_1} = \frac{Q_1}{v_{\text{out}}} = \frac{150 \dfrac{\text{ft}^3}{\text{min}}}{13.95 \dfrac{\text{ft}^3}{\text{lbm air}}} = 10.75 \text{ lbm air/min}$$

The air conditioner capacity is given by

$$q = \dot{m}_{\text{air}}(h_{T,2} - h_{T,1}) = \dot{m}_1(h_{\text{out}} - h_{\text{co}})$$

$$= \frac{\left(10.75 \ \dfrac{\text{lbm air}}{\text{min}}\right)\left(36.2 \ \dfrac{\text{Btu}}{\text{lbm air}} - 20.3 \ \dfrac{\text{Btu}}{\text{lbm air}}\right)}{200 \ \dfrac{\text{Btu}}{\text{min-ton}}}$$

$$= 0.85 \text{ tons} \quad (0.90 \text{ tons})$$

The answer is (A).

9. Use the illustration shown.

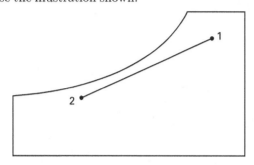

At point 1, from the psychrometric chart, $T_{\text{db}} = 95°\text{F}$ and $\phi = 70\%$. [ASHRAE Psychrometric Chart No. 1 - Normal Temperature at Sea Level]

$$h_1 = 50.7 \text{ Btu/lbm air}$$
$$v_1 = 14.56 \text{ ft}^3/\text{lbm air}$$
$$W_1 = 0.0253 \text{ lbm water/lbm air}$$

At point 2, from the psychrometric chart, $T_{\text{db}} = 60°\text{F}$ and $\phi = 95\%$. [ASHRAE Psychrometric Chart No. 1 - Normal Temperature at Sea Level]

$$h_2 = 25.8 \text{ Btu/lbm air}$$
$$W_2 = 0.0105 \text{ lbm water/lbm air}$$

The air mass flow rate (of dry air) is

$$\dot{m}_a = \frac{Q}{v_1} = \frac{5000 \ \dfrac{\text{ft}^3}{\text{min}}}{14.56 \ \dfrac{\text{ft}^3}{\text{lbm air}}} = 343.4 \text{ lbm air/min}$$

Arrange the heat equation to find the rate of heat removed from the air. The heat change associated with dehumidification is negligible compared to the heat change due to the cooling and can be neglected.

Moist-Air Cooling and Dehumidification

$$q = \dot{m}_{\text{da}}(h_1 - h_2)$$

$$= \left(343.4 \ \dfrac{\text{lbm air}}{\text{min}}\right)\left(50.7 \ \dfrac{\text{Btu}}{\text{lbm air}} - 25.8 \dfrac{\text{Btu}}{\text{lbm air}}\right)$$

$$= 8551 \text{ Btu/min} \quad (8550 \text{ Btu/min})$$

The answer is (D).

10. The T-s and h-s diagrams are shown for a refrigeration cycle using refrigerant-12 (R-12) and operating at saturated conditions at 100°F.

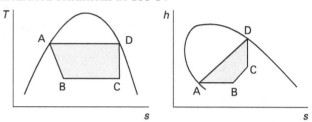

A to B is the throttling of refrigerant from high pressure to low pressure. The internal energy does not change, so enthalpy is constant. From the ideal gas law, the reduction in pressure has a proportional reduction in temperature.

B to C is the evaporation of the low-pressure gas of the air. This absorbs energy from the air and decreases the temperature of the air.

C to D is the compression process from low pressure to high pressure with a proportional increase in temperature. Treat this as a reversible adiabatic process so that entropy is constant (isentropic). The work done on the refrigerant leads to an increase in entropy. The question states that the cycle is saturated at the high temperature condition and not superheated, so point D is located at the saturation line of the T-s and h-s diagrams.

D to A is the condensation of the refrigerant from a saturated vapor to saturated liquid with a decrease in enthalpy. Only the latent heat of vaporization is removed, so no temperature change occurs.

11. *Method 1*: Using a high-temperature psychrometric chart, locate the point where the 180°F line intersects the 20% relative humidity line. From that point, move horizontally to the left to intersect the saturation curve. [ASHRAE Psychrometric Chart No. 3 - High Temperature at Sea Level]

The dew-point temperature is 116°F (115°F).

Method 2: Interpolating from steam tables, the saturation pressure, p_{sat}, of 180°F water vapor is 7.52 psia. [Properties of Saturated Water and Steam (Temperature) - I-P Units]

Use p_{sat} to find the partial pressure of the water vapor, p_w. The relative humidity, ϕ, is given as 20%.

$$\phi = \frac{p_w}{p_{sat}}$$

$$p_w = \phi p_{sat} = (0.20)\left(7.52 \frac{\text{lbf}}{\text{in}^2}\right)$$

$$= 1.5 \text{ lbf/in}^2$$

From steam tables, the dew-point temperature corresponding to 1.5 lbf/in² is approximately 115.8°F (115°F). [Properties of Saturated Water and Steam (Pressure) - I-P Units]

The answer is (B).

12. With a pressure of 19 psia, standard conditions do not apply, so a high-temperature psychrometric chart cannot be used. There are 7000 grains per pound. The humidity ratio expressed in lbm/lbm is

$$W = \frac{560 \frac{\text{gr}}{\text{lbm}}}{7000 \frac{\text{gr}}{\text{lbm}}} = 0.08 \frac{\text{lbm}}{\text{lbm}}$$

The total pressure, p_t, is 19 psia. Using the humidity ratio, find the partial pressure of the water vapor, p_w, in the combustion products. [Psychrometric Properties]

$$p_w = \frac{p_t W}{0.622 + W} = \frac{(19 \text{ psia})\left(0.08 \frac{\text{lbm}}{\text{lbm}}\right)}{0.622 + 0.08 \frac{\text{lbm}}{\text{lbm}}}$$

$$= 2.17 \text{ psia}$$

The dew point is found from steam tables as the saturation temperature corresponding to the partial pressure of the water vapor. Use 2.0 psia for convenience. [Properties of Saturated Water and Steam (Temperature) - I-P Units]

The saturation temperature of 2.0 psia water vapor is 126°F (130°F).

The answer is (D).

13. On a psychrometric chart, plot the outside air (state 1: 91°F dry-bulb and 77°F wet-bulb) and the room air (state 2: 75°F dry-bulb and 50% relative humidity). [ASHRAE Psychrometric Chart No. 1 - Normal Temperature at Sea Level]

$$h_{oa} = h_1 = 40 \frac{\text{BTU}}{\text{hr}}$$

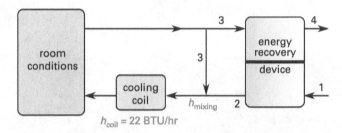

(a) energy recovery device

(b) energy recovery ventilation schematic

This problem asks to calculate the cooling load of the AC system. That is the energy absorbed by the air and dissipated by the given cooling coil, with efficiency, as the air flows through. In this case, the airflow is mixed. This is fresh air in addition to recirculated air.

Then, obtain the cooling load from the difference in enthalpies before and after the cooling coil. Thus, we need to find h_2.

From given outside air and room conditions,

$$h_{oa} = h_1 = 40 \frac{\text{Btu}}{\text{hr}}$$

$$h_{\text{roomcondition}} = h_3 = 28 \frac{\text{Btu}}{\text{hr}}$$

The total effectiveness of an ERV is

$$\epsilon_t = \frac{\dot{m}_s(h_2 - h_1)}{\dot{m}_{\min}(h_3 - h_1)} = \frac{(h_2 - 40)}{(28 - 40)}$$

$$h_2 = 31 \frac{\text{Btu}}{\text{hr}}$$

Because of the assumptions permitted in the description, the term with the mass flow rate is equal to 1.

This h_2 is enthalpy of air leaving the ERV which then mixes with the 7000 CFM of return air. The mixing of these two air streams gives the enthalpy of air before the cooling coil.

$$h_{\text{mixing}} = \frac{\left(31\ \dfrac{\text{Btu}}{\text{hr}}\right)(3000\text{CFM}) + \left(28\ \dfrac{\text{Btu}}{\text{hr}}\right)(7000\text{CFM})}{10{,}000\text{CFM}}$$

$$= 28.9\ \dfrac{\text{Btu}}{\text{hr}}$$

Finally, having obtained h_{mixing}, calculate the cooling load corresponding to the proposed air mixture. The simplified cooling load equation can be used because of the assumptions in the introduction. [Heat Gain Calculations Using Standard Air and Water Values]

$$Q_t = 4.5\text{CFM}(\Delta h)$$

$$= 4.5(10{,}000\text{CFM})\left(28.9\ \dfrac{\text{Btu}}{\text{hr}} - 22\ \dfrac{\text{Btu}}{\text{hr}}\right)$$

$$= 310{,}500\ \text{Btu/hr}$$

The answer is (A).

14. According to the figure for airstream numbering conventions, in a Heat Recovery Ventilator (*NCEES Handbook*), the supply air is cold and gains heat from the HRV resulting in an enthalpy increase from h_1 (entrance) to h_2 (exit).

The exhaust air is warm and transfers heat to the supply air resulting in an enthalpy decrease from h_3 (entrance) to h_4 (exit).

In this problem, the outside air loses heat since it is cooled by the return air, thus reducing the cooling load on the system.

The maximum possible heat transfer is driven the enthalpy difference between the warmest stream (outside air entering, h_1) and the coldest stream (return air entering, h_3) in the HRV.

Using a psychrometric chart, find the enthalpies of the outside and return air entering the ERV. [ASHRAE Psychrometric Chart No. 1 - Normal Temperature at Sea Level]

For outside air at 95°F dry-bulb temperature and 81°F wet-bulb temperature,

$$h_1 = 44.6\ \text{Btu/lbm}$$

For the return air at 75°F dry-bulb and 63°F wet-bulb,

$$h_3 = 28.5\ \text{Btu/lbm}$$

Find the maximum energy transfer rate by using the formula for maximum latent heat transfer. Multiply the volumetric flow rate by the given density to convert to mass flow rate.

Energy-Recovery Ventilator (ERV)

$$q_{t,\max} = 60\dot{m}_{\min}(h_1 - h_3)$$

$$= \left(60\ \dfrac{\min}{\text{hr}}\right)\left(9000\ \dfrac{\text{ft}^3}{\min}\right)$$

$$\times \left(44.6\ \dfrac{\text{Btu}}{\text{lbm}} - 28.5\ \dfrac{\text{Btu}}{\text{lbm}}\right)\left(0.069\ \dfrac{\text{lbm}}{\text{ft}^3}\right)$$

$$= 599{,}886\ \text{Btu/hr}$$

Use the equation for determining the total effectiveness, ϵ_t, of an ERV to find the actual energy transfer rate.

Energy-Recovery Ventilator (ERV)

$$\epsilon_t = \dfrac{q_t}{q_{t,\max}}$$

$$q_t = \epsilon_t q_{t,\max}$$

$$= (0.55)\left(599{,}886\ \dfrac{\text{Btu}}{\text{hr}}\right)$$

$$= 329{,}937\ \text{Btu/hr}\quad (330{,}000\ \text{Btu/hr})$$

The answer is (C).

38 Cooling Towers and Fluid Coolers

Content in blue refers to the *NCEES Handbook*.

PRACTICE PROBLEMS

1. A counterflow air cooling tower removes 1,000,000 Btu/hr from a water flow through evaporation. The temperature of the water passing through the cooling tower is reduced from 120°F to 110°F. Air enters the cooling tower at 91°F dry-bulb temperature and 60% relative humidity, and air leaves the cooling tower at 100°F and 82% relative humidity. The flow rate of the air is most nearly

(A) 39,500 lbm/hr
(B) 42,000 lbm/hr
(C) 50,500 lbm/hr
(D) 55,000 lbm/hr

2. A counterflow air cooling tower cools a water flow through evaporation. The temperature of the water passing through the cooling tower is reduced from 120°F to 110°F. Air enters the cooling tower at 91°F dry-bulb temperature and 60% relative humidity, and air leaves the cooling tower at 100°F and 82% relative humidity. Assume the mass flow rate of the dry air through the tower is 50,000 lbm/hr. The needed rate of makeup water is most nearly

(A) 650 lbm/hr
(B) 780 lbm/hr
(C) 900 lbm/hr
(D) 1000 lbm/hr

3. An evaporative cooling tower is used by a large power-generating facility. The circulating water enters the tower at 1000 gal/min at 90°F. The dry-bulb temperature of the air at the inlet is 70°F and the wet-bulb temperature is 50°F. The exit temperature of the air is 80°F at 80% relative humidity. The cooling tower is designed to cool the water to 70°F. The needed rate of makeup water is most nearly

(A) 12 gal/min
(B) 14 gal/min
(C) 16 gal/min
(D) 18 gal/min

4. An ideal vapor compression refrigeration system using refrigerant-134a is used in a facility where the sensible heat load is 100,000 Btu/hr and the sensible heat ratio is 0.70. The refrigeration cycle operates between an evaporator pressure of 20 psia and a condenser pressure of 100 psia. The temperature rise of the cooling water is limited to 10°F. The water flow rate needed in the condenser is most nearly

(A) 29 gpm
(B) 34 gpm
(C) 41 gpm
(D) 45 gpm

SOLUTIONS

1. The dry-bulb temperature, T_{db}, of the air entering the cooling tower is 91°F and the relative humidity, ϕ, is 60%. From the psychrometric chart, find the enthalpy of the air entering the cooling tower. [ASHRAE Psychrometric Chart No. 1 - Normal Temperature at Sea Level]

$$h_{air,in} \approx 42.7 \text{ Btu/lbm}$$

From the psychrometric chart, find the enthalpy of the air leaving the cooling tower (100 F, dry bulb temperature and 82% relative humidity). [ASHRAE Psychrometric Chart No. 3 - High Temperature at Sea Level]

$$h_{air,out} = 62.5 \text{ Btu/lbm air}$$

Find the mass flow rate of dry air.

Moist-Air Sensible Heating or Cooling

$$q = \dot{m}_a(h_2 - h_1)$$

$$\dot{m}_{air} = \frac{q}{h_2 - h_1} = \frac{1{,}000{,}000 \frac{\text{Btu}}{\text{hr}}}{62.5 \frac{\text{Btu}}{\text{lbm air}} - 42.7 \frac{\text{Btu}}{\text{lbm air}}}$$

$$= 50{,}505 \text{ lbm air/hr} \quad (50{,}500 \text{ lbm air/hr})$$

The answer is (C).

2. The dry-bulb temperature, T_{db}, of the air entering the cooling tower is 91°F and the relative humidity, ϕ, is 60%. From the psychrometric chart, find the humidity ratio of the air entering the cooling tower. [ASHRAE Psychrometric Chart No. 1 - Normal Temperature at Sea Level]

$$W_{air,in} = 0.0190 \text{ lbm moisture/lbm air}$$

From the psychrometric chart, find the humidity ratio of the air leaving the cooling tower. [ASHRAE Psychrometric Chart No. 3 - High Temperature at Sea Level]

$$W_{air,out} = 0.0345 \text{ lbm moisture/lbm air}$$

The moisture removed by the air from the cooling tower is replaced with makeup water, so the mass flow rate of the makeup water is a function of the air's change in humidity ratio. Find the needed rate of makeup water.

Moist-Air Cooling and Dehumidification

$$\dot{m}_{makeup} = \dot{m}_{air}(W_1 - W_2)$$

$$= \left(50{,}000 \frac{\text{lbm air}}{\text{hr}}\right)$$

$$\times \left(0.0345 \frac{\text{lbm moisture}}{\text{lbm air}} - 0.0190 \frac{\text{lbm moisture}}{\text{lbm air}}\right)$$

$$= 775 \text{ lbm water/hr} \quad (780 \text{ lbm water/hr})$$

The answer is (B).

3. The following subscripts are used for the different streams:

inlet air – 1

outlet air – 2

inlet water – 3

outlet water (after the addition of makeup water) – 4

makeup water – 5

Find the mass flow rate of the circulating water entering the tower in pounds per minute. [Commonly Used Equivalents]

$$\dot{m}_3 = \left(1000 \frac{\text{gal}}{\text{min}}\right)\left(8.34 \frac{\text{lbm}}{\text{gal}}\right) = 8340 \text{ lbm/min} = \dot{m}_4$$

The inlet air conditions are $T_{db} = 70°F$ and $T_{wb} = 50°F$. Use the psychrometric chart to locate the intersection of these points. [ASHRAE Psychrometric Chart No. 1 - Normal Temperature at Sea Level]

$$W_1 = 0.0032 \text{ lbm H}_2\text{O/lbm dry air}$$
$$h_1 = 20.3 \text{ Btu/lbm dry air}$$

The outlet air conditions are 80°F and 80% relative humidity. At these conditions,

$$W_2 = 0.0178 \text{ lbm H}_2\text{O/lbm dry air}$$
$$h_2 = 39.0 \text{ Btu/lbm dry air}$$

Using steam tables, determine the enthalpies of water. [Properties of Saturated Water and Steam (Temperature) - I-P Units]

$$h_3 = h_f \text{ at } 90°F = 58.05 \text{ Btu/lbm H}_2\text{O}$$

$$h_4 = h_5 = h_f \text{ at } 70°F = 38.08 \text{ Btu/lbm H}_2\text{O}$$

The mass flow rate of dry air entering and leaving the tower is the same.

$$\dot{m}_1 = \dot{m}_2 = \dot{m}_{air}$$

Find the needed mass flow of dry air. The energy balance around the cooling tower results in the following equation.

$$\dot{m}_{air}h_1 + \dot{m}_3 h_3 + \dot{m}_5 h_5 = \dot{m}_4 h_4 + \dot{m}_{air} h_2$$

The mass balance for the water is

$$\dot{m}_3 + \dot{m}_5 + \dot{m}_{air} W_1 = \dot{m}_4 + \dot{m}_{air} W_2$$

Since $\dot{m}_3 = \dot{m}_4$,

$$\dot{m}_5 = \dot{m}_{air}(W_2 - W_1)$$

Substitute the values into the energy balance equation and solve for mass flow rate of the air.

$$\dot{m}_{air}h_1 + \dot{m}_3 h_3 + \dot{m}_{air}(W_2 - W_1)h_5 = \dot{m}_4 h_4 + \dot{m}_{air} h_2$$

$$\dot{m}_{air} = \frac{\dot{m}_3(h_3 - h_4)}{h_2 - h_1 - h_5(W_2 - W_1)}$$

$$= \frac{\left(8340 \dfrac{\text{lbm H}_2\text{O}}{\text{min}}\right) \times \left(58.05 \dfrac{\text{Btu}}{\text{lbm H}_2\text{O}} - 38.08 \dfrac{\text{Btu}}{\text{lbm H}_2\text{O}}\right)}{39 \dfrac{\text{Btu}}{\text{lbm dry air}} - 20.3 \dfrac{\text{Btu}}{\text{lbm dry air}} - \left(38.08 \dfrac{\text{Btu}}{\text{lbm H}_2\text{O}}\right) \times \left(0.0178 \dfrac{\text{lbm H}_2\text{O}}{\text{lbm dry air}} - 0.0032 \dfrac{\text{lbm H}_2\text{O}}{\text{lbm dry air}}\right)}$$

$$= 9179 \text{ lbm dry air/min}$$

Determine the needed makeup water.

$$\dot{m}_5 = \dot{m}_{air}(W_2 - W_1)$$

$$= \left(9179 \dfrac{\text{lbm dry air}}{\text{min}}\right) \times \left(0.0178 \dfrac{\text{lbm H}_2\text{O}}{\text{lbm dry air}} - 0.0032 \dfrac{\text{lbm H}_2\text{O}}{\text{lbm dry air}}\right)$$

$$= 134.01 \text{ lbm H}_2\text{O/min}$$

Convert the needed makeup water to gallons per minute.

$$Q_{water} = \frac{134.01 \dfrac{\text{lbm H}_2\text{O}}{\text{min}}}{8.34 \dfrac{\text{lbm H}_2\text{O}}{\text{gal}}} = 16.07 \text{ gal/min} \quad (16 \text{ gal/min})$$

The answer is (C).

4. Calculate the total cooling load for the system.

Psychrometric Properties

$$\text{SHR} = \frac{\text{sensible heat gain}}{\text{total heat gain}} = \frac{q_s}{q_t}$$

$$q_t = \frac{q_s}{\text{SHR}}$$

$$= \frac{100{,}000 \dfrac{\text{Btu}}{\text{hr}}}{0.70}$$

$$= 142{,}857 \text{ Btu/hr}$$

The enthalpies at each state of the refrigeration system are determined from a pressure versus enthalpy diagram for R-134a.

Refrigeration Cycle - Single Stage

The refrigerant at the compressor inlet is state 1; at the compressor outlet is state 2; at the condenser outlet is state 3; and at the evaporator inlet is state 4. [Pressure Versus Enthalpy Curves for Refrigerant 134a]

Find the specific enthalpy and the specific entropy of the refrigerant at the compressor inlet.

$$h_1 = h_g \text{ at 20 psia} = 102 \text{ Btu/lbm}$$

$$s_1 = s_g \text{ at 20 psia} = 0.225 \text{ Btu/lbm-°R}$$

For isentropic compression, $s_1 = s_2 = 0.225$ Btu/lbm-°R. The pressure after compression is $p_2 = 100$ psia. At $s_2 = 0.225$ and $p_2 = 100$ psia, find the enthalpy of the refrigerant at the compressor outlet. [Pressure Versus Enthalpy Curves for Refrigerant 134a]

$$h_2 = 115 \text{ Btu/lbm}$$

Find the enthalpy of the refrigerant at the condenser outlet.

$$h_3 = h_f \text{ at 100 psia} = 39 \text{ Btu/lbm}$$

The enthalpy of the refrigerant at the outlet of the condenser equals the enthalpy of the refrigerant at the evaporator inlet.

$$h_3 = h_4 = 39 \text{ Btu/lbm}$$

Find the needed mass flow rate of the refrigerant.

$$\dot{m}_{ref} = \frac{q_t}{h_1 - h_4} = \frac{142{,}857 \; \frac{\text{Btu}}{\text{hr}}}{102 \; \frac{\text{Btu}}{\text{lbm}} - 39 \; \frac{\text{Btu}}{\text{lbm}}}$$

$$= 2268 \; \text{lbm/hr}$$

Find the rate of heat rejection in the condenser.

$$\begin{aligned} q_{condenser,\,out} &= \dot{m}_{ref}(h_2 - h_3) \\ &= \left(2268 \; \frac{\text{lbm}}{\text{hr}}\right)\left(115 \; \frac{\text{Btu}}{\text{lbm}} - 39 \; \frac{\text{Btu}}{\text{lbm}}\right) \\ &= 172{,}368 \; \text{Btu/hr} \end{aligned}$$

Find the needed flow rate of the cooling water.

Water-Cooled Condensers

$$Q = \frac{q_{condenser,\,out}}{\rho c_p(T_2 - T_1)} = \frac{172{,}368 \; \frac{\text{Btu}}{\text{hr}}}{\left(62.4 \; \frac{\text{lbm}}{\text{ft}^3}\right)\left(1.0 \; \frac{\text{Btu}}{\text{lbm-}°\text{F}}\right)(10°\text{F})}$$

$$= 276.23 \; \text{ft}^3/\text{hr}$$

Multiply ft³/hr by 0.125 to get gpm.

$$\begin{aligned} Q_{gpm} &= (276.23)(0.125) \\ &= 34.53 \; \text{gal/min} \quad (34 \; \text{gpm}) \end{aligned}$$

The answer is (B).

39 Ventilation

Content in blue refers to the NCEES Handbook.

PRACTICE PROBLEMS

1. An auditorium is designed to seat 4500 people. The ventilation rate is 1.62×10^7 ft³/hr (4.54×10^5 m³/h) of outside air. The outside temperature is 0°F (−18°C) dry-bulb, and the outside pressure is 14.6 psia (100.6 kPa). Air leaves the auditorium at 70°F (21°C) dry-bulb. Use 225 Btu/hr for the sensible heat of a person. There is no recirculation. The temperature at which the air enters the auditorium is most nearly

(A) 52°F (11°C)
(B) 55°F (13°C)
(C) 62°F (17°C)
(D) 67°F (19°C)

2. A room is maintained at design conditions of 75°F (23.9°C) dry-bulb and 50% relative humidity. The air outside is at 95°F (35°C) dry-bulb temperature and 75°F (23.9°C) wet-bulb temperature. The outside air is conditioned and mixed with some room exhaust air. The mixed, conditioned air enters the room and increases 20°F (11.1°C) in temperature before being removed from the room. The sensible and latent loads are 200,000 Btu/hr (60 kW) and 50,000 Btu/hr (15 kW), respectively. Air leaves the coil at 50.8°F (10°C). The volume of air flowing through the coil is most nearly

(A) 4200 ft³/min (120 m³/min)
(B) 5800 ft³/min (160 m³/min)
(C) 7500 ft³/min (210 m³/min)
(D) 9300 ft³/min (260 m³/min)

3. Assuming atmospheric air is a mixture of oxygen and nitrogen only, above what approximate ambient volumetric nitrogen concentration should supplemental oxygen be provided by an employer to employees?

(A) 77% nitrogen
(B) 79% nitrogen
(C) 81% nitrogen
(D) 88% nitrogen

4. What is the appropriate inlet face velocity of return air adjacent to an office desk that is usually occupied?

(A) 300 ft/min
(B) 500 ft/min
(C) 700 ft/min
(D) 900 ft/min

5. From ASHRAE Standard 62.1-2007, a minimum separation distance of at least 10 m is recommended for the installation of an air intake in a space with

(A) significantly contaminated exhaust air
(B) cooling tower exhaust
(C) noxious or dangerous exhaust
(D) exhaust from buses or trucks

SOLUTIONS

1. *Customary U.S. Solution*

Determine the density of outside air at 0°F from a table of properties for air. [Properties of Air at Atmospheric Pressure]

$$\rho = 0.0862 \text{ lbm/ft}^3$$

Calculate the mass flow rate of the air.

$$\dot{m}_{\text{air}} = Q_{\text{air}}\rho = \left(1.62 \times 10^7 \frac{\text{ft}^3}{\text{hr}}\right)\left(0.0862 \frac{\text{lbm}}{\text{ft}^3}\right)$$
$$= 1.396 \times 10^6 \text{ lbm/hr}$$

For a conservative estimate, assume that there is no latent heat from people in the auditorium.

Calculate the rate of heat input to the space from the generation of sensible heat.

$$q_{\text{in}} = \left(225 \frac{\frac{\text{Btu}}{\text{hr}}}{\text{person}}\right)(4500 \text{ persons})$$
$$= 1.01 \times 10^6 \text{ Btu/hr}$$

This heat picked up by the air entering the auditorium can be represented by the following equation. Solve the equation for the change in temperature for the air entering and leaving the auditorium, ΔT_{air}. [Thermal and Physical Properties of Ideal Gases (at Room Temperature)]

$$q_{\text{in}} = \dot{m}_{\text{air}} c_{p,\text{air}} \Delta T_{\text{air}}$$
$$\Delta T_{\text{air}} = \frac{q_{\text{in}}}{\dot{m}_{\text{air}} c_{p,\text{air}}}$$
$$= \frac{1.01 \times 10^6 \frac{\text{Btu}}{\text{hr}}}{\left(1.396 \times 10^6 \frac{\text{lbm}}{\text{hr}}\right)\left(0.240 \frac{\text{Btu}}{\text{lbm-°F}}\right)}$$
$$= 3.01\text{°F}$$

The temperature of the air entering the auditorium is

$$\Delta T_{\text{air}} = T_{\text{air,out}} - T_{\text{air,in}}$$
$$T_{\text{air,in}} = T_{\text{air,out}} - \Delta T_{\text{air}}$$
$$= 70\text{°F} - 3.01\text{°F}$$
$$= 66.99\text{°F} \quad (67\text{°F})$$

SI Solution

Determine the density of outside air at −18°C from a table of properties for air. [Properties of Air at Atmospheric Pressure]

$$\rho = \frac{\left(0.0862 \frac{\text{lbm}}{\text{ft}^3}\right)\left(3.281 \frac{\text{ft}}{\text{m}}\right)^3}{2.205 \frac{\text{lbm}}{\text{kg}}}$$
$$= 1.381 \text{ kg/m}^3$$

Calculate the mass flow rate of air.

$$\dot{m}_{\text{air}} = Q_{\text{air}}\rho = \left(4.54 \times 10^5 \frac{\text{m}^3}{\text{h}}\right)\left(1.381 \frac{\text{kg}}{\text{m}^3}\right)$$
$$= 6.270 \times 10^5 \text{ kg/h}$$

For a conservative estimate, assume that there is no latent heat from people in the auditorium. From a table of rates of heat given off by people in different states of activity, find the sensible heat generated from a person seated in a theatre. [Representative Rates at Which Heat and Moisture Are Given Off by People in Different States of Activity]

$$\text{sensible heat per person} = 225 \text{ Btu/hr}$$

Calculate the rate of heat input to the space from the generation of sensible heat. [Measurement Relationships]

$$q_{\text{in}} = \left(225 \frac{\frac{\text{Btu}}{\text{hr}}}{\text{person}}\right)(4500 \text{ persons})\left(1.055 \frac{\text{kJ}}{\text{Btu}}\right)$$
$$= 1.0656 \text{ kJ/h}$$

This heat picked up by the air entering the auditorium can be represented by the following equation. Solve the equation for the change in temperature for the air entering and leaving the auditorium, ΔT_{air}. [Thermal and Physical Properties of Ideal Gases (at Room Temperature)]

$$q_{\text{in}} = \dot{m}_{\text{air}} c_{p,\text{air}} \Delta T_{\text{air}}$$
$$\Delta T_{\text{air}} = \frac{q_{\text{in}}}{\dot{m}_{\text{air}} c_{p,\text{air}}}$$
$$= \frac{1.0656 \times 10^6 \frac{\text{kJ}}{\text{h}}}{\left(6.270 \times 10^5 \frac{\text{kg}}{\text{h}}\right)\left(1.004 \frac{\text{kJ}}{\text{kg-°C}}\right)}$$
$$= 1.69\text{°C}$$

The temperature of the air entering the auditorium is

$$\Delta T_{air} = T_{air,out} - T_{air,in}$$
$$T_{air,in} = T_{air,out} - \Delta T_{air}$$
$$= 21°C - 1.69°C$$
$$= 19.31°C \quad (19°C)$$

The answer is (D).

2. *Customary U.S. Solution*

The mixed, conditioned air undergoes a temperature increase of 20°F before leaving the room at 75°F. Therefore, the mixed conditioned air enters the room at

$$T_{mixed} = 75°F - 20°F = 55°F$$

Calculate the volumetric flow rate of air entering the room, Q_{mixed}, using the equation for determining the sensible heat load.

Heat Gain Calculations Using Standard Air and Water Values
$$q_s = 1.10 Q_{mixed} \Delta T$$
$$Q_{mixed} = \frac{q_s}{1.10 \Delta T}$$
$$= \frac{200{,}000 \, \frac{\text{Btu}}{\text{hr}}}{\left(1.10 \, \frac{\text{Btu-min}}{\text{ft}^3\text{-hr-°F}}\right)(20°F)}$$
$$= 9091 \, \text{ft}^3/\text{min}$$

The air entering the room is a mix of return air from the room and air from the coil. Use the equation for the mixing of air streams. Since the mixed air consists of both a fraction of return air and a fraction of coil air, the fraction of the return air, f_{return}, can be expressed as $1 - f_{coil}$.

Air-Handling Unit Mixed-Air Plenums
$$T_{mixed} = (\text{fraction coil air}) T_{coil}$$
$$+ (\text{fraction return air}) T_{return}$$
$$= f_{coil} T_{coil} + f_{return} T_{return}$$
$$= f_{coil} T_{coil} + (1 - f_{coil}) T_{return}$$

Solving for the fraction of the coil air gives

$$f_{coil} = \frac{T_{mixed} - T_{return}}{T_{coil} - T_{return}}$$
$$= \frac{55°F - 75°F}{50.8°F - 75°F}$$
$$= 0.8264$$

Find the volume of the air flowing through the coil.

$$Q_{coil} = f_{coil} Q_{mixed}$$
$$= (0.8264)\left(9091 \, \frac{\text{ft}^3}{\text{min}}\right)$$
$$= 7513 \, \text{ft}^3/\text{min} \quad (7500 \, \text{ft}^3/\text{min})$$

SI Solution

The mixed conditioned air undergoes a temperature increase of 11.1°C before leaving the room at 23.9°C. Therefore, the mixed conditioned air enters the room at

$$T_{mixed} = 23.9°C - 11.1°C = 12.8°C$$

Convert the standard density of air to kg/m³. [Measurement Relationships] [Standard Dry Air Conditions at Sea Level]

$$\rho = \frac{\left(0.075 \, \frac{\text{lbm}}{\text{ft}^3}\right)\left(3.281 \, \frac{\text{ft}}{\text{m}}\right)^3}{2.205 \, \frac{\text{lbm}}{\text{kg}}}$$
$$= 1.201 \, \text{kg/m}^3$$

Find the specific heat of air at room temperature using a table of thermal properties of ideal gas. [Thermal and Physical Properties of Ideal Gases (at Room Temperature)]

$$c_p = 1.004 \, \text{kJ/kg·K} = 1.004 \, \text{kJ/kg·°C}$$

Calculate the volumetric flow rate of air entering the room using the equation for determining the sensible heat load.

$$q_s = \dot{m} c_p \Delta T = (\rho Q_{mixed}) c_p \Delta T$$
$$Q_{mixed} = \frac{q_s}{\rho c_p \Delta T}$$
$$= \frac{60 \, \text{kW}}{\left(1.201 \, \frac{\text{kg}}{\text{m}^3}\right)\left(1.004 \, \frac{\text{kJ}}{\text{kg·°C}}\right)(11.1°C)}$$
$$= 4.483 \, \text{m}^3/\text{s}$$

The air entering the room is a mix of return air from the room and air from the coil. Use the equation for the mixing of air streams. Since the mixed air consists of both a fraction of return air and a fraction of coil air, the fraction of the return air, f_{return}, can be expressed as $1 - f_{coil}$.

Air-Handling Unit Mixed-Air Plenums

$$T_{\text{mixed}} = (\text{fraction coil air})\, T_{\text{coil}}$$
$$+ (\text{fraction return air})\, T_{\text{return}}$$
$$= f_{\text{coil}} T_{\text{coil}} + f_{\text{return}} T_{\text{return}}$$
$$= f_{\text{coil}} T_{\text{coil}} + (1 - f_{\text{coil}})\, T_{\text{return}}$$

Solving for the fraction of the coil air gives

$$f_{\text{coil}} = \frac{T_{\text{mixed}} - T_{\text{return}}}{T_{\text{coil}} - T_{\text{return}}}$$
$$= \frac{12.8°C - 23.9°C}{10°C - 23.9°C}$$
$$= 0.7986$$

Find the volume of the air flowing through the coil.

$$Q_{\text{coil}} = f_{\text{coil}} Q_{\text{mixed}}$$
$$= (0.7986)\left(4.483\ \frac{m^3}{s}\right)\left(60\ \frac{s}{\text{min}}\right)$$
$$= 214.8\ m^3/\text{min} \quad (210\ m^3/\text{min})$$

The answer is (C).

3. OSHA defines an oxygen-deficient atmosphere as one having less than 19.5% oxygen. In this scenario, the maximum nitrogen content would be

$$100\% - 19.5\% = 80.5\% \quad (81\%)$$

The answer is (C).

4. The space adjacent to an office desk is considered to be an occupied zone near a seat. From a table of return air face velocities, the velocity should be in the range of 400 fpm to 600 fpm. [Recommended Return Inlet Face Velocity]

The answer is (B).

5. From ASHRAE Standard 62.1-2007, the minimum separation distance from noxious and dangerous exhaust should be at least 10 m (30 ft). [Air Intake Minimum Separation Distance, Based on ANSI/ASHRAE Standard 62.1-2007]

The answer is (C).

40 Fans, Ductwork, and Terminal Devices

Content in blue refers to the *NCEES Handbook*.

PRACTICE PROBLEMS

(Note: Round all duct dimensions to the next larger whole inch or multiples of 25 mm.)

1. A round 18 in (457 mm) duct is to be replaced by a rectangular duct with an aspect ratio of 4:1. The dimensions of the rectangular duct are most nearly

(A) 8 in × 30 in (200 mm × 760 mm)
(B) 9 in × 35 in (220 mm × 880 mm)
(C) 12 in × 18 in (300 mm × 460 mm)
(D) 18 in × 72 in (460 mm × 1800 mm)

2. A fan moves 10,000 SCFM (4700 L/s) through ductwork that has a total pressure of 4 in wg (1 kPa). If the fan speed is decreased so that the flow rate becomes 8000 SCFM (3700 L/s), the total pressure in the duct after the fan will most nearly be

(A) 2.2 in wg (0.53 kPa)
(B) 2.4 in wg (0.58 kPa)
(C) 2.6 in wg (0.62 kPa)
(D) 2.8 in wg (0.68 kPa)

3. Air flows in an 18 in diameter round duct at a flow rate of 1500 SCFM. After a branch reduction of 300 SCFM, the fitting reduces to 14 in in the through direction. The equation for determining the static regain in the through direction is

$$SR_{actual} = (1.1)\left[\frac{v_{up}^2 - v_{down}^2}{\left((\sqrt{in\ wg})\left(4005\ \frac{ft}{min}\right)\right)^2}\right]$$

The static regain in the through direction is most nearly

(A) −0.037 in wg loss
(B) −0.073 in wg loss
(C) −0.11 in wg loss
(D) −0.19 in wg loss

4. Which of the following parameters is normally sensed in order to control a variable frequency drive (VFD) in a variable air volume (VAV) system?

(A) current of the fan
(B) duct velocity pressure
(C) motor flow rate through the supply air fan
(D) flow rate through the VAV box

5. Fan A and fan B are identical variable-speed fans operating in parallel. They are of the same rating, and they have the same efficiency. Both fans draw from the same source. Fan A always turns at 2000 rev/min and has a volumetric flow rate of 4000 ft³/min. The speed of Fan B varies according to demand. When fan B turns at 1800 rev/min, its volumetric flow rate is 5000 ft³/min. At any particular moment, the total flow rate supplied by fans A and B is 10,000 ft³/min. When operating in parallel with fan A, the speed of fan B is most nearly

(A) 1500 rev/min
(B) 2100 rev/min
(C) 2200 rev/min
(D) 4200 rev/min

6. Which of the following terminal units is best suited for zones where cooling loads fluctuate during occupied hours?

(A) single duct constant volume system
(B) single duct variable air volume (VAV) system
(C) fan-powered parallel VAV system
(D) fan-powered series VAV system

SOLUTIONS

1. Customary U.S. Solution

Find the dimensions of the rectangular duct using the equation for the equivalent diameter. Since the aspect ratio of the rectangular duct is 4:1, $b = 4a$. Therefore,

Rectangular Ducts

$$D_{eq} = \frac{1.30(ab)^{0.625}}{(a+b)^{0.250}}$$

$$= \frac{1.30(4a^2)^{0.625}}{(5a)^{0.250}}$$

$$18 \text{ in} = 2.0677a$$
$$a = 8.71 \text{ in} \quad (9 \text{ in})$$

Solving for the other dimension gives

$$b = 4a = (4)(8.71 \text{ in}) = 34.84 \text{ in} \quad (35 \text{ in})$$

SI Solution

Find the dimensions of the rectangular duct using the equation for equivalent diameter. Since the aspect ratio of the rectangular duct is 4:1, $b = 4a$. Therefore,

Rectangular Ducts

$$D_{eq} = \frac{1.30(ab)^{0.625}}{(a+b)^{0.250}}$$

$$= \frac{1.30(4a^2)^{0.625}}{(5a)^{0.250}}$$

$$457 \text{ mm} = 2.0677a$$
$$a = 221 \text{ mm} \quad (220 \text{ mm})$$

Solving for the other dimension gives

$$b = 4a = (4)(220 \text{ mm}) = 880 \text{ mm}$$

The answer is (B).

2. Customary U.S. Solution

Use the fan affinity laws to calculate the pressure in the duct after the fan. Since the same fan is used, $D_2 = D_1$. The same fluid is being moved, so $\rho_2 = \rho_1$. Therefore,

Fan Laws

$$p_1 = p_2 \times \left(\frac{D_2}{D_1}\right)^4 \left(\frac{Q_1}{Q_2}\right)^2 \left(\frac{\rho_1}{\rho_2}\right) \quad [\text{law 3b}]$$

$$= (4 \text{ in wg})\left(\frac{8000 \text{ SCFM}}{10{,}000 \text{ SCFM}}\right)^2$$

$$= 2.56 \text{ in wg} \quad (2.6 \text{ in wg})$$

SI Solution

Use the fan affinity laws to calculate the pressure in the duct after the fan. Since the same fan is used, $D_2 = D_1$. The same fluid is being moved, so $\rho_2 = \rho_1$. Therefore,

Fan Laws

$$p_1 = p_2 \times \left(\frac{D_2}{D_1}\right)^4 \left(\frac{Q_1}{Q_2}\right)^2 \left(\frac{\rho_1}{\rho_2}\right) \quad [\text{law 3b}]$$

$$= (1 \text{ kPa})\left(\frac{3700 \, \frac{L}{s}}{4700 \, \frac{L}{s}}\right)^2$$

$$= 0.6197 \text{ kPa} \quad (0.62 \text{ kPa})$$

The answer is (C).

3. Find the area of the 18 in duct.

$$A_1 = \frac{\pi}{4}D^2 = \left(\frac{\pi}{4}\right)\left(\frac{18 \text{ in}}{12 \, \frac{\text{in}}{\text{ft}}}\right)^2 = 1.767 \text{ ft}^2$$

Calculate the velocity in the 18 in duct using the continuity equation.

Continuity Equation

$$Q = Av$$

$$v_{up} = \frac{Q_1}{A_1}$$

$$= \frac{1500 \, \frac{\text{ft}^3}{\text{min}}}{1.767 \text{ ft}^2}$$

$$= 848.9 \text{ ft/min}$$

Find the area of the 14 in duct.

$$A_2 = \frac{\pi}{4}D^2 = \left(\frac{\pi}{4}\right)\left(\frac{14 \text{ in}}{12 \frac{\text{in}}{\text{ft}}}\right)^2 = 1.069 \text{ ft}^2$$

Calculate the velocity in the 14 in duct using the continuity equation.

Continuity Equation

$$Q = A\text{v}$$

$$\text{v}_{\text{down}} = \frac{Q_2}{A_2}$$

$$= \frac{\left(1500 \frac{\text{ft}^3}{\text{min}} - 300 \frac{\text{ft}^3}{\text{min}}\right)}{1.069 \text{ ft}^2}$$

$$= 1122.5 \text{ ft/min}$$

Since $\text{v}_2 > \text{v}_1$, there will be a pressure loss in the direction of flow. This pressure loss will be regained as static pressure. The static regain is

$$\text{SR}_{\text{actual}} = (1.1)\left(\frac{\text{v}_{\text{up}}^2 - \text{v}_{\text{down}}^2}{\left((\sqrt{\text{in wg}})\left(4005 \frac{\text{ft}}{\text{min}}\right)\right)^2}\right)$$

$$= (1.1)\left(\frac{\left(848.9 \frac{\text{ft}}{\text{min}}\right)^2 - \left(1122.5 \frac{\text{ft}}{\text{min}}\right)^2}{\left((\sqrt{\text{in wg}})\left(4005 \frac{\text{ft}}{\text{min}}\right)\right)^2}\right)$$

$$= -0.037 \text{ in wg loss}$$

The answer is (A).

4. A pitot tube placed at the entrance of the variable air volume (VAV) box can measure the velocity of air. The pressure taps of the pitot tube can measure the difference between the total (or stagnation) pressure and the static pressure. This difference is the duct velocity pressure. A differential pressure transmitter sends the signal to the controller, which can determine the fan speed needed.

The answer is (B).

5. The volumetric flow rate of fan A is 4000 ft³/min. Therefore, the volumetric flow rate of fan B is

$$Q_B = Q_{\text{total}} - Q_A = 10{,}000 \frac{\text{ft}^3}{\text{min}} - 4000 \frac{\text{ft}^3}{\text{min}}$$

$$= 6000 \text{ ft}^3/\text{min}$$

Fan B delivers 5000 ft³/min at 1800 rpm. Calculate the fan speed needed to deliver 6000 ft³/min by using the fan affinity laws. Since the fan size is constant, $D_1 = D_2$.

Fan Laws

$$Q_1 = Q_2\left(\frac{D_1}{D_2}\right)^3\left(\frac{N_1}{N_2}\right) \quad \text{[law 1a]}$$

$$N_2 = N_1\left(\frac{Q_2}{Q_1}\right)$$

$$= \left(1800 \frac{\text{rev}}{\text{min}}\right)\left(\frac{6000 \frac{\text{ft}^3}{\text{min}}}{5000 \frac{\text{ft}^3}{\text{min}}}\right)$$

$$= 2160 \text{ rev/min} \quad (2200 \text{ rev/min})$$

The answer is (C).

6. In a fan-powered parallel variable air volume (VAV) system, the fan operates in parallel with the VAV damper for the primary supply air. The fan typically controls the flow of the return or plenum air and can operate intermittently to respond to fluctuating cooling loads. [Parallel Fan-Powered VAV Terminal Unit]

The answer is (C).

41 Heating Load

Content in blue refers to the *NCEES Handbook*.

PRACTICE PROBLEMS

1. A building is located in New York City. There are 4772 (2651 in SI) degree-days (basis of 65°F (18°C)) during the October 15 to May 15 period for this area. The building is heated by fuel oil that has a heating value of 153,600 Btu/gal (42 800 MJ/m³) and costs $2.00/gal ($0.54/L). The calculated design heat loss is 3.5×10^6 Btu/hr (1 MW) based on 70°F (21.1°C) inside and 0°F (−17.8°C) outside design temperatures. The boiler has an efficiency of 85%. The cost of heating this building during the winter (October 15 through May 15) is most nearly

(A) $70,000
(B) $80,000
(C) $90,000
(D) $100,000

2. A flat roof consists of 1½ in of expanded perlite insulation installed over 3 in of spruce-pine-fir sheathing, and a ¾ in wood-fiberboard acoustical tile ceiling suspended 3.5 in below the spruce-pine-fir. (The surface emissivity of soft pine is $\epsilon_{wood} = 0.90$.) The heat flow direction is downward. The outside wind velocity is 15 mi/hr. The interior design temperature is 80°F. The exterior design temperature is 95°F. The overall coefficient of heat transfer is most nearly

(A) 0.08 Btu/hr-ft²-°F
(B) 1.0 Btu/hr-ft²-°F
(C) 4.1 Btu/hr-ft²-°F
(D) 18 Btu/hr-ft²-°F

3. A 12 ft × 12 ft (3.6 m × 3.6 m) floor is constructed as a concrete slab with two exposed edges. The slab edge heat loss coefficient for these two edges is 0.55 Btu/hr-ft-°F (0.95 W/m·°C). The other two edges form part of a basement wall exposed to 70°F air. The inside design temperature is 70°F (21.1°C). The outdoor design temperature is −10°F (−23.3°C). Using the slab edge method, the heat loss from the slab is most nearly

(A) 500 Btu/hr (140 W)
(B) 800 Btu/hr (240 W)
(C) 1100 Btu/hr (300 W)
(D) 4200 Btu/hr (1300 W)

4. An outdoor solarium is heated to 68°F by a small electric cabinet heater. The solarium contains 160 ft² of fixed, vertical double-pane window units with a ½ in air gap and unbroken metal framing. In addition to the fixed window units, there is a ⅛ in sliding glass door unit with an area of 52 ft² and reinforced vinyl framing. The outdoor design temperature is 10°F (23.3°C). The capacity of the cabinet electric heater required to maintain the space temperature of 68°F is most nearly

(A) 8600 Btu/hr
(B) 9100 Btu/hr
(C) 9400 Btu/hr
(D) 9800 Btu/hr

SOLUTIONS

1. *Customary U.S. Solution*

Estimate the fuel consumption during the heating season.

$$\frac{\left(24 \frac{\text{hr}}{\text{day}}\right) q_{\text{Btu/hr}}(\text{HDD})}{(T_i - T_o)(\text{HV}_{\text{Btu/gal}})\eta_{\text{furnace}}}$$

$$= \frac{\left(24 \frac{\text{hr}}{\text{day}}\right)\left(3.5 \times 10^6 \frac{\text{Btu}}{\text{hr}}\right)(4772°\text{F-days})}{(70°\text{F} - 0°\text{F})\left(153{,}600 \frac{\text{Btu}}{\text{gal}}\right)(0.85)}$$

$$= 43{,}860 \text{ gal}$$

The total cost of 43,860 gal fuel at $2.00/gal is

$$(43{,}860 \text{ gal})\left(\frac{\$2.00}{\text{gal}}\right) = \$87{,}720 \quad (\$90{,}000)$$

SI Solution

Estimate the fuel consumption during the heating season.

$$\frac{\left(86\,400 \frac{\text{s}}{\text{d}}\right) q_{\text{kW}}(\text{HDD})}{(T_i - T_o)(\text{HV}_{\text{kJ/L}})\eta_{\text{furnace}}}$$

$$= \frac{\left(86\,400 \frac{\text{s}}{\text{d}}\right)(1 \text{ MW})\left(1000 \frac{\text{kW}}{\text{MW}}\right)}{(21.1°\text{C} - (-17.8°\text{C}))\left(42\,800 \frac{\text{MJ}}{\text{m}^3}\right)}$$

$$\times \left(1000 \frac{\text{kJ}}{\text{MJ}}\right)(0.85)$$

$$= 161\,849 \text{ L}$$

The total cost at $0.54/L is

$$(161\,849 \text{ L})\left(\frac{\$0.54}{\text{L}}\right) = \$87{,}398 \quad (\$90{,}000)$$

The answer is (C).

2. From a table of surface film coefficients for air, the surface film coefficient for outdoor air (horizontal, heat flow downward 15 mph) is $h_o = 6.00$ Btu/hr-ft²-°F. [Surface Film Coefficients/Resistances for Air]

The film thermal resistance is

$$R_1 = \frac{1}{h_o} = \frac{1}{6.00 \frac{\text{Btu}}{\text{hr-ft}^2\text{-°F}}}$$

$$= 0.167 \text{ hr-ft}^2\text{-°F/Btu}$$

From a table of thermal resistances of building materials, for expanded perlite roof insulation, the thermal resistance per inch is 2.78 hr-ft²-°F/Btu-in. [Thermal Resistance of Building Materials]

$$R_2 = \left(2.78 \frac{\text{hr-ft}^2\text{-°F}}{\text{Btu-in}}\right)(1.5 \text{ in}) = 4.17 \text{ hr-ft}^2\text{-°F/Btu}$$

The thermal resistance per inch for softwood spruce-pine-fir is between 1.11 hr-ft²-°F/Btu-in and 1.35 hr-ft²-°F/Btu-in. Use the weighted average of 1.25 hr-ft²-°F/Btu-in. [Thermal Resistance of Building Materials]

$$R_3 = \left(1.25 \frac{\text{hr-ft}^2\text{-°F}}{\text{Btu-in}}\right)(3 \text{ in}) = 3.75 \text{ hr-ft}^2\text{-°F/Btu}$$

Most building materials, such as wood and paper, have an emissivity around 0.90. For an air gap bordered by two such surfaces, each with an emissivity of 0.90, the effective emittance is 0.82. [Emissivity of Surfaces and Effective Emittances of Facing Spaces]

From a table of thermal resistances of plane air spaces, the thermal resistance of 3.5 in horizontal planar air space with an effective emittance of 0.82 and mean temperature of 90°F with heat flow down ward is 1.00 hr-ft²-°F/Btu. [Thermal Resistances of Plane Air Spaces]

$$R_4 = 1.00 \text{ hr-ft}^2\text{-°F/Btu}$$

Using a table of thermal resistances of building materials, find the thermal resistance for 0.75 in wood-fiberboard acoustical tile. [Thermal Resistance of Building Materials]

$$R_5 = 1.89 \text{ hr-ft}^2\text{-°F/Btu}$$

From a table of surface film coefficients, the surface film coefficient for indoor air (v = 0) (horizontal, heat flow downward) is $h_i = 1.08$ Btu/hr-ft²-°F. [Surface Film Coefficients/Resistances for Air]

The film thermal resistance is

$$R_6 = \frac{1}{h_i} = \frac{1}{1.08 \frac{\text{Btu}}{\text{hr-ft}^2\text{-°F}}} = 0.926 \text{ hr-ft}^2\text{-°F/Btu}$$

The total resistance is

$$R_{\text{total}} = R_1 + R_2 + R_3 + R_4 + R_5 + R_6$$
$$= 0.167 \frac{\text{hr-ft}^2\text{-}°\text{F}}{\text{Btu}} + 4.17 \frac{\text{hr-ft}^2\text{-}°\text{F}}{\text{Btu}}$$
$$+ 3.75 \frac{\text{hr-ft}^2\text{-}°\text{F}}{\text{Btu}} + 1.00 \frac{\text{hr-ft}^2\text{-}°\text{F}}{\text{Btu}}$$
$$+ 1.89 \frac{\text{hr-ft}^2\text{-}°\text{F}}{\text{Btu}} + 0.926 \frac{\text{hr-ft}^2\text{-}°\text{F}}{\text{Btu}}$$
$$= 11.903 \text{ hr-ft}^2\text{-}°\text{F/Btu}$$

The overall coefficient of heat transfer is [Composite Plane Wall]

$$U = \frac{1}{R_{\text{total}}} = \frac{1}{11.903 \frac{\text{hr-ft}^2\text{-}°\text{F}}{\text{Btu}}}$$
$$= 0.084 \text{ Btu/hr-ft}^2\text{-}°\text{F} \quad (0.08 \text{ Btu/hr-ft}^2\text{-}°\text{F})$$

The answer is (A).

3. *Customary U.S. Solution*

The slab edge coefficient, F_p, is = 0.55 Btu/hr-ft-°F.

The perimeter length is

$$P = 12 \text{ ft} + 12 \text{ ft} \quad \text{[2 edges only]}$$
$$= 24 \text{ ft}$$

The heat loss from the slab is

$$q = F_p P(T_i - T_o)$$
$$= \left(0.55 \frac{\text{Btu}}{\text{hr-ft-}°\text{F}}\right)(24 \text{ ft})\left(70°\text{F} - (-10°\text{F})\right)$$
$$= 1056 \text{ Btu/hr} \quad (1100 \text{ Btu/hr})$$

SI Solution

The slab edge coefficient, F_p, is = 0.95 W/m·°C.

The perimeter length is

$$P = 3.6 \text{ m} + 3.6 \text{ m} \quad \text{[2 edges only]}$$
$$= 7.2 \text{ m}$$

The heat loss from the slab is

$$q = F_p P(T_i - T_o)$$
$$= \left(0.95 \frac{\text{W}}{\text{m·°C}}\right)(7.2 \text{ m})\left(21.1°\text{C} - (-23.3°\text{C})\right)$$
$$= 303.7 \text{ W} \quad (300 \text{ W})$$

The answer is (C).

4. Using a table of *U*-factors for various fenestration products, find the overall window unit heat transfer coefficient. [U-Factors for Various Fenestration Products]

$$U_{\text{windows}} = \left(0.62 \frac{\text{Btu}}{\text{hr-ft}^2\text{-}°\text{F}}\right)$$

The heat loss through the windows is [Heat Gain Through Interior Surfaces]

$$q_{\text{windows}} = U_{\text{windows}} A_{\text{windows}} (T_i - T_o)$$
$$= \left(0.62 \frac{\text{Btu}}{\text{hr-ft}^2\text{-}°\text{F}}\right)(160 \text{ ft}^2)(68°\text{F} - 10°\text{F})$$
$$= 5754 \text{ Btu/hr}$$

From a table of design *U*-factors of swinging doors, for sliding glass door units, use the fenestration tables for "operable." [Design U-Factors of Swinging Doors]

Using a table of *U*-factors for various fenestration products, find the overall unit heat transfer coefficient for the sliding glass door. [U-Factors for Various Fenestration Products]

$$U_{\text{door}} = \left(0.93 \frac{\text{Btu}}{\text{hr-ft}^2\text{-}°\text{F}}\right)$$

The heat loss through the sliding glass door is [Heat Gain Through Interior Surfaces]

$$q_{\text{door}} = U_{\text{door}} A_{\text{door}} (T_i - T_o)$$
$$= \left(0.93 \frac{\text{Btu}}{\text{hr-ft}^2\text{-}°\text{F}}\right)(52 \text{ ft}^2)(68°\text{F} - 10°\text{F})$$
$$= 2805 \text{ Btu/hr}$$

The total heat loss for the solarium is

$$q_{\text{solarium}} = q_{\text{windows}} + q_{\text{door}}$$
$$= 5754 \frac{\text{Btu}}{\text{hr}} + 2805 \frac{\text{Btu}}{\text{hr}}$$
$$= 8559 \text{ Btu/hr} \quad (8600 \text{ Btu/hr})$$

The answer is (A).

42 Cooling Load

Content in blue refers to the NCEES Handbook.

PRACTICE PROBLEMS

1. An air conditioning unit has a SEER-13 rating and a cooling load of 8000 Btu/hr. The unit operates eight hours per day for 140 days each year. The air conditioning unit's annual power usage is most nearly

(A) 450,000 W-hr/yr

(B) 690,000 W-hr/yr

(C) 9,000,000 W-hr/yr

(D) 120,000,000 W-hr/yr

2. An air conditioning unit has a SEER-13 rating and a cooling load of 10,000 Btu/hr. The average cost of electricity is \$0.25/kW-hr. What is most nearly the hourly cost of operating the air conditioning unit?

(A) \$0.19/hr

(B) \$0.47/hr

(C) \$1.10/hr

(D) \$2.00/hr

3. Multiple blower door leakage tests performed on a building are used to develop a CFM50 building leakage curve correlation with coefficient $C = 110.2$ and exponent $n = 0.702$. What airflow is needed to create a 5 Pa pressure difference in the building?

(A) 210 ft³/min

(B) 340 ft³/min

(C) 390 ft³/min

(D) 430 ft³/min

4. The north facing wall of a building has the characteristics shown here.

4 in brick facing (resistance = 0.11 hr-ft²-°F/Btu-in)

3 in concrete block (resistance = 0.13 hr-ft²-°F/Btu-in)

1 in mineral wool (resistance = 3.38 hr-ft²-°F/Btu-in)

1.5 in furring (assume as air space, effective emittance = 0.82)

0.375 in drywall gypsum (emissivity = 0.95)

0.50 in cement plaster

The wall has an area of 1600 ft². The inside design temperature is 78°F. The outside design temperature is 95°F. The heat gain from the wall is most nearly

(A) 3400 Btu/hr

(B) 4200 Btu/hr

(C) 5100 Btu/hr

(D) 6500 Btu/hr

5. The composite roof of a building has the components shown here.

4 in concrete (density = 130 lbm/ft^3)

2 in insulation (resistance = 2.28 hr-ft^2-°F/Btu-in)

single permeable felt layer

1.5 in air gap (assume emissivity = 0.82, downward heat flow, a mean temperature of 50°F, and a temperature difference of 20°F.)

0.50 in acoustical ceiling tile

The area of the roof is 6000 ft^2. The maximum temperature differences occur between 4 p.m. and 6 p.m. The equivalent maximum temperature difference for the concrete roof is 70°F. The heat gain through the roof is most nearly

(A) 23,000 Btu/hr

(B) 32,000 Btu/hr

(C) 44,000 Btu/hr

(D) 49,000 Btu/hr

6. A single-glazed glass window in a building is 1/8 in thick and has an area of 100 ft^2. At the window's location, the incident irradiance is 68 Btu/hr-ft^2 and the solar heat gain coefficient (SHGC) is 0.75. The inside temperature is 70°F and the outside temperature is 95°F. The heat gain through the window is most nearly

(A) 6900 Btu/hr

(B) 7700 Btu/hr

(C) 8200 Btu/hr

(D) 9500 Btu/hr

7. A space is supplied with 3000 cfm of primary air at the time of changeover. The sensible heat gain from internal sources is 90,000 Btu/hr and the external heat gain is 22,000 Btu/hr. The outside air temperature is 85°F. The room temperature at the time of changeover is 72°F, and the primary air temperature of the unit after changeover is 56°F. The temperature at the changeover point is most nearly

(A) 35°F

(B) 44°F

(C) 53°F

(D) 61°F

8. Which of the following statements about active chilled beam systems is NOT correct?

(A) use natural convective currents to cool a space

(B) operate as sensible cooling units only

(C) use induction nozzles that capture room air to mix with primary air

(D) can be two-pipe or four-pipe systems

9. The average occupancy in an office space during a workday is 25. The space contains lighting equivalent to a total of 2 kW, 30 computers with an average power consumption of 60 W each, including the display, a coffee maker, and a reach-in refrigerator with a small freezer.

A reasonable estimate of heat gain in the office space while occupied is most nearly

(A) 16,000 Btu/hr

(B) 19,000 Btu/hr

(C) 24,000 Btu/hr

(D) 28,000 Btu/hr

SOLUTIONS

1. Calculate the total cooling load per year.

$$q_{annual} = \left(8000 \, \frac{\text{Btu}}{\text{hr}}\right)\left(140 \, \frac{\text{days}}{\text{yr}}\right)\left(8 \, \frac{\text{hr}}{\text{day}}\right)$$
$$= 8{,}960{,}000 \text{ Btu/yr}$$

Use the seasonal energy efficiency ratio (SEER), and solve for the annual energy needed to achieve the total cooling load per year.

$$\text{SEER} = \frac{\text{total seasonal cooling output (Btu)}}{\text{total seasonal input energy (W-hr)}}\overset{\text{Efficiency}}{=} \frac{q_{annual}}{E_{annual}}$$

$$E_{annual} = \frac{q_{annual}}{\text{SEER}}$$
$$= \frac{8{,}960{,}000 \, \frac{\text{Btu}}{\text{yr}}}{13 \, \frac{\text{Btu}}{\text{W-hr}}}$$
$$= 689{,}231 \text{ W-hr} \quad (690{,}000 \text{ W-hr/yr})$$

The answer is (B).

2. Use the seasonal energy efficiency ratio (SEER) and solve for the total input energy in kilowatts, P_{kW}, needed to operate the unit.

$$\text{SEER} = \frac{\text{total seasonal cooling output (Btu)}}{\text{total seasonal input energy (W-hr)}}\overset{\text{Efficiency}}{=} \frac{q_{cooling,\text{Btu/hr}}}{P_{kW}}$$

$$P_{kW} = \frac{q_{cooling}}{\text{SEER}} = \frac{10{,}000 \, \frac{\text{Btu}}{\text{hr}}}{\left(13 \, \frac{\text{Btu}}{\text{W-hr}}\right)\left(1000 \, \frac{\text{W}}{\text{kW}}\right)}$$
$$= 0.769 \text{ kW}$$

Calculate the hourly cost to operate the unit.

$$C_{hr} = (P_{kW})(C_{kW\text{-}hr})$$
$$= (0.769 \text{ kW})\left(\frac{\$0.25}{\text{kW-hr}}\right)$$
$$= \$0.192/\text{hr} \quad (\$0.19/\text{hr})$$

The answer is (A).

3. Building leakage is a function of the pressure difference between the building air and the outdoor air. The air flow can be calculated using the correlation for building leakage.

$$Q_{\text{ft}^3/\text{min}} = C\Delta p_{\text{Pa}}^n = (110.2)(5 \text{ Pa})^{0.702}$$
$$= 341.1 \text{ ft}^3/\text{min} \quad (340 \text{ ft}^3/\text{min})$$

The answer is (B).

4. The heat gain through the wall is the overall temperature difference divided by the sum of the thermal resistances of the components of the wall. Calculate the total thermal resistance of the wall.

Composite Plane Wall

$$R_{total} = \frac{1}{h_i} + R_1 + R_2 + \ldots\ldots + R_n + \frac{1}{h_o}$$

$$R_{i,conv} = \frac{1}{h_i}$$

$$R_{o,conv} = \frac{1}{h_o}$$

The components of the wall include 4 in brick facing (BF), 3 in concrete block (CB), 1 in mineral wool insulation (INS), 1.5 in furring (FRG), 0.375 in drywall gypsum (DW), 0.50 in plaster (PL). The convection resistances of the inside and outside air will be included in the total thermal resistance.

$$R_{total} = R_{i,conv} + R_{BF} + R_{CB} + R_{INS}$$
$$+ R_{DW} + R_{PL} + R_{FRG} + R_{o,conv}$$

Calculate the resistances of the components of the wall from the data provided for building materials. [Thermal Resistance of Building Materials]

Use the lower end of the range of resistances for the concrete.

$$R_{BF} = \left(0.11 \, \frac{\text{hr-ft}^2\text{-}°\text{F}}{\text{Btu-in}}\right)(4 \text{ in})$$
$$= 0.44 \text{ hr-ft}^2\text{-}°\text{F/Btu}$$

$$R_{CB} = \left(0.13 \, \frac{\text{hr-ft}^2\text{-}°\text{F}}{\text{Btu-in}}\right)(3 \text{ in})$$
$$= 0.39 \text{ hr-ft}^2\text{-}°\text{F/Btu}$$

$$R_{INS} = 3.38 \text{ hr-ft}^2\text{-}°\text{F/Btu}$$

$$R_{DW} = 0.32 \text{ hr-ft}^2\text{-}°\text{F/Btu}$$

Find the thermal resistance for cement plaster and calculate for 0.50 in thickness.

$$R_{PL} = \left(0.2 \; \frac{\text{hr-ft}^2\text{-}°F}{\text{Btu-in}}\right)(0.50 \text{ in})$$
$$= 0.10 \text{ hr-ft}^2\text{-}°F/\text{Btu}$$

The furring can be treated as an air space with a thickness of 1.5 inch and an effective emittance of 0.82. [Thermal Resistances of Plane Air Spaces]

The furring in the wall is similar to a vertical air space. Interpolate the thermal resistance value for a vertical air space with horizontal heat flow, a mean temperature of 50°F, and a temperature difference of 20°F.

$$R_{FRG} = \frac{0.90 \; \frac{\text{hr-ft}^2\text{-}°F}{\text{Btu}} + 1.02 \; \frac{\text{hr-ft}^2\text{-}°F}{\text{Btu}}}{2}$$
$$= 0.96 \text{ hr-ft}^2\text{-}°F/\text{Btu}$$

Determine the convection resistances for the inside air and the outside air from a table of surface film coefficients and resistances for air. Use the values for nonreflective surface emittance. [Surface Film Coefficients/Resistances for Air]

For a vertical surface with horizontal heat flow,

$$R_{i,\text{conv}} = 0.68 \text{ hr-ft}^2\text{-}°F/\text{Btu}$$

For outdoor air in summer,

$$R_{o,\text{conv}} = 0.25 \text{ hr-ft}^2\text{-}°F/\text{Btu}$$

Calculate the sum of thermal resistances for the composite wall.

$$R_{\text{total}} = \sum R$$
$$= R_{i,\text{conv}} + R_{BF} + R_{CB} + R_{INS}$$
$$\quad + R_{DW} + R_{PL} + R_{FRG} + R_{o,\text{conv}}$$
$$= 0.68 \; \frac{\text{hr-ft}^2\text{-}°F}{\text{Btu}} + 0.44 \; \frac{\text{hr-ft}^2\text{-}°F}{\text{Btu}}$$
$$\quad + 0.39 \; \frac{\text{hr-ft}^2\text{-}°F}{\text{Btu}} + 3.38 \; \frac{\text{hr-ft}^2\text{-}°F}{\text{Btu}}$$
$$\quad + 0.32 \; \frac{\text{hr-ft}^2\text{-}°F}{\text{Btu}} + 0.10 \; \frac{\text{hr-ft}^2\text{-}°F}{\text{Btu}}$$
$$\quad + 0.96 \; \frac{\text{hr-ft}^2\text{-}°F}{\text{Btu}} + 0.25 \; \frac{\text{hr-ft}^2\text{-}°F}{\text{Btu}}$$
$$= 6.52 \text{ hr-ft}^2\text{-}°F/\text{Btu}$$

Calculate the overall heat transfer coefficient.

Composite Plane Wall

$$U = \frac{1}{R_{\text{total}}}$$
$$= \frac{1}{6.52 \; \frac{\text{hr-ft}^2\text{-}°F}{\text{Btu}}}$$
$$= 0.1534 \text{ hr-ft}^2\text{-}°F/\text{Btu}$$

Calculate the heat gain through the wall.

Heat Gain Through Interior Surfaces

$$q = UA(T_{\text{outside}} - T_{\text{inside}})$$
$$= \left(0.1534 \; \frac{\text{Btu}}{\text{hr-ft}^2\text{-}°F}\right)(1600 \text{ ft}^2)(95°F - 78°F)$$
$$= 4172.48 \text{ Btu/hr-ft}^2\text{-}°F \quad (4200 \text{ Btu/hr})$$

The answer is (B).

5. The heat gain through the wall is the overall temperature difference divided by the sum of the thermal resistances of the components of the wall.

Composite Plane Wall

$$R_{\text{total}} = \frac{1}{h_i} + R_1 + R_2 + \ldots\ldots + R_n + \frac{1}{h_o}$$

$$R_{i,\text{conv}} = \frac{1}{h_i}$$

$$R_{o,\text{conv}} = \frac{1}{h_o}$$

The components of the roof include 4 in concrete (CON), 2 in insulation (INS), single layer felt (FLT), 1 in air space (ASP), and 0.50 in acoustical ceiling tile (ACT). Therefore,

$$R_{\text{total}} = R_{i,\text{conv}} + R_{CON} + R_{INS} + R_{FLT}$$
$$\quad + R_{ACT} + R_{ASP} + R_{o,\text{conv}}$$

Use the provided data and a table of thermal resistances of building materials to find the individual resistances of the components of the roof. [Thermal Resistance of Building Materials]

Use the lower end of the range of resistances for the concrete.

$$R_{CON} = \left(0.08 \; \frac{\text{hr-ft}^2\text{-}°F}{\text{Btu-in}}\right)(4 \text{ in})$$
$$= 0.32 \text{ hr-ft}^2\text{-}°F/\text{Btu}$$

$$R_{\text{INS}} = \left(2.28 \ \frac{\text{hr-ft}^2\text{-°F}}{\text{Btu-in}}\right)(2 \text{ in})$$
$$= 4.56 \ \text{hr-ft}^2\text{-°F/Btu}$$

$$R_{\text{FLT}} = 0.06 \ \text{hr-ft}^2\text{-°F/Btu}$$

$$R_{\text{ACT}} = 1.25 \ \text{hr-ft}^2\text{-°F/Btu}$$

Interpolate the thermal resistance value for a 1.5 in horizontal air space (effective emittance 0.82) with downward heat flow, a mean temperature of 50°F, and a temperature difference of 20°F. [Thermal Resistances of Plane Air Spaces]

$$R_{\text{ASP}} = \frac{1.14 \ \frac{\text{hr-ft}^2\text{-°F}}{\text{Btu}} + 1.15 \ \frac{\text{hr-ft}^2\text{-°F}}{\text{Btu}}}{2}$$
$$= 1.145 \ \text{hr-ft}^2\text{-°F/Btu}$$

Determine the convection resistances for the inside air and the outside air. Use the values for nonreflective surface emittance. [Surface Film Coefficients/Resistances for Air]

For a horizontal surface with downward heat flow,

$$R_{i,\text{conv}} = 0.92 \ \text{hr-ft}^2\text{-°F/Btu}$$

For outdoor air in summer,

$$R_{o,\text{conv}} = 0.25 \ \text{hr-ft}^2\text{-°F/Btu}$$

Calculate the sum of thermal resistances for the composite roof.

$$R_{\text{total}} = \sum R$$
$$= R_{i,\text{conv}} + R_{\text{CON}} + R_{\text{INS}} + R_{\text{FLT}}$$
$$\quad + R_{\text{ACT}} + R_{\text{ASP}} + R_{o,\text{conv}}$$
$$= 0.92 \ \frac{\text{hr-ft}^2\text{-°F}}{\text{Btu}} + 0.32 \ \frac{\text{hr-ft}^2\text{-°F}}{\text{Btu}} + 4.56 \ \frac{\text{hr-ft}^2\text{-°F}}{\text{Btu}}$$
$$\quad + 0.06 \ \frac{\text{hr-ft}^2\text{-°F}}{\text{Btu}} + 1.25 \ \frac{\text{hr-ft}^2\text{-°F}}{\text{Btu}}$$
$$\quad + 1.145 \ \frac{\text{hr-ft}^2\text{-°F}}{\text{Btu}} + 0.25 \ \frac{\text{hr-ft}^2\text{-°F}}{\text{Btu}}$$
$$= 8.51 \ \text{hr-ft}^2\text{-°F/Btu}$$

Calculate the overall heat transfer coefficient.

Composite Plane Wall

$$U = \frac{1}{R_{\text{total}}}$$
$$= \frac{1}{8.51 \ \frac{\text{hr-ft}^2\text{-°F}}{\text{Btu}}}$$
$$= 0.1175 \ \text{hr-ft}^2\text{-°F/Btu}$$

Calculate the heat gain through the roof.

Heat Gain Through Interior Surfaces

$$q = UA(T_{\text{outside}} - T_{\text{inside}})$$
$$= UA \Delta T_{\text{eq}}$$
$$= \left(0.1175 \ \frac{\text{Btu}}{\text{hr-ft}^2\text{-°F}}\right)(6000 \ \text{ft}^2)(70°\text{F})$$
$$= 49{,}350 \ \text{Btu/hr-ft}^2\text{-°F} \quad (49{,}000 \ \text{Btu/hr-ft}^2\text{-°F})$$

The answer is (D).

6. Use the equation for fenestration to calculate the heat gain through the window.

Fenestration

$$q = UA_{\text{pf}}(T_{\text{out}} - T_{\text{in}}) + (\text{SHGC})A_{\text{pf}}E_t$$
$$\quad + (C)(AL)A_{\text{pf}}\rho c_p(T_{\text{out}} - T_{\text{in}})$$

From a table of U-factors for various fenestration products, for $1/8$ in thick, single-glazed glass window, $U = 1.04 \ \text{Btu/hr-ft}^2\text{-°F}$. [U-Factors for Various Fenestration Products]

Assuming no air leakage, the revised equation for fenestration is

$$q = UA_{\text{pf}}(T_{\text{out}} - T_{\text{in}}) + (\text{SHGC})A_{\text{pf}}E_t$$
$$= \left(1.04 \ \frac{\text{Btu}}{\text{hr-ft}^2\text{-°F}}\right)(100 \ \text{ft}^2)(95°\text{F} - 70°\text{F})$$
$$\quad + (0.75)(100 \ \text{ft}^2)\left(68 \ \frac{\text{Btu}}{\text{hr-ft}^2}\right)$$
$$= 7700 \ \text{Btu/hr}$$

The answer is (B).

7. Using a chart of primary air temperature versus outdoor air temperature, find the approximate air-to-transmission ratio at the intersection of a primary air temperature of 56°F and an outside air temperature of 85°F. [Primary-Air Temperature Versus Outdoor Air Temperature]

$$\frac{A}{T} = 1.40$$

Determine the heat transmission per degree of temperature difference between the room air and the outside air.

Transmission of Heat in a Space

$$\frac{\text{primary airflow}}{\text{transmission per degree}} = \frac{A}{T} = 1.40$$

$$T = \frac{A}{1.40}$$

$$= \frac{3000 \frac{\text{ft}^3}{\text{min}}}{1.40}$$

$$= 2143 \text{ Btu/hr-}°\text{F}$$

Use the equation for changeover temperature. The change in heat transmission per degree, Δq_{td}, is equal to the transmission per degree, T, in the previous equation. The value of the primary air quantity, Q_p, is equal to the primary airflow, A, in the previous equation.

In-Room Terminal Systems

$$T_{co} = T_r - \frac{q_{is} + q_{es} - 1.1 Q_p(T_r - T_p)}{\Delta q_{td}}$$

$$= 72°\text{F} - \left(\frac{\left(90{,}000 \frac{\text{Btu}}{\text{hr}} + 22{,}000 \frac{\text{Btu}}{\text{hr}} - \left(1.1 \frac{\text{Btu-min}}{\text{hr-ft}^3\text{-}°\text{F}}\right) \times \left(3000 \frac{\text{ft}^3}{\text{min}}\right)(72°\text{F} - 56°\text{F})\right)}{2143 \frac{\text{Btu}}{\text{hr-}°\text{F}}} \right)$$

$$= 44.37°\text{F} \quad (44°\text{F})$$

The answer is (B).

8. Active chilled beam systems can be two-pipe or four-pipe systems, are designed to be operated as sensible cooling units only, and use induction nozzles that entrain room air and mix it with primary air. Passive chilled beam systems are designed to use natural convective currents to cool a space. Therefore, option A is incorrect. [Chilled Beam Systems] [Passive and Active Chilled-Beam Operation]

The answer is (A).

9. Convert the heat gain due to lighting from kW to Btu/hr.

Measurement Relationships

$$Q_{\text{lighting}} = (2 \text{ kW})\left(1000 \frac{\text{W}}{\text{kW}}\right)\left(3.413 \frac{\text{Btu}}{\text{hr-W}}\right)$$

$$= 6826 \text{ Btu/hr}$$

For moderately active office work, the adjusted heat gain is 450 Btu/hr-person. [Representative Rates at Which Heat and Moisture Are Given Off by People in Different States of Activity]

Therefore, the average total heat gain due to people is

$$Q_{\text{People}} = \left(450 \frac{\text{Btu}}{\text{hr-person}}\right)(25 \text{ persons}) = 11{,}250 \text{ Btu/hr}$$

The total heat gain due to computer equipment is

Measurement Relationships

$$Q_{\text{computers}} = \left(60 \frac{\text{W}}{\text{computer}}\right)(30 \text{ computers})\left(3.413 \frac{\text{Btu}}{\text{hr-W}}\right)$$

$$= 6144 \text{ Btu/hr}$$

The total heat gain from each of the appliances is 1200 Btu/hr each for the coffee maker and reach-in refrigerator, and 1100 Btu/hr for the small freezer. [Recommended Rates of Radiant and Convective Heat Gain: Unhooded Appliances During Idle (Ready-to-Cook) Conditions]

The total heat gain from all appliances is 1200 Btu/hr + 1200 Btu/hr + 1100 Btu/hr = 3500 Btu/hr.

The total heat gain in the office space is

$$Q_{\text{office space}} = Q_{\text{lighting}} + Q_{\text{people}} + Q_{\text{computers}} + Q_{\text{appliances}}$$

$$= 6826 \frac{\text{Btu}}{\text{hr}} + 11250 \frac{\text{Btu}}{\text{hr}} + 6144 \frac{\text{Btu}}{\text{hr}} + 3500 \frac{\text{Btu}}{\text{hr}}$$

$$= 27{,}720 \text{ Btu/hr} \quad (28{,}000 \text{ Btu/hr})$$

The answer is (D).

43 Air Conditioning Systems and Controls

Content in blue refers to the NCEES Handbook.

PRACTICE PROBLEMS

1. What is the typical approximate supply air pressure in a pneumatic HVAC control system?

(A) 7 psig
(B) 18 psig
(C) 35 psig
(D) 140 psig

2. What kind of device is a pneumatic thermostat?

(A) reverse acting
(B) normally closed
(C) normally open
(D) direct acting

3. What is the term for a device that converts a pneumatic signal of 15 psig to a control pressure of 20 psig?

(A) pneumatic transformer
(B) pneumatic amplifier
(C) pneumatic relay
(D) pneumatic booster

4. Which Boolean truth table describes the control of motor M in the relay logic diagram shown?

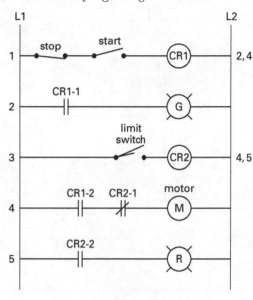

(A)

CR1	CR2	M
0	0	1
0	1	0
1	0	0
1	1	0

(B)

CR1	CR2	M
0	0	0
0	1	1
1	0	0
1	1	0

(C)

CR1	CR2	M
0	0	0
0	1	0
1	0	1
1	1	0

(D)

CR1	CR2	M
0	0	0
0	1	0
1	0	0
1	1	1

5. Given that the stop switch is closed as shown and motor M is not running, what are the functions of switches PB1 and PB2 in the relay logic diagram?

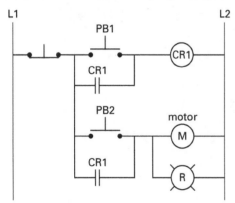

(A) PB1 starts the motor running continuously, and PB2 stops the motor.

(B) PB1 starts the motor for as long as PB1 is pushed, and PB2 starts the motor running continuously.

(C) PB1 stops the motor, and PB2 starts the motor running continuously.

(D) PB1 starts the motor running continuously, and PB2 starts the motor for as long as PB2 is pushed.

SOLUTIONS

1. Pneumatic air pressure is typically supplied at 18 psig to 20 psig, but may be as high as 25 psig.

The answer is (B).

2. A pneumatic thermostat is direct acting because it can turn on an air conditioning system or a heating system based on the temperature of the controlled space. The thermostat is connected to a series of tubes and can sense the temperature of the environment based on the pressure in the tubes. [Control System Types]

In summer, when a room is getting too hot, the air pressure in the tubes connected to the thermostat increases, causing the air conditioning system to turn on. Once the air pressure in the tubes returns to normal, the air conditioner will automatically switch off. Similarly, in winter, the pneumatic thermostat senses when a room gets too cold as the air pressure in the connected tubes dissipates. It will then turn on the heating system until the room temperature rises to the original temperature.

The answer is (D).

3. Pneumatic relays use compressed air as an input signal. In a pneumatic relay, the output control pressure is proportional to the input signal pressure.

The answer is (C).

4. When the start switch in rung 1 is pushed, the coil in control relay CR1 is energized. The coil controls the contacts in rungs 2 and 4. In rung 4, the normally open (NO) contacts CR1-2 also close, and power is supplied through the normally closed (NC) contacts CR2-1 to motor M. If the limit switch is closed by overtravel, relay coil CR2 will be energized, opening the NC CR2-1 contacts in rung 4, and stopping the motor, M. Therefore, in order for the motor to run, CR1 must be energized (state 1), and CR2 must not be energized (state 0). When the motor M is running, it is in state 1 (same as On). [Digital Controllers]

The answer is (C).

5. Initially, the motor is not running and both the CR1 relay contacts are in the normally open (NO) position. Once PB1 is pushed, it energizes the coil CR1 (circled) and both the CR1 relay contacts will be actuated to the closed position. The motor will start running continuously due to the power supply from the sub-rung of the motor location. Similarly, when the motor is not running, pushing PB2 starts the motor. However, PB2 does not control a relay coil and has no effect on the CR1 contacts. Therefore, PB1 starts the motor and keeps it running continuously and pushing PB2 starts the motor for as long as PB2 is pushed. [Digital Controllers]

The answer is (D).

Determinate Statics

Content in blue refers to the NCEES Handbook.

PRACTICE PROBLEMS

1. Two legs of a tripod are mounted on a vertical wall, as shown. Both legs are horizontal. The apex is 12 ft from the wall. The right leg is 13.4 ft long. The wall mounting points are 10 ft apart. A third leg is mounted on the wall 6 ft to the left of the right upper leg and 9 ft below the two top legs. A vertical downward load of 200 lbf is supported at the apex. What is most nearly the reaction at the lowest mounting point?

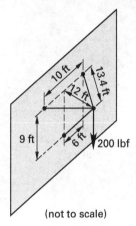

(not to scale)

(A) 120

(B) 170

(C) 250

(D) 330

2. The ideal truss shown is supported by a pinned connection at point D and a roller connection at point C. Loads are applied at points A and F. What is most nearly the force in member DE?

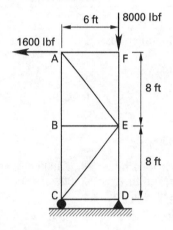

(A) 1200 lbf

(B) 2700 lbf

(C) 3300 lbf

(D) 3700 lbf

3. A pin-connected tripod is loaded at the apex by a horizontal force of 1200, as shown. What is most nearly the magnitude of the force in member AD?

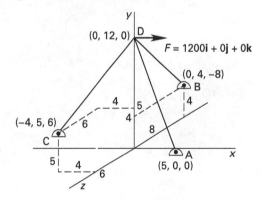

(A) 1100

(B) 1300

(C) 1800

(D) 2500

4. A truss is loaded by forces of 4000 at each upper connection point and forces of 60,000 at each lower connection point, as shown.

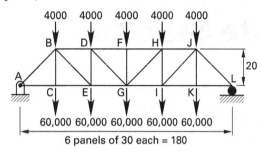

The force in member DE is most nearly

(A) 36,000

(B) 45,000

(C) 60,000

(D) 160,000

5. A truss is loaded by forces of 4000 at each upper connection point and forces of 60,000 at each lower connection point, as shown.

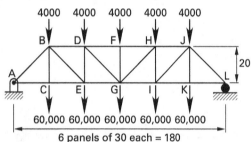

The force in member HJ is most nearly

(A) 24,000

(B) 60,000

(C) 160,000

(D) 380,000

6. A truss is shown. What is most nearly the force in member FC?

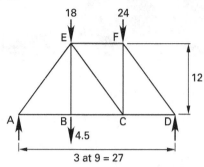

(A) 0.5

(B) 17

(C) 23

(D) 29

7. A truss is shown. The force in member CE is most nearly

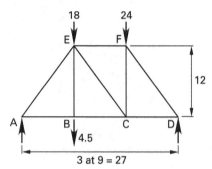

(A) 0.50

(B) 0.63

(C) 11

(D) 17

8. The rigid rod AO is supported by guy wires BO and CO, as shown. (Points A, B, and C are all in the same vertical plane. Points A, O, and C are all in the same horizontal plane. Points A, O, and B are all in the same vertical plane.) Vertical and horizontal forces are 12,000 and 6000, respectively, as carried at the end of the rod. What are most nearly the x-, y-, and z-components of the reactions at point C?

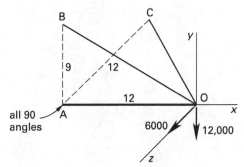

(A) $(C_x, C_y, C_z) = (0, 6000, 0)$

(B) $(C_x, C_y, C_z) = (6000, 0, 6000)$

(C) $(C_x, C_y, C_z) = (6000, 0, 4200)$

(D) $(C_x, C_y, C_z) = (4200, 0, 8500)$

9. The lever shown is pinned at and free to rotate about point O. The lever is acted upon by forces F_1 and F_2, but not by gravity. What is the algebraic expression for force F_2 in terms of F_1 that will keep the lever stationary?

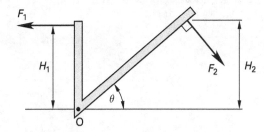

(A) $F_2 = \dfrac{F_1 H_1 \sin\theta}{H_2}$

(B) $F_2 = \dfrac{F_1 H_1}{H_2 \sin\theta}$

(C) $F_2 = \dfrac{F_1 H_1}{H_2}$

(D) $F_2 = \dfrac{F_1 H_1 \cos\theta}{H_2}$

10. A power pole loaded horizontally by electrical wires T_1 and T_2 is braced by a guy wire, as shown. The guy wire is 25 ft long, is attached 20 ft up the pole, and has an internal tensile force of 12,500 lbf. The loads are perpendicular to the vertical power pole, and the bases of the guy wire and power pole are both even and at ground level. All forces are in equilibrium. What are most nearly the magnitudes of the horizontal loads T_1 and T_2, respectively?

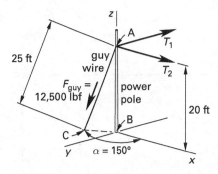

(A) $T_1 = 3500$ lbf; $T_2 = 5800$ lbf

(B) $T_1 = 3800$ lbf; $T_2 = 6500$ lbf

(C) $T_1 = 3900$ lbf; $T_2 = 6800$ lbf

(D) $T_1 = 6300$ lbf; $T_2 = 6300$ lbf

11. A 30 ft × 10 ft, 9000 lbf, box-shaped piece of machinery is lifted using a three-cable hoist ring, as shown. Two cables of the hoist are connected to opposite corners of the 10 ft side, and the other cable is connected to the center of the other 10 ft side. The hoist ring is 20 ft directly above the center of the machinery. What are most nearly the forces in hoist cables C_1, C_2, and C_3, respectively?

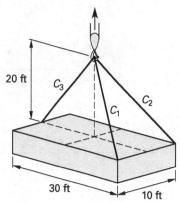

(A) 2300 lbf, 2300 lbf, and 6800 lbf

(B) 2500 lbf, 2700 lbf, and 3900 lbf

(C) 2900 lbf, 2900 lbf, and 5600 lbf

(D) 3000 lbf, 3000 lbf, and 3000 lbf

SOLUTIONS

1. *step 1:* Draw the tripod with the origin at the apex.

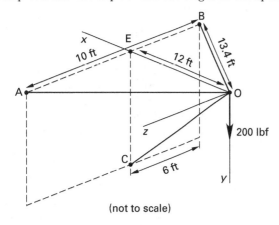

(not to scale)

step 2: By inspection, the force components are $F_x = 0$, $F_y = 200$ lbf, and $F_z = 0$.

step 3: First, from triangle EBO, length BE is

$$BE = \sqrt{(13.4 \text{ ft})^2 - (12 \text{ ft})^2}$$
$$= 5.96 \text{ ft}$$

The (x, y, z) coordinates of the three support points are

Point A: (12 ft, 0 ft, 4 ft)
Point B: (12 ft, 0 ft, −6 ft)
Point C: (12 ft, 9 ft, 0 ft)

step 4: Find the lengths of the legs.

$$AO = \sqrt{(x_A - x_o)^2 + (y_A - y_o)^2 + (z_A - z_o)^2}$$
$$= \sqrt{(12 \text{ ft})^2 + (0 \text{ ft})^2 + (4 \text{ ft})^2}$$
$$= 12.65 \text{ ft}$$

$$BO = \sqrt{(x_B - x_o)^2 + (y_B - y_o)^2 + (z_B - z_o)^2}$$
$$= \sqrt{(12 \text{ ft})^2 + (0 \text{ ft})^2 + (-6 \text{ ft})^2}$$
$$= 13.4 \text{ ft}$$

$$CO = \sqrt{(x_C - x_o)^2 + (y_C - y_o)^2 + (z_C - z_o)^2}$$
$$= \sqrt{(12 \text{ ft})^2 + (9 \text{ ft})^2 + (0 \text{ ft})^2}$$
$$= 15 \text{ ft}$$

step 5: Find the direction cosines for each leg.
For leg AO,

$$\cos\theta_{A,x} = \frac{x_A}{AO} = \frac{12 \text{ ft}}{12.65 \text{ ft}} = 0.949$$

$$\cos\theta_{A,y} = \frac{y_A}{AO} = \frac{0 \text{ ft}}{12.65 \text{ ft}} = 0$$

$$\cos\theta_{A,z} = \frac{z_A}{AO} = \frac{4 \text{ ft}}{12.65 \text{ ft}} = 0.316$$

For leg BO,

$$\cos\theta_{B,x} = \frac{x_B}{BO} = \frac{12 \text{ ft}}{13.4 \text{ ft}} = 0.896$$

$$\cos\theta_{B,y} = \frac{y_B}{BO} = \frac{0 \text{ ft}}{13.4 \text{ ft}} = 0$$

$$\cos\theta_{B,z} = \frac{z_B}{BO} = \frac{-6 \text{ ft}}{13.4 \text{ ft}} = -0.448$$

For leg CO,

$$\cos\theta_{C,x} = \frac{x_C}{CO} = \frac{12 \text{ ft}}{15 \text{ ft}} = 0.80$$

$$\cos\theta_{C,y} = \frac{y_C}{CO} = \frac{9 \text{ ft}}{15 \text{ ft}} = 0.60$$

$$\cos\theta_{C,z} = \frac{z_C}{CO} = \frac{0 \text{ ft}}{15 \text{ ft}} = 0$$

steps 6 and 7: Use the direction cosines to calculate the x, y, and z component of each internal leg's force and sum the x, y, and z components of each leg's forces with respect to the tripod apex.

Resolution of a Force

$$F_x = F\cos(\theta_x)$$
$$F_y = F\cos(\theta_y)$$
$$F_z = F\cos(\theta_z)$$

Systems of n Forces

$$\sum F_x = 0 = F_{A,x} + F_{B,x} + F_{C,x} + F_x$$
$$\sum F_y = 0 = F_{A,y} + F_{B,y} + F_{C,y} + F_y$$
$$\sum F_z = 0 = F_{A,z} + F_{B,z} + F_{C,z} + F_z$$

$$F_A\cos(\theta_{A,x}) + F_B\cos(\theta_{B,x}) + F_C\cos(\theta_{C,x}) + F_x = 0$$
$$F_A\cos(\theta_{A,y}) + F_B\cos(\theta_{B,y}) + F_C\cos(\theta_{C,y}) + F_y = 0$$
$$F_A\cos(\theta_{A,z}) + F_B\cos(\theta_{B,z}) + F_C\cos(\theta_{C,z}) + F_z = 0$$

$$0.949F_A + 0.896F_B + 0.80F_C + 0 = 0$$
$$0F_A + 0F_B + 0.60F_C + 200 \text{ lbf} = 0$$
$$0.316F_A - 0.448F_B + 0F_C + 0 = 0$$

Solve the three equations simultaneously.

$$F_A = 168.6 \text{ lbf}$$
$$F_B = 118.9 \text{ lbf}$$
$$F_C = -333.3 \text{ lbf} \quad (330)$$

The answer is (D).

2. First, find the vertical reaction at point D.

Systems of n Forces

$$\sum M_C = 0$$
$$= (CD)D_y - (AF)(8000 \text{ lbf}) + (AC)(1600 \text{ lbf})$$
$$= (6 \text{ ft})D_y - (6 \text{ ft})(8000 \text{ lbf}) + (16 \text{ ft})(1600 \text{ lbf})$$
$$= 0$$

Solve for $D_y = 3733.3$.

The free-body diagram of pin D is as follows.

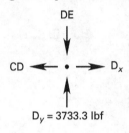

$$\sum F_y = D_y - DE = 0$$

Therefore,

$$DE = D_y = 3733.3 \text{ lbf} \quad (3700 \text{ lbf})$$

The answer is (D).

3. *step 1:* Move the origin to the apex of the tripod. Call this point O.

step 2: By inspection, the force components are $F_x = 1200$, $F_y = 0$, and $F_z = 0$.

step 3: The (x, y, z) coordinates of the three support points are

point A: $(5, -12, 0)$
point B: $(0, -8, -8)$
point C: $(-4, -7, 6)$

step 4: Find the lengths of the legs.

$$AO = \sqrt{(x_A - x_O)^2 + (y_A - y_O)^2 + (z_A - z_O)^2}$$
$$= \sqrt{(5)^2 + (-12)^2 + (0)^2}$$
$$= 13.0$$
$$BO = \sqrt{(x_B - x_O)^2 + (y_B - y_O)^2 + (z_B - z_O)^2}$$
$$= \sqrt{(0)^2 + (-8)^2 + (-8)^2}$$
$$= 11.31$$
$$CO = \sqrt{(x_C - x_O)^2 + (y_C - y_O)^2 + (z_C - z_O)^2}$$
$$= \sqrt{(-4)^2 + (-7)^2 + (6)^2}$$
$$= 10.05$$

step 5: Use the law of cosines to find the direction cosines for each leg. [Trigonometry: Basics]

For leg AO,

$$\cos\theta_{A,x} = \frac{x_A}{AO} = \frac{5}{13.0} = 0.385$$
$$\cos\theta_{A,y} = \frac{y_A}{AO} = \frac{-12}{13.0} = -0.923$$
$$\cos\theta_{A,z} = \frac{z_A}{AO} = \frac{0}{13.0} = 0$$

For leg BO,

$$\cos\theta_{B,x} = \frac{x_B}{BO} = \frac{0}{11.31} = 0$$
$$\cos\theta_{B,y} = \frac{y_B}{BO} = \frac{-8}{11.31} = -0.707$$
$$\cos\theta_{B,z} = \frac{z_B}{BO} = \frac{-8}{11.31} = -0.707$$

For leg CO,

$$\cos\theta_{C,x} = \frac{x_C}{CO} = \frac{-4}{10.05} = -0.398$$
$$\cos\theta_{C,y} = \frac{y_C}{CO} = \frac{-7}{10.05} = -0.697$$
$$\cos\theta_{C,z} = \frac{z_C}{CO} = \frac{6}{10.05} = 0.597$$

steps 6 and 7: Use the direction cosines to calculate the x, y, and z component of each internal leg's force and sum the x, y, and z components of each leg's forces with respect to the tripod apex.

Resolution of a Force

$$F_x = F\cos(\theta_x)$$
$$F_y = F\cos(\theta_y)$$
$$F_z = F\cos(\theta_z)$$

Systems of n Forces

$$\sum F_x = 0 = F_{A,x} + F_{B,x} + F_{C,x} + F_x$$
$$\sum F_y = 0 = F_{A,y} + F_{B,y} + F_{C,y} + F_y$$
$$\sum F_z = 0 = 0 = F_{A,z} + F_{B,z} + F_{C,z} + F_z$$

$$\sum F_x = 0 = F_A\cos\theta_{A,x} + F_B\cos\theta_{B,x} + F_C\cos\theta_{C,x} + F_x$$
$$\sum F_y = 0 = F_A\cos\theta_{A,y} + F_B\cos\theta_{B,y} + F_C\cos\theta_{C,y} + F_y$$
$$\sum F_z = 0 = F_A\cos\theta_{A,z} + F_B\cos\theta_{B,z} + F_C\cos\theta_{C,z} + F_z$$
$$0.385F_A + 0F_B - 0.398F_C + 1200 = 0$$
$$-0.923F_A - 0.707F_B - 0.697F_C + 0 = 0$$
$$0F_A - 0.707F_B + 0.597F_C + 0 = 0$$

Solve the three equations simultaneously.

$$F_A = -1794 \quad (C) \quad (1800)$$
$$F_B = 1081 \quad (T)$$
$$F_C = 1280 \quad (T)$$

The answer is (C).

4. First, find the vertical reactions.

$$\sum F_y = A_y + L_y - (5)(4000) - (5)(60{,}000) = 0$$

By symmetry, $A_y = L_y$.

$$2A_y = (5)(4000) + (5)(60{,}000)$$
$$A_y = 160{,}000$$
$$L_y = 160{,}000$$

Make a cut in members BD, DE, and EG, as shown.

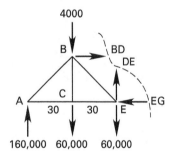

The force is

Systems of n Forces

$$F = \sum F_n$$

$$\sum F_y = 160{,}000 - 60{,}000 - 60{,}000$$
$$+ DE - 4000$$
$$= 0$$
$$DE = -36{,}000 \quad (36{,}000 \text{ in the downward direction.})$$

The answer is (A).

5. First, find the vertical reactions.

Systems of n Forces

$$F = \sum F_n$$

$$\sum F_y = A_y + L_y - (5)(4000) - (5)(60{,}000) = 0$$

By symmetry, $A_y = L_y$.

$$2A_y = (5)(4000) + (5)(60{,}000)$$
$$A_y = 160{,}000$$
$$L_y = 160{,}000$$

Make a cut in members HJ, HI, and GI.

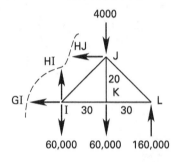

Systems of n Forces

$$M = \sum M_n$$

$$\sum M_I = (160{,}000)(60) - (60{,}000)(30)$$
$$- (4000)(30) + HJ(20)$$
$$= 0$$
$$HJ = -384{,}000 \quad (C) \quad (380{,}000)$$

The answer is (D).

6. First, find the reactions. Take clockwise moments about A as positive.

Systems of n Forces

$$\sum M_A = 0 = (27)(-D_y) + (18)(24) + (9)(22.5)$$
$$D_y = 23.5$$
$$A_y = 18 + 24 + 4.5 - 23.5 = 23$$

Either the method of sections (easiest) or a member-by-member analysis can be used.

The general force triangle is

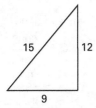

At pin A,

$$AE_y = 23$$
$$AE_x = \left(\frac{9}{12}\right)(23) = 17.25$$
$$AE = \left(\frac{15}{12}\right)(23) = 28.75 \quad (C)$$
$$AB = AE_x = 17.25 \quad (T)$$

At pin B,

$$BE = 4.5 \quad (T)$$
$$BC = AB = 17.25 \quad (T)$$

At pin D,

$$DF_y = 23.5$$
$$DF_x = \left(\frac{9}{12}\right)(23.5) = 17.63$$
$$DF = \left(\frac{15}{12}\right)(23.5) = 29.38 \quad (C)$$
$$DC = DF_x = 17.63 \quad (T)$$

At pin F,

$$FE = DF_x = 17.63 \quad (C)$$
$$FC = 24 - DF_y = 24 - 23.5 = 0.5 \quad (C)$$

The answer is (A).

7. First, find the reactions. Take clockwise moments about A as positive.

Systems of n Forces

$$\sum M_A = 0 = (27)(-D_y) + (18)(24) + (9)(22.5)$$
$$D_y = 23.5$$
$$A_y = 18 + 24 + 4.5 - 23.5 = 23$$

Either the method of sections (easiest) or a member-by-member analysis can be used.

The general force triangle is

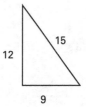

Sum the forces at in the y-direction point D.

$$\sum F_{D,y} = 0 = \left(\frac{12}{15}\right)FD + D_y$$
$$FD = -D_y\left(\frac{15}{12}\right) = -29.38$$

Sum the forces in the y-direction at point F. At point F, FD acts in the opposite direction as it does at point D.

$$\sum F_{F,y} = 0 = \left(\frac{12}{15}\right)FD + FC - 24$$
$$FC = 24 - \left(\frac{12}{15}\right)29.38 = 0.5$$

Sum the forces in the y-direction at point C. At point C, FC acts in the opposite direction as point F.

$$\sum F_{C,y} = 0 = FC + \left(\frac{12}{15}\right)CE$$
$$CE = \left(\frac{15}{12}\right)0.5 = 0.63$$

The answer is (B).

8. First, consider a free-body diagram at point O in the vertical (x-y) plane.

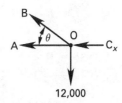

θ is obtained from triangle AOB.

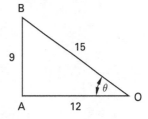

C_x is obtained from triangle AOC.

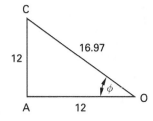

Equilibrium in the vertical (x-y) plane at point O requires

$\sum F_y = 0$

$$B \sin \theta = 12{,}000$$
$$B = \frac{12{,}000}{\sin \theta} = \frac{12{,}000}{\frac{9}{15}}$$
$$= 20{,}000$$

$$B_x = B \cos \theta = (20{,}000)\left(\frac{12}{15}\right) = 16{,}000$$

$$B_y = B \sin \theta = (20{,}000)\left(\frac{9}{15}\right) = 12{,}000$$

Since BO is in the x-y plane, $B_z = 0$.

Next, consider a free-body diagram at point O in the horizontal (x-z) plane.

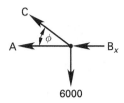

$\sum F_x = 0$:

$$A + B_x + C \cos \phi = 0$$
$$A + C \cos \phi = -B_x = -B \cos \theta$$
$$= (-20{,}000)\left(\frac{12}{15}\right)$$
$$= -16{,}000$$

$\sum F_y = 0$:

$$C \sin \phi = 6000$$
$$C = \frac{6000}{\sin \phi}$$
$$= \frac{6000}{\frac{12}{16.97}}$$
$$= 8485$$

Therefore,

$$A = -C \cos \phi - 16{,}000$$
$$= (-8485)\left(\frac{12}{16.97}\right) - 16{,}000$$
$$= -22{,}000$$

Since AO is on the x-axis,

$$A_x = -22{,}000$$
$$A_y = 0$$
$$A_z = 0$$

Solve for C reactions.

$$C_x = C \cos \phi = (8485)\left(\frac{12}{16.97}\right)$$
$$= 6000$$
$$C_z = C \sin \phi = (8485)\left(\frac{12}{16.97}\right)$$
$$= 6000$$

Since CO is in the horizontal (x-z) plane, $C_y = 0$.

The answer is (B).

9. For the lever to be stationary, the sum of the moments about point O must be zero.

$$\sum M_O = 0 = F_1 H_1 - \frac{F_2 H_2}{\sin \theta}$$

Solve for F_2.

$$F_2 = \frac{F_1 H_1 \sin \theta}{H_2}$$

The answer is (A).

10. *step 1:* Establish point A as the origin.

step 2: By inspection, length AC = 25 ft and length AB = 20 ft. Use the Pythagorean theorem to solve for length BC.

$$BC = \sqrt{(AC)^2 - (AB)^2} = \sqrt{(25 \text{ ft})^2 - (20 \text{ ft})^2}$$
$$= 15 \text{ ft}$$

Because loads T_1 and T_2 are horizontal (i.e., in the x-y plane), neglect all z components. Use trigonometry to find the x- and y-coordinates for point C. [Trigonometry: Basics]

$$x_C = BC \cos \alpha = (15 \text{ ft}) \cos 150° = -13 \text{ ft}$$
$$y_C = BC \sin \alpha = (15 \text{ ft}) \sin 150° = 7.5 \text{ ft}$$

step 3: Find the x- and y-direction cosines for the guy wire.

$$\cos \theta_{\text{guy},x} = \frac{d_x}{AC} = \frac{-13 \text{ ft} - 0 \text{ ft}}{25 \text{ ft}} = -0.52 \text{ ft}$$
$$\cos \theta_{\text{guy},y} = \frac{d_y}{AC} = \frac{7.5 \text{ ft} - 0 \text{ ft}}{25 \text{ ft}} = 0.3 \text{ ft}$$

The x- and y-components of the guy wire's force are

$$F_{\text{guy},x} = -0.52 F_{\text{guy}}$$
$$F_{\text{guy},y} = 0.3 F_{\text{guy}}$$

Resolution of a Force

$$F_x = F \cos \theta_x = F_{\text{guy},x} = F_{\text{guy}} \cos \theta_{\text{guy},x} = F_{\text{guy}}(-0.52)$$
$$F_y = F \cos \theta_y = F_{\text{guy},y} = F_{\text{guy}} \cos \theta_{\text{guy},y} = F_{\text{guy}}(0.3)$$

steps 4 and 5: Write the equilibrium equations for the forces in the vertical direction.

$$F_{\text{guy},y} - T_1 = 0$$
$$F_{\text{guy},x} + T_2 = 0$$

$$F = \sum F_n = F_{\text{guy},y} - T_1 = 0$$
$$F = \sum F_n = F_{\text{guy},x} + T_2 = 0$$

Substitute and solve for the magnitudes of T_1 and T_2.

$$0.3 F_{\text{guy}} - T_1 = 0$$
$$(0.3)(12{,}500 \text{ lbf}) - T_1 = 0$$
$$T_1 = 3750 \text{ lbf} \quad (3800 \text{ lbf})$$

$$-0.52 F_{\text{guy}} + T_2 = 0$$
$$(-0.52)(12{,}500 \text{ lbf}) + T_2 = 0$$
$$T_2 = 6500 \text{ lbf}$$

The answer is (B).

11. The load in each cable can be found by the following steps.

step 1: Establish the apex as the origin, O.

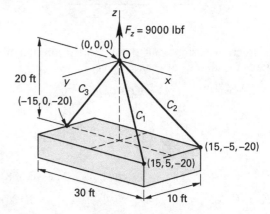

step 2: Determine the x-, y-, and z-components for cables C_1, C_2, and C_3. By inspection, the coordinates are $C_1 =$ (15 ft, 5 ft, −20 ft), $C_2 = $ (15 ft, −5 ft, −20 ft), and $C_3 =$ (−15 ft, 0 ft, −20 ft).

step 3: Find the cable lengths, L_{C_i}. By inspection, $L_{C_1} = L_{C_2}$.

$$L_{C_1} = L_{C_2} = \sqrt{x^2 + y^2 + z^2}$$
$$= \sqrt{(x_O - x_{C_1})^2 + (y_O - y_{C_1})^2 + (z_O - z_{C_1})^2}$$
$$= \sqrt{\begin{array}{c}(0 \text{ ft} - 15 \text{ ft})^2 + (0 \text{ ft} - 5 \text{ ft})^2 \\ + \left(0 \text{ ft} - (-20 \text{ ft})\right)^2\end{array}}$$
$$= 25.5 \text{ ft}$$

$$L_{C_3} = \sqrt{x^2 + y^2 + z^2}$$
$$= \sqrt{(x_O - x_{C_3})^2 + (y_O - y_{C_3})^2 + (z_O - z_{C_3})^2}$$
$$= \sqrt{\begin{array}{c}\left(0 \text{ ft} - (-15 \text{ ft})\right)^2 + (0 \text{ ft} - 0 \text{ ft})^2 \\ + \left(0 \text{ ft} - (-20 \text{ ft})\right)^2\end{array}}$$
$$= 25 \text{ ft}$$

steps 4 and 5: Find the direction cosine for each cable.

For cable C_1,

$$\cos\theta_{C_1,x} = \frac{x_{C_1}}{L_{C_1}} = \frac{0 \text{ ft} - 15 \text{ ft}}{25.5 \text{ ft}} = -0.588$$

$$\cos\theta_{C_1,y} = \frac{y_{C_1}}{L_{C_1}} = \frac{0 \text{ ft} - 5 \text{ ft}}{25.5 \text{ ft}} = -0.196$$

$$\cos\theta_{C_1,z} = \frac{z_{C_1}}{L_{C_1}} = \frac{0 \text{ ft} - (-20 \text{ ft})}{25.5 \text{ ft}} = 0.784$$

The direction cosines and x-, y-, and z-components of the force in cable C_1 are

$$F_{C_1,x} = -0.588 F_{C_1}$$
$$F_{C_1,y} = -0.196 F_{C_1}$$
$$F_{C_1,z} = 0.784 F_{C_1}$$

$$F_x = F\cos\theta_x = F_{C_1,x} = F_{C_1}\cos\theta_{C_1,x} = F_{C_1}(-0.588)$$
$$F_y = F\cos\theta_y = F_{C_1,y} = F_{C_1}\cos\theta_{C_1,y} = F_{C_1}(-0.196)$$
$$F_z = F\cos\theta_z = F_{C_1,z} = F_{C_1}\cos\theta_{C_1,z} = F_{C_1}(0.784)$$

For cable C_2,

$$\cos\theta_{C_2,x} = \frac{x_{C_2}}{L_{C_2}} = \frac{0 \text{ ft} - 15 \text{ ft}}{25.5 \text{ ft}} = -0.588$$

$$\cos\theta_{C_2,y} = \frac{y_{C_2}}{L_{C_2}} = \frac{0 \text{ ft} - (-5 \text{ ft})}{25.5 \text{ ft}} = 0.196$$

$$\cos\theta_{C_2,z} = \frac{z_{C_2}}{L_{C_2}} = \frac{0 \text{ ft} - (-20 \text{ ft})}{25.5 \text{ ft}} = 0.784$$

The direction cosines and x-, y-, and z-components of the force in cable C_2 are

$$F_{C_2,x} = -0.588 F_{C_2}$$
$$F_{C_2,y} = 0.196 F_{C_2}$$
$$F_{C_2,z} = 0.784 F_{C_2}$$

$$F_x = F\cos\theta_x = F_{C_2,x} = F_{C_2}\cos\theta_{C_2,x} = F_{C_2}(-0.588)$$
$$F_y = F\cos\theta_y = F_{C_2,y} = F_{C_2}\cos\theta_{C_2,y} = F_{C_2}(0.196)$$
$$F_z = F\cos\theta_z = F_{C_2,z} = F_{C_2}\cos\theta_{C_2,z} = F_{C_2}(0.784)$$

For cable C_3,

$$\cos\theta_{C_3,x} = \frac{x_{C_3}}{L_{C_3}} = \frac{0 \text{ ft} - (-15 \text{ ft})}{25 \text{ ft}} = 0.6$$

$$\cos\theta_{C_3,y} = \frac{y_{C_3}}{L_{C_3}} = \frac{0 \text{ ft} - 0 \text{ ft}}{25 \text{ ft}} = 0$$

$$\cos\theta_{C_3,z} = \frac{z_{C_3}}{L_{C_3}} = \frac{0 \text{ ft} - (-20 \text{ ft})}{25 \text{ ft}} = 0.8$$

The direction cosines and x-, y-, and z-components of the force in cable C_3 are

$$F_{C_3,x} = 0.6 F_{C_3}$$
$$F_{C_3,y} = 0 F_{C_3}$$
$$F_{C_3,z} = 0.8 F_{C_3}$$

$$F_x = F\cos\theta_x = F_{C_3,x} = F_{C_3}\cos\theta_{C_3,x} = F_{C_3}(0.6)$$
$$F_y = F\cos\theta_y = F_{C_3,y} = F_{C_3}\cos\theta_{C_3,y} = F_{C_3}(0)$$
$$F_z = F\cos\theta_z = F_{C_3,z} = F_{C_3}\cos\theta_{C_3,z} = F_{C_3}(0.8)$$

step 7: Write the three sum-of-forces equilibrium equations for the apex.

$$-0.588 F_{C_1} - 0.588 F_{C_2} + 0.6 F_{C_3} = 0$$
$$-0.196 F_{C_1} + 0.196 F_{C_2} + 0 F_{C_3} = 0$$
$$0.784 F_{C_1} + 0.784 F_{C_2} + 0.8 F_{C_3} = -9000$$

Systems of n Forces

$$F_x = \sum F_x = 0 = -0.588 F_{C_1} - 0.588 F_{C_2} + 0.6 F_{C_3}$$
$$F_y = \sum F_y = 0 = -0.196 F_{C_1} + 0.196 F_{C_2} + 0 F_{C_3}$$
$$F_z = \sum F_z = -9000 = 0.784 F_{C_1} + 0.784 F_{C_2} + 0.8 F_{C_3}$$

The solution to these simultaneous equations is

$$F_{C_1} = -2870 \text{ lbf} \quad (2900 \text{ lbf})$$
$$F_{C_2} = -2870 \text{ lbf} \quad (2900 \text{ lbf})$$
$$F_{C_3} = -5625 \text{ lbf} \quad (5600 \text{ lbf})$$

The answer is (C).

45 Indeterminate Statics

Content in blue refers to the NCEES Handbook.

PRACTICE PROBLEMS

1. What is the degree of indeterminacy of the structure shown?

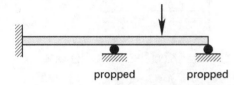

(A) 1

(B) 2

(C) 3

(D) 4

2. A 1 in × 2 in × 10 in (2 cm × 5 cm × 30 cm) copper tie rod experiences a 55°F (30°C) increase from the no-stress temperature. The rod has a modulus of elasticity of 18×10^6 lbf/in² (12×10^4 MPa). The coefficient of linear thermal expansion for copper is 8.9×10^{-6} 1/°F (16.0×10^{-6} 1/°C). The compressive load is most nearly

(A) 8000 lbf (26 kN)

(B) 12,000 lbf (38 kN)

(C) 15,000 lbf (48 kN)

(D) 18,000 lbf (58 kN)

3. A 2 in (5 cm) diameter steel rod supports a 2250 lbf (10 000 N) compressive load. The stress is most nearly

(A) 570 lbf/in² (4.0 MPa)

(B) 720 lbf/in² (5.1 MPa)

(C) 1100 lbf/in² (7.8 MPa)

(D) 1400 lbf/in² (9.9 MPa)

4. A 15 in (40 cm) long steel rod supports a 2250 lbf (10 000 N) compressive load. The stress is 716.2 lbf/in² (5.093 MPa). The rod has a modulus of elasticity of 29×10^6 lbf/in² (20×10^4 MPa). The decrease in length is most nearly

(A) 8.7×10^{-5} in (2.3×10^{-6} m)

(B) 1.1×10^{-4} in (3.0×10^{-6} m)

(C) 3.7×10^{-4} in (1.0×10^{-5} m)

(D) 9.3×10^{-4} in (2.6×10^{-5} m)

5. For the structure shown, the beam is made of steel, with an area moment of inertia of 10 in⁴ (4.17×10^6 mm⁴). The cable cross section is 0.0124 in² (8 mm²). Before the 270 lbf load is applied, the cable is taut but carries no load.

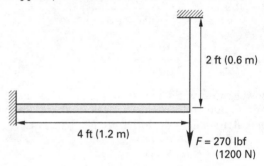

What is most nearly the load carried by the steel cable?

(A) 180 lbf (780 N)

(B) 200 lbf (870 N)

(C) 220 lbf (950 N)

(D) 240 lbf (1000 N)

6. A beam is simply supported at its ends and by a column of area A at its center, as shown. The beam and column are of the same material. The beam is subject to a uniform load, w.

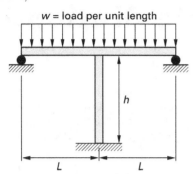

What are each of the left and right support reactions?

(A) $R = wL + \dfrac{5AwL^4}{48hI + 8AL^3}$

(B) $R = wL\left(\dfrac{48hI + 3AL^3}{48hI + 8AL^3}\right)$

(C) $R = wL + wL\left(\dfrac{5AL^3}{48hI + 8AL^3}\right)$

(D) $R = wL + \dfrac{10AwL^4}{24hI + 4AL^3}$

7. A weightless rigid bar is supported by a hinge and two aluminum rods as shown. Deformations are small, and angular geometry changes are negligible.

$a = 6$ ft (2 m)
$b = 3$ ft (1 m)
$c = 3$ ft (1 m)
$d = 12$ ft (4 m)
$P = 4500$ lbf (20 kN)
$A_{\text{rod}} = 0.124$ in^2 (80 mm^2)
$E_{\text{rod}} = 10 \times 10^6$ lbf/in^2 (70 × 10^3 MPa)

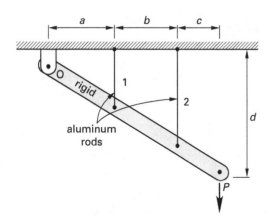

The force in each aluminum rod is most nearly

(A) 3200 lbf (14 kN)
(B) 3600 lbf (16 kN)
(C) 4000 lbf (18 kN)
(D) 4300 lbf (19 kN)

8. What are most nearly the reactions R_1, R_2, and R_3, respectively, in the structure shown?

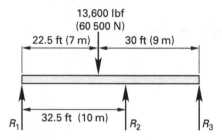

(A) 2000 lbf; 15,000 lbf; −3600 lbf (8500 N; 68,000 N; −16,000 N)

(B) 2700 lbf; 13,000 lbf; −2500 lbf (11,000 N; 60,000 N; −11,000 N)

(C) 3900 lbf; 10,000 lbf; −540 lbf (16,000 N; 46,000 N; −2500 N)

(D) 4700 lbf; 8200 lbf; 770 lbf (20,000 N; 37,000 N; −3500 N)

9. What are most nearly the vertical reactions, R_2 and R_1, respectively, at the ends of the structure?

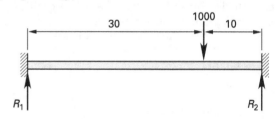

(A) 510, 490
(B) 560, 440
(C) 670, 330
(D) 845, 155

10. A rigid bar is supported by a hinge and two aluminum rods as shown. Deformations are small, and angular geometry changes are negligible.

$$a = 6 \text{ ft } (2 \text{ m})$$
$$b = 3 \text{ ft } (1 \text{ m})$$
$$c = 3 \text{ ft } (1 \text{ m})$$
$$d = 12 \text{ ft } (4 \text{ m})$$
$$P = 4500 \text{ lbf } (20 \text{ kN})$$
$$A_{\text{rod}} = 0.124 \text{ in}^2 \ (80 \text{ mm}^2)$$
$$E_{\text{rod}} = 10 \times 10^6 \text{ lbf/in}^2 \ (70 \times 10^3 \text{ MPa})$$

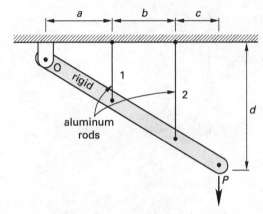

The force in each aluminum rod is 3600 lbf (16 kN). The elongations of the two aluminum rods are most nearly

(A) 0.13 in and 0.26 in (3.3 mm and 6.6 mm)

(B) 0.21 in and 0.31 in (5.7 mm and 8.6 mm)

(C) 0.21 in and 0.47 in (5.7 mm and 12 mm)

(D) 0.26 in and 0.62 in (6.6 mm and 16 mm)

11. The truss shown carries a moving uniform live load of 2 kips/ft and a moving concentrated live load of 15 kips.

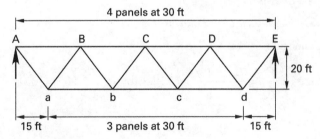

The maximum force in member Bb is most nearly

(A) 40 kips

(B) 51 kips

(C) 59 kips

(D) 73 kips

12. The truss shown carries a moving uniform live load of 2 kips/ft and a moving concentrated live load of 15 kips.

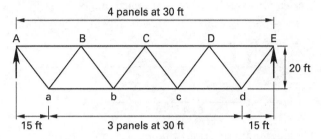

The maximum force in member BC is most nearly

(A) 15 kips

(B) 80 kips

(C) 95 kips

(D) 250 kips

13. A moving load, consisting of two 30 kip forces separated by a constant 6 ft, travels over a two-span bridge as shown. The bridge has an interior expansion joint that can be considered to be an ideal hinge.

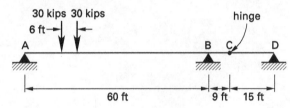

The maximum moment at point B is most nearly

(A) 250 ft-kips

(B) 310 ft-kips

(C) 380 ft-kips

(D) 430 ft-kips

14. A moving load, consisting of two 30 kip forces separated by a constant 6 ft, travels over a two-span bridge as shown. The bridge has an interior expansion joint that can be considered to be an ideal hinge.

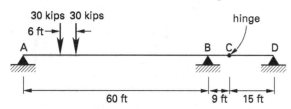

The maximum reaction at point B is most nearly

(A) 30 kips

(B) 60 kips

(C) 70 kips

(D) 90 kips

SOLUTIONS

1. The degree of indeterminacy is 2. Remove the two props (vertical reactions) in order to make the structure statically determinate.

The answer is (B).

2. *Customary U.S. Solution*

The thermal strain is

$$\epsilon_{\text{th}} = \alpha \Delta T = \left(8.9 \times 10^{-6} \, \frac{1}{°\text{F}}\right)(55°\text{F}) = 0.00049$$

The strain must be counteracted to maintain the rod in its original position.

$$\sigma = E\epsilon_{\text{th}} = \left(18 \times 10^6 \, \frac{\text{lbf}}{\text{in}^2}\right)(0.00049) = 8820 \, \text{lbf/in}^2$$

$$F = \sigma A = \left(8820 \, \frac{\text{lbf}}{\text{in}^2}\right)(1 \, \text{in})(2 \, \text{in})$$

$$= 17{,}640 \, \text{lbf} \quad (18{,}000 \, \text{lbf})$$

SI Solution

The thermal strain is

$$\epsilon_{\text{th}} = \alpha \Delta T = \left(16.0 \times 10^{-6} \, \frac{1}{°\text{C}}\right)(30°\text{C}) = 0.00048$$

The strain must be counteracted to maintain the rod in its original position.

$$\sigma = E\epsilon_{\text{th}} = (12 \times 10^4 \, \text{MPa})(0.00048)$$

$$= 57.6 \, \text{MPa}$$

$$F = \sigma A = (57.6 \times 10^6 \, \text{Pa})(0.02 \, \text{m})(0.05 \, \text{m})$$

$$= 57{,}600 \, \text{N} \quad (58 \, \text{kN})$$

The answer is (D).

3. *Customary U.S. Solution*

The rod's stress is

$$\sigma = \frac{F}{A} = \frac{2250 \, \text{lbf}}{\frac{\pi(2 \, \text{in})^2}{4}} = 716.2 \, \text{lbf/in}^2 \quad (720 \, \text{lbf/in}^2)$$

SI Solution

The rod's stress is

$$\sigma = \frac{F}{A} = \frac{10\,000 \text{ N}}{\frac{\pi (0.05 \text{ m})^2}{4}}$$

$$= 5.093 \times 10^6 \text{ Pa} \quad (5.1 \text{ MPa})$$

The answer is (B).

4. *Customary U.S. Solution*

Solve for the decrease in length.

$$\epsilon = \frac{\sigma}{E} = \frac{\delta}{L_o}$$

$$\delta = L_o\left(\frac{\sigma}{E}\right) = (15 \text{ in})\left(\frac{716.2 \frac{\text{lbf}}{\text{in}^2}}{29 \times 10^6 \frac{\text{lbf}}{\text{in}^2}}\right)$$

$$= 3.7 \times 10^{-4} \text{ in}$$

SI Solution

Solve for the decrease in length.

$$\epsilon = \frac{\sigma}{E} = \frac{\delta}{L_o}$$

$$\delta = L_o\left(\frac{\sigma}{E}\right) = (0.4 \text{ m})\left(\frac{5.093 \text{ MPa}}{20 \times 10^4 \text{ MPa}}\right)$$

$$= 1.02 \times 10^{-5} \text{ m} \quad (1.0 \times 10^{-5} \text{ m})$$

The answer is (C).

5. *Customary U.S. Solution*

The deflection of the beam is $\delta_b = P_b L^3/3EI$, where P_b is the net load at the beam tip. If P_c is the tension in the cable, the elongation of the cable is

$$\delta_c = \frac{P_c L_c}{AE}$$

$\delta_c = \delta_b$ is the constraint on the deformation. Therefore,

$$\frac{P_b L_b^3}{3EI} = \frac{P_c L_c}{AE} \quad [\text{Eq. I}]$$

Another equation is the equilibrium equation.

$$F - P_c = P_b \quad [\text{Eq. II}]$$

Solving Eq. I and Eq. II simultaneously,

$$P_c = \frac{\dfrac{FL_b^3}{3I}}{\dfrac{L_c}{A} + \dfrac{L_b^3}{3I}}$$

$$= \frac{(270 \text{ lbf})\left((4 \text{ ft})\left(12 \dfrac{\text{in}}{\text{ft}}\right)\right)^3}{(3)(10 \text{ in}^4)}$$

$$\overline{\dfrac{(2 \text{ ft})\left(12 \dfrac{\text{in}}{\text{ft}}\right)}{0.0124 \text{ in}^2} + \dfrac{\left((4 \text{ ft})\left(12 \dfrac{\text{in}}{\text{ft}}\right)\right)^3}{(3)(10 \text{ in}^4)}}$$

$$= 177 \text{ lbf} \quad (180 \text{ lbf})$$

SI Solution

The deflection of the beam is $\delta_b = P_b L^3/3EI$, where P_b is the net load at the beam tip. If P_c is the tension in the cable, the elongation of the cable is

$$\delta_c = \frac{P_c L_c}{AE}$$

$\delta_c = \delta_b$ is the constraint on the deformation. Therefore,

$$\frac{P_b L_b^3}{3EI} = \frac{P_c L_c}{AE} \quad [\text{Eq. I}]$$

Another equation is the equilibrium equation.

$$F - P_c = P_b \quad [\text{Eq. II}]$$

Solving Eq. I and Eq. II simultaneously,

$$P_c = \frac{\dfrac{FL_b^3}{3I}}{\dfrac{L_c}{A} + \dfrac{L_b^3}{3I}} = \frac{\dfrac{(1200 \text{ N})(1.2 \text{ m})^3 \left(1000 \dfrac{\text{mm}}{\text{m}}\right)^4}{(3)(4.17 \times 10^6 \text{ mm}^4)}}{\dfrac{(0.6 \text{ m})\left(1000 \dfrac{\text{mm}}{\text{m}}\right)^2}{8 \text{ mm}^2}}$$

$$+ \dfrac{(1.2 \text{ m})^3 \left(1000 \dfrac{\text{mm}}{\text{m}}\right)^4}{(3)(4.17 \times 10^6 \text{ mm}^4)}$$

$$= 778 \text{ N} \quad (780 \text{ N})$$

The answer is (A).

6. Let deflection downward be positive.

The deflection at the center of the beam is

$$\delta_b = \frac{5w(2L)^4}{384EI} - \frac{F(2L)^3}{48EI}$$

F is the force applied by the column at the beam center. The beam deflection is equal to the shortening of the column.

$$\delta_c = \frac{Fh}{EA}$$

Since $\delta_b = \delta_c$,

$$\frac{Fh}{EA} = \frac{5w(2L)^4}{384EI} - \frac{F(2L)^3}{48EI}$$

$$F = \frac{5AwL^4}{24hI + 4AL^3} \quad \text{[Eq. I]}$$

Another equation is the equilibrium equation. Let R_1 and R_2 be the left and right support reactions, respectively, on the beam. By symmetry,

$$R_1 = R_2 = R$$
$$2R + F - 2wL = 0 \quad \text{[Eq. II]}$$

Solving Eq. I and Eq. II simultaneously,

$$R = \frac{2wL - F}{2}$$
$$= wL\left(\frac{48hI + 3AL^3}{48hI + 8AL^3}\right)$$

The answer is (B).

7. *Customary U.S. Solution*

Let F_1 and F_2 and δ_1 and δ_2 be the tensions and the deformations in the rods, respectively. The moment equilibrium equation is taken at the hinge of the rigid bar and is

$$\sum M_o = aF_1 + (a+b)F_2 - (a+b+c)P = 0 \quad \text{[Eq. I]}$$

Since the bar is rigid and angle changes are negligible, the elongations of the two aluminum rods are proportional to distances from the hinge at point O. The relationship between the elongations is

$$\frac{\delta_1}{a} = \frac{\delta_2}{a+b}$$

This can be rewritten as

$$\frac{F_1 L_1}{AEa} = \frac{F_2 L_2}{AE(a+b)}$$

Since $L_1/a = L_2/(a+b)$,

$$F_1 = F_2 \quad \text{[Eq. II]}$$

Solving Eq. I and Eq. II,

$$aF + (a+b)F - (a+b+c)P = 0$$

$$F = F_1 = F_2 = \left(\frac{a+b+c}{2a+b}\right)P$$
$$= \left(\frac{6 \text{ ft} + 3 \text{ ft} + 3 \text{ ft}}{(2)(6 \text{ ft}) + 3 \text{ ft}}\right)(4500 \text{ lbf})$$
$$= 3600 \text{ lbf}$$

SI Solution

Let F_1 and F_2 and δ_1 and δ_2 be the tensions and the deformations in the rods, respectively. The moment equilibrium equation is taken at the hinge of the rigid bar and is

$$\sum M_o = aF_1 + (a+b)F_2 - (a+b+c)P = 0 \quad \text{[Eq. I]}$$

Since the bar is rigid and angle changes are negligible, the elongations of the two aluminum rods are proportional to distances from the hinge at point O. The relationship between the elongations is

$$\frac{\delta_1}{a} = \frac{\delta_2}{a+b}$$

This can be rewritten as

$$\frac{F_1 L_1}{AEa} = \frac{F_2 L_2}{AE(a+b)}$$

Since $L_1/a = L_2/(a+b)$,

$$F_1 = F_2 \quad \text{[Eq. II]}$$

Solving Eq. I and Eq. II,

$$aF + (a+b)F - (a+b+c)P = 0$$

$$F = F_1 = F_2 = \left(\frac{a+b+c}{2a+b}\right)P$$

$$= \left(\frac{2\text{ m} + 1\text{ m} + 1\text{ m}}{(2)(2\text{ m}) + 1\text{ m}}\right)(20\text{ kN})$$

$$= 16\text{ kN}$$

The answer is (B).

8. *Customary U.S. Solution*

Use the three-moment method. The first moment of the area is

$$A_1 a = \tfrac{1}{6} Fc(L^2 - c^2)$$

$$= \left(\frac{1}{6}\right)(13{,}600\text{ lbf})(22.5\text{ ft})\big((32.5\text{ ft})^2 - (22.5\text{ ft})^2\big)$$

$$= 28{,}050{,}000\text{ ft}^3\text{-lbf}$$

Since there is no force between R_2 and R_3, $A_2 b = 0$.

The left and right ends of the beam are simply supported; M_1 and M_3 are zero. Therefore, the three-moment equation becomes

$$2M_2(32.5\text{ ft} + 20\text{ ft}) = (-6)\left(\frac{28{,}050{,}000\text{ ft}^3\text{-lbf}}{32.5\text{ ft}}\right)$$

$$M_2 = -49{,}318.7\text{ ft-lbf}$$

M_2 can be written in terms of the load and reactions to the left of support 2.

$$M_2 = (-13{,}600\text{ lbf})(10\text{ ft}) + (32.5\text{ ft})R_1$$

$$= -49{,}318.7\text{ ft-lbf}$$

$$R_1 = 2667.1\text{ lbf} \quad (2700\text{ lbf})$$

Now that R_1 is known, moments can be taken about support 3 to the left.

$$\sum M_3 = (2667.1\text{ lbf})(52.5\text{ ft}) - (13{,}600\text{ lbf})(30\text{ ft})$$
$$+ R_2(20\text{ ft}) = 0$$
$$R_2 = 13{,}398.9\text{ lbf} \quad (13{,}000\text{ lbf})$$

R_3 can be obtained by taking moments about support 1.

$$\sum M_1 = (22.5\text{ ft})(13{,}600\text{ lbf}) - (32.5\text{ ft})(13{,}398.9\text{ lbf})$$
$$-(52.5)R_3 = 0$$
$$R_3 = -2466.0\text{ lbf} \quad (-2500\text{ lbf})$$

SI Solution

Use the three-moment method. The first moment of the area is

$$A_1 a = \tfrac{1}{6} Fc(L^2 - c^2)$$

$$= \left(\frac{1}{6}\right)(60{,}500\text{ N})(7\text{ m})\big((10\text{ m})^2 - (7\text{ m})^2\big)$$

$$= 3.60 \times 10^6\text{ N}\cdot\text{m}^3$$

Since there is no force between R_2 and R_3, $A_2 b = 0$.

The left and right ends of the beam are simply supported; M_1 and M_3 are zero. Therefore, the three-moment equation becomes

$$2M_2(10\text{ m} + 6\text{ m}) = (-6)\left(\frac{3.6 \times 10^6\text{ N}\cdot\text{m}^3}{10\text{ m}}\right)$$

$$M_2 = -67{,}500\text{ N}\cdot\text{m}$$

M_2 can be written in terms of the load and reactions to the left of support 2.

$$M_2 = (-60{,}500\text{ N})(10\text{ m} - 7\text{ m}) + (10\text{ m})R_1$$
$$= -67{,}500\text{ N}\cdot\text{m}$$
$$R_1 = 11{,}400\text{ N} \quad (11{,}000\text{ N})$$

Now that R_1 is known, moments can be taken about support 3 to the left.

$$\sum M_3 = (11{,}400\text{ N})(16\text{ m}) - (60{,}500\text{ N})(9\text{ m})$$
$$+ R_2(6\text{ m}) = 0$$
$$R_2 = 60{,}350\text{ N} \quad (60{,}000\text{ N})$$

R_3 can be obtained by taking moments about support 1.

$$\sum M_1 = (7\text{ m})(60{,}500\text{ N}) - (10\text{ m})(60{,}350\text{ N})$$
$$-(16\text{ m})R_3 = 0$$
$$R_3 = -11{,}250\text{ N} \quad (-11{,}000\text{ N})$$

The answer is (B).

9. The equilibrium requirement is

$$R_1 + R_2 = 1000$$

The vertical reactions are

$$R_1 = \frac{Pb^2(l+2a)}{l^3}$$

$$= \frac{(1000)(10)^2(40 + 2(30))}{40^3}$$

$$= 156.25$$

$$R_2 = \frac{Pa^2(l+2b)}{l^3}$$
$$= \frac{(1000)(30)^2(40+2(10))}{40^3}$$
$$= 843.75$$

The equilibrium requirement is

$$R_1 + R_2 = 1000$$
$$156.25 + 843.75 = 1000 \quad \text{[check]}$$

The answer is (D).

10. *Customary U.S. Solution*

Let deflection downward be positive. The slope of the rigid member is

$$\frac{d}{a+b+c}$$

$$\delta_1 = \left(\frac{F_1}{AE}\right)L_1 = \left(\frac{F_1}{AE}\right)\left(\frac{d}{a+b+c}\right)a$$
$$= \left(\frac{3600 \text{ lbf}}{(0.124 \text{ in}^2)\left(10\times 10^6 \frac{\text{lbf}}{\text{in}^2}\right)}\right)$$
$$\times \left(\frac{12 \text{ ft}}{6 \text{ ft}+3 \text{ ft}+3 \text{ ft}}\right)(6 \text{ ft})\left(12 \frac{\text{in}}{\text{ft}}\right)$$
$$= 0.21 \text{ in}$$

$$\delta_2 = \left(\frac{F_2}{AE}\right)L_2 = \left(\frac{F_2}{AE}\right)\left(\frac{d}{a+b+c}\right)(a+b)$$
$$= \left(\frac{3600 \text{ lbf}}{(0.124 \text{ in}^2)\left(10\times 10^6 \frac{\text{lbf}}{\text{in}^2}\right)}\right)$$
$$\times \left(\frac{12 \text{ ft}}{6 \text{ ft}+3 \text{ ft}+3 \text{ ft}}\right)(6 \text{ ft}+3 \text{ ft})\left(12 \frac{\text{in}}{\text{ft}}\right)$$
$$= 0.31 \text{ in}$$

SI Solution

Let deflection downward be positive. The slope of the rigid member is

$$\frac{d}{a+b+c}$$

$$\delta_1 = \left(\frac{F_1}{AE}\right)L_1 = \left(\frac{F_1}{AE}\right)\left(\frac{d}{a+b+c}\right)a$$
$$= \left(\frac{(16{,}000 \text{ N})\left(1000 \frac{\text{mm}}{\text{m}}\right)^2}{(80 \text{ mm}^2)(70\times 10^9 \text{ Pa})}\right)$$
$$\times \left(\frac{4 \text{ m}}{2 \text{ m}+1 \text{ m}+1 \text{ m}}\right)(2 \text{ m})\left(1000 \frac{\text{mm}}{\text{m}}\right)$$
$$= 5.7 \text{ mm}$$

$$\delta_2 = \left(\frac{F_2}{AE}\right)L_2 = \left(\frac{F_2}{AE}\right)\left(\frac{d}{a+b+c}\right)(a+b)$$
$$= \left(\frac{(16{,}000 \text{ N})\left(1000 \frac{\text{mm}}{\text{m}}\right)^2}{(80 \text{ mm}^2)(70\times 10^9 \text{ Pa})}\right)$$
$$\times \left(\frac{4 \text{ m}}{2 \text{ m}+1 \text{ m}+1 \text{ m}}\right)(2 \text{ m}+1 \text{ m})\left(1000 \frac{\text{mm}}{\text{m}}\right)$$
$$= 8.6 \text{ mm}$$

The answer is (B).

11. The force in member Bb depends on the shear, V, across the cut shown.

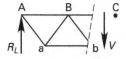

Influence diagram for shear across panel Bb:

If the unit load is to the right of point C, the reaction, R_L, will be

$$R_L = \frac{x}{120 \text{ ft}}$$

(x is the distance from the right reaction to the unit load.)

$$V_L = R_L = \frac{x}{120 \text{ ft}}$$

If the unit load is to the left of point B,

$$R_L = \frac{x}{120 \text{ ft}}$$
$$V = R_L - 1 = \frac{x}{120 \text{ ft}} - 1$$

At points B and C,

$$V_B = \frac{90 \text{ ft}}{120 \text{ ft}} - 1 = -0.25$$
$$V_C = \frac{60 \text{ ft}}{120 \text{ ft}} = 0.5$$

The influence diagram for shear in member Bb is

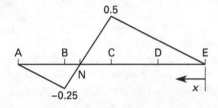

The neutral point, N, is located at

$$x = (2)(30 \text{ ft}) + \left(\frac{0.5}{0.5 + 0.25}\right)(30 \text{ ft}) = 80 \text{ ft}$$

Maximum shear due to moving uniform load:

The moving load, perhaps representing a stream of cars, is allowed to be over any part or all of the bridge deck. The shear will be maximum in member Bb if the load is distributed from N to E.

The area under the influence line from N to E is

$$\left(\tfrac{1}{2}\right)\left((20 \text{ ft})(0.5) + (2)(30 \text{ ft})(0.5)\right) = 20 \text{ ft}$$

The maximum shear, V, is

$$(20 \text{ ft})\left(2 \ \frac{\text{kips}}{\text{ft}}\right) = 40 \text{ kips}$$

Maximum shear due to moving concentrated load:

From the influence diagram, maximum shear will occur when the concentrated load is at point C. The shear in panel Bb is

$$(0.5)(15 \text{ kips}) = 7.5 \text{ kips}$$

Tension in member Bb:
The force triangle is

The total maximum shear across panel Bb is

$$40 \text{ kips} + 7.5 \text{ kips} = 47.5 \text{ kips}$$

The total maximum tension in member Bb is

$$(47.5 \text{ ft})\left(\frac{25 \text{ ft}}{20 \text{ ft}}\right) = 59.375 \text{ kips} \quad (59 \text{ kips})$$

The answer is (C).

12. *Influence diagram for moment at point b:*

The horizontal member BC cannot resist vertical shear.

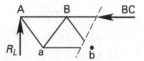

With no loads between A and C, the force in member BC can be found by summing the moments about point b. Taking clockwise moments as positive,

$$\sum M_b = (45 \text{ ft})R_L - (20 \text{ ft})(BC)$$
$$= 0$$
$$BC = \frac{45 R_L}{20 \text{ ft}}$$

$45 R_L$ is the moment that the moment from force BC opposes. In general,

$$BC = \frac{M_b}{20 \text{ ft}}$$

If the load is between C and E,

$$R_L = \frac{x}{120 \text{ ft}} \quad [x \text{ is measured from E}]$$

The moment caused by R_L is

$$M_b = (45 \text{ ft})\left(\frac{x}{120 \text{ ft}}\right)$$
$$= 0.375x$$

If the load is between A and B, the reaction is

$$R_L = \frac{x}{120 \text{ ft}}$$

The moment at b is also affected by the load between A and B.

$$M_b = 0.375x - (1)(x - 75 \text{ ft})$$
$$= 75 \text{ ft} - 0.625x$$

Plotting these values versus x,

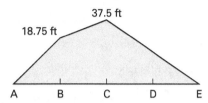

Maximum moment due to uniform load:

The moment at b is maximum when the entire truss is loaded from A to E. The area under the curve is

$$\left(\frac{1}{2}\right)(30 \text{ ft})(18.75 \text{ ft}) + \left(\frac{1}{2}\right)(60 \text{ ft})(37.5 \text{ ft})$$
$$+ (30 \text{ ft})(18.75 \text{ ft})$$
$$+ \left(\frac{1}{2}\right)(30 \text{ ft})(37.5 \text{ ft} - 18.75 \text{ ft})$$
$$= 2250 \text{ ft}^2$$

The maximum moment is

$$M_b = \left(2 \frac{\text{kips}}{\text{ft}}\right)(2250 \text{ ft}^2) = 4500 \text{ ft-kips}$$

Maximum moment due to concentrated load:

Maximum moment will occur when the load is at C.

$$M_b = (15 \text{ kips})(37.5 \text{ ft})$$
$$= 562.5 \text{ ft-kips}$$

Total maximum moment:

$$M_b = 562.5 \text{ ft-kips} + 4500 \text{ ft-kips}$$
$$= 5062.5 \text{ ft-kips}$$

Compression in BC:

$$BC = \frac{M_b}{20 \text{ ft}} = \frac{5062.5 \text{ ft-kips}}{20 \text{ ft}}$$
$$= 253.1 \text{ kips} \quad (250 \text{ kips})$$

The answer is (D).

13. Put a hinge at point B and rotate.

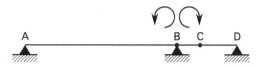

The moment influence diagram is

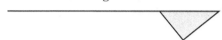

One of the loads should be at point C.

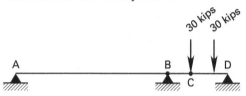

Since the slope of the influence line is less between C and D, the ordinate 6 ft to the right of point C will be larger than the ordinate 6 ft to the left of point C. The reaction at the hinge due to the 30 kip load 6 ft to the right of point C is

$$R = (30 \text{ kips})\left(\frac{15 \text{ ft} - 6 \text{ ft}}{15 \text{ ft}}\right) = 18 \text{ kips}$$

The moment at point B is

$$M_B = (9 \text{ ft})(30 \text{ kips} + 18 \text{ kips})$$
$$= 432 \text{ ft-kips} \quad (430 \text{ ft-kips})$$

The answer is (D).

14. Use the method of virtual displacement to draw the shear influence diagram. Since the point is a reaction point, lift the point a distance of 1. The shear diagram is

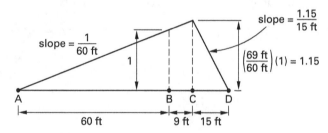

For maximum shear, one load or the other must be at point C.

By inspection, the effect of having both loads to the left of C is greater than having them to the right.

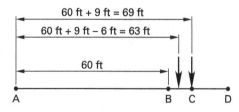

The maximum shear is

$$V_{\max} = (30 \text{ kips})\left(\frac{63 \text{ ft}}{60 \text{ ft}} + \frac{69 \text{ ft}}{60 \text{ ft}}\right) = 66 \text{ kips} \quad (70 \text{ kips})$$

The answer is (C).

46 Engineering Materials

Content in blue refers to the *NCEES Handbook*.

PRACTICE PROBLEMS

1. A composite polymer consists of silicon carbide fiber and polyvinyl chloride. The density of the composite is 2.22 g/cm³. The mass fraction of silicon carbide fiber in the composite is most nearly

(A) 0.20
(B) 0.35
(C) 0.55
(D) 0.70

2. A composite material consists of 40% boron fibers by volume. The fibers are oriented axially in a polymer matrix. The modulus of elasticity of the polymer is 0.55 Mpsi. A tensile load is applied along the direction of the fibers. Most nearly, the ratio of the load carried by the fiber to the load carried by the polymer is

(A) 30:1
(B) 50:1
(C) 70:1
(D) 90:1

SOLUTIONS

1. Use the formula for calculating the density of a composite to determine the volume fraction of each individual material.

Composite Materials

$$\rho_c = \sum f_i \rho_i = f_{PVC}\rho_{PVC} + f_{SCF}\rho_{SCF}$$
$$= (1 - f_{SCF})\rho_{PVC} + f_{SCF}\rho_{SCF}$$

The density of polyvinyl chloride is 1.3 Mg/m³, and the density of silicon carbide fiber is 3.2 Mg/m³. [Typical Material Properties]

Convert the density of each material to units of grams per cubic centimeter.

$$\rho_{PVC} = 1.3 \frac{Mg}{m^3} = \frac{\left(1.3 \frac{Mg}{m^3}\right)\left(10^6 \frac{g}{Mg}\right)}{10^6 \frac{cm^3}{m^3}} = 1.3 \text{ g/cm}^3$$

$$\rho_{SCF} = 3.2 \frac{Mg}{m^3} = \frac{\left(3.2 \frac{Mg}{m^3}\right)\left(10^6 \frac{g}{Mg}\right)}{10^6 \frac{cm^3}{m^3}} = 3.2 \text{ g/cm}^3$$

Substitute the known values into the equation.

$$2.22 \frac{g}{cm^3} = (1 - f_{SCF})\left(1.3 \frac{g}{cm^3}\right) + f_{SCF}\left(3.2 \frac{g}{cm^3}\right)$$

Solving for the volume fraction of the silicon carbide fiber through trial and error, $f_{SCF} = 0.4842$.

Find the volume fraction of polyvinyl chloride.

$$f_{PVC} = 1 - f_{SCF} = 1 - 0.4842 = 0.5158$$

The volume of silicon carbide fiber in 1 cm³ of composite is

$$V_{SCF} = f_{SCF}V_C = (0.4842)(1 \text{ cm}^3) = 0.4842 \text{ cm}^3$$

The mass of silicon carbide fiber is

$$m_{SCF} = V_{SCF}\rho_{SCF} = (0.4842 \text{ cm}^3)\left(3.2 \frac{\text{g}}{\text{cm}^3}\right) = 1.5494 \text{ g}$$

The mass of the composite is

$$m_C = V_C\rho_C = (1 \text{ cm}^3)\left(2.22 \frac{\text{g}}{\text{cm}^3}\right) = 2.22 \text{ g}$$

The mass fraction of silicon carbide fiber is

$$y_i = \frac{m_{SCF}}{m_C} = \frac{1.5494 \text{ g}}{2.22 \text{ g}} = 0.6979 \quad (0.70)$$

The answer is (D).

2. For a composite material, the strains of the individual materials are equal in the axial direction.

Composite Materials

$$\left(\frac{\Delta L}{L}\right)_1 = \left(\frac{\Delta L}{L}\right)_2$$

The equation for the engineering strain is

Engineering Strain

$$\epsilon = \frac{\Delta L}{L_o}$$

Therefore, $\epsilon_{fiber} = \epsilon_{matrix}$. The equation for the modulus of elasticity is

Uniaxial Loading and Deformation

$$E = \frac{\sigma}{\epsilon} = \frac{\frac{P}{A_o}}{\frac{\delta}{L}} = \frac{\frac{P}{A_o}}{\epsilon}$$

$$= \frac{P}{A_o \epsilon}$$

Therefore,

$$\epsilon_{fiber} = \frac{P_{fiber}}{A_{o,fiber} E_{fiber}}$$

$$\epsilon_{matrix} = \frac{P_{matrix}}{A_{o,matrix} E_{matrix}}$$

Because $\epsilon_{fiber} = \epsilon_{matrix}$,

$$\frac{P_{fiber}}{A_{o,fiber} E_{fiber}} = \frac{P_{matrix}}{A_{o,matrix} E_{matrix}}$$

The cross-sectional areas are proportional to the volumes of the individual materials.

$$\frac{P_{fiber}}{V_{fiber} E_{fiber}} = \frac{P_{matrix}}{V_{matrix} E_{matrix}}$$

The modulus of elasticity of boron fiber is 58 Mpsi. [Typical Material Properties]

Calculate the required ratio of the loads carried by the boron fiber and the polymer.

$$\frac{P_{fiber}}{P_{matrix}} = \frac{V_{fiber} E_{fiber}}{V_{matrix} E_{matrix}}$$

$$= \frac{(0.40)(58 \text{ Mpsi})}{(0.60)(0.55 \text{ Mpsi})}$$

$$= 70.30 \quad (70:1)$$

The answer is (C).

47 Material Properties and Testing

Content in blue refers to the NCEES Handbook.

PRACTICE PROBLEMS

1. At a particular instant during a tensile test, a 0.5 in diameter steel cylinder 2 in long experienced a length of 2.6 in at a load of 4200 lbf. The cylinder was prepared according to ASTM specifications. Uniform narrowing of the cylinder should be assumed. The true stress in the steel cylinder at that instant was most nearly

(A) 28,000 lbf/in^2
(B) 32,000 lbf/in^2
(C) 36,000 lbf/in^2
(D) 40,000 lbf/in^2

2. A copper specimen has an engineering stress of 20,000 lbf/in^2 (140 MPa) and an engineering strain of 0.0200 in/in (0.020 mm/mm). Poisson's ratio for the specimen is 0.3. The true strain is most nearly

(A) 0.0182 in/in (0.0182 mm/mm)
(B) 0.0189 in/in (0.0189 mm/mm)
(C) 0.0194 in/in (0.0194 mm/mm)
(D) 0.0198 in/in (0.0198 mm/mm)

3. A graph of engineering stress-strain for a material is shown. Poisson's ratio for the material is 0.3.

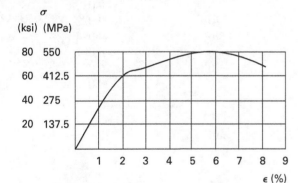

The 0.5% yield strength is most nearly

(A) 70,000 lbf/in^2 (480 MPa)
(B) 76,000 lbf/in^2 (530 MPa)
(C) 84,000 lbf/in^2 (590 MPa)
(D) 98,000 lbf/in^2 (690 MPa)

4. A graph of engineering stress-strain for a material is shown. Poisson's ratio for the material is 0.3.

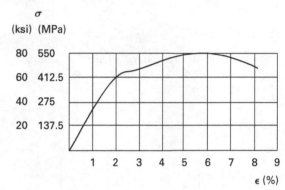

The elastic modulus is most nearly

(A) 2.4×10^6 lbf/in^2 (17 GPa)
(B) 2.7×10^6 lbf/in^2 (19 GPa)
(C) 2.9×10^6 lbf/in^2 (20 GPa)
(D) 3.0×10^6 lbf/in^2 (21 GPa)

5. A graph of engineering stress-strain for a material is shown. Poisson's ratio for the material is 0.3. The ultimate strength is most nearly

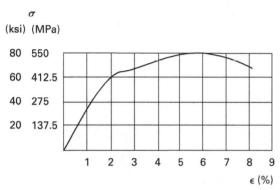

(A) 72,000 lbf/in² (500 MPa)

(B) 76,000 lbf/in² (530 MPa)

(C) 80,000 lbf/in² (550 MPa)

(D) 84,000 lbf/in² (590 MPa)

6. A graph of engineering stress-strain for a material is shown. Poisson's ratio for the material is 0.3. The fracture strength is most nearly

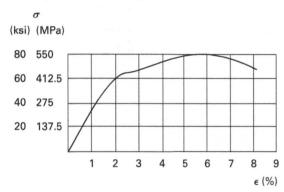

(A) 62,000 lbf/in² (430 MPa)

(B) 70,000 lbf/in² (480 MPa)

(C) 76,000 lbf/in² (530 MPa)

(D) 84,000 lbf/in² (590 MPa)

7. A graph of engineering stress-strain for a material is shown. Poisson's ratio for the material is 0.3.

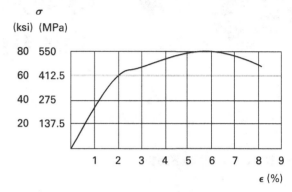

The percentage of elongation at fracture is most nearly

(A) 6%

(B) 8%

(C) 10%

(D) 12%

8. A specimen with an unstressed cross-sectional area of 4.0 in² (25 cm²) necks down to 3.42 in² (22 cm²) before breaking in a standard tensile test. The percentage reduction in area of the material is most nearly

(A) 9.4% (8.9%)

(B) 10% (9.4%)

(C) 13% (10%)

(D) 15% (12%)

9. A constant 15,000 lbf/in² (100 MPa) tensile stress is applied to a specimen. The stress is known to be less than the material's yield strength. The strain is measured at various times, as shown.

time (hr)	strain (in/in)
5	0.018
10	0.022
20	0.026
30	0.031
40	0.035
50	0.040
60	0.046

What is most nearly the steady-state creep rate for the material?

(A) 0.00037 hr^{-1}

(B) 0.00041 hr^{-1}

(C) 0.00051 hr^{-1}

(D) 0.00075 hr^{-1}

10. At a particular instant during a tensile test, a 0.5 in diameter steel cylinder 2 in long experiences an instantaneous length of 2.6 in at a load of 4200 lbf. The cylinder is prepared according to ASTM specifications. The true stress in the steel cylinder at that instant is most nearly

(A) 28,000 lbf/in^2

(B) 32,000 lbf/in^2

(C) 36,000 lbf/in^2

(D) 40,000 lbf/in^2

11. True stress and true strain from a tensile test performed on a metal bar are plotted on a log-log graph in accordance with ASTM E646. If the slope of the curve is between 0.5 and 1.0, which conclusion can be drawn about the metal?

(A) The metal is brittle.

(B) The metal is hard.

(C) The metal is highly malleable and ductile.

(D) The metal has a high strain-hardening capacity.

12. Which nondestructive testing method would be best suited to detect a subsurface crack in a part made of 300 series stainless steel?

(A) liquid penetrant inspection

(B) magnetic particle inspection

(C) D-sight inspection

(D) ultrasonic inspection

13. According to ASTM standards, which test is used to identify the ductile-to-brittle transition temperature of metals?

(A) Jominy

(B) Vickers

(C) Charpy

(D) thermography

14. A 52100 steel plate containing a 1.5 inch long crack is 8 inches wide and 0.2 inches thick. The plate is pulled with a uniform tensile force of 5000 lbf. What is most nearly the stress intensity factor at the end of the crack?

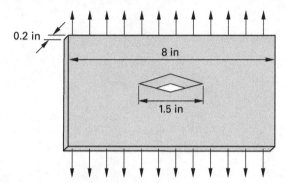

(A) $4.2 \text{ ksi} \cdot \sqrt{\text{in}}$

(B) $4.8 \text{ ksi} \cdot \sqrt{\text{in}}$

(C) $5.3 \text{ ksi} \cdot \sqrt{\text{in}}$

(D) $6.8 \text{ ksi} \cdot \sqrt{\text{in}}$

SOLUTIONS

1. As volume is constant, calculate instantaneous area, A.

$$AL = A_o L_o$$

$$A = \frac{A_o L_o}{L} = \frac{\pi(0.25 \text{ in})^2 (2 \text{ in})}{2.6 \text{ in}} = 0.15 \text{ in}^2$$

Solve for true stress, σ.

$$\sigma = \frac{F}{A} = \frac{4200 \text{ lbf}}{0.15 \text{ in}^2} = \boxed{28{,}000 \text{ lbf/in}^2}$$

The answer is (A).

2. *Customary U.S. Solution*

The true strain is

True Strain

$$\epsilon_T = \ln(1 + \epsilon) = \ln\left(1 + 0.020 \, \frac{\text{in}}{\text{in}}\right)$$
$$= 0.0198 \text{ in/in}$$

SI Solution

The true strain is

True Strain

$$\epsilon_T = \ln(1 + \epsilon) = \ln\left(1 + 0.020 \, \frac{\text{mm}}{\text{mm}}\right)$$
$$= 0.0198 \text{ mm/mm}$$

The answer is (D).

3. *Customary U.S. Solution*

Extend a line from the 0.5% offset strain value parallel to the linear portion of the curve.

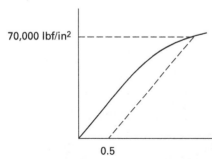

The 0.5% yield strength is 70,000 lbf/in².

SI Solution

Extend a line from the 0.5% offset strain value parallel to the linear portion of the curve.

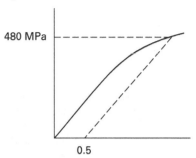

The 0.5% yield strength is 480 MPa.

The answer is (A).

4. *Customary U.S. Solution*

At the elastic limit, the stress is 60,000 lbf/in², and the percent strain is 2. The elastic modulus is

Uniaxial Loading and Deformation

$$E = \frac{\sigma}{\epsilon} = \frac{60{,}000 \, \frac{\text{lbf}}{\text{in}^2}}{0.02}$$
$$= 3.0 \times 10^6 \text{ lbf/in}^2$$

SI Solution

At the elastic limit, the stress is 410 MPa, and the percent strain is 2. The elastic modulus is

Uniaxial Loading and Deformation

$$E = \frac{\sigma}{\epsilon} = \frac{410 \text{ MPa}}{(0.02)\left(1000 \, \frac{\text{MPa}}{\text{GPa}}\right)}$$
$$= 20.5 \text{ GPa} \quad (21 \text{ GPa})$$

The answer is (D).

5. *Customary U.S. Solution*

The ultimate strength is the highest point of the curve. This value is 80,000 lbf/in².

SI Solution

The ultimate strength is the highest point of the curve. This value is 550 MPa.

The answer is (C).

6. *Customary U.S. Solution*

The fracture strength is at the end of the curve. This value is 70,000 lbf/in².

SI Solution

The fracture strength is at the end of the curve. This value is 480 MPa.

The answer is (B).

7. The percent elongation at fracture is determined by extending a straight line parallel to the initial strain line from the fracture point. This gives an approximate value of 6%.

The answer is (A).

8. *Customary U.S. Solution*

Use Eq. 47.8.

Percent Reduction in Area (RA)

$$\%\text{RA} = \frac{A_o - A_f}{A_o} \times 100\%$$

$$= \frac{4.0 \text{ in}^2 - 3.42 \text{ in}^2}{4.0 \text{ in}^2} \times 100\%$$

$$= 14.5\% \quad (15\%)$$

SI Solution

Use Eq. 47.8.

Percent Reduction in Area (RA)

$$\%\text{RA} = \frac{A_o - A_f}{A_o} \times 100\%$$

$$= \frac{25 \text{ cm}^2 - 22 \text{ cm}^2}{25 \text{ cm}^2} \times 100\%$$

$$= 12\%$$

The answer is (D).

9. Plot the data and draw a best fit straight line through the data points. Calculate the slope of the line to obtain the steady-state creep rate.

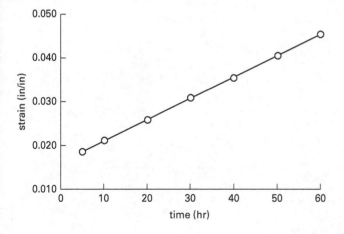

The creep rate is

Mechanical Properties

$$\frac{d\varepsilon}{dt} = \frac{0.046 \frac{\text{in}}{\text{in}} - 0.018 \frac{\text{in}}{\text{in}}}{60 \text{ hr} - 5 \text{ hr}}$$

$$= 0.0005091 \text{ hr}^{-1} \quad (0.00051 \text{ hr}^{-1})$$

The answer is (C).

10. Volume is constant, so the instantaneous area at the time the load is applied is

$$AL = A_o L_o$$

$$A = \frac{A_o L_o}{L} = \frac{\pi (0.25 \text{ in})^2 (2 \text{ in})}{2.6 \text{ in}} = 0.15 \text{ in}^2$$

The true stress is

Uniaxial Loading and Deformation

$$\sigma_T = \frac{P}{A} = \frac{4200 \text{ lbf}}{0.15 \text{ in}^2} = 28{,}000 \text{ lbf/in}^2$$

The answer is (A).

11. The true stress-true strain curve of many metals in the region of uniform plastic deformation can be expressed by the power curve relationship $\sigma_T = K\epsilon_T^n$. When data following this equation is plotted in log-log format, the result will be a straight line with linear slope n. The slope is known as the strain-hardening exponent. The strain-hardening exponent may have values from 0 (perfectly plastic solid) to 1 (perfectly elastic solid). For most metals, n is between 0.10 and 0.50. Only copper-based alloys have values in the vicinity of 0.50. Annealed steels, for example, have a strain-hardening exponent between 0.10 and 0.25. A strain-hardening exponent between 0.5 and 1.0 would indicate that the material becomes stronger and harder as it is strained. That is, it has a high strain-hardening capacity.

The answer is (D).

12. The most effective inspection method must be able to effectively detect subsurface cracking in the part material. Since the crack is subsurface, inspection techniques for surface flaws will not be effective. This rules out liquid penetrant inspection and D-sight inspection. Since the part is made of 300 series stainless steel, which is non-magnetic, magnetic particle inspection will not be a viable technique. Ultrasonic inspection is well suited to inspecting internal flaws, such as cracks, in all metals. [Nondestructive Test Methods]

The answer is (D).

13. In order to measure a metal's ductile-to-brittle transition temperature, a Charpy impact test is used to determine the amount of energy required to cause failure in standardized test samples. The test is repeated at various specimen temperatures to determine the ductile-to-brittle transition temperature. [Impact Test]

The answer is (C).

14. The nominal engineering stress is

Uniaxial Loading and Deformation

$$\sigma = \frac{P}{A_o} = \frac{(5000 \text{ lbf})}{(8 \text{ in})(0.2 \text{ in})} = 3125 \text{ lbf/in}^2 \quad (3125 \text{ psi})$$

Since this is an interior crack, $Y = 1$ and the crack length for an internal crack is $2a$. We know that $a = 1.5/2 = 0.75$. The stress intensity factor is

Mechanical Properties

$$K_{LC} = Y\sigma\sqrt{\pi a} = (1)(3125 \text{ psi})\sqrt{\pi(0.75 \text{ in})}$$
$$= 4796.8 \text{ psi} \cdot \sqrt{\text{in}} \quad (4.8 \text{ ksi} \cdot \sqrt{\text{in}})$$

The answer is (B).

48 Thermal Treatment of Metals

Content in blue refers to the *NCEES Handbook*.

PRACTICE PROBLEMS

1. Refer to the following equilibrium diagram for an alloy of elements A and B.

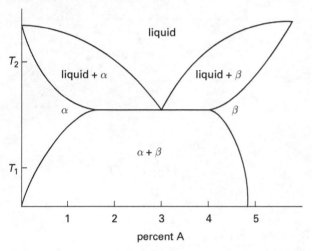

For a 4%A alloy at temperature T_1, what is the percentage of α and β?

(A) $\alpha = 7.0\%, \beta = 30.0\%$
(B) $\alpha = 10.0\%, \beta = 43.0\%$
(C) $\alpha = 14.0\%, \beta = 86.0\%$
(D) $\alpha = 20.0\%, \beta = 76.0\%$

2. Refer to the equilibrium diagram shown for an alloy of elements A and B.

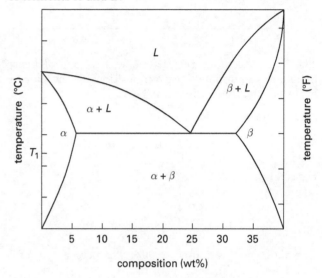

For a 20%A alloy at temperature T_1, the weight percentage of α in the solid phase is most nearly

(A) 5%
(B) 20%
(C) 35%
(D) 48%

3. A low-carbon steel is to be quickly quenched from 1355°F to 400°F. Which quenchant will result in the most rapid cooling rate?

(A) air
(B) brine
(C) oil
(D) water

SOLUTIONS

1. For temperature T_1, the equilibrium diagram is [Lever Rule]

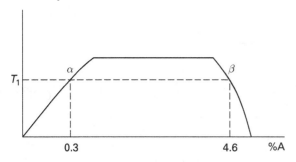

The intersections of the tie-line with α and β phase boundaries determines the compositions of the phases. From the diagram, solid α is 0.3%A.

The %B for solid α is

$$\%B = 100\% - 0.3\% = 99.7\%$$

For solid β,

$$\%A = 4.6\%$$
$$\%B = 100\% - 4.6\% = 95.4\%$$

The alloy composition is $x = 4\%$. The percentages of α and β are

Lever Rule

$$\text{wt}\%\alpha = \frac{x_\beta - x}{x_\beta - x_\alpha} \times 100\%$$

$$\%\alpha = \frac{4.6\% - 4\%}{4.6\% - 0.3\%}(100\%)$$
$$= 0.14 \quad (14.0\%)$$
$$\%\beta = 100\% - \%\alpha$$
$$= 100\% - 14.0\%$$
$$= 86.0\%$$

The answer is (C).

2. Use the diagram to find the composition of α and β in the solid phase at T_1, and solve for the weight percentage of α using the lever rule.

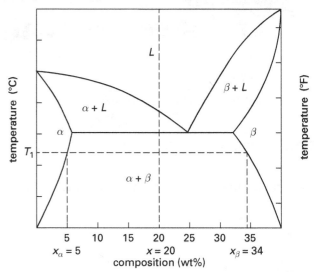

From the lever rule,

Lever Rule

$$\text{wt}\%\alpha = \frac{x_\beta - x}{x_\beta - x_\alpha} \times 100\%$$
$$= \left(\frac{34 - 20}{34 - 5}\right)(100\%)$$
$$= 48.3\% \quad (48\%)$$

The answer is (D).

3. Brine has the fastest quenching rate of the quench media identified.

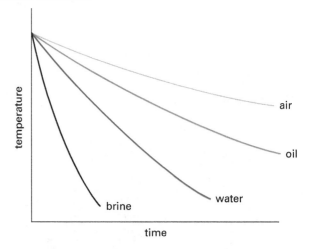

The answer is (B).

49 Properties of Areas

Content in blue refers to the *NCEES Handbook*.

PRACTICE PROBLEMS

1. Where is the x-coordinate of the centroid of the area most nearly located?

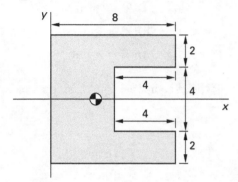

(A) 2.7 units

(B) 2.9 units

(C) 3.1 units

(D) 3.3 units

2. Most nearly, what is the centroidal moment of inertia about an axis parallel to the x-axis shown?

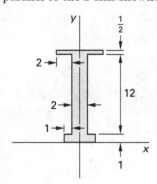

(A) 160 units^4

(B) 290 units^4

(C) 570 units^4

(D) 740 units^4

3. A rectangular 4 in × 10 in area has a 2 in diameter hole in its geometric center, as shown. What is most nearly the moment of inertia about the y-axis?

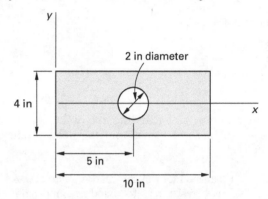

(A) 1030 in^4

(B) 1150 in^4

(C) 1250 in^4

(D) 1370 in^4

4. In the graph shown, what is most nearly the moment of inertia about the x-axis of the area OAB?

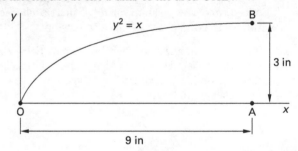

(A) 32 in^4

(B) 67 in^4

(C) 70 in^4

(D) 76 in^4

5. An annular flat ring has an outer diameter of 4 in and an inner diameter of 2 in, as shown. What is most nearly the radius of gyration of the ring about the y-axis?

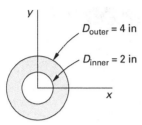

(A) 0.89 in

(B) 1.1 in

(C) 1.3 in

(D) 2.2 in

6. Which option best describes the difference between the mass moment of inertia and the polar moment of inertia?

(A) The mass moment of inertia characterizes the object's resistance to bending, and the polar moment of inertia characterizes the object's resistance to changes in rotational speed.

(B) The mass moment of inertia characterizes the object's resistance to torsion, and the polar moment of inertia characterizes the object's resistance to changes in rotational speed.

(C) The mass moment of inertia characterizes the object's resistance to changes in rotational speed, and the polar moment of inertia characterizes the object's resistance to torsion.

(D) The mass moment of inertia characterizes the object's resistance to torsion, and the polar moment of inertia characterizes the object's resistance changes in bending.

SOLUTIONS

1. The area is divided into three basic shapes.

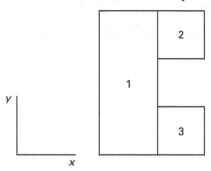

First, calculate the areas of the basic shapes.

$$A_1 = (4)(8) = 32 \, \text{units}^2$$
$$A_2 = (4)(2) = 8 \, \text{units}^2$$
$$A_3 = (4)(2) = 8 \, \text{units}^2$$

Next, find the x-components of the centroids of the basic shapes.

$$x_{c,1} = 2 \, \text{units}$$
$$x_{c,2} = 6 \, \text{units}$$
$$x_{c,3} = 6 \, \text{units}$$

Finally, use Eq. 49.5.

$$\begin{aligned} x_c &= \frac{\sum A_i x_{c,i}}{\sum A_i} \\ &= \frac{(32 \, \text{units}^2)(2 \, \text{units}) + (8 \, \text{units}^2)(6 \, \text{units})}{32 \, \text{units}^2 + 8 \, \text{units}^2 + 8 \, \text{units}^2} \\ &\quad + (8 \, \text{units}^2)(6 \, \text{units}) \\ &= 3.33 \, \text{units} \quad (3.3 \, \text{units}) \end{aligned}$$

The answer is (D).

2. The area is divided into three basic shapes.

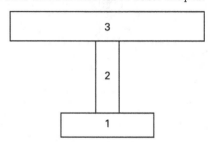

First, calculate the areas of the basic shapes.

$$A_1 = (4)(1) = 4 \text{ units}^2$$
$$A_2 = (2)(12) = 24 \text{ units}^2$$
$$A_3 = (6)(0.5) = 3 \text{ units}^2$$

Next, find the y-components of the centroids of the basic shapes.

$$y_{ac,1} = 0.5 \text{ units}$$
$$y_{ac,2} = 7 \text{ units}$$
$$y_{ac,3} = 13.25 \text{ units}$$

From Eq. 49.6, the centroid of the area is

$$y_{ac} = \frac{\sum A_n y_{ac,i}}{\sum A_n} = \frac{(4)(0.5)+(24)(7)+(3)(13.25)}{4+24+3}$$
$$= 6.77 \text{ units}$$

The moment of inertia of basic shape 1 about its own centroid is

Properties of Various Shapes

$$I_{cx,1} = \frac{bh^3}{12} = \frac{(4)(1)^3}{12} = 0.33 \text{ units}^4$$

The moment of inertia of basic shape 2 about its own centroid is

Properties of Various Shapes

$$I_{cx,2} = \frac{bh^3}{12} = \frac{(2)(12)^3}{12} = 288 \text{ units}^4$$

The moment of inertia of basic shape 3 about its own centroid is

Properties of Various Shapes

$$I_{cx,3} = \frac{bh^3}{12} = \frac{(6)(0.5)^3}{12} = 0.063 \text{ units}^4$$

From the parallel axis theorem, Eq. 49.20, the moment of inertia of basic shape 1 about the centroidal axis of the section is

Moment of Inertia Parallel Axis Theorem

$$I' = I_c + d^2 A = 0.33 + (4)(6.77 - 0.5)^2$$
$$= 157.6 \text{ units}^4$$

The moment of inertia of basic shape 2 about the centroidal axis of the section is

$$I_{x,2} = I_{cx,2} + A_2 d_2^2 = 288 + (24)(7.0 - 6.77)^2$$
$$= 289.3 \text{ units}^4$$

The moment of inertia of basic shape 3 about the centroidal axis of the section is

$$I_{x,3} = I_{cx,3} + A_3 d_3^2 = 0.063 + (3)(13.25 - 6.77)^2$$
$$= 126.0 \text{ units}^4$$

The total moment of inertia about the centroidal axis of the section is

$$I_x = I_{x,1} + I_{x,2} + I_{x,3}$$
$$= 157.6 \text{ units}^4 + 289.3 \text{ units}^4 + 126.0 \text{ units}^4$$
$$= 572.9 \text{ units}^4 \quad (570 \text{ units}^4)$$

The answer is (C).

3. This is a composite area. Divide the composite area into two basic shapes, a rectangle and a circle.

The moment of inertia of the rectangle with respect to an edge (in this case, the y-axis) is

Properties of Various Shapes

$$I_{y,\text{rectangle}} = \frac{b^3 h}{3} = \frac{(4 \text{ in})(10 \text{ in})^3}{3} = 1333.33 \text{ in}^4$$

The centroidal moment of inertia of the circle is

Properties of Various Shapes

$$I_{c,\text{circle}} = \frac{\pi r^4}{4} = \frac{\pi (1 \text{ in})^4}{4} = 0.785 \text{ in}^4$$

From the parallel axis theorem, the moment of inertia of the circle with respect to the y-axis is

Moment of Inertia Parallel Axis Theorem

$$I'_y = I_{y_c} + d_x^2 A = 0.785 \text{ in}^4 + (5 \text{ in})^2 \pi (1 \text{ in})^2$$
$$= 79.285 \text{ in}^4$$

The moment of inertia about the y-axis is

$$I_{y,\text{composite area}} = I_{y,\text{rectangle}} - I_{y,\text{circle}}$$
$$= 1333.33 \text{ in}^4 - 79.285 \text{ in}^4$$
$$= 1254.05 \text{ in}^4 \quad (1250 \text{ in}^4)$$

The answer is (C).

4. Determine dA, which is the shaded area within the curve parallel to the x-axis.

$$dA = (9-x)\,dy$$

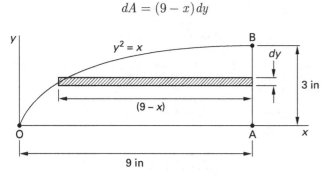

Using Eq. 49.17, calculate the moment of inertia from $y=0$ in to $y=3$ in.

Moment of Inertia

$$\begin{aligned}
I_x &= \int y^2\,dA = \int_{0\text{ in}}^{3\text{ in}} y^2(9-x)\,dy \\
&= \int_{0\text{ in}}^{3\text{ in}} y^2(9-y^2)\,dy \\
&= 9\int_{0\text{ in}}^{3\text{ in}} y^2\,dy - \int_{0\text{ in}}^{3\text{ in}} y^4\,dy \\
&= (9)\left(\frac{y^3}{3}\bigg|_{0\text{ in}}^{3\text{ in}}\right) - \frac{y^5}{5}\bigg|_{0\text{ in}}^{3\text{ in}} \\
&= 32.4 \text{ in}^4 \quad (32 \text{ in}^4)
\end{aligned}$$

The answer is (A).

5. Due to symmetry, the moment of inertia and radius of gyration about the y-axis will be the same as for the x-axis. The moment of inertia of the annular ring is

Properties of Various Shapes

$$\begin{aligned}
I &= \frac{\pi(a^4 - b^4)}{4} \\
&= \frac{\pi\left(\frac{4\text{ in}}{2}\right)^4}{4} - \frac{\pi\left(\frac{2\text{ in}}{2}\right)^4}{4} \\
&= 11.78 \text{ in}^4
\end{aligned}$$

The area of the composite area a is

$$A = \pi r_{\text{outer}}^2 - \pi r_{\text{inner}}^2 = \pi\left(\frac{4\text{ in}}{2}\right)^2 - \pi\left(\frac{2\text{ in}}{2}\right)^2$$
$$= 9.42 \text{ in}^2$$

From Eq. 49.28, the radius of gyration of the annular ring about the y-axis is

Radius of Gyration

$$\begin{aligned}
r_y &= \sqrt{\frac{I_y}{A}} = \sqrt{\frac{11.78 \text{ in}^4}{9.42 \text{ in}^2}} \\
&= 1.12 \text{ in} \quad (1.1 \text{ in})
\end{aligned}$$

The answer is (B).

6. The mass moment of inertia is the limit of the product of the mass of the solid body and the square of the distance from a given axis. It measures the object's resistance to changes in rotational speed. Typical use is in the equation $T = I\alpha$. A solid body with a small amount of inertia will be less likely to resist a change in rotational speed.

The polar moment of inertia is measured about an axis perpendicular to the plane of the area. It characterizes the object's resistance to torsion. Typical use is in the equation $\theta = TL/JG$. A solid body with a small area polar moment inertia will twist more than a solid body with greater inertia.

The answer is (C).

50 Strength of Materials

Content in blue refers to the NCEES Handbook.

PRACTICE PROBLEMS

1. A beam 14 ft (4.2 m) long is simply supported at the left end and 2 ft (0.6 m) from the right end. The beam has a mass of 20 lbm/ft (30 kg/m). A 100 lbf (450 N) load is applied 2 ft (0.6 m) from the left end. An 80 lbf (350 N) load is applied at the right end.

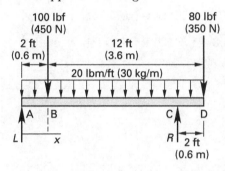

The maximum moment is most nearly

(A) 150 ft-lbf (200 N·m)

(B) 250 ft-lbf (340 N·m)

(C) 390 ft-lbf (520 N·m)

(D) 830 ft-lbf (1100 N·m)

2. A 0.25% carbon steel supporting strap in an oven furnace carries a constant tensile load while the oven temperature is increased from 400°F to 800°F. The modulus of elasticity of steel at 400°F is 27,000,000 lbf/in². The modulus of elasticity of steel at 800°F is 24,200,000 lbf/in². Compared to the strain at 400°F, what is most nearly the strain at 800°F?

(A) $1.1\epsilon_{400°F}$

(B) $1.3\epsilon_{400°F}$

(C) $1.4\epsilon_{400°F}$

(D) $2.6\epsilon_{400°F}$

3. A 1 in thick steel plate and 1 in thick aluminum plate are held together with a ³⁄₄ in-10 UNC bolt and nut. The bolt is snug, with no initial preload and is made of steel. The modulus of elasticity of the steel and aluminum are 30×10^6 lbf/in² and 10×10^6 lbf/in², respectively. The coefficients of linear thermal expansion of steel and aluminum are 6.5×10^{-6} 1/°F and 12.8×10^{-6} 1/°F, respectively. The temperature is increased by 250°F. The tensile stress in the bolt is most nearly

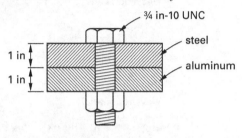

(A) 24,000 psi

(B) 26,000 psi

(C) 28,000 psi

(D) 31,000 psi

4. A 1 in (25 mm) diameter solid rod is held firmly in a chuck. A wrench with a 12 in (300 mm) moment arm applies 60 lbf (270 N) of force 8 in (200 mm) up from the chuck.

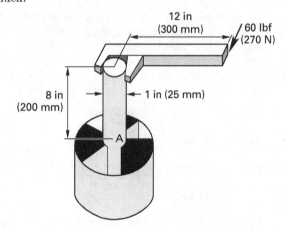

The maximum normal stress in the rod is most nearly

(A) 5400 lbf/in² (38 MPa)

(B) 6900 lbf/in² (49 MPa)

(C) 7500 lbf/in² (54 MPa)

(D) 11,000 lbf/in² (77 MPa)

5. The offset wrench handle shown is constructed from a ⅝ in (16 mm) diameter round bar. The handle cross-section stays the same as it bends. The bar's modulus of elasticity is 29.6×10^6 lbf/in² (204 GPa). The 3 in (75 mm) rise is in the plane of the socket.

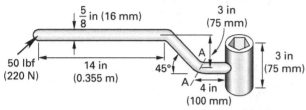

elevation view

view from handle

The maximum bending stress at the extreme fiber at section A-A is most nearly

(A) 30,000 lbf/in² (190 MPa)

(B) 35,000 lbf/in² (240 MPa)

(C) 58,000 lbf/in² (390 MPa)

(D) 71,000 lbf/in² (470 MPa)

6. A brass tube 6 ft (1.8 m) long, with a 2.0 in (50 mm) outside diameter, a 1.0 in (25 mm) inside diameter, and a modulus of elasticity of 1.5×10^7 lbf/in² (100 GPa) is used as a cantilever beam. When a concentrated load of 50 lbf (220 N) is applied at the free end, the tip deflection is found to be excessive. To reduce the deflection, a tight-fitting 1.0 in (25 mm) outside diameter soft steel rod with a modulus of elasticity of 2.9×10^7 lbf/in² (200 GPa) is inserted into the entire length of the brass tube.

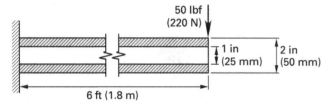

Neglecting self-weights, the percentage change in the tip deflection is most nearly

(A) 5.7%

(B) 7.1%

(C) 9.6%

(D) 11%

SOLUTIONS

1. *Customary U.S. Solution*

First, determine the reactions. The uniform load can be assumed to be concentrated at the center of the beam.

Sum the moments about A.

Systems of n Forces
$$\sum M_n = 0$$

$$(100 \text{ lbf})(2 \text{ ft}) + (80 \text{ lbf})(14 \text{ ft})$$
$$+ \left(20 \frac{\text{lbm}}{\text{ft}}\right)\left(\frac{32.2 \frac{\text{ft}}{\text{sec}^2}}{32.2 \frac{\text{ft-lbm}}{\text{lbf-sec}^2}}\right)$$
$$\times (14 \text{ ft})(7 \text{ ft}) - R(12 \text{ ft}) = 0$$
$$R = 273.3 \text{ lbf}$$

Sum the forces in the vertical direction.

$$L + 273.3 \text{ lbf} = 100 \text{ lbf} + 80 \text{ lbf}$$
$$+ \left(20 \frac{\text{lbm}}{\text{ft}}\right)\left(\frac{32.2 \frac{\text{ft}}{\text{sec}^2}}{32.2 \frac{\text{ft-lbm}}{\text{lbf-sec}^2}}\right)(14 \text{ ft})$$
$$L = 186.7 \text{ lbf}$$

The shear diagram starts at +186.7 lbf at the left reaction and decreases linearly at a rate of 20 lbf/ft to 146.7 lbf at point B. The concentrated load reduces the shear to 46.7 lbf. The shear then decreases linearly at a rate of 20 lbf/ft to point C. Measuring x from the left, the shear line goes through zero at

$$x = 2 \text{ ft} + \frac{46.7 \text{ lbf}}{20 \frac{\text{lbf}}{\text{ft}}} = 4.3 \text{ ft}$$

The shear at the right of the beam at point D is 80 lbf and increases linearly at a rate of 20 lbf/ft to 120 lbf at point C. The reaction, R, at point C decreases the shear to -153.3 lbf. This is sufficient to draw the shear diagram.

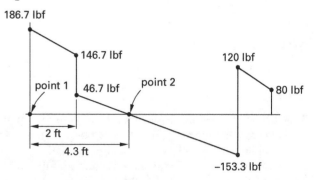

From the shear diagram, the maximum moment occurs when the shear is zero. Call this point 2. The moment at the left reaction is zero. Call this point 1. Use Eq. 50.43

Beams

$$M_2 = M_1 + \int_{x_1}^{x_2} V(x)\, dx$$

The integral is the area under the curve from $x_1 = 0$ to $x_2 = 4.3$ ft.

$$\begin{aligned}M_2 &= 0 + (146.7\ \text{lbf})(2\ \text{ft}) \\ &\quad + \left(\frac{1}{2}\right)(186.7\ \text{lbf} - 146.7\ \text{lbf})(2\ \text{ft}) \\ &\quad + \left(\frac{1}{2}\right)(46.7\ \text{lbf})(4.3\ \text{ft} - 2\ \text{ft}) \\ &= 387.1\ \text{ft-lbf}\quad (390\ \text{ft-lbf})\end{aligned}$$

SI Solution

First, determine the reactions. The uniform load can be assumed to be concentrated at the center of the beam. Sum the moments about A.

Systems of n Forces

$$\sum M_n = 0$$

$$(450\ \text{N})(0.6\ \text{m}) + (350\ \text{N})(4.2\ \text{m})$$
$$+ \left(30\ \frac{\text{kg}}{\text{m}}\right)(4.2\ \text{m})\left(9.81\ \frac{\text{N}}{\text{kg}}\right)(2.1\ \text{m})$$
$$- R(3.6\ \text{m}) = 0$$
$$R = 1204.4\ \text{N}$$

Sum the forces in the vertical direction.

$$L + 1204.4\ \text{N} = 450\ \text{N} + 350\ \text{N}$$
$$+ \left(30\ \frac{\text{kg}}{\text{m}}\right)(4.2\ \text{m})\left(9.81\ \frac{\text{N}}{\text{kg}}\right)$$
$$L = 831.7\ \text{N}$$

The shear diagram starts at +831.7 N at the left reaction and decreases linearly at a rate of 294.3 N/m to 655.1 N at point B. The concentrated load reduces the shear to 205.1 N. The shear then decreases linearly at a rate of 294.3 N/m to point C. Measuring x from the left, the shear line goes through zero at

$$x = 0.6\ \text{m} + \frac{205.1\ \text{N}}{294.3\ \frac{\text{N}}{\text{m}}} = 1.3\ \text{m}$$

The shear at the right of the beam at point D is 350 N and increases linearly at a rate of 294.3 N/m to 526.3 N at point C. The reaction, R, at point C decreases the shear to −677.8 N. This is sufficient to draw the shear diagram.

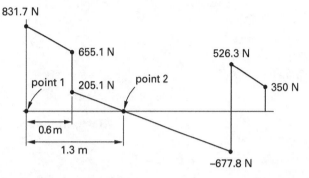

From the shear diagram, the maximum moment occurs when the shear is zero. Call this point 2. The moment at the left reaction is zero. Call this point 1.

Beams

$$M_2 = M_1 + \int_{x_1}^{x_2} V(x)\, dx$$

The integral is the area under the curve from $x_1 = 0$ to $x_2 = 1.3$ m.

$$\begin{aligned}M_2 &= 0 + (655.1\ \text{N})(0.6\ \text{m}) \\ &\quad + \left(\frac{1}{2}\right)(831.7\ \text{N} - 655.1\ \text{N})(0.6\ \text{m}) \\ &\quad + \left(\frac{1}{2}\right)(205.1\ \text{N})(1.3\ \text{m} - 0.6\ \text{m}) \\ &= 517.8\ \text{N·m}\quad (520\ \text{N·m})\end{aligned}$$

The answer is (C).

2. From the equation for the modulus of elasticity, the strain at 400°F is

Uniaxial Loading and Deformation

$$E = \frac{\sigma}{\epsilon}$$

$$\epsilon_{400°\text{F}} = \frac{\sigma}{E_{400°\text{F}}}$$

The strain at 800°F is

$$\epsilon_{800°\text{F}} = \frac{\sigma}{E_{800°\text{F}}}$$

Since the stress is constant, the strain at 800°F is

$$\epsilon_{800°F} = \left(\frac{E_{400°F}}{E_{800°F}}\right)\epsilon_{400°F} = \left(\frac{27{,}000{,}000\ \frac{\text{lbf}}{\text{in}^2}}{24{,}200{,}000\ \frac{\text{lbf}}{\text{in}^2}}\right)\epsilon_{400°F}$$

$$= 1.12\epsilon_{400°F}\quad (1.1\epsilon_{400°F})$$

The answer is (A).

3. The unconstrained thermal deformations are

Thermal Deformations
$$\Delta_{th} = \alpha L(T - T_o)$$
$$\Delta_{th,\text{bolt}} = \alpha L_o(T_2 - T_1)$$
$$= \left(6.5 \times 10^{-6}\ \frac{1}{°F}\right)(2\ \text{in})(250°F)$$
$$= 0.00325\ \text{in}$$
$$\Delta_{th,\text{steel}} = \alpha L_o(T_2 - T_1)$$
$$= \left(6.5 \times 10^{-6}\ \frac{1}{°F}\right)(1\ \text{in})(250°F)$$
$$= 0.001625\ \text{in}$$
$$\Delta_{th,\text{aluminum}} = \alpha L_o(T_2 - T_1)$$
$$= \left(12.8 \times 10^{-6}\ \frac{1}{°F}\right)(1\ \text{in})(250°F)$$
$$= 0.0032\ \text{in}$$

The unrealized elongation of the bolt is

$$\Delta = \delta_{th,\text{aluminum}} + \delta_{th,\text{steel}} - \delta_{th,\text{bolt}}$$
$$= 0.0032\ \text{in} + 0.001625\ \text{in} - 0.00325\ \text{in}$$
$$= 0.001575\ \text{in}$$

Rearrange the equation for thermal deformation, and solve for the mechanical strain due to the unrealized elongation of the bolt.

Thermal Deformations
$$\Delta_{th} = \alpha L(T - T_o)$$
$$\epsilon_{\text{bolt}} = \alpha(T - T_o) = \frac{\Delta_{th}}{L}$$
$$= \frac{0.001575\ \text{in}}{2\ \text{in}}$$
$$= 0.0007875$$

Use Hooke's law to calculate the tensile stress in the bolt.

$$\sigma = (\epsilon_{\text{bolt}})(E)$$
$$= \left(0.0007875\ \frac{\text{in}}{\text{in}}\right)\left(30 \times 10^6\ \frac{\text{lbf}}{\text{in}^2}\right)$$
$$= 23{,}625\ \text{lbf/in}^2\quad (24{,}000\ \text{psi})$$

The answer is (A).

4. *Customary U.S. Solution*

First, find the properties of the rod cross section. The area is

$$A = \frac{\pi a^2}{4} = \frac{\pi(1\ \text{in})^2}{4} = 0.7854\ \text{in}^2$$

The moment of inertia is

Properties of Various Shapes
$$I_c = \frac{\pi a^4}{4} = \frac{\pi\left(\frac{1\ \text{in}}{2}\right)^4}{4} = 0.04909\ \text{in}^4$$

The polar moment of inertia is

Properties of Various Shapes
$$J = \frac{\pi a^4}{2} = \frac{\pi\left(\frac{1\ \text{in}}{2}\right)^4}{2} = 0.09817\ \text{in}^4$$

The moment at the chuck is

$$M = (60\ \text{lbf})(8\ \text{in}) = 480\ \text{in-lbf}$$

The maximum bending stress will occur where the moment along the length of the rod is maximum. The maximum bending stress at the extreme fiber of the rod is

Stresses in Beams
$$\sigma_x \pm \frac{Mc}{I} = \frac{(480\ \text{in-lbf})\left(\frac{1\ \text{in}}{2}\right)}{0.04909\ \text{in}^4} = 4889\ \text{lbf/in}^2$$

The torque applied to the rod is

$$T = (60\ \text{lbf})(12\ \text{in}) = 720\ \text{in-lbf}$$

The maximum torsional shear stress at the extreme fiber of the rod is

$$\tau = \frac{Tr}{J} = \frac{(720 \text{ in-lbf})\left(\dfrac{1 \text{ in}}{2}\right)}{0.09817 \text{ in}^4} = 3667 \text{ lbf/in}^2 \quad \textit{Torsion}$$

The direct shear stress is zero at the surface of the rod. The maximum shear stress in the rod is

$$\tau_1 = R = \sqrt{\left(\frac{\sigma_x - \sigma_y}{2}\right)^2 + \tau_{xy}^2} \quad \textit{Mohr's Circle—Stress, 2D}$$

$$= \sqrt{\left(\frac{4889 \dfrac{\text{lbf}}{\text{in}^2} - 0 \dfrac{\text{lbf}}{\text{in}^2}}{2}\right)^2 + \left(3667 \dfrac{\text{lbf}}{\text{in}^2}\right)^2}$$

$$= 4407 \text{ lbf/in}^2$$

The maximum normal stress in the rod is

$$\sigma_1 = C + R = \frac{\sigma_x + \sigma_y}{2} + \tau_1 \quad \textit{Mohr's Circle—Stress, 2D}$$

$$= \frac{4889 \dfrac{\text{lbf}}{\text{in}^2} + 0 \dfrac{\text{lbf}}{\text{in}^2}}{2} + 4407 \dfrac{\text{lbf}}{\text{in}^2}$$

$$= 6852 \text{ lbf/in}^2 \quad (6900 \text{ lbf/in}^2)$$

SI Solution

First, find the properties of the rod cross section. The area is

$$A = \frac{\pi a^2}{4} = \frac{\pi (0.025 \text{ m})^2}{4} = 4.909 \times 10^{-4} \text{ m}^2 \quad \textit{Properties of Various Shapes}$$

The moment of inertia is

$$I_c = \frac{\pi a^4}{4} = \frac{\pi\left(\dfrac{0.025 \text{ m}}{2}\right)^4}{4} = 1.917 \times 10^{-8} \text{ m}^4 \quad \textit{Properties of Various Shapes}$$

The polar moment of inertia is

$$J = \frac{\pi a^4}{2} = \frac{\pi\left(\dfrac{0.025 \text{ m}}{2}\right)^4}{2} = 3.835 \times 10^{-8} \text{ m}^4 \quad \textit{Properties of Various Shapes}$$

The moment at the chuck is

$$M = (270 \text{ N})(0.2 \text{ m}) = 54 \text{ N·m}$$

The maximum bending stress will occur where the moment along the length of the rod is maximum. The maximum bending stress at the extreme fiber of the rod is

$$\sigma \pm \frac{Mc}{I_c} = \frac{(54 \text{ N·m})\left(\dfrac{0.025 \text{ m}}{2}\right)}{1.917 \times 10^{-8} \text{ m}^4} \quad \textit{Stresses in Beams}$$

$$= 3.52 \times 10^7 \text{ Pa} \quad (35.2 \text{ MPa})$$

The torque applied to the rod is

$$T = (270 \text{ N})(0.3 \text{ m}) = 81 \text{ N·m}$$

The maximum torsional shear stress at the extreme fiber of the rod is

$$\tau = \frac{Tr}{J} = \frac{(81 \text{ N·m})\left(\dfrac{0.025 \text{ m}}{2}\right)}{3.835 \times 10^{-8} \text{ m}^4} \quad \textit{Torsion}$$

$$= 2.64 \times 10^7 \text{ Pa} \quad (26.4 \text{ MPa})$$

The direct shear stress is zero at the surface of the rod. The maximum shear stress in the rod is

$$\tau_1 = R = \sqrt{\left(\frac{\sigma_x - \sigma_y}{2}\right)^2 + \tau_{xy}^2} \quad \textit{Mohr's Circle—Stress, 2D}$$

$$= \sqrt{\left(\frac{35.2 \text{ MPa} - 0 \text{ MPa}}{2}\right)^2 + (26.4 \text{ MPa})^2}$$

$$= 31.7 \text{ MPa}$$

The maximum normal stress in the rod is

$$\sigma_1 = C + R = \frac{\sigma_x + \sigma_y}{2} + \tau_1 \quad \textit{Mohr's Circle—Stress, 2D}$$

$$= \frac{35.2 \text{ MPa} + 0 \text{ MPa}}{2} + 31.7 \text{ MPa}$$

$$= 49.3 \text{ MPa} \quad (49 \text{ MPa})$$

The answer is (B).

5. Customary U.S. Solution

Find the properties of the handle cross section. The moment of inertia is

Properties of Various Shapes

$$I_c = \frac{\pi a^4}{4} = \frac{\pi \left(\frac{0.625 \text{ in}}{2}\right)^4}{4}$$
$$= 0.00749 \text{ in}^4$$

The moment at section A-A is

$$M = (50 \text{ lbf})(14 \text{ in} + 3 \text{ in})$$
$$= 850 \text{ in-lbf}$$

The maximum bending stress at the extreme fiber at section A-A is

Stresses in Beams

$$\sigma = \pm \frac{Mc}{I_c} = \frac{(850 \text{ in-lbf})\left(\frac{0.625 \text{ in}}{2}\right)}{0.00749 \text{ in}^4}$$
$$= 35{,}464 \text{ lbf/in}^2 \quad (35{,}000 \text{ lbf/in}^2)$$

SI Solution

Find the properties of the handle cross section. The moment of inertia is

Properties of Various Shapes

$$I_c = \frac{\pi a^4}{4} = \frac{\pi \left(\frac{0.016 \text{ m}}{2}\right)^4}{4}$$
$$= 3.22 \times 10^{-9} \text{ m}^4$$

The moment at section A-A is

$$M = (220 \text{ N})(0.355 \text{ m} + 0.075 \text{ m})$$
$$= 94.6 \text{ N·m}$$

The maximum bending stress at the extreme fiber at section A-A is

Stresses in Beams

$$\sigma = \pm \frac{Mc}{I_c} = \frac{(94.6 \text{ N·m})\left(\frac{0.016 \text{ m}}{2}\right)}{3.22 \times 10^{-9} \text{ m}^4}$$
$$= 2.35 \times 10^8 \text{ Pa} \quad (240 \text{ MPa})$$

The answer is (B).

6. Customary U.S. Solution

The moment of inertia of the brass tube annular cross section is

Properties of Various Shapes

$$I_{\text{brass}} = \frac{\pi(a^4 - b^4)}{4} = \frac{\left(\pi\left(\frac{2.0 \text{ in}}{2}\right)^4 - \left(\frac{1.0 \text{ in}}{2}\right)^4\right)}{4}$$
$$= 0.736 \text{ in}^4$$

The moment of inertia of the steel rod insert circular cross section is

Properties of Various Shapes

$$I_{\text{steel}} = \frac{\pi a^4}{4} = \frac{\pi \left(\frac{1.0 \text{ in}}{2}\right)^4}{4}$$
$$= 0.0491 \text{ in}^4$$

The product EI for the brass tube is

$$E_{\text{brass}} I_{\text{brass}} = \left(1.5 \times 10^7 \, \frac{\text{lbf}}{\text{in}^2}\right)(0.736 \text{ in}^4)$$
$$= 1.104 \times 10^7 \text{ lbf-in}^2$$

The product EI for the steel rod insert is

$$E_{\text{steel}} I_{\text{steel}} = \left(2.9 \times 10^7 \, \frac{\text{lbf}}{\text{in}^2}\right)(0.0491 \text{ in}^4)$$
$$= 0.142 \times 10^7 \text{ lbf-in}^2$$

The total EI for the composite is

$$E_c I_c = E_{\text{brass}} I_{\text{brass}} + E_{\text{steel}} I_{\text{steel}}$$
$$= 1.104 \times 10^7 \text{ lbf-in}^2 + 0.142 \times 10^7 \text{ lbf-in}^2$$
$$= 1.246 \times 10^7 \text{ lbf-in}^2$$

The equation for the tip deflection of the tube is

Bending Moment, Vertical Shear, and Deflection of Beams of Uniform Cross Section, Under Various Conditions of Loading

$$y_{\text{tip}} = \frac{PL^3}{3EI}$$

The percent change in tip deflection is

$$\text{percent} = \frac{y_{\text{tip}_{\text{brass}}} - y_{\text{tip}_{\text{brass + steel}}}}{y_{\text{tip}_{\text{brass}}}} \times 100\%$$

$$= \frac{\dfrac{PL^3}{3E_{\text{brass}}I_{\text{brass}}} - \dfrac{PL^3}{3E_c I_c}}{\dfrac{PL^3}{3E_{\text{brass}}I_{\text{brass}}}} \times 100\%$$

Simplify.

$$\text{percent} = \frac{\dfrac{1}{E_{\text{brass}}I_{\text{brass}}} - \dfrac{1}{E_c I_c}}{\dfrac{1}{E_{\text{brass}}I_{\text{brass}}}} \times 100\%$$

$$= \frac{\dfrac{1}{1.104 \times 10^7 \text{ lbf-in}^2} - \dfrac{1}{1.246 \times 10^7 \text{ lbf-in}^2}}{\dfrac{1}{1.104 \times 10^7 \text{ lbf-in}^2}} \times 100\%$$

$$= 11.4\% \quad (11\%)$$

SI Solution

The moment of inertia of the brass tube annular cross section is

Properties of Various Shapes

$$I_{\text{brass}} = \frac{\pi(a^4 - b^4)}{4} = \frac{\pi\left(\left(\dfrac{0.050 \text{ m}}{2}\right)^4 - \left(\dfrac{0.025 \text{ m}}{2}\right)^4\right)}{4}$$

$$= 2.876 \times 10^{-7} \text{ m}^4$$

The moment of inertia of the steel rod insert circular cross section is

Properties of Various Shapes

$$I_{\text{steel}} = \frac{\pi a^4}{4} = \frac{\pi\left(\dfrac{0.025 \text{ m}}{2}\right)^4}{4}$$

$$= 1.917 \times 10^{-8} \text{ m}^4$$

The product EI for the brass tube is

$$E_{\text{brass}}I_{\text{brass}} = (100 \times 10^9 \text{ Pa})(2.876 \times 10^{-7} \text{ m}^4)$$

$$= 28\,760 \text{ N·m}^2$$

The product EI for the steel rod insert is

$$E_{\text{steel}}I_{\text{steel}} = (200 \times 10^9 \text{ Pa})(1.917 \times 10^{-8} \text{ m}^4)$$

$$= 3834 \text{ N·m}^2$$

The total EI for the composite is

$$E_c I_c = E_{\text{brass}}I_{\text{brass}} + E_{\text{steel}}I_{\text{steel}}$$

$$= 28\,760 \text{ N·m}^2 + 3834 \text{ N·m}^2$$

$$= 32\,594 \text{ N·m}^2$$

The equation for the tip deflection of the tube is

Bending Moment, Vertical Shear, and Deflection of Beams of Uniform Cross Section, Under Various Conditions of Loading

$$y_{\text{tip}} = \frac{PL^3}{3EI}$$

The percent change in tip deflection is

$$\text{percent} = \frac{y_{\text{tip}_{\text{brass}}} - y_{\text{tip}_{\text{brass + steel}}}}{y_{\text{tip}_{\text{brass}}}} \times 100\%$$

$$= \frac{\dfrac{PL^3}{3E_{\text{brass}}I_{\text{brass}}} - \dfrac{PL^3}{3E_c I_c}}{\dfrac{PL^3}{3E_{\text{brass}}I_{\text{brass}}}} \times 100\%$$

Simplify.

$$\text{percent} = \frac{\dfrac{1}{E_{\text{brass}}I_{\text{brass}}} - \dfrac{1}{E_c I_c}}{\dfrac{1}{E_{\text{brass}}I_{\text{brass}}}} \times 100\%$$

$$= \frac{\dfrac{1}{28\,760 \text{ N·m}^2} - \dfrac{1}{32\,594 \text{ N·m}^2}}{\dfrac{1}{28\,760 \text{ N·m}^2}} \times 100\%$$

$$= 11.8\% \quad (11\%)$$

The answer is (D).

Failure Theories

Content in blue refers to the NCEES Handbook.

PRACTICE PROBLEMS

1. A shaft with a 1.125 in (28.6 mm) diameter receives 400 in-lbf (45 N·m) of torque through a pinned sleeve. The pin is manufactured from steel with a tensile yield strength of 73.9 ksi (510 MPa). Using a factor of safety of 2.5, what is most nearly the pin diameter needed?

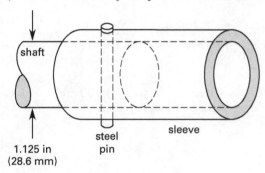

- (A) 0.16 in (0.0041 m)
- (B) 0.19 in (0.0048 m)
- (C) 0.26 in (0.0066 m)
- (D) 0.37 in (0.0094 m)

2. The pressure in a small pressure vessel operating at room temperature, 70°F (21°C), varies continually between the extremes of 50 psig and 350 psig (340 kPa and 2400 kPa). The vessel is closed by a 1/2 in (12 mm) plate. The plate is attached to the vessel flange with six 3/8-24 UNF bolts evenly spaced around a 9 1/2 in (240 mm) circle. Each bolt is tightened to an initial preload of 3700 lbf (16.4 kN). The bolts and nuts are constructed of cold-rolled steel with a 90 ksi (620 MPa) yield strength and 110 ksi (760 MPa) ultimate strength. The plate, flange, and vessel are constructed of steel with a 30 ksi (205 MPa) yield strength and a 50 ksi (345 MPa) ultimate strength. The stress concentration factor for the bolt threads is 2 and the maximum stress in the bolts is 42,680 psi (293.2 MPa), and the minimum stress in the bolts is 42,218 psi (289.9 MPa). The vessel is intended to be used indefinitely. Neglect the effects of the gasket (not shown). Neglect bending of the plate.

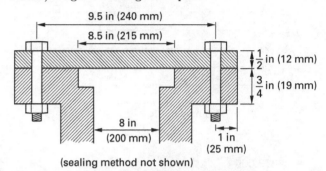

(sealing method not shown)

Use appropriate endurance strength derating factors and disregarding miscellaneous effects such as residual stresses and corrosion. The mean stress is the average of the maximum and minimum stresses in the bolt. The normal alternating stress is half the stress range (maximum - minimum). Based on the modified Goodman failure theory, the factor of safety used in the design of the bolt is most nearly

- (A) 1.2
- (B) 1.8
- (C) 2.1
- (D) 2.5

3. A structural member with the S-N curve shown is subjected to repeated loadings. 10% of the time, the member experiences cycles at 117% of the endurance strength; 15% of the time, the cycles are at 110% of the endurance strength; and 20% of the time, the cycles are at 105% of the endurance strength. The rest of the time, the stress is below the endurance limit. Most nearly, how many cycles can the member experience before failure?

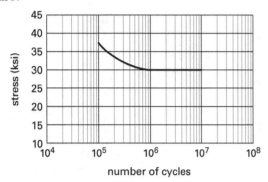

(A) 312,000

(B) 423,000

(C) 536,000

(D) 625,000

SOLUTIONS

1. *Customary U.S. Solution*

Use the distortion energy failure theory. The yield strength in shear is

$$S_{ys} = 0.577 S_{yt} = (0.577)\left(73.9 \times 10^3 \,\frac{\text{lbf}}{\text{in}^2}\right)$$
$$= 42.64 \times 10^3 \,\text{lbf/in}^2$$

For a safety factor of 2.5, the maximum allowable shear stress is

$$\tau_{\max} = \frac{S_{ys}}{\text{FS}} = \frac{42.64 \times 10^3 \,\frac{\text{lbf}}{\text{in}^2}}{2.5} = 17.06 \times 10^3 \,\text{lbf/in}^2$$

The total shear at the pin is

$$V = \frac{\text{shaft torque}}{\text{shaft radius}} = \frac{400 \text{ in-lbf}}{\frac{1.125 \text{ in}}{2}} = 711.1 \text{ lbf}$$

This is not a case of biaxial stress. There is no bending. The direct shear stress in the pin is

$$\tau_{\max} = \frac{V}{A}$$

Solve for the required total pin area.

$$A = \frac{V}{\tau_{\max}} = \frac{711.1 \text{ lbf}}{17.06 \times 10^3 \,\frac{\text{lbf}}{\text{in}^2}} = 0.04168 \text{ in}^2$$

Since two surfaces of the pin resist the shear,

$$A = (2)\left(\frac{\pi D^2}{4}\right) = \frac{\pi D^2}{2}$$

Solve for the pin diameter.

$$D = \sqrt{\frac{2A}{\pi}} = \sqrt{\frac{(2)(0.04168 \text{ in}^2)}{\pi}}$$
$$= 0.163 \text{ in} \quad (0.16 \text{ in})$$

SI Solution

Use the distortion energy failure theory. The yield strength in shear is

$$S_{ys} = 0.577 S_{yt} = (0.577)(510 \times 10^6 \text{ Pa})$$
$$= 294.3 \times 10^6 \text{ Pa}$$

For a safety factor of 2.5, the maximum allowable shear stress is

$$\tau_{\max} = \frac{S_{ys}}{\text{FS}} = \frac{294.3 \times 10^6 \text{ Pa}}{2.5} = 117.7 \times 10^6 \text{ Pa}$$

The total shear at the pin is

$$V = \frac{\text{shaft torque}}{\text{shaft radius}} = \frac{45 \text{ N·m}}{\frac{0.0286 \text{ m}}{2}} = 3146.9 \text{ N}$$

This is not a case of biaxial stress. There is no bending. The direct shear stress in the pin is

$$\tau_{\max} = \frac{V}{A}$$

Solve for the required total pin area.

$$A = \frac{V}{\tau_{\max}} = \frac{3146.9 \text{ N}}{117.7 \times 10^6 \text{ Pa}} = 2.674 \times 10^{-5} \text{ m}^2$$

Since two surfaces of the pin resist the shear, the total pin area is

$$A = (2)\left(\frac{\pi D^2}{4}\right) = \frac{\pi D^2}{2}$$

Solve for the pin diameter.

$$D = \sqrt{\frac{2A}{\pi}} = \sqrt{\frac{(2)(2.674 \times 10^{-5} \text{ m}^2)}{\pi}}$$
$$= 0.00413 \text{ m} \quad (0.0041 \text{ m})$$

The answer is (A).

2. *Customary U.S. Solution*

Calculate the mean stress

$$\sigma_m = \frac{\sigma_{\max} + \sigma_{\min}}{2} = \frac{42{,}680 \frac{\text{lbf}}{\text{in}^2} + 42{,}218 \frac{\text{lbf}}{\text{in}^2}}{2}$$
$$= 42{,}449 \text{ lbf/in}^2$$

The actual alternating stress for a thread stress concentration factor of 2 is twice the normal alternating stress

$$\sigma_a = (2)\left(\frac{\sigma_{\max} - \sigma_{\min}}{2}\right) = \sigma_{\max} - \sigma_{\min}$$
$$= 42{,}680 \frac{\text{lbf}}{\text{in}^2} - 42{,}218 \frac{\text{lbf}}{\text{in}^2}$$
$$= 462 \text{ lbf/in}^2$$

The ideal endurance strength of the bolt is considered to be one-half the ultimate strength.

Variable Loading Failure Theories

$$S'_e = 0.5 S_{ut} = (0.5)\left(110{,}000 \frac{\text{lbf}}{\text{in}^2}\right) = 55{,}000 \text{ lbf/in}^2$$

From the rotary beam method, the modified endurance limit, S_e, is

Variable Loading Failure Theories

$$S_e = k_a k_b k_c k_d k_e S'_e$$

The surface factor is

Variable Loading Failure Theories

$$k_a = a S_{ut}^b = (2.70)(110 \text{ ksi})^{-0.265}$$
$$= 0.777 \quad [\text{cold-rolled bolt}]$$

The diameter is nominally 0.315 in (8 mm), so the size modification factor is 1. [Variable Loading Failure Theories] [Basic Dimensions for Fine Thread Series (UNF/UNRF)]

The beam is under axial loading and has an ultimate tensile strength of less than 220.6 ksi (1520 MPa), so the load factor is 0.923. [Variable Loading Failure Theories]

The temperature is less than 842°F (450°C), so the temperature factor is 1. [Variable Loading Failure Theories]

Effects like corrosion and residual stresses are to be ignored, so use a miscellaneous effects factor of 1. [Variable Loading Failure Theories]

The derated endurance strength of the bolt is

Variable Loading Failure Theories

$$S_e = k_a k_b k_c k_d k_e S'_e$$
$$= (0.777)(1.00)(0.923)(1.00)(1.00)$$
$$\times \left(55{,}000 \frac{\text{lbf}}{\text{in}^2}\right)$$
$$= 39{,}444 \text{ lbf/in}^2$$

Calculate Goodman equivalent stress

Variable Loading Failure Theories

$$\sigma_{eq} = \sigma_a + \left(\frac{S_e}{S_{ut}}\right)\sigma_m$$

$$= 462 \ \frac{\text{lbf}}{\text{in}^2} + \left(\frac{39{,}444 \ \frac{\text{lbf}}{\text{in}^2}}{110{,}000 \ \frac{\text{lbf}}{\text{in}^2}}\right)\left(42{,}449 \ \frac{\text{lbf}}{\text{in}^2}\right)$$

$$= 15{,}683 \ \text{lbf/in}^2$$

Based on the Modified Goodman Failure Theory, the factor of safety in the bolt design is

$$\text{FS} = \frac{S_e}{\sigma_{eq}} = \frac{39{,}444 \ \frac{\text{lbf}}{\text{in}^2}}{15{,}683 \ \frac{\text{lbf}}{\text{in}^2}} = 2.5$$

SI Solution

Calculate the mean stress is

$$\sigma_m = \frac{\sigma_{max} + \sigma_{min}}{2} = \frac{2.932 \times 10^8 \ \text{Pa} + 2.899 \times 10^8 \ \text{Pa}}{2}$$

$$= 2.916 \times 10^8 \ \text{Pa} \quad (291.6 \ \text{MPa})$$

Based on a stress concentration factor of 2 for the threads, the actual alternating stress is twice the normal alternating stress.

$$\sigma_a = (2)\left(\frac{\sigma_{max} - \sigma_{min}}{2}\right) = \sigma_{max} - \sigma_{min}$$

$$= 2.932 \times 10^8 \ \text{Pa} - 2.899 \times 10^8 \ \text{Pa}$$

$$= 0.033 \times 10^8 \ \text{Pa} \quad (3.3 \ \text{MPa})$$

The ideal endurance strength of the bolt is considered to be one-half the ultimate strength.

Variable Loading Failure Theories

$$S_e' = 0.5 S_{ut} = (0.5)(760 \ \text{MPa}) = 380 \ \text{MPa}$$

From the rotary beam method, the modified endurance limit, S_e, is

Variable Loading Failure Theories

$$S_e = k_a k_b k_c k_d k_e S_e'$$

The surface factor is

Variable Loading Failure Theories

$$k_a = a S_{ut}^b = (4.51)(760 \ \text{MPa})^{-0.265}$$

$$= 0.777 \quad [\text{cold-rolled bolt}]$$

The diameter is nominally 8 mm, so the size modification factor is 1. [Variable Loading Failure Theories] [Basic Dimensions for Fine Thread Series (UNF/UNRF)]

The beam is under axial loading and has an ultimate tensile strength of less than 1520 MPa, so the load factor is 0.923. [Variable Loading Failure Theories]

The temperature is less than 450°C, so the temperature factor is 1. [Variable Loading Failure Theories]

Effects like corrosion and residual stresses are to be ignored, so use a miscellaneous effects factor of 1. [Variable Loading Failure Theories]

The derated endurance strength of the bolt is

Variable Loading Failure Theories

$$S_e = k_a k_b k_c k_d k_e S_e'$$

$$= (0.777)(1.00)(0.923)(1.00)(1.00)$$

$$\times (380 \ \text{MPa})$$

$$= 272.5 \ \text{MPa}$$

Calculate the Goodman equivalent stress

Variable Loading Failure Theories

$$\sigma_{eq} = \sigma_a + \left(\frac{S_e}{S_{ut}}\right)\sigma_m$$

$$= 3.3 \ \text{MPa} + \left(\frac{272.5 \ \text{MPa}}{760 \ \text{MPa}}\right)(291.6 \ \text{MPa})$$

$$= 107.9 \ \text{MPa}$$

Based on the Modified Goodman Failure Theory, the factor of safety used in the bolt design is

$$\text{FS} = \frac{S_e}{\sigma_{eq}} = \frac{272.5 \ \text{MPa}}{107.9 \ \text{MPa}} = 2.5$$

The answer is (D).

3. The endurance limit is approximately 30 ksi. Determine the fatigue life for each stress level.

For a stress level of 117% of S_e, for example,

$$\sigma = 1.17 S_e = (1.17)\left(30 \ \frac{\text{kips}}{\text{in}^2}\right) = 35 \ \text{ksi}$$

Tabulate the results as shown.

amount of time (%)	stress level (% of S_e)	stress (ksi)	fatigue life, N_i (cycles)
10%	117%	35	1.5×10^5
15%	110%	33	2.4×10^5
20%	105%	31.5	3.5×10^5

Use Miner's rule. Let N^* represent the number of cycles at failure. Then, the number of cycles experienced at the 117% stress level is $1.1N^*$, and so on.

Variable Loading Failure Theories

$$\sum \frac{n_i}{N_i} = C = \frac{0.1N^*}{1.5 \times 10^5} + \frac{0.15N^*}{2.4 \times 10^5} + \frac{0.2N^*}{3.5 \times 10^5} = 1$$

$$N^* = 536{,}170 \quad (536{,}000)$$

The answer is (C).

Basic Machine Design

Content in blue refers to the *NCEES Handbook*.

PRACTICE PROBLEMS

Note: Unless instructed otherwise in a problem, use the following properties:

steel: $E = 30 \times 10^6$ lbf/in² (20×10^4 MPa)
$G = 11.5 \times 10^6$ lbf/in² (8.0×10^4 MPa)
$\alpha = 6.5 \times 10^{-6}$ 1/°F (1.2×10^{-5} 1/°C)
$\nu = 0.3$
aluminum: $E = 10 \times 10^6$ lbf/in² (70×10^3 MPa)
copper: $E = 17.5 \times 10^6$ lbf/in² (12×10^4 MPa)

1. The yield strength of a structural steel member is 36,000 lbf/in² (250 MPa). The tensile stress is 8240 lbf/in² (57 MPa). The factor of safety in tension is most nearly

(A) 2.5
(B) 3.1
(C) 3.6
(D) 4.4

2. A structural steel member 50 ft (15 m) long is used as a long column to support 75,000 lbf (330 kN). Both ends are built-in, and there are no intermediate supports. A factor of safety of 2.5 is used. The required moment of inertia is most nearly

(A) 48 in⁴ (2.0×10^{-5} m⁴)
(B) 72 in⁴ (3.0×10^{-5} m⁴)
(C) 96 in⁴ (4.0×10^{-5} m⁴)
(D) 130 in⁴ (5.5×10^{-5} m⁴)

3. A shell with an outside diameter of 16 in (406 mm) and a wall thickness of 0.10 in (2.54 mm) is subjected to a 40,000 lbf/in² (280 MPa) tensile load and a 400,000 in-lbf torque (45 k·N m). What is the approximate maximum shear stress?

(A) 12,400 psi (86 MPa)
(B) 18,400 psi (126 MPa)
(C) 22,400 psi (156 MPa)
(D) 28,400 psi (196 MPa)

4. The bracket shown is attached to a column with three 0.75 in (19 mm) bolts arranged in an equilateral triangular layout. A force is applied with a moment arm of 20 in (500 mm) measured to the centroid of the bolt group. The maximum shear stress in the bolts is limited to 15,000 lbf/in² (100 MPa). Perform an elastic analysis to determine the approximate maximum force that the connection can support.

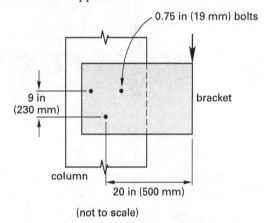

(A) 4,700 lbf (20 kN)
(B) 6,100 lbf (26 kN)
(C) 7,200 lbf (33 kN)
(D) 9,700 lbf (42 kN)

5. Two brackets are connected by two bolts with individual cross-sectional areas of A_{bolt}. Each bracket carries a load of $2F$, as shown.

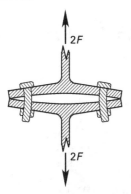

What best describes the bolt configuration shown?

(A) prying action

(B) complex double shear

(C) complex double tension

(D) plastic bending

6. Two brackets are connected by two bolts with individual cross-sectional areas of A_{bolt}. Each bracket carries a load of $2F$, as shown.

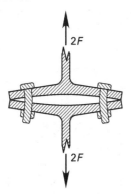

Neglecting secondary effects, what is the equation for the tensile stress in each bolt?

(A) $F/2A_{\text{bolt}}$

(B) F/A_{bolt}

(C) $2F/A_{\text{bolt}}$

(D) $4F/A_{\text{bolt}}$

7. An American Unified Standard Threaded bolt is preloaded with a tensile force of 80,000 lbf. The bolt's modulus of elasticity is 20×10^6 lbf/in^2, the original length is 4 in, and the maximum allowable elongation due to preloading is 0.05 in. The minimum appropriate coarse series nominal size is

(A) $1/4$ in

(B) $3/8$ in

(C) $7/16$ in

(D) $3/4$ in

8. A $9/16$ in UNF bolt is loaded with a tensile proof load. Under a proof stress of 70,000 lbf/in^2, the bolt elongates 0.0147 in. The spring constant for the bolt is most nearly

(A) 6.2×10^5 lbf/in

(B) 9.7×10^5 lbf/in

(C) 17×10^5 lbf/in

(D) 22×10^5 lbf/in

9. A bolt is preloaded with a force of 500,000 lbf. The bolt has a cross-sectional area of 1 in^2, a spring constant of 10×10^6 lbf/in, and a modulus of elasticity of 20×10^6 lbf/in^2. The elongation of the bolt as a percentage of the original length is most nearly

(A) 1.0%

(B) 1.5%

(C) 1.7%

(D) 2.5%

SOLUTIONS

1. *Customary U.S. Solution*

The factor of safety is

$$\text{FS} = \frac{S_{yt}}{\sigma} = \frac{36{,}000 \ \frac{\text{lbf}}{\text{in}^2}}{8240 \ \frac{\text{lbf}}{\text{in}^2}} = 4.37 \quad (4.4)$$

SI Solution

The factor of safety is

$$\text{FS} = \frac{S_{yt}}{\sigma} = \frac{250 \ \text{MPa}}{57 \ \text{MPa}} = 4.39 \quad (4.4)$$

The answer is (D).

2. *Customary U.S. Solution*

The design load for a factor of safety of 2.5 is

$$P_{cr} = (\text{FS})P = (2.5)(75{,}000 \ \text{lbf}) = 187{,}500 \ \text{lbf}$$

Substitute the equation for the radius of gyration into the critical buckling stress equation for long columns.

Radius of Gyration

$$r = \sqrt{\frac{I}{A}}$$

Long Columns

$$\sigma_{cr} = \frac{P_{cr}}{A} = \frac{\pi^2 E}{\left(\frac{KL}{r}\right)^2} = \frac{\pi^2 E}{\left(\frac{KL}{\sqrt{\frac{I}{A}}}\right)^2} = \frac{I\pi^2 E}{A(KL)^2}$$

Rearrange and solve for moment of inertia. The theoretical end restraint coefficient for built-in ends is $K = 0.5$, and the recommended design value is 0.65. [Approximate Values of Effective Length Factor, K]

$$I = \frac{P_{cr}(KL)^2}{\pi^2 E}$$

$$= \frac{(187{,}500 \ \text{lbf})\left((0.65)\left((50 \ \text{ft})\left(12 \ \frac{\text{in}}{\text{ft}}\right)\right)\right)^2}{\pi^2 \left(30 \times 10^6 \ \frac{\text{lbf}}{\text{in}^2}\right)}$$

$$= 96.32 \ \text{in}^4 \quad (96 \ \text{in}^4)$$

SI Solution

The design load for a factor of safety of 2.5 is

$$P_{cr} = (\text{FS})P = (2.5)(330 \times 10^3 \ \text{N}) = 825 \times 10^3 \ \text{N}$$

Substitute the equation for the radius of gyration into the critical buckling stress equation for long columns.

Radius of Gyration

$$r = \sqrt{\frac{I}{A}}$$

Long Columns

$$\sigma_{cr} = \frac{P_{cr}}{A} = \frac{\pi^2 E}{\left(\frac{KL}{r}\right)^2} = \frac{\pi^2 E}{\left(\frac{KL}{\sqrt{\frac{I}{A}}}\right)^2} = \frac{I\pi^2 E}{A(KL)^2}$$

Rearrange and solve for moment of inertia. The theoretical end restraint coefficient for built-in ends is $K = 0.5$, and the recommended design value is 0.65. [Approximate Values of Effective Length Factor, K]

$$I = \frac{P_{cr}(KL)^2}{\pi^2 E}$$

$$= \frac{(825 \times 10^3 \ \text{N})\big((0.65)(15 \ \text{m})\big)^2}{\pi^2 \left(20 \times 10^{10} \ \frac{\text{N}}{\text{m}^2}\right)}$$

$$= 3.9673 \times 10^{-5} \ \text{m}^4 \quad (4.0 \times 10^{-5} \ \text{m}^4)$$

The answer is (C).

3. *Customary U.S. Solution*

The mean radius of the shell is

$$r_m = \frac{D_o - t}{2} = \frac{16 \ \text{in} - 0.10 \ \text{in}}{2} = 7.95 \ \text{in}$$

The mean area enclosed by the centerline of the shell is

Hollow, Thin-Walled Shafts

$$A_m = \pi r_m^2 = \pi (7.95 \ \text{in})^2 = 198.6 \ \text{in}^2$$

The shear stress is

Hollow, Thin-Walled Shafts

$$\tau = \frac{T}{2 A_m t}$$

$$= \frac{(400{,}000 \ \text{in-lbf})}{(2)(198.6 \ \text{in}^2)(0.10 \ \text{in})}$$

$$= 10{,}072 \ \text{lbf/in}^2$$

The maximum shear stress is

Mohr's Circle—Stress, 2D

$$r = \sqrt{\left(\frac{\sigma_x - \sigma_y}{2}\right)^2 + \tau_{xy}^2}$$

$$= \sqrt{\left(\frac{40{,}000 \frac{\text{lbf}}{\text{in}^2} - 0}{2}\right)^2 + \left(10{,}072 \frac{\text{lbf}}{\text{in}^2}\right)^2}$$

$$= 22{,}393 \text{ psi} \quad (22{,}400 \text{ psi})$$

SI Solution

The mean radius of the shell is

$$r_m = \frac{D_o - t}{2} = \frac{406 \text{ mm} - 2.54 \text{ mm}}{2} = 201.73 \text{ mm}$$

The mean area enclosed by the centerline of the shell is

Hollow, Thin-Walled Shafts

$$A_m = \pi r_m^2 = \frac{\pi (201.73 \text{ mm})^2}{\left(1000 \frac{\text{mm}}{\text{m}}\right)^2} = 0.12785 \text{ m}^2$$

The shear stress is

Hollow, Thin-Walled Shafts

$$\tau = \frac{T}{2A_m t}$$

$$= \frac{(45 \times 10^3 \text{ N·m})\left(1000 \frac{\text{mm}}{\text{m}}\right)}{(2)(0.12785 \text{ m}^2)(2.54 \text{ mm})}$$

$$= 69.29 \times 10^6 \text{ N/m}^2$$

$$= 69.3 \text{ MPa}$$

The maximum shear stress is

Mohr's Circle—Stress, 2D

$$r = \sqrt{\left(\frac{\sigma_x + \sigma_y}{2}\right)^2 + \tau_{xy}^2}$$

$$= \sqrt{\left(\frac{280 \text{ MPa} + 0}{2}\right)^2 + (69.3 \text{ MPa})^2}$$

$$= 156.2 \text{ MPa} \quad (156 \text{ MPa})$$

The answer is (C).

4. *Customary U.S. Solution*

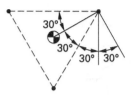

Find the properties of the bolt area. For an equilateral triangular layout, the distance from the centroid of the bolt group to the center of each bolt is

$$r = a = \frac{2}{3}(9 \text{ in}) = 6 \text{ in}$$

The area of each bolt is

$$A = \frac{\pi D^2}{4} = \frac{\pi (0.75 \text{ in})^2}{4}$$

$$= 0.442 \text{ in}^2$$

The polar moment of inertia of each bolt about the centroid is

$$J_B = Ar^2 = (0.442 \text{ in}^2)(6 \text{ in})^2$$

$$= 15.91 \text{ in}^4$$

Using the parallel axis theorem, the polar moment of inertia of the three bolts at the centroid of the group is,

$$J = 3J_B = 3(15.91 \text{ in}^4)$$

$$= 47.73 \text{ in}^4$$

The vertical shear load at each bolt is

$$F_v = \frac{F}{3} = 0.333F \quad (\text{lbf})$$

The moment applied to each bolt is

$$M = 20 \text{ in} \times F = 20F \quad (\text{in-lbf})$$

The vertical shear stress in each bolt is

$$\tau_v = \frac{F_v}{A} = \frac{0.333F}{0.442 \text{ in}^2} = 0.753F \quad (\text{lbf / in}^2)$$

The torsional shear stress in each bolt is

Torsion

$$\tau = \frac{Fe}{J} = \frac{Mr}{J} = \frac{(20.0F)(6 \text{ in})}{47.73 \text{ in}^4} = 2.51F \quad (\text{lbf / in}^2)$$

The most highly stressed bolt is the rightmost bolt. The shear stress configuration is

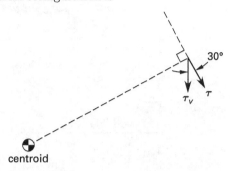

The stresses are combined to find the maximum stress.

$$\tau_{\max} = \sqrt{(\tau\sin 30)^2 + (\tau_v + \tau\cos 30)^2}$$

$$15000\frac{\text{lbf}}{\text{in}^2} = \sqrt{((2.51\text{F})\sin 30)^2 + (0.753\,\text{F} + (2.51\text{F})\cos 30)^2}$$

$$= 3.18\text{F}$$

$$\text{F} = \frac{15000}{3.18} = 4{,}716 \text{ lbf} \quad (4{,}700 \text{ lbf})$$

SI Solution

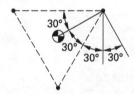

Find the properties of the bolt area. For an equilateral triangular layout, the distance from the centroid of the bolt group to each bolt is

$$r = a = \frac{2}{3}(230 \text{ mm}) = 153 \text{ mm}$$

The area of each bolt is

$$A = \frac{\pi D^2}{4} = \frac{\pi\left(19 \text{ mm} \times \dfrac{1 \text{ m}}{1000 \text{ mm}}\right)^2}{4}$$

$$= 2.835 \times 10^{-4} \text{ m}^2$$

The polar moment of inertia of each bolt about the centroid is

Properties of Various Shapes

$$J_B = Ar^2 = (2.835 \times 10^{-4} \text{ m}^2)(0.153 \text{ m})^2$$

$$= 6.636 \times 10^{-6} \text{ m}^4$$

Using the parallel axis theorem, the polar moment of inertia of the three bolts at the centroid of the group is,

$$J = 3J_B = 3(6.636 \times 10^{-6} \text{ m}^4)$$

$$= 1.991 \times 10^{-5} \text{ m}^4$$

The vertical shear load at each bolt is

$$F_v = \frac{F}{3} = 0.333F \quad (\text{N})$$

The moment applied to each bolt is

$$M = 0.50 \text{ m} \times F = 0.50F \quad (\text{N-m})$$

The vertical shear stress in each bolt is

$$\tau_v = \frac{F_v}{A} = \frac{0.333F}{2.835 \times 10^{-4} \text{ m}^2} = 1175F \quad (\text{N}/\text{m}^2 \text{ or Pa})$$

The torsional shear stress in each bolt is

Torsion

$$\tau = \frac{Fe}{J} = \frac{Mr}{J} = \frac{(0.50F)(0.153 \text{ m})}{1.991 \times 10^{-5} \text{ m}^4} = 3842F \quad (\text{Pa})$$

The most highly stressed bolt is the rightmost bolt. The shear stress configuration is

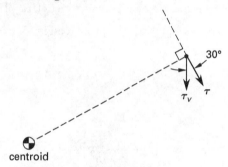

The stresses are combined to find the maximum stress.

$$\tau_{\max} = \sqrt{(\tau\sin 30)^2 + (\tau_v + \tau\cos 30)^2}$$

$$(100 \text{ MPa})\left(\frac{10^6 \text{ Pa}}{\text{MPa}}\right) = \sqrt{\begin{array}{l}((3842F)\sin 30)^2 \\ +(1175\,F + (3842F)\cos 30)^2\end{array}}$$

$$= 4895F$$

$$F = \frac{100 \times 10^6}{4895} = 20\,430 \text{ N} \quad (20 \text{ kN})$$

The answer is (A).

5. The bolts are experiencing prying action. The deformation may or may not be plastic.

The answer is (A).

6. Either load can be considered the "applied" load, while the other load can be considered the "resisting" load. The two loads are not additive. The bending (secondary effects) are disregarded.

Uniaxial Loading and Deformation

$$\sigma_t = \frac{P_{\text{total}}}{A_{\text{total}}} = \frac{2F}{2A_{\text{bolt}}} = \frac{F}{A_{\text{bolt}}}$$

The answer is (B).

7. Since the elongation can be calculated as PL/AE, the cross-sectional area of the major diameter is

Uniaxial Loading and Deformation

$$\delta = \frac{PL}{AE}$$

$$A = \frac{PL}{\delta E} = \frac{(80{,}000 \text{ lbf})(4 \text{ in})}{(0.05 \text{ in})\left(20{,}000{,}000 \frac{\text{lbf}}{\text{in}^2}\right)} = 0.32 \text{ in}^2$$

For this tensile stress area, the appropriate nominal size is ¾ in. [Basic Dimensions for Coarse Thread Series (UNC/UNRC)]

The answer is (D).

8. For a ⁹⁄₁₆ in UNF bolt, the tensile stress area is 0.203 in². [Basic Dimensions for Fine Thread Series (UNF/UNRF)]

From the stress equation, the proof load is

Uniaxial Loading and Deformation

$$\sigma_{\text{proof}} = \frac{P_{\text{proof}}}{A}$$

$$P_{\text{proof}} = \sigma_{\text{proof}} A$$
$$= \left(70{,}000 \frac{\text{lbf}}{\text{in}^2}\right)(0.203 \text{ in}^2)$$
$$= 14{,}210 \text{ lbf}$$

The spring constant is

Spring Energy

$$F_s = kx$$

$$k = \frac{F}{x} = \frac{P_{\text{proof}}}{\delta}$$
$$= \frac{14{,}210 \text{ lbf}}{0.0147 \text{ in}}$$
$$= 9.66 \times 10^5 \text{ lbf/in} \quad (9.7 \times 10^5 \text{ lbf/in})$$

The answer is (B).

9. The elongation of the bolt is

Uniaxial Loading and Deformation

$$E = \frac{\sigma}{\epsilon}$$

$$\epsilon = \frac{\sigma}{E}$$
$$= \frac{F}{AE}$$
$$= \frac{500{,}000 \text{ lbf}}{(1 \text{ in}^2)\left(20{,}000{,}000 \frac{\text{lbf}}{\text{in}^2}\right)}$$
$$= 0.025 \quad (2.5\%)$$

The answer is (D).

53 Advanced Machine Design

Content in blue refers to the *NCEES Handbook*.

PRACTICE PROBLEMS

1. A shaft with a keyway transmits 300 hp at 1200 rev/min. The shaft is 2 ft long with a diameter of 3 in. The keyway has a width of 1 in and a length of 3 in. The shear stress on the keyway is most nearly

(A) 440 lbf/in^2

(B) 1200 lbf/in^2

(C) 1800 lbf/in^2

(D) 3500 lbf/in^2

2. A spring with 12 active coils and a spring index of 9 supports a static load of 50 lbf (220 N) with a deflection of 0.5 in (12 mm). The shear modulus of the spring material is 1.2×10^7 lbf/in^2 (83 GPa). The theoretical wire diameter is most nearly

(A) 0.58 in (15 mm)

(B) 0.63 in (16 mm)

(C) 0.69 in (18 mm)

(D) 0.74 in (19 mm)

3. A spring with 11 active coils and a spring index of 9 supports a static load of 60 lbf (267 N) with a deflection of 0.75 in (19 mm). The shear modulus of the spring material is 1.2×10^7 lbf/in^2 (83 GPa). The mean spring diameter is most nearly

(A) 3.9 in (98 mm)

(B) 6.4 in (160 mm)

(C) 7.1 in (180 mm)

(D) 7.9 in (200 mm)

4. A severe service valve spring is to be manufactured from unpeened ASTM A230 steel wire in standard W&M sizes. The valve spring will operate continuously between 20 lbf and 30 lbf (100 N and 150 N). The valve lift is 0.3 in (8 mm). The force to compress the spring to its solid height is 46 lbf (247 N). The deflection at solid height is most nearly

(A) 1.4 in (39 mm)

(B) 1.7 in (43 mm)

(C) 2.1 in (53 mm)

(D) 2.4 in (61 mm)

5. A severe service valve spring is to be manufactured from unpeened ASTM A230 steel wire in standard W&M sizes. The valve spring will operate continuously between 25 lbf and 32 lbf (111 N and 142 N). The valve lift is 0.35 in (9 mm). The force to compress the spring to its solid height of 1.5 in (38 mm) is 49 lbf (218 N). Including the deflection due to difference in applied forces, the minimum free height is most nearly

(A) 2.1 in (53 mm)

(B) 4.0 in (101 mm)

(C) 3.2 in (81 mm)

(D) 4.5 in (110 mm)

6. The material in a spring wire has a shear modulus of 1.2×10^7 lbf/in^2 (83 GPa). The maximum allowable stress under design conditions is 50,000 lbf/in^2 (350 MPa). The spring index is 7. The maximum stress is experienced when a 700 lbm (320 kg) object falls from a height of 46 in (120 cm) above the tip of the spring, impacts squarely on the spring, and deflects the spring 10 in (26 cm). The minimum wire diameter for the spring is most nearly

(A) 0.58 in (15 mm)

(B) 0.88 in (23 mm)

(C) 1.4 in (36 mm)

(D) 1.8 in (47 mm)

7. The material in a spring wire has a shear modulus of 1.2×10^7 lbf/in^2 (83 GPa). The maximum allowable stress under design conditions is 45,000 lbf/in^2 (310 MPa). The spring index is 8. The maximum stress is experienced when a 650 lbm (295 kg) object falls from a height of 42 in (107 cm) above the tip of the spring, impacts squarely on the spring, and deflects the spring 9.4 in (24 cm). The mean coil diameter is most nearly

(A) 8.6 in (220 mm)

(B) 10 in (250 mm)

(C) 15.5 in (390 mm)

(D) 16 in (410 mm)

8. The material in a spring wire has a shear modulus of 1.2×10^7 lbf/in^2 (83 GPa). The spring index is 6. Assume a wire diameter of 1.63 in (41.4 mm) is required to withstand the maximum stress experienced when a 815 lbm (370 kg) object falls from a height of 38 in (96.5 cm) above the tip of the spring, impacts squarely on the spring, and deflects the spring 11.4 in (29 cm). The number of active coils is most nearly

(A) 8

(B) 18

(C) 12

(D) 14

9. A 6 in (150 mm) wide, 24 in (610 mm) long cantilever steel spring supports an 800 lbf (3.5 kN) load at its tip. The deflection is to be less than 1 in (25 mm), and the bending stress is limited to 50,000 lbf/in^2 (345 MPa). The minimum thickness as limited by deflection alone is most nearly

(A) 0.37 in (9.4 mm)

(B) 0.45 in (11 mm)

(C) 0.63 in (16 mm)

(D) 0.77 in (20 mm)

10. A 6 in (150 mm) wide, 26 in (660 mm) long cantilever steel spring supports an 650 lbf (2.9 kN) load at its tip. The deflection is to be less than 1.125 in (29 mm). The bending stress is limited to 45,000 lbf/in^2 (310 MPa). The minimum thickness as limited by bending stress is most nearly

(A) 0.61 in (16 mm)

(B) 0.66 in (17 mm)

(C) 0.72 in (18 mm)

(D) 0.78 in (20 mm)

11. A bathroom scale design makes use of four cantilever flat steel springs (beams) located at the corners of a rectangular load plate, as shown. Each beam is made from high-strength plate and is 1¾ in wide by 0.313 in thick. The four beams are equally loaded. Strain gauges are mounted on the upper surface of each of the four beams. A digital readout is scaled to convert the resulting voltage due to load to a person's weight. All pieces other than the beam springs are rigid. The modulus of elasticity, E, is 30×10^6 lbf/in^2.

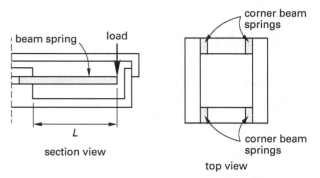

Given a spring constant of 25,758 lbf/in for each beam spring, the user weight that would produce an engineering strain of 1.2×10^{-3} in/in on the upper surface of one of the beams is most nearly

(A) 77 lbf

(B) 150 lbf

(C) 410 lbf

(D) 1600 lbf

12. A 2 in NPS schedule-80 steel pipe is rigidly attached to a wall at one end and is connected to a solid 2 in × 2 in square aluminum rod at the other end, as shown. The effective length of each piece is 12 in. The assembly is part of a torque-limiting system and experiences torsion.

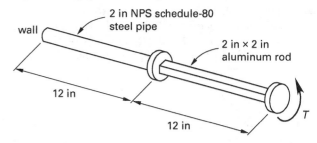

Neglecting the connection points, the equivalent torsional spring constant for a torque applied to the end of the aluminum rod is most nearly

(A) 5.0×10^5 in-lbf/rad

(B) 7.0×10^5 in-lbf/rad

(C) 17×10^5 in-lbf/rad

(D) 24×10^5 in-lbf/rad

13. A gear train is to have a speed reduction of 600:1. The gears used can have no fewer than 12 teeth and no more than 96 teeth. The pinion gears in the first three stages have the same number of teeth. The number of stages needed is most nearly

(A) 2
(B) 3
(C) 4
(D) 5

14. The power transmission system shown consists of spur gears and AISI 1045 cold-drawn steel shafting. The gears have a 20° normal pressure angle, and a diametral pitch of 5 (module of 5 mm). The yield strength of the 1045 steel is 69,000 lbf/in² (480 MPa). Loading is slow and steady. Use a factor of safety of 2 and the maximum shear stress failure theory.

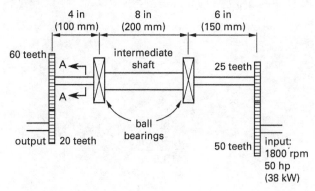

Assume static loading and a worst-case scenario where all moments on the shaft are additive. The safe minimum shaft diameter at section A-A considering torsional stress only is

(A) 0.4 in (10.2 mm)
(B) 0.45 in (11.4 mm)
(C) 0.55 in (14 mm)
(D) 0.65 in (16.5 mm)

15. An American standard full-depth involute 20° gear has a circular pitch of 1 in and a pitch circle diameter of 8 in. The number of teeth on the gear, N, is most nearly

(A) 20
(B) 25
(C) 28
(D) 32

16. Which of the following angles is the angle between mating bevel-gear axes?

(A) shaft angle
(B) root angle
(C) spiral angle
(D) pitch angle

17. A bevel gear has a pitch angle of 22.5° and a normal pressure angle of 20°. The gear is exposed to an axial force of 1000 lbf. The transmitted load on the gear is most nearly

(A) 1200 lbf
(B) 2800 lbf
(C) 4400 lbf
(D) 7200 lbf

18. A bevel gear with a pitch angle of 40° experiences an axial force of 350 lbf and a tangential force of 1500 lbf. The resultant force is most nearly

(A) 1200 lbf
(B) 1450 lbf
(C) 1550 lbf
(D) 1600 lbf

19. A 1 in pump journal bearing 2 in long is exposed to a 2500 lbf lateral shaft load. The pressure on the journal bearing is most nearly

(A) 1050 lbf/in²
(B) 1250 lbf/in²
(C) 1310 lbf/in²
(D) 1550 lbf/in²

20. A plain cylindrical journal bearing is 3 in long and has a radial clearance of 0.01 in. The bearing load is 5500 lbf and the operating speed is 205 rpm. The oil used as a lubricant has a dynamic viscosity of 1.5×10^{-6} lbf-sec/in². The bearing radius required to keep the torque on the bearing below 0.15 ft-lbf is most nearly

(A) 3.1 in
(B) 4.6 in
(C) 6.3 in
(D) 8.5 in

21. A roller bearing is exposed to an equivalent radial load of 1000 lbf. The basic load rating of the bearing is 3480 lbf. The rotational speed is 200 rev/min. The rated life of the bearing is most nearly

(A) 3200 hr

(B) 5300 hr

(C) 5400 hr

(D) 5500 hr

22. A ball bearing is to carry an equivalent radial load of 1000 lbf. The design life is 500,000,000 revolutions. The basic load rating is most nearly

(A) 6500 lbf

(B) 7100 lbf

(C) 7900 lbf

(D) 8000 lbf

SOLUTIONS

1. The torque of the shaft, T, is

$$T = \frac{63{,}025 P}{n} = \frac{(63{,}025)(300 \text{ hp})}{1200 \frac{\text{rev}}{\text{min}}} = 15{,}756 \text{ in-lbf}$$

The keyway's shear stress, τ, is

$$\tau = \frac{F}{A} = \frac{\dfrac{T}{D/2}}{wL} = \frac{2T}{DwL} = \frac{(2)(15{,}756 \text{ in-lbf})}{(3 \text{ in})(1 \text{ in})(3 \text{ in})}$$
$$= 3501 \text{ lbf/in}^2 \quad (3500 \text{ lbf/in}^2)$$

The answer is (D).

2. *Customary U.S. Solution*

From the equation for applied force, the spring constant is

Spring Energy

$$F_s = kx$$
$$k = \frac{F_s}{x}$$
$$= \frac{50 \text{ lbf}}{0.5 \text{ in}}$$
$$= 100 \text{ lbf/in}$$

Substitute the equation for the spring index into the equation for the spring constant and rearrange the equation to solve for the wire diameter.

Mechanical Springs

$$C = \frac{D}{d}$$
$$k = \frac{d^4 G}{8 D^3 N}$$
$$= \frac{dG}{8 C^3 N}$$
$$d = \frac{8 k C^3 N}{G}$$
$$= \frac{(8)\left(100 \dfrac{\text{lbf}}{\text{in}}\right)(9)^3 (12)}{\left(1.2 \times 10^7 \dfrac{\text{lbf}}{\text{in}^2}\right)}$$
$$= 0.583 \text{ in} \quad (0.58 \text{ in})$$

SI Solution

From the equation for applied force, the spring constant is

Spring Energy
$$F_s = kx$$
$$k = \frac{F_s}{x}$$
$$= \frac{220 \text{ N}}{0.012 \text{ m}}$$
$$= 18{,}333 \text{ N/m}$$

Substitute the equation for the spring index into the equation for the spring constant and rearrange the equation to solve for the wire diameter.

Mechanical Springs
$$C = \frac{D}{d}$$
$$k = \frac{d^4 G}{8 D^3 N}$$
$$= \frac{dG}{8 C^3 N}$$
$$d = \frac{8 k C^3 N}{G}$$
$$= \left(\frac{(8)\left(18{,}333\ \frac{\text{N}}{\text{m}}\right)(9)^3 (12)}{(83 \times 10^9 \text{ Pa})}\right)\left(1000\ \frac{\text{mm}}{\text{m}}\right)$$
$$= 15.46 \text{ mm} \quad (15 \text{ mm})$$

The answer is (A).

3. *Customary U.S. Solution*

From the equation for applied force, the spring constant is

Spring Energy
$$F_s = kx$$
$$k = \frac{F_s}{x}$$
$$= \frac{60 \text{ lbf}}{0.75 \text{ in}}$$
$$= 80 \text{ lbf/in}$$

Substitute the equation for the spring index into the equation for the spring constant and rearrange the equation to solve for the wire diameter.

Mechanical Springs
$$C = \frac{D}{d}$$
$$k = \frac{d^4 G}{8 D^3 N}$$
$$= \frac{dG}{8 C^3 N}$$
$$d = \frac{8 k C^3 N}{G}$$
$$= \frac{(8)\left(80\ \frac{\text{lbf}}{\text{in}}\right)(9)^3 (11)}{\left(1.2 \times 10^7\ \frac{\text{lbf}}{\text{in}^2}\right)}$$
$$= 0.428 \text{ in}$$

Rearrange the equation for the spring index and solve for the mean spring diameter.

Mechanical Springs
$$C = \frac{D}{d}$$
$$D = Cd$$
$$= (9)(0.428 \text{ in})$$
$$= 3.852 \text{ in} \quad (3.9 \text{ in})$$

SI Solution

From the equation for applied force, the spring constant is

Spring Energy
$$F_s = kx$$
$$k = \frac{F_s}{x}$$
$$= \frac{267 \text{ N}}{0.019 \text{ m}}$$
$$= 14{,}053 \text{ N/m}$$

Substitute the equation for the spring index into the equation for the spring constant and rearrange the equation to solve for the wire diameter.

Mechanical Springs

$$C = \frac{D}{d}$$

$$k = \frac{d^4 G}{8D^3 N}$$

$$= \frac{dG}{8C^3 N}$$

$$d = \frac{8kC^3 N}{G}$$

$$= \left(\frac{(8)\left(14{,}053 \ \frac{\text{N}}{\text{m}}\right)(9)^3(11)}{(83 \times 10^9 \ \text{Pa})}\right)\left(1000 \ \frac{\text{mm}}{\text{m}}\right)$$

$$= 10.86 \ \text{mm}$$

Rearrange the equation for the spring index and solve for the mean spring diameter.

Mechanical Springs

$$C = \frac{D}{d}$$

$$D = Cd$$

$$= (9)(10.86 \ \text{mm})$$

$$= 97.75 \ \text{mm} \quad (98 \ \text{mm})$$

The answer is (A).

4. *Customary U.S. Solution*

From the equation for applied force, the spring constant is

Spring Energy

$$F_s = kx$$

$$k = \frac{F_s}{x} = \frac{\Delta F}{\Delta x}$$

$$k = \frac{F_1 - F_2}{x_1 - x_2} = \frac{30 \ \text{lbf} - 20 \ \text{lbf}}{0.3 \ \text{in}}$$

$$= 33.33 \ \text{lbf/in}$$

The deflection at solid height is

$$x_s = \frac{F_s}{k} = \frac{46 \ \text{lbf}}{33.33 \ \frac{\text{lbf}}{\text{in}}} = 1.38 \ \text{in} \quad (1.4 \ \text{in})$$

SI Solution

From the equation for applied force, the spring constant is

Spring Energy

$$F_s = kx$$

$$k = \frac{F_s}{x} = \frac{\Delta F}{\Delta x}$$

$$k = \frac{F_1 - F_2}{x_1 - x_2} = \frac{(150 \ \text{N} - 100 \ \text{N})\left(1000 \ \frac{\text{mm}}{\text{m}}\right)}{8 \ \text{mm}}$$

$$= 6.25 \times 10^3 \ \text{N/m}$$

The deflection at solid height is

$$x_s = \frac{F_s}{k} = \left(\frac{247 \ \text{N}}{6.25 \times 10^3 \ \frac{\text{N}}{\text{m}}}\right)\left(1000 \ \frac{\text{mm}}{\text{m}}\right)$$

$$= 39.52 \ \text{mm} \quad (39 \ \text{mm})$$

The answer is (A).

5. *Customary U.S. Solution*

To determine the free height of the spring, the deflection at the solid height must be calculated. First, calculate the spring constant from the equation for applied force.

Spring Energy

$$F_s = kx$$

$$k = \frac{F_s}{x} = \frac{\Delta F}{\Delta x}$$

$$k = \frac{F_1 - F_2}{x_1 - x_2} = \frac{32 \ \text{lbf} - 25 \ \text{lbf}}{0.35 \ \text{in}}$$

$$= 20 \ \text{lbf/in}$$

The deflection at solid height is

Spring Energy

$$x_s = \frac{F_s}{k} = \frac{49 \ \text{lbf}}{20 \ \frac{\text{lbf}}{\text{in}}} = 2.45 \ \text{in}$$

The minimum free height is the solid height plus the deflection at the solid height.

$$h_f = h_s + x_s = 1.5 \ \text{in} + 2.45 \ \text{in} = 3.95 \ \text{in} \quad (4.0 \ \text{in})$$

SI Solution

To determine the free height of the spring, the spring deflection at the solid height must be calculated. First, calculate the spring constant from the equation for applied force.

Spring Energy

$$F_s = kx$$

$$k = \frac{F_s}{x} = \frac{\Delta F}{\Delta x}$$

$$k = \frac{F_1 - F_2}{x_1 - x_2} = \frac{(142 \text{ N} - 111 \text{ N})\left(1000 \dfrac{\text{mm}}{\text{m}}\right)}{9 \text{ mm}}$$

$$= 3.44 \times 10^3 \text{ N/m}$$

The deflection at solid height is

$$x_s = \frac{F_s}{k} = \left(\frac{218 \text{ N}}{3.44 \times 10^3 \dfrac{\text{N}}{\text{m}}}\right)\left(1000 \dfrac{\text{mm}}{\text{m}}\right)$$

$$= 63.37 \text{ mm}$$

The minimum free height is the solid height plus the deflection to the solid height.

$$h_f = h_s + x_s = 38 \text{ mm} + 63.37 \text{ mm}$$

$$= 101.37 \text{ mm} \quad (101 \text{ mm})$$

The answer is (B).

6. *Customary U.S. Solution*

The potential energy absorbed is

Potential Energy

$$PE = \frac{mgh}{g_c} = \frac{(700 \text{ lbm})\left(32.174 \dfrac{\text{ft}}{\text{sec}^2}\right)(46 \text{ in} + 10 \text{ in})}{32.174 \dfrac{\text{lbm-ft}}{\text{lbf-sec}^2}}$$

$$= 39{,}200 \text{ in-lbf}$$

The work done by the spring is equal to the potential energy.

Spring Energy

$$U = k\frac{x^2}{2} = PE$$

$$k = \frac{2(PE)}{x^2}$$

$$= \frac{(2)(39{,}200 \text{ in-lbf})}{(10 \text{ in})^2}$$

$$= 784 \text{ lbf/in}$$

The equivalent spring force is

Spring Energy

$$F_s = kx = \left(784 \dfrac{\text{lbf}}{\text{in}}\right)(10 \text{ in}) = 7840 \text{ lbf}$$

The stress correction factor is

Mechanical Springs

$$K_s = \frac{4C + 2}{4C - 3} = \frac{(4)(7) + 2}{(4)(7) - 3} = 1.2$$

Substitute the equation for the spring index into the equation for shear stress in a helical compression spring and solve for the wire diameter.

Mechanical Springs

$$\tau_{\text{allowable}} = K_s \frac{8FD}{\pi d^3}$$

$$C = \frac{D}{d}$$

$$\tau_{\text{allowable}} = K_s \frac{8F_s C}{\pi d^2}$$

$$d = \sqrt{\frac{K_s 8 F_s C}{\pi \tau_{\text{allowable}}}}$$

$$= \sqrt{\frac{(1.2)(8)(7840 \text{ lbf})(7)}{\pi\left(50{,}000 \dfrac{\text{lbf}}{\text{in}^2}\right)}}$$

$$= 1.83 \text{ in} \quad (1.8 \text{ in})$$

SI Solution

The potential energy absorbed is

Potential Energy

$$PE = mgh = (320 \text{ kg})\left(9.807 \dfrac{\text{m}}{\text{s}^2}\right)(1.2 \text{ m} + 0.26 \text{ m})$$

$$= 4582 \text{ J}$$

The work done by the spring is equal to the potential energy.

Spring Energy

$$U = k\frac{x^2}{2} = PE$$

$$k = \frac{2(PE)}{x^2}$$

$$= \frac{(2)(4582 \text{ J})}{(0.26 \text{ m})^2}$$

$$= 135\,562 \text{ N/m}$$

The equivalent spring force is

Spring Energy
$$F_s = kx = \left(135\,562\,\frac{\text{N}}{\text{m}}\right)(0.26\text{ m}) = 35\,246\text{ N}$$

The stress correction factor is

Spring Energy
$$K_s = \frac{4C+2}{4C-3} = \frac{(4)(7)+2}{(4)(7)-3} = 1.2$$

Substitute the equation for spring index into the equation for shear stress in a helical compression spring and solve for the wire diameter.

$$\tau_{\text{allowable}} = K_s \frac{8FD}{\pi d^3}$$
$$C = \frac{D}{d}$$
$$\tau_{\text{allowable}} = K_s \frac{8F_s C}{\pi d^2}$$
$$d = \sqrt{\frac{K_s 8 F_s C}{\pi \tau_{\text{allowable}}}}$$
$$= \sqrt{\frac{(1.2)(8)(35\,246\text{ N})(7)\left(10^3\,\frac{\text{mm}}{\text{m}}\right)^2}{\pi(350 \times 10^6\text{ Pa})}}$$
$$= 46.4\text{ mm} \quad (47\text{ mm})$$

The answer is (D).

7. *Customary U.S. Solution*

The potential energy absorbed is

Potential Energy
$$\text{PE} = \frac{mgh}{g_c} = \frac{(650\text{ lbm})\left(32.174\,\frac{\text{ft}}{\text{sec}^2}\right)(42\text{ in} + 9.4\text{ in})}{32.174\,\frac{\text{lbm-ft}}{\text{lbf-sec}^2}}$$
$$= 33{,}410\text{ in-lbf}$$

The work done by the spring is equal to the potential energy.

Spring Energy
$$U = k\frac{x^2}{2} = \text{PE}$$
$$k = \frac{2(\text{PE})}{x^2}$$
$$= \frac{(2)(33{,}410\text{ in-lbf})}{(9.4\text{ in})^2}$$
$$= 756\text{ lbf/in}$$

The equivalent spring force is

Spring Energy
$$F_s = kx = \left(756\,\frac{\text{lbf}}{\text{in}}\right)(9.4\text{ in}) = 7106\text{ lbf}$$

The stress correction factor is

Mechanical Springs
$$K_s = \frac{4C+2}{4C-3} = \frac{(4)(8)+2}{(4)(8)-3} = 1.17$$

Substitute the equation for spring index into the equation for shear stress in a helical compression spring and solve for the wire diameter.

Mechanical Springs
$$\tau_{\text{allowable}} = K_s \frac{8FD}{\pi d^3}$$
$$C = \frac{D}{d}$$
$$\tau_{\text{allowable}} = K_s \frac{8F_s C}{\pi d^2}$$
$$d = \sqrt{\frac{K_s 8 F_s C}{\pi \tau_{\text{allowable}}}}$$
$$= \sqrt{\frac{(1.17)(8)(7106\text{ lbf})(8)}{\pi\left(45{,}000\,\frac{\text{lbf}}{\text{in}^2}\right)}}$$
$$= 1.94\text{ in}$$

Rearranging the equation for the spring index, the mean coil diameter is

$$C = \frac{D}{d}$$
$$D = Cd$$
$$= (8)(1.94\text{ in})$$
$$= 15.52\text{ in} \quad (15.5\text{ in})$$

SI Solution

The potential energy absorbed is

Potential Energy
$$\text{PE} = mgh = (295\text{ kg})\left(9.807\,\frac{\text{m}}{\text{s}^2}\right)(1.07\text{ m} + 0.24\text{ m})$$
$$= 3790\text{ J}$$

The work done by the spring is equal to the potential energy.

Spring Energy

$$U = k\frac{x^2}{2} = \text{PE}$$

$$k = \frac{2(\text{PE})}{x^2}$$

$$= \frac{(2)(3790 \text{ J})}{(0.24 \text{ m})^2}$$

$$= 131\,597 \text{ N/m}$$

The equivalent spring force is

Spring Energy

$$F_s = kx = \left(131\,597 \ \frac{\text{N}}{\text{m}}\right)(0.24 \text{ m}) = 31\,583 \text{ N}$$

The stress correction factor is

Spring Energy

$$K_s = \frac{4C+2}{4C-3} = \frac{(4)(8)+2}{(4)(8)-3} = 1.17$$

Substitute the equation for spring index into the equation for shear stress in a helical compression spring and solve for the wire diameter.

$$\tau_{\text{allowable}} = K_s \frac{8FD}{\pi d^3}$$

$$C = \frac{D}{d}$$

$$\tau_{\text{allowable}} = K_s \frac{8F_s C}{\pi d^2}$$

$$d = \sqrt{\frac{K_s 8 F_s C}{\pi \tau_{\text{allowable}}}}$$

$$= \sqrt{\frac{(1.17)(8)(31\,583 \text{ N})(8)\left(10^3 \ \frac{\text{mm}}{\text{m}}\right)^2}{\pi(310 \times 10^6 \text{ Pa})}}$$

$$= 49.3 \text{ mm} \quad (49 \text{ mm})$$

Rearranging the equation for the spring index, the mean coil diameter is

$$C = \frac{D}{d}$$

$$D = Cd$$

$$= (8)(49.3 \text{ mm})$$

$$= 394 \text{ mm} \quad (390 \text{ mm})$$

The answer is (C).

8. *Customary U.S. Solution*

The potential energy absorbed is

Potential Energy

$$PE = \frac{mgh}{g_c} = \frac{(815 \text{ lbm})\left(32.174 \ \frac{\text{ft}}{\text{sec}^2}\right)(38 \text{ in} + 11.4 \text{ in})}{32.174 \ \frac{\text{lbm-ft}}{\text{lbf-sec}^2}}$$

$$= 40{,}261 \text{ in-lbf}$$

The work done by the spring is equal to the potential energy.

Spring Energy

$$U = k\frac{x^2}{2} = \text{PE}$$

$$k = \frac{2(\text{PE})}{x^2}$$

$$= \frac{(2)(40{,}261 \text{ in-lbf})}{(11.4 \text{ in})^2}$$

$$= 619.6 \text{ lbf/in}$$

Substituting in the equation for spring index into the equation for spring stiffness, the number of active coils is

Spring Energy

$$C = \frac{D}{d}$$

$$k = \frac{d^4 G}{8D^3 N} = \frac{dG}{8C^3 N}$$

$$N = \frac{dG}{k 8 C^3} = \frac{(1.63 \text{ in})\left(1.2 \times 10^7 \ \frac{\text{lbf}}{\text{in}^2}\right)}{(6)^3 \left(619.6 \ \frac{\text{lbf}}{\text{in}}\right)(8)}$$

$$= 18.3 \quad (18)$$

SI Solution

The potential energy absorbed is

Potential Energy

$$PE = mgh = (370 \text{ kg})\left(9.807 \ \frac{\text{m}}{\text{s}^2}\right)(0.965 \text{ m} + 0.29 \text{ m})$$

$$= 4554 \text{ J}$$

The work done by the spring is equal to the potential energy.

Spring Energy

$$U = k\frac{x^2}{2} = \text{PE}$$

$$k = \frac{2(\text{PE})}{x^2} = \frac{(2)(4554 \text{ J})}{(0.29 \text{ m})^2} = 108\,297 \text{ N/m}$$

Substituting in the equation for spring index into the equation for spring stiffness the number of active coils is

Mechanical Springs

$$C = \frac{D}{d}$$

$$k = \frac{d^4 G}{8D^3 N} = \frac{d\,G}{8C^3 N}$$

$$N = \frac{dG}{k8C^3} = \frac{(41.4 \text{ mm})(83 \times 10^9 \text{ Pa})}{(6)^3 \left(108,297\,\frac{\text{N}}{\text{m}}\right)(8)\left(10^3\,\frac{\text{mm}}{\text{m}}\right)}$$

$$= 18.4 \quad (18)$$

The answer is (B).

9. *Customary U.S. Solution*

The moment of inertia of the beam cross section is

Properties of Various Shapes

$$I = \frac{bh^3}{12} = \frac{(6 \text{ in})h^3}{12} = 0.5h^3 \quad [\text{in}^4]$$

The modulus of elasticity for a steel spring is 29.0×10^6 lbf/in^2. [Typical Material Properties]

The tip deflection for a cantilevered beam is

Bending Moment, Vertical Shear, and Deflection of Beams of Uniform Cross Section, Under Various Conditions of Loading

$$y = \frac{PL^3}{3EI} = \frac{PL^3}{3E(0.5)h^3}$$

$$h = \left(\frac{PL^3}{1.5yE}\right)^{1/3} = \left(\frac{(800 \text{ lbf})(24 \text{ in})^3}{(1.5 \text{ in})(1.0 \text{ in})\left(29.0 \times 10^6\,\frac{\text{lbf}}{\text{in}^2}\right)}\right)^{1/3}$$

$$= 0.633 \text{ in} \quad (0.63 \text{ in})$$

SI Solution

The moment of inertia of the beam cross section is

Properties of Various Shapes

$$I = \frac{bh^3}{12} = \frac{(150 \text{ mm})h^3}{12} = 12.5h^3 \quad [\text{mm}^4]$$

The modulus of elasticity for a steel spring is 200 GPa. [Typical Material Properties]

The tip deflection for a cantilevered beam is

Bending Moment, Vertical Shear, and Deflection of Beams of Uniform Cross Section, Under Various Conditions of Loading

$$y = \frac{PL^3}{3EI} = \frac{PL^3}{3E(12.5)h^3}$$

$$h = \left(\frac{PL^3}{37.5yE}\right)^{1/3}$$

$$= \left(\frac{(3.5 \text{ kN})\left(1000\,\frac{\text{N}}{\text{kN}}\right) \times (610 \text{ mm})^3\left(1000\,\frac{\text{mm}}{\text{m}}\right)^2}{(37.5)(25 \text{ mm})(200 \times 10^9 \text{ Pa})}\right)^{1/3}$$

$$= 16.18 \text{ mm} \quad (16 \text{ mm})$$

The answer is (C).

10. *Customary U.S. Solution*

The moment of inertia of the beam cross section is

Properties of Various Shapes

$$I = \frac{bh^3}{12} = \frac{(6 \text{ in})h^3}{12} = 0.5h^3$$

The moment at the fixed end is

Bending Moment, Vertical Shear, and Deflection of Beams of Uniform Cross Section, Under Various Conditions of Loading

$$M = FL = (650 \text{ lbf})(26 \text{ in}) = 16,900 \text{ in-lbf}$$

The bending stress is

Stresses in Beams

$$\sigma \pm \frac{Mc}{I} = \frac{M\left(\frac{h}{2}\right)}{\left(\frac{h^3}{2}\right)} = \frac{M}{h^2}$$

$$h = \sqrt{\frac{M}{\sigma}} = \sqrt{\frac{16,900 \text{ in-lbf}}{(1.125 \text{ in})\left(45,000\,\frac{\text{lbf}}{\text{in}^2}\right)}} = 0.61 \text{ in}$$

SI Solution

The moment of inertia of the beam cross section is

Properties of Various Shapes

$$I = \frac{bh^3}{12} = \frac{(150 \text{ mm})h^3}{12} = (12.5 \text{ mm})h^3$$

The moment at the fixed end is

Bending Moment, Vertical Shear, and Deflection of Beams of Uniform Cross Section, Under Various Conditions of Loading

$$M = FL = \frac{(2.9 \text{ kN})\left(1000 \dfrac{\text{N}}{\text{kN}}\right)(660 \text{ mm})}{1000 \dfrac{\text{mm}}{\text{m}}}$$

$$= 1.914 \times 10^3 \text{ N·m}$$

The bending stress is

$$\sigma \pm \frac{Mc}{I} = \frac{M\left(\dfrac{h}{2}\right)}{(12.5h^3)} = \frac{M}{25h^2}$$

$$h = \sqrt{\frac{M}{25\sigma}}$$

$$= \sqrt{\frac{(1.914 \times 10^3 \text{ N·m})\left(1000 \dfrac{\text{mm}}{\text{m}}\right)^2}{(29 \text{ mm})(310 \text{ MPa})\left(10^6 \dfrac{\text{Pa}}{\text{MPa}}\right)}}$$

$$= 15.6 \text{ mm} \quad (16 \text{ mm})$$

The answer is (A).

11. The moment of inertia, I, for one beam is

Properties of Various Shapes

$$I = \frac{1}{12}bh^3 = \left(\frac{1}{12}\right)(1.75 \text{ in})(0.313 \text{ in})^3$$

$$= 4.47 \times 10^{-3} \text{ in}^4$$

The deflection equation for a cantilevered beam with an end load is

Bending Moment, Vertical Shear, and Deflection of Beams of Uniform Cross Section, Under Various Conditions of Loading

$$y = \frac{PL^3}{3EI}$$

The equation for the resulting spring constant of each beam is

Spring Energy

$$F_s = kx$$

$$k = \frac{F_s}{x} = \frac{P}{y} = \frac{3EI}{L^3}$$

Solving this equation for length,

$$L = \sqrt[3]{\frac{3EI}{k}} = \sqrt[3]{\frac{(3)\left(30 \times 10^6 \dfrac{\text{lbf}}{\text{in}^2}\right)(4.47 \times 10^{-3} \text{ in}^4)}{25{,}758 \dfrac{\text{lbf}}{\text{in}}}}$$

$$= 2.5 \text{ in}$$

The longitudinal strain at the surface is related to the stress, moment, and force.

Hooke's Law

$$\sigma = E\epsilon = \frac{Mc}{I} = \frac{PLc}{I}$$

Solve for the tip force, P.

$$P = \frac{E\epsilon I}{Lc} = \frac{2E\epsilon I}{Lh}$$

$$= \frac{(2)\left(30 \times 10^6 \dfrac{\text{lbf}}{\text{in}^2}\right)\left(1.2 \times 10^{-3} \dfrac{\text{in}}{\text{in}}\right)}{(2.5 \text{ in})(0.313 \text{ in})}$$

$$\times (4.47 \times 10^{-3} \text{ in}^4)$$

$$= 411.3 \text{ lbf}$$

With the beams equally loaded, the user weight, W, is the load per beam multiplied by 4.

$$W = 4P$$
$$= (4)(411.3 \text{ lbf})$$
$$= 1645.2 \text{ lbf} \quad (1600 \text{ lbf})$$

The answer is (D).

12. The angle of twist, ϕ, of a shaft due to an applied torque is given by

Torsional Strain

$$\phi = \frac{TL}{GJ}$$

Combining with the equation for the relationship between deflection and moment, the torsional spring constant, k, is given by

Mechanical Springs

$$Fr = k\theta$$
$$T = k\phi$$
$$k = \frac{T}{\phi} = \frac{GJ}{L}$$

The modulus of rigidity of steel is 11.5×10^6 lbf/in², and the modulus of rigidity of aluminum is 3.8×10^6 lbf/in² [Typical Material Properties]

2 in NPS schedule-80 steep pipe has an inner diameter of 1.939 in, and a wall thickness of 0.218 in. [Pipe and Tube Data]

The inner and outer radii of the pipe are

$$b_{steel} = \frac{D_i}{2} = \frac{1.939 \text{ in}}{2} = 0.9695 \text{ in}$$

$$a_{steel} = \frac{D_i + 2t}{2} = \frac{1.939 \text{ in} + (2)(0.218 \text{ in})}{2} = 1.1875 \text{ in}$$

The polar moment of inertia for the steel circular shaft, J_{steel}, is

Properties of Various Shapes

$$J_{steel} = \frac{\pi(a_{steel}^4 - b_{steel}^4)}{2}$$

$$= \frac{\pi((1.1875 \text{ in})^4 - (0.9695 \text{ in})^4)}{2}$$

$$= 1.733 \text{ in}^4$$

The polar moment of inertia for the aluminum solid section, $J_{aluminum}$, is.

Properties of Various Shapes

$$J_{aluminum} = \frac{bh(b^2 + h^2)}{12}$$

$$= \frac{(2 \text{ in})(2 \text{ in})((2 \text{ in})^2 + (2 \text{ in})^2)}{12}$$

$$= 2.67 \text{ in}^4$$

The torsional spring constants for the steel and aluminum sections are

$$k_{steel} = \frac{T}{\phi} = \frac{T}{\dfrac{TL_{steel}}{G_{steel}J_{steel}}} = \frac{G_{steel}J_{steel}}{L_{steel}}$$

$$= \frac{\left(11.5 \times 10^6 \dfrac{\text{lbf}}{\text{in}^2}\right)(1.733 \text{ in}^4)}{12 \text{ in}}$$

$$= 16.6 \times 10^5 \text{ in-lbf/rad}$$

$$k_{aluminum} = \frac{T}{\phi} = \frac{T}{\dfrac{TL_{aluminum}}{G_{aluminum}J_{aluminum}}} = \frac{G_{aluminum}J_{aluminum}}{L_{aluminum}}$$

$$= \frac{\left(3.8 \times 10^6 \dfrac{\text{lbf}}{\text{in}^2}\right)(2.67 \text{ in}^4)}{12 \text{ in}}$$

$$= 8.45 \times 10^5 \text{ in-lbf/rad}$$

The equivalent torsional spring constant, k_{eq}, is found from the equation for springs in series.

Equivalent Springs

$$\frac{1}{k_{eq}} = \left(\frac{1}{k_1} + \frac{1}{k_2} + \ldots + \frac{1}{k_n}\right)$$

$$k_{eq} = \left(\frac{1}{k_{steel}} + \frac{1}{k_{aluminum}}\right)^{-1}$$

$$= \left(\frac{1}{16.6 \times 10^5 \dfrac{\text{in-lbf}}{\text{rad}}} + \frac{1}{8.45 \times 10^5 \dfrac{\text{in-lbf}}{\text{rad}}}\right)^{-1}$$

$$= 5.6 \times 10^5 \text{ in-lbf/rad} \quad (5.0 \times 10^5 \text{ in-lbf/rad})$$

The answer is (A).

13. The maximum speed ratio is

$$\frac{N_{max}}{N_{min}} = \frac{96 \text{ teeth}}{12 \text{ teeth}} = 8$$

For three stages,

$$(8)^3 = 512$$

Since this is less than 600, four stages are required.

The answer is (C).

14. *Customary U.S. Solution*

To determine the transmitted load, the pitch circle diameter, intermediate shaft speed, and pitch line velocity must be calculated. The pitch circle diameter is

Involute Gear Tooth Nomenclature

$$p_D = \frac{N}{D}$$

$$D = \frac{N}{P_D}$$

$$= \frac{60 \text{ teeth}}{5 \dfrac{\text{teeth}}{\text{in}}}$$

$$= 12 \text{ in}$$

The gear speed is

$$n_{int} = n_{input}\left(\frac{N_{input}}{N_{int\,at\,output}}\right) = \left(1800 \frac{\text{rev}}{\text{min}}\right)\left(\frac{50 \text{ teeth}}{25 \text{ teeth}}\right)$$

$$= 3600 \text{ rev/min}$$

The pitch-line velocity is

Spur Gears

$$v = \frac{\pi D n}{12}$$

$$= \frac{\pi(12 \text{ in})\left(3600 \dfrac{\text{rev}}{\text{min}}\right)}{12 \dfrac{\text{in}}{\text{ft}}}$$

$$= 11{,}310 \text{ ft/min}$$

The transmitted load is

Spur Gears

$$W_t = \frac{33{,}000 H}{v} = \frac{(33{,}000)(50 \text{ hp})}{11{,}310 \dfrac{\text{ft}}{\text{min}}}$$

$$= 145.9 \text{ lbf}$$

The torsional moment on the shaft is

$$T = W_t\left(\frac{D}{2}\right)$$

$$= (145.9 \text{ lbf})\left(\frac{12 \text{ in}}{2}\right)$$

$$= 875 \text{ in-lbf}$$

According to the maximum shear stress failure theory, the shear stress is limited to

Ductile Materials

$$\tau_{\max} \geq \frac{S_y}{2}$$

$$\tau_{\max} = \frac{69{,}000 \dfrac{\text{lbf}}{\text{in}^2}}{2}$$

$$= 34{,}500 \text{ lbf/in}^2$$

With a factor of safety of 2, the allowable shear stress is

$$\tau_a = \frac{\tau_{\max}}{2}$$

$$= \frac{34{,}500 \dfrac{\text{lbf}}{\text{in}^2}}{2}$$

$$= 17{,}250 \text{ lbf/in}^2$$

The minimum shaft diameter due to torsional stress alone is

Torsion

$$\tau = \frac{Tr}{J}$$

Properties of Various Shapes

$$J = \frac{\pi a^4}{2}$$

$$\tau = \frac{Tr}{J} = \frac{Tr}{\left(\dfrac{\pi r^4}{2}\right)}$$

$$= \frac{2T}{\pi r^3}$$

$$r = \sqrt[3]{\frac{2T}{\pi \tau}}$$

$$D_{\text{shaft}} = 2r$$

$$= 2\left(\frac{2T}{\pi \tau}\right)^{\frac{1}{3}}$$

$$= (2)\left(\frac{(2)(875 \text{ in-lbf})}{\pi\left(17{,}250 \dfrac{\text{lbf}}{\text{in}^2}\right)}\right)^{\frac{1}{3}}$$

$$= 0.637 \text{ in} \quad (0.65 \text{ in})$$

SI Solution

To determine the transmitted load, the pitch circle diameter and intermediate shaft speed must be calculated. The pitch circle diameter is

Involute Gear Tooth Nomenclature

$$m = \frac{d}{N}$$

$$d = mN$$

$$= (5 \text{ mm})(60 \text{ teeth})$$

$$= 300 \text{ mm}$$

The intermediate speed is

$$n_{\text{int}} = n_{\text{input}}\left(\frac{N_{\text{input}}}{N_{\text{int at output}}}\right) = \left(1800 \dfrac{\text{rev}}{\text{min}}\right)\left(\dfrac{50 \text{ teeth}}{25 \text{ teeth}}\right)$$

$$= 3600 \text{ rev/min}$$

The transmitted load is

Spur Gears

$$W_t = \frac{60{,}000 H}{\pi D n} = \frac{(60{,}000)(38 \text{ kW})}{\pi(300 \text{ mm})\left(3600 \dfrac{\text{rev}}{\text{min}}\right)}$$

$$= 0.67 \text{ kN} \quad (670 \text{ N})$$

The torsional moment on the shaft is

$$T = W_t\left(\frac{D}{2}\right)$$

$$= (670 \text{ N})\left(\frac{300 \text{ mm}}{(2)\left(1000 \frac{\text{mm}}{\text{m}}\right)}\right)$$

$$= 100.5 \text{ N·m}$$

According to the maximum shear stress failure theory, the shear stress is limited to

Ductile Materials

$$\tau_{\max} = 0.5 S_{yt}$$

$$= \frac{480 \text{ MPa}}{2}$$

$$= 240 \text{ MPa}$$

With a factor of safety of 2, the allowable shear stress is

$$\tau_a = \frac{\tau_{\max}}{2}$$

$$= \frac{240 \text{ MPa}}{2}$$

$$= 120 \text{ MPa}$$

The minimum shaft diameter due to torsional stress alone is

Torsion

$$\tau = \frac{Tr}{J}$$

Properties of Various Shapes

$$J = \frac{\pi a^4}{2}$$

$$\tau = \frac{Tr}{J} = \frac{Tr}{\left(\frac{\pi r^4}{2}\right)}$$

$$= \frac{2T}{\pi r^3}$$

$$r = \sqrt[3]{\frac{2T}{\pi \tau}}$$

$$D_{\text{shaft}} = 2r$$

$$= 2\left(\frac{2T}{\pi \tau}\right)^{\frac{1}{3}}$$

$$= (2)\left(\frac{(2)(100.5 \text{ N·m})}{\pi(120 \text{ MPa})\left(1 \times 10^6 \frac{\text{Pa}}{\text{MPa}}\right)}\right)^{\frac{1}{3}}\left(1000 \frac{\text{mm}}{\text{m}}\right)$$

$$= 16.21 \text{ mm} \quad (16.5 \text{ mm})$$

The answer is (D).

15. From the equation for the circular pitch, the number of teeth is

Involute Gear Tooth Nomenclature

$$p_c = \frac{\pi D}{N}$$

$$N = \frac{\pi D}{p_c}$$

$$= \frac{\pi(8 \text{ in})}{1 \text{ in}}$$

$$= 25.13 \quad (25)$$

The answer is (B).

16. The spiral angle is the angle between the tooth trace and an element of the pitch cone. The root angle is the angle formed between a tooth root element and the axis of the bevel gear. The shaft angle is the angle between the mating bevel-gear axes; it is also the sum of the two pitch angles.

The answer is (A).

17. The transmitted load is

Bevel Gears

$$W_a = W_t \tan\phi \sin\gamma$$

$$W_t = \frac{W_a}{\tan\phi \sin\gamma}$$

$$= \frac{1000 \text{ lbf}}{(\tan 20°)(\sin 22.5°)}$$

$$= 7179.5 \text{ lbf} \quad (7200 \text{ lbf})$$

The answer is (D).

18. The pressure angle of the bevel gear is

Bevel Gears

$$W_a = W_t \tan\phi \sin\gamma$$

$$\phi = \tan^{-1}\left(\frac{W_a}{W_t \sin\gamma}\right)$$

$$= \tan^{-1}\left(\frac{350 \text{ lbf}}{(1500 \text{ lbf})\sin 40°}\right)$$

$$= 20°$$

The resultant load is

$$W = \frac{W_t}{\cos\phi} = \frac{1500 \text{ lbf}}{\cos 20°}$$

$$= 1596.3 \text{ lbf} \quad (1600 \text{ lbf})$$

The answer is (D).

19. The projected area, A, of the journal bearing that is exposed to the shaft load is

$$A = LD = (2 \text{ in})(1 \text{ in}) = 2 \text{ in}^2$$

The pressure, p, on the journal bearing is a function of load and projected area.

$$p = \frac{\text{lateral shaft load}}{A} = \frac{2500 \text{ lbf}}{2 \text{ in}^2} = 1250 \text{ lbf/in}^2$$

The answer is (B).

20. From Petroff's law, the required bearing radius is

Ball/Roller Bearing Selection

$$T = \frac{4\pi^2 r^3 L \mu N}{c}$$

$$r = \left(\frac{Tc}{4\pi^2 L \mu N}\right)^{\frac{1}{3}}$$

$$= \left(\frac{(0.15 \text{ ft-lbf})\left(12 \frac{\text{in}}{\text{ft}}\right)(0.01 \text{ in})\left(60 \frac{\text{sec}}{\text{min}}\right)}{4\pi^2(3 \text{ in})\left(1.5 \times 10^{-6} \frac{\text{lbf-sec}}{\text{in}^2}\right)\left(205 \frac{\text{rev}}{\text{min}}\right)}\right)^{\frac{1}{3}}$$

$$= 3.09 \text{ in} \quad (3.1 \text{ in})$$

The answer is (A).

21. Rated life, L_R, is a function of the design life. The design life is

Ball/Roller Bearing Selection

$$C = PL^{\frac{1}{a}}$$

$$L = \left(\frac{C}{P}\right)^{\frac{10}{3}} \times 10^6 \text{ rev}$$

$$= \left(\frac{3480 \text{ lbf}}{1000 \text{ lbf}}\right)^{\frac{10}{3}} \times 10^6 \text{ rev}$$

$$= 63.9 \times 10^6 \text{ rev}$$

The lifetime for the roller bearing based on a shaft speed of 200 rpm is

$$L_R = \frac{L}{n_{\text{rpm}}}$$

$$= \left(\frac{63.9 \times 10^6 \text{ rev}}{\left(200 \frac{\text{rev}}{\text{min}}\right)\left(60 \frac{\text{min}}{\text{hr}}\right)}\right)$$

$$= 5325 \text{ hr} \quad (5300 \text{ hr})$$

The answer is (B).

22. The basic load rating is

Ball/Roller Bearing Selection

$$C = PL^{\frac{1}{a}}$$

$$= (1000 \text{ lbf})\left(\frac{500{,}000{,}000 \text{ rev}}{10^6 \text{ rev}}\right)^{\frac{1}{3}}$$

$$= 7937 \text{ lbf} \quad (7900 \text{ lbf})$$

The answer is (C).

Pressure Vessels

Content in blue refers to the *NCEES Handbook*.

PRACTICE PROBLEMS

1. What is the major distinction between pressure vessels marked with "U" and "UM" stamps?

(A) UM-stamped pressure vessels may be used in corrosive environments.

(B) UM-stamped pressure vessels are not intended for installation in European Commonwealth countries.

(C) UM stamps are for pressure vessels intended for U.S. military installations.

(D) UM stamps are for miniature pressure vessels.

2. Which pressure range of pressure vessels is NOT covered by Section VIII of the ASME *Boiler and Pressure Vessel Code* (BPVC)?

(A) less than 3000 psig

(B) 3000 psig to 10,000 psig

(C) greater than 10,000 psig

(D) none of the above

3. The pressure vessel shown is intended for general service. No radiography is performed. The corrosion allowance is $1/8$ in. The vessel is seamless, operates at 350 psig and 500°F, and is made from SA-106 steel. The shell has a 48 in outside diameter and a nominal thickness of 0.750 in. At 500°F, SA-106 steel has an allowable stress of 17,100 psi and a code weld-joint efficiency of 0.85. The sump is a seamless pipe with a 16 in outside diameter and 0.625 in thickness.

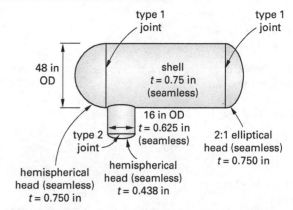

What is most nearly the required minimum thickness of the shell?

(A) 0.34 in

(B) 0.40 in

(C) 0.57 in

(D) 0.70 in

4. A pressure vessel previously used in a refinery has been out of service for a number of years. Its pressure must be checked using a new equation in order to ensure it is still viable. The equation to find the hydrostatic pressure is shown.

$$P = \frac{1.3 P_{max} S_{allowable\,at\,room\,temp}}{S_{allowable\,at\,operating\,temp}}$$

The vessel is known to be constructed from 0.625 in SA-516, grade 65 steel plate, but no other documentation is available. The longitudinal seam joint in the vessel is type 1, and the heads are attached with type 2 joints. The vessel is spot radiographed and operates at 350 psig and 850°F. The corrosion allowance is 1/16 in. SA-516 grade 65 steel plate has a maximum allowable stress of 18,600 lbf/in² at room temperature, and a maximum allowable stress of 8700 lbf/in² at 850°F.

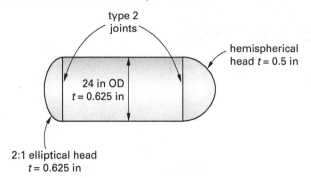

If the vessel is tested at room temperature, the hydrostatic test pressure is most nearly

(A) 600 psig

(B) 800 psig

(C) 900 psig

(D) 1000 psig

5. A large pressure vessel with a circular, flat unstayed integral head is to operate at 20 psig. The minimum head thickness for this vessel is found from the equation shown.

$$t = D\sqrt{\frac{Cp}{SE}}$$

The shell and head are constructed from material with a maximum allowable stress of 20,000 psi. The inside diameter of the head is 147 in. The flange attachment factor is 0.2. The weld efficiency is 1.00. Exclusive of any corrosion allowance, what is most nearly the minimum head thickness?

(A) 2.1 in

(B) 2.3 in

(C) 2.5 in

(D) 2.7 in

SOLUTIONS

1. "UM" stamps are for miniature pressure vessels (inside diameters no greater than 16 in, working pressures no greater than 100 psig, and internal volumes no greater than 5 ft³).

The answer is (D).

2. BPVC Sec. VIII is divided into three divisions. Division 1 covers pressure vessels that operate at less than 3000 psig; Division 2 covers the pressure range of 3000 psig to 10,000 psig; and Division 3 covers pressures greater than 10,000 psig.

The answer is (D).

3. The minimum required thickness of the shell is [Cylindrical Pressure Vessel]

$$r_i = 24.000 \text{ in} - 0.750 \text{ in}$$
$$= 23.25 \text{ in}$$

$$t_{\text{circumferential}} = \frac{P_i r_i}{Se - 0.6P_i} + c$$

$$= \frac{\left(350 \frac{\text{lbf}}{\text{in}^2}\right)(23.25 \text{ in})}{\left(17,100 \frac{\text{lbf}}{\text{in}^2}\right)(0.85) - (0.6)\left(350 \frac{\text{lbf}}{\text{in}^2}\right)} + \frac{1}{8} \text{ in}$$

$$= 0.693 \text{ in} \quad (0.70 \text{ in})$$

The answer is (D).

4. The hydrostatic pressure at room temperature is

$$p = \frac{1.3 p_{\max} S_{\text{allowable at room temp}}}{S_{\text{allowable at operating temp}}} = \frac{(1.3)\left(350 \frac{\text{lbf}}{\text{in}^2}\right)\left(18,600 \frac{\text{lbf}}{\text{in}^2}\right)}{8700 \frac{\text{lbf}}{\text{in}^2}}$$

$$= 973 \text{ lbf/in}^2 \quad (1000 \text{ psig})$$

The answer is (D).

5. The thickness is

$$t = D\sqrt{\frac{Cp}{SE}}$$

$$= (147 \text{ in})\sqrt{\frac{(0.2)\left(20 \frac{\text{lbf}}{\text{in}^2}\right)}{\left(20,000 \frac{\text{lbf}}{\text{in}^2}\right)(1.0)}}$$

$$= 2.08 \text{ in} \quad (2.1 \text{ in})$$

The answer is (A).

55 Properties of Solid Bodies

Content in blue refers to the NCEES Handbook.

PRACTICE PROBLEMS

1. A spoked flywheel has an outside diameter of 60 in (1500 mm). The rim thickness is 6 in (150 mm), and the width is 12 in (300 mm). The cylindrical hub has an outside diameter of 12 in (300 mm), a thickness of 3 in (75 mm), and a width of 12 in (300 mm). The rim and hub are connected by six equally spaced cylindrical radial spokes, each having a diameter of 4.25 in (110 mm). All parts of the flywheel are ductile cast iron with a density of 0.256 lbm/in^3 (7080 kg/m^3). What is the rotational mass moment of inertia of the entire flywheel?

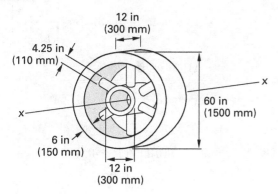

(A) 16,000 lbm-ft^2 (625 kg·m^2)

(B) 16,200 lbm-ft^2 (631 kg·m^2)

(C) 16,800 lbm-ft^2 (654 kg·m^2)

(D) 193,000 lbm-ft^2 (7500 kg·m^2)

2. Which option best describes the difference between the mass moment of inertia for a solid body and the area moment of inertia for an area?

(A) The mass moment of inertia characterizes the object's resistance to changes in rotational speed, and the area moment of inertia characterizes the object's resistance to bending.

(B) The mass moment of inertia characterizes the object's resistance to bending, and the area moment of inertia characterizes the object's resistance changes in rotational speed.

(C) The mass moment of inertia characterizes the object's resistance to torsion, and the area moment of inertia characterizes the object's resistance changes in bending.

(D) The mass moment of inertia characterizes the object's resistance to bending, and the area moment of inertia characterizes the object's resistance changes in torsion.

3. A solid, cylindrical rod is 9 ft long and has a diameter of 0.5 ft. The rod has a density of 280 lbm/ft^3, and the centroidal axis is perpendicular to the rod's length. The mass radius of gyration is most nearly

(A) 3.6 ft

(B) 4.5 ft

(C) 2.6 ft

(D) 4.1 ft

SOLUTIONS

1. *Customary U.S. Solution*

From the hollow circular cylinder in App. 55.A, the mass moment of inertia of the rim about the x-axis (i.e., the rotational axis) is

Mass and Mass Moments of Inertia of Geometric Shapes

$$I_{x,1} = \frac{m(r_o^2 + r_i^2)}{2}$$

$$= \left(\frac{\pi \rho L}{2}\right)(r_o^4 - r_i^4)$$

$$= \frac{\left(\frac{\pi}{2}\right)\left(0.256 \ \frac{\text{lbm}}{\text{in}^3}\right)(12 \text{ in})}{\left(12 \ \frac{\text{in}}{\text{ft}}\right)^2}$$

$$\times \left[\left(\frac{60 \text{ in}}{2}\right)^4 - \left(\frac{60 \text{ in} - 12 \text{ in}}{2}\right)^4\right]$$

$$= 16{,}025 \text{ lbm-ft}^2$$

From the hollow circular cylinder in App. 55.A, the mass moment of inertia of the hub is

Mass and Mass Moments of Inertia of Geometric Shapes

$$I_{x,2} = \frac{m(r_o^2 + r_i^2)}{2}$$

$$= \left(\frac{\pi \rho L}{2}\right)(r_o^4 - r_i^4)$$

$$= \frac{\left(\frac{\pi}{2}\right)\left(0.256 \ \frac{\text{lbm}}{\text{in}^3}\right)(12 \text{ in})}{\left(12 \ \frac{\text{in}}{\text{ft}}\right)^2}$$

$$\times \left[\left(\frac{12 \text{ in}}{2}\right)^4 - \left(\frac{12 \text{ in} - 6 \text{ in}}{2}\right)^4\right]$$

$$= 41 \text{ lbm-ft}^2$$

The length of a cylindrical spoke is

$$L = 24 \text{ in} - 6 \text{ in} = 18 \text{ in}$$

The mass of a spoke is

Mass and Mass Moments of Inertia of Geometric Shapes

$$m = \rho A L = \rho \pi r^2 L$$

$$= \left(0.256 \ \frac{\text{lbm}}{\text{in}^3}\right) \pi \left(\frac{4.25 \text{ in}}{2}\right)^2 (18 \text{ in})$$

$$= 65.37 \text{ lbm}$$

Find mass moment of inertia of a spoke about its own centroidal axis. Use the value of the length, L, from the previous equation for the height, h.

Mass and Mass Moments of Inertia of Geometric Shapes

$$I_{c,x} = \frac{m(3r^2 + L^2)}{12}$$

$$= \frac{(65.37 \text{ lbm})\left[(3)\left(\frac{4.25 \text{ in}}{2}\right)^2 + (18 \text{ in})^2\right]}{(12)\left(12 \ \frac{\text{in}}{\text{ft}}\right)^2}$$

$$= 13 \text{ lbm-ft}^2$$

Use the parallel axis theorem, to find the mass moment of inertia of a spoke about the axis of the flywheel.

$$d = \frac{12 \text{ in}}{2} + \frac{18 \text{ in}}{2} = 15 \text{ in}$$

Parallel-Axis Theorem

$$I_{x,3\,\text{per spoke}} = I_{c,x} + md^2$$

$$= 13 \text{ lbm-ft}^2 + (65.37 \text{ lbm})\left(\frac{15 \text{ in}}{12 \ \frac{\text{in}}{\text{ft}}}\right)^2$$

$$= 115 \text{ lbm-ft}^2$$

The total for six spokes is

$$I_{x,3} = 6 I_{x,3\,\text{per spoke}}$$

$$= (6)(115 \text{ lbm-ft}^2)$$

$$= 690 \text{ lbm-ft}^2$$

Finally, the total rotational mass moment of inertia of the flywheel is the sum of the entirety of the object.

$$I = I_{x,1} + I_{x,2} + I_{x,3}$$

$$= 16{,}025 \text{ lbm-ft}^2 + 41 \text{ lbm-ft}^2 + 690 \text{ lbm-ft}^2$$

$$= 16{,}756 \text{ lbm-ft}^2 \quad (16{,}800 \text{ lbm-ft}^2)$$

SI Solution

From the hollow circular cylinder in App. 55.A, the mass moment of inertia of the rim about the x-axis (i.e., the rotational axis) is

Mass and Mass Moments of Inertia of Geometric Shapes

$$I_{x,1} = \frac{m(r_o^2 + r_i^2)}{2}$$

$$= \left(\frac{\pi \rho L}{2}\right)(r_o^4 - r_i^4)$$

$$= \left(\frac{\pi (7080\ \frac{\text{kg}}{\text{m}^3})(0.3\ \text{m})}{2}\right)$$

$$\times \left(\left(\frac{1.5\ \text{m}}{2}\right)^4 - \left(\frac{1.5\ \text{m} - 0.30\ \text{m}}{2}\right)^4\right)$$

$$= 623.3\ \text{kg} \cdot \text{m}^2$$

From the hollow circular cylinder in App. 55.A, the mass moment of inertia of the hub is

Mass and Mass Moments of Inertia of Geometric Shapes

$$I_{x,2} = \frac{m(r_o^2 + r_i^2)}{2}$$

$$= \left(\frac{\pi \rho L}{2}\right)(r_o^4 - r_i^4)$$

$$= \left(\frac{\pi (7080\ \frac{\text{kg}}{\text{m}^3})(0.3\ \text{m})}{2}\right)$$

$$\times \left(\left(\frac{0.3\ \text{m}}{2}\right)^4 - \left(\frac{0.3\ \text{m} - 0.15\ \text{m}}{2}\right)^4\right)$$

$$= 1.6\ \text{kg} \cdot \text{m}^2$$

The length of a cylindrical spoke is

$$L = 0.6\ \text{m} - 0.15\ \text{m} = 0.45\ \text{m}$$

The mass of a spoke is

Mass and Mass Moments of Inertia of Geometric Shapes

$$m = \rho A L = \rho \pi r^2 L$$

$$= \left(7080\ \frac{\text{kg}}{\text{m}^3}\right)\pi \left(\frac{0.11\ \text{m}}{2}\right)^2 (0.45\ \text{m})$$

$$= 30.3\ \text{kg}$$

Find mass moment of inertia of a spoke about its own centroidal axis. Use the value of the length, L, from the previous equation for the height, h.

Mass and Mass Moments of Inertia of Geometric Shapes

$$I_{c,x} = \frac{m(3r^2 + h^2)}{12}$$

$$= \frac{(30.3\ \text{kg})\left((3)\left(\frac{0.11\ \text{m}}{2}\right)^2 + (0.45\ \text{m})^2\right)}{12}$$

$$= 0.53\ \text{kg} \cdot \text{m}^2$$

Use the parallel axis theorem to find the mass moment of inertia of a spoke about the axis of the flywheel.

$$d = \frac{0.30\ \text{m}}{2} + \frac{0.45\ \text{m}}{2} = 0.375\ \text{m}$$

Parallel-Axis Theorem

$$I_{x,3\,\text{per spoke}} = I_{c,x} + md^2$$

$$= 0.53\ \text{kg} \cdot \text{m}^2 + (30.3\ \text{kg})(0.375\ \text{m})^2$$

$$= 4.8\ \text{kg} \cdot \text{m}^2$$

The total for six spokes is

$$I_{x,3} = 6 I_{x,3\,\text{per spoke}} = (6)(4.8\ \text{kg} \cdot \text{m}^2)$$

$$= 28.8\ \text{kg} \cdot \text{m}^2$$

The total rotational mass moment of inertia of the flywheel is

$$I = I_{x,1} + I_{x,2} + I_{x,3}$$

$$= 623.3\ \text{kg} \cdot \text{m}^2 + 1.6\ \text{kg} \cdot \text{m}^2 + 28.8\ \text{kg} \cdot \text{m}^2$$

$$= 653.7\ \text{kg} \cdot \text{m}^2 \quad (654\ \text{kg} \cdot \text{m}^2)$$

The answer is (C).

2. The mass moment of inertia is the limit of the product of the mass of the solid body and the square of the distance from a given axis. It measures the object's resistance to changes in rotational speed. Typical use is in the equation $T = I\alpha$. A solid body with a small amount of inertia will be less likely to resist a change in rotational speed.

The area moment of inertia, I, is the limit of the product of the area of the solid body and the square of the distance from a given axis. It measures the object's resistance to bending. Typical use is in the equation $\sigma = Mc/I$. A solid body with a small amount of inertia will bend more than a solid body with greater inertia. [Stresses in Beams]

The answer is (A).

3. The cross-sectional area of the rod is

$$A = \frac{\pi D^2}{4} = \frac{\pi (0.5 \text{ ft})^2}{4} = 0.196 \text{ ft}^2$$

The mass of the rod is

Mass and Mass Moments of Inertia of Geometric Shapes

$$m = \rho L A$$
$$= \left(280 \, \frac{\text{lbm}}{\text{ft}^3}\right)(9 \text{ ft})(0.196 \text{ ft}^2)$$
$$= 493.92 \text{ lbm}$$

Use the formula for the mass moment of inertia of a cylindrical rod.

Mass and Mass Moments of Inertia of Geometric Shapes

$$I_{y,c} = \frac{mL^2}{12}$$
$$= \frac{(493.92 \text{ lbm})(9 \text{ ft})^2}{12}$$
$$= 3333.96 \text{ lbm-ft}^2$$

Find the mass radius of gyration.

Mass Radius of Gyration

$$r_m = \sqrt{\frac{I_{y,c}}{m}}$$
$$= \sqrt{\frac{3333.96 \text{ lbm-ft}^2}{493.92 \text{ lbm}}}$$
$$= 2.598 \text{ ft} \quad (2.6 \text{ ft})$$

The answer is (C).

Kinematics

Content in blue refers to the NCEES Handbook.

PRACTICE PROBLEMS

1. Consider a particle that moves horizontally according to the formula $s = 2t^2 - 8t + 3$. When $t = 2$, what are most nearly the position, velocity, and acceleration, respectively?

(A) $s = -5$, v $= 0$, $a = 4$

(B) $s = 0$, v $= 0$, $a = 4$

(C) $s = 3$, v $= -4$, $a = 4$

(D) $s = 3$, v $= -8$, $a = 4$

2. Consider a particle that moves horizontally according to the formula $s = 2t^2 - 8t + 3$. What are most nearly the linear displacement and total distance traveled between $t = 1$ and $t = 3$, respectively?

(A) 0, 0

(B) 0, 4

(C) 6, 6

(D) 6, 10

3. A projectile is fired at 45° from the horizontal with an initial velocity of 2700 ft/sec (820 m/s). Neglecting air friction, what are most nearly the maximum altitude and range, respectively?

(A) $H = 56{,}600$ ft, $R = 226{,}500$ ft ($H = 17\,140$ m, $R = 68\,550$ m)

(B) $H = 81{,}800$ ft, $R = 203{,}400$ ft ($H = 24\,560$ m, $R = 62\,710$ m)

(C) $H = 56{,}600$ ft, $R = 113{,}300$ ft ($H = 17\,140$ m, $R = 34\,300$ m)

(D) $H = 81{,}800$ ft, $R = 113{,}300$ ft ($H = 24\,560$ m, $R = 34\,300$ m)

4. A projectile is launched with an initial velocity of 900 ft/sec (270 m/s). The target is 12,000 ft (3600 m) away and 2000 ft (600 m) higher than the launch point. Air friction is to be neglected. Most nearly, at what angle should the projectile be launched?

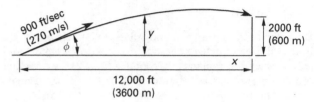

(A) 75°, 15°

(B) 75°, 24°

(C) 1.3°, 0.25°

(D) 1.3°, 0.40°

5. A baseball is hit at 60 ft/sec (20 m/s) and 36.87° from the horizontal. It strikes a fence 72 ft (22 m) away. What are most nearly the velocity components and the elevation, respectively, above the origin at impact?

(A) $v_x = 16$ ft/sec, $v_y = -36$ ft/sec, $y = -18$ ft
($v_x = 5$ m/s, $v_y = -11$ m/s, $y = 6$ m)

(B) $v_x = 48$ ft/sec, $v_y = -12$ ft/sec, $y = 18$ ft
($v_x = 16$ m/s, $v_y = -1.5$ m/s, $y = 7$ m)

(C) $v_x = 41$ ft/sec, $v_y = -93$ ft/sec, $y = -103$ ft
($v_x = 12$ m/s, $v_y = -28$ m/s, $y = -31$ m)

(D) $v_x = 130$ ft/sec, $v_y = -190$ ft/sec, $y = 230$ ft
($v_x = 41$ m/s, $v_y = -57$ m/s, $y = 70$ m)

6. A bomb is dropped from a plane that is climbing at 30° and 600 ft/sec (180 m/s) while passing through 12,000 ft (3600 m) altitude. Approximately how long will it take for the bomb to reach the ground from the release point?

(A) 24 sec

(B) 38 sec

(C) 43 sec

(D) 55 sec

7. A point starts from rest and travels in a circle according to the equation $\omega = 6t^2 - 10t$, where t is time measured in seconds, and ω is measured in radians per second. At $t = 2$, the direction of motion is clockwise. The minimum displacement between the point's positions at $t=1$ and $t=3$ is most nearly

(A) 0.57 rad

(B) 0.82 rad

(C) 12 rad

(D) 15 rad

8. A point starts from rest and travels in a circle according to $\omega = 6t^2 - 10t$, where t is time measured in seconds, and ω is measured in radians per second. At $t = 2$, the direction of motion is clockwise. Assume $\theta(0) = 0$. The total angle turned through between $t=1$ and $t=3$ is most nearly

(A) 6 rad

(B) 8 rad

(C) 12 rad

(D) 15 rad

9. A balloon is 200 ft (60 m) above the ground and is rising at a constant 15 ft/sec (4.5 m/s). An automobile passes under it, traveling along a straight and level road at 45 mph (72 km/h). Approximately how fast is the distance between them changing 1 sec later?

(A) 13 ft/sec (4.0 m/s)

(B) 34 ft/sec (10 m/s)

(C) 37 ft/sec (11 m/s)

(D) 51 ft/sec (15 m/s)

10. The center of a wheel with an outer diameter of 24 in (610 mm) is moving at 28 mph (12.5 m/s). There is no slippage between the wheel and surface. A valve stem is mounted 6 in (150 mm) from the center. What is most nearly the velocity of the valve stem when it is 45° from the horizontal?

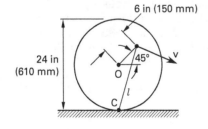

(A) 30 ft/sec (9 m/s)

(B) 57 ft/sec (17 m/s)

(C) 61 ft/sec (19 m/s)

(D) 97 ft/sec (30 m/s)

11. The axle that the wheel shown is attached to moves to the right at 10 ft/sec (3 m/s). What are most nearly the velocities of points A and B with respect to point O if the wheel rolls without slipping?

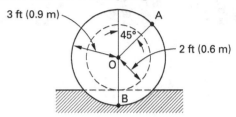

(A) $v_{A/O} = 5$ ft/sec, 45° below horizontal, to the right; $v_{B/O} = 5$ ft/sec, horizontal, to the left ($v_{A/O} = 1.5$ m/s, 45° below horizontal, to the right; $v_{B/O} = 1.5$ m/s, horizontal, to the left)

(B) $v_{A/O} = 6.7$ ft/sec, 45° below horizontal, to the right; $v_{B/O} = 6.7$ ft/sec, horizontal, to the left ($v_{A/O} = 2.0$ m/s, 45° below horizontal, to the right; $v_{B/O} = 2.0$ m/s, horizontal, to the left)

(C) $v_{A/O} = 10$ ft/sec, 45° below horizontal, to the right; $v_{B/O} = 10$ ft/sec, horizontal, to the left ($v_{A/O} = 3.3$ m/s, 45° below horizontal, to the right; $v_{B/O} = 3.3$ m/s, horizontal, to the left)

(D) $v_{A/O} = 15$ ft/sec, 45° below horizontal, to the right; $v_{B/O} = 15$ ft/sec, horizontal, to the left ($v_{A/O} = 4.5$ m/s, 45° below horizontal, to the right; $v_{B/O} = 4.5$ m/s, horizontal, to the left)

12. The axle that the wheel shown is attached to moves to the right at 10 ft/sec (3 m/s).

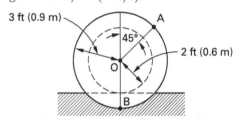

The velocity of point B with respect to point A is most nearly

(A) 28 ft/sec, 23° above horizontal, to the left (8.3 m/s, 23° above horizontal, to the left)

(B) 30.0 ft/sec, 22.5° above horizontal, to the left (9.1 m/s, 22.5° above horizontal, to the left)

(C) 41.3 ft/sec, 22.5° above horizontal, to the left (12.5 m/s, 22.5° above horizontal, to the left)

(D) 42.3 ft/sec, 22.5° above horizontal, to the left (12.8 m/s, 22.5° above horizontal, to the left)

SOLUTIONS

1. From the problem statement,

$$s(t) = 2t^2 - 8t + 3$$

The equations for the instantaneous velocity and acceleration are

$$v(t) = \frac{ds(t)}{dt} = 4t - 8$$
$$a(t) = \frac{dv(t)}{dt} = 4$$

At $t = 2$,

$$s = (2)(2)^2 - (8)(2) + 3 = -5$$
$$v = (4)(2) - 8 = 0$$
$$a = 4$$

The answer is (A).

2. At $t = 1$,

$$s = (2)(1)^2 - (8)(1) + 3 = -3$$

At $t = 3$,

$$s = (2)(3)^2 - (8)(3) + 3 = -3$$

The displacement is

$$\text{displacement} = s(t_2) - s(t_1) = -3 - (-3) = 0$$

The total distance traveled from $t = 1$ to $t = 3$ is

$$s = \int_1^3 |v(t)|\, dt = \int_1^3 |4t - 8|\, dt$$
$$= \int_1^2 (8 - 4t)\, dt + \int_2^3 (4t - 8)\, dt$$
$$= 2 + 2$$
$$= 4$$

The answer is (B).

3. From the equations for projectile motion,

Projectile Motion

$$v_y = -gt + v_0 \sin\theta$$

Solve for t.

$$t = \frac{v_0 \sin\theta - v_y}{g}$$

At the time the projectile reaches the apex of its arc, $v_y = 0$. The equation simplifies to

$$t = \frac{v_0 \sin\theta}{g}$$

Insert the equation for t into the equation for y.

Projectile Motion

$$y = -\frac{gt^2}{2} + (v_0 \sin\theta)t + y_0$$
$$= -\frac{g}{2}\left(\frac{v_0 \sin\theta}{g}\right)^2 + (v_0 \sin\theta)\left(\frac{v_0 \sin\theta}{g}\right) + y_0$$
$$= \frac{v_0^2 \sin^2\theta}{2g}$$

The total time of travel for the projectile is twice the time required to reach the maximum height. Therefore the equation for the range is

Projectile Motion

$$x = (v_0 \cos\theta)t + x_0$$
$$= (v_0 \cos\theta)\left(\frac{2v_0 \sin\theta}{g}\right)$$
$$= \frac{v_0^2 \sin 2\theta}{g}$$

Customary U.S. Solution

The altitude and range are

$$H = \frac{v_0^2 \sin^2\theta}{2g} = \frac{\left(2700\, \frac{\text{ft}}{\text{sec}}\right)^2 (\sin 45°)^2}{(2)\left(32.2\, \frac{\text{ft}}{\text{sec}^2}\right)}$$
$$= 56{,}599\text{ ft}\quad (56{,}600\text{ ft})$$

$$R = \frac{v_0^2 \sin 2\theta}{g} = \frac{\left(2700\, \frac{\text{ft}}{\text{sec}}\right)^2 (\sin(2)(45°))}{32.2\, \frac{\text{ft}}{\text{sec}^2}}$$
$$= 226{,}398\text{ ft}\quad (226{,}500\text{ ft})$$

SI Solution

The altitude and range are

$$H = \frac{v_0^2 \sin^2 \theta}{2g} = \frac{\left(820 \, \frac{\text{m}}{\text{s}}\right)^2 (\sin 45°)^2}{(2)\left(9.81 \, \frac{\text{m}}{\text{s}^2}\right)}$$

$$= 17\,136 \text{ m} \quad (17\,140 \text{ m})$$

$$R = \frac{v_0^2 \sin 2\theta}{g} = \frac{\left(820 \, \frac{\text{m}}{\text{s}}\right)^2 (\sin(2)(45°))}{9.81 \, \frac{\text{m}}{\text{s}^2}}$$

$$= 68\,542 \text{ m} \quad (68\,550 \text{ m})$$

The answer is (A).

4. *Customary U.S. Solution*

The initial x-distance is zero, so the equation for the x-distance is

Projectile Motion

$$x = (v_0 \cos \theta)t + x_0 = (v_0 \cos \theta)t$$

Solve for t.

$$t = \frac{x}{v_0 \cos \theta} = \frac{12{,}000 \text{ ft}}{\left(900 \, \frac{\text{ft}}{\text{sec}}\right)\cos \theta} = \frac{13.33}{\cos \theta}$$

The initial y-distance is 0, so the equation for the y-distance is

Projectile Motion

$$y = -\frac{gt^2}{2} + (v_0 \sin \theta)t + y_0 = (v_0 \sin \theta)t - \frac{gt^2}{2}$$

Substitute t and the given value of y.

$$2000 \text{ ft} = \left(900 \, \frac{\text{ft}}{\text{sec}}\right)\sin \theta\left(\frac{13.33}{\cos \theta}\right)$$
$$- \left(\frac{1}{2}\right)\left(32.2 \, \frac{\text{ft}}{\text{sec}^2}\right)\left(\frac{13.33}{\cos \theta}\right)^2$$

Simplify.

$$1 = 6.0 \tan \theta - \frac{1.43}{\cos^2 \theta}$$

Use the trigonometric identity for the secant.

Identities

$$\sec \theta = \frac{1}{\cos \theta}$$

$$\frac{1}{\cos^2 \theta} = \sec^2 \theta = 1 + \tan^2 \theta$$

$$1 = 6.0 \tan \theta - 1.43 - 1.43 \tan^2 \theta$$

Simplify.

$$\tan^2 \theta - 4.20 \tan \theta + 1.70 = 0$$

Use the quadratic formula.

$$\tan \theta = \frac{4.20 \pm \sqrt{(4.20)^2 - (4)(1)(1.70)}}{(2)(1)}$$

$$= 3.75, \, 0.454 \quad [\text{radians}]$$

$$\theta = \tan^{-1} 3.75, \, \tan^{-1} 0.454$$

$$= 75.1°, \, 24.4° \quad (75°, \, 24°)$$

SI Solution

The initial x-distance is zero, so the equation for the x-distance is

Projectile Motion

$$x = (v_0 \cos \theta)t + x_0 = (v_0 \cos \theta)t$$

Solve for t.

$$t = \frac{x}{v_0 \cos \phi} = \frac{3600 \text{ m}}{\left(270 \, \frac{\text{m}}{\text{s}}\right)\cos \theta} = \frac{13.33}{\cos \theta}$$

The initial y-distance is 0, so the equation for the y-distance is

Projectile Motion

$$y = -\frac{gt^2}{2} + (v_0 \sin \theta)t + y_0 = (v_0 \sin \theta)t - \frac{gt^2}{2}$$

Substitute t and the given value of y.

$$600 \text{ m} = \left(270 \, \frac{\text{m}}{\text{s}}\right)\sin \theta\left(\frac{13.33}{\cos \theta}\right)$$
$$- \left(\frac{1}{2}\right)\left(9.81 \, \frac{\text{m}}{\text{s}^2}\right)\left(\frac{13.33}{\cos \theta}\right)^2$$

Simplify.

$$1 = 6.0 \tan \theta - \frac{1.45}{\cos^2 \theta}$$

Use the trigonometric identity for the secant.

Identities

$$\sec\theta = \frac{1}{\cos\theta}$$

$$\frac{1}{\cos^2\theta} = \sec^2\theta = 1 + \tan^2\theta$$

$$1 = 6.0\tan\theta - 1.45 - 1.45\tan^2\theta$$

Simplify.

$$\tan^2\theta - 4.14\tan\theta + 1.69 = 0$$

Use the quadratic formula.

$$\tan\theta = \frac{4.14 \pm \sqrt{(4.14)^2 - (4)(1)(1.69)}}{(2)(1)}$$

$$= 3.68, \, 0.459 \quad \text{[radians]}$$

$$\theta = \tan^{-1} 3.68, \, \tan^{-1} 0.459$$

$$= 74.8°, \, 24.7° \quad (75°, \, 24°)$$

The answer is (B).

5. *Customary U.S. Solution*

Neglecting air friction, the x-component of the velocity is

Projectile Motion

$$v_x = v_0 \cos\theta = \left(60 \, \frac{\text{ft}}{\text{sec}}\right)(\cos 36.87°)$$

$$= 48 \text{ ft/sec} \quad \text{[constant]}$$

$$t = \frac{s}{v_x} = \frac{72 \text{ ft}}{48 \, \frac{\text{ft}}{\text{sec}}} = 1.5 \text{ sec}$$

Neglecting air friction, the y-component of the velocity is

Projectile Motion

$$v_0 \sin\theta - gt$$

$$= \left(60 \, \frac{\text{ft}}{\text{sec}}\right)(\sin 36.87°) - \left(32.2 \, \frac{\text{ft}}{\text{sec}^2}\right)(1.5 \text{ sec})$$

$$= -12.3 \text{ ft/sec} \quad (-12 \text{ ft/sec})$$

$$y = (v_0 \sin\theta)t - \frac{1}{2}gt^2$$

$$= \left(60 \, \frac{\text{ft}}{\text{sec}}\right)(\sin 36.87°)(1.5 \text{ sec})$$

$$- \left(\frac{1}{2}\right)\left(32.2 \, \frac{\text{ft}}{\text{sec}^2}\right)(1.5 \text{ sec})^2$$

$$= 17.78 \text{ ft} \quad (18 \text{ ft})$$

SI Solution

Neglecting air friction, the x-component of the velocity is

Projectile Motion

$$v_0 \cos\theta = \left(20 \, \frac{\text{m}}{\text{s}}\right)(\cos 36.87°)$$

$$= 16 \text{ m/s} \quad \text{[constant]}$$

$$t = \frac{s}{v_x} = \frac{22 \text{ m}}{16 \, \frac{\text{m}}{\text{s}}} = 1.375 \text{ s}$$

Neglecting air friction, the y-component of the velocity is

Projectile Motion

$$v_y = v_0 \sin\theta - gt$$

$$= \left(20 \, \frac{\text{m}}{\text{s}}\right)(\sin 36.87°) - \left(9.81 \, \frac{\text{m}}{\text{s}^2}\right)(1.375 \text{ s})$$

$$= -1.49 \text{ m/s} \quad (-1.5 \text{ m/s})$$

$$y = (v_0 \sin\theta)t - \frac{1}{2}gt^2$$

$$= \left(20 \, \frac{\text{m}}{\text{s}}\right)(\sin 36.87°)(1.375 \text{ s})$$

$$- \left(\frac{1}{2}\right)\left(9.81 \, \frac{\text{m}}{\text{s}^2}\right)(1.375 \text{ s})^2$$

$$= 7.227 \text{ m} \quad (7 \text{ m})$$

The answer is (B).

6. First find the maximum height above the reference height ($z = 12{,}000$ ft) that the bomb reaches.

Projectile Motion

$$v_y = -gt + v_0 \sin\theta$$

Solve for t.

$$t = \frac{v_0 \sin\theta - v_y}{g}$$

At the time the projectile reaches the apex of its arc, $v_y = 0$. Call this time t_1.

Find the bomb's maximum altitude by adding

$$t_1 = \frac{v_0 \sin\theta}{g}$$

Insert the equation for t_1 into the equation for y and simplify.

$$y = -\frac{gt^2}{2} + (v_0 \sin\theta)t + y_0$$

$$= -\left(\frac{g}{2}\right)\left(\frac{v_0 \sin\theta}{g}\right)^2 + (v_0 \sin\theta)\left(\frac{v_0 \sin\theta}{g}\right) + 0$$

$$= \frac{v_0^2 \sin^2\theta}{2g}$$

Let t_2 be the time the bomb takes to fall from H. Rearrange the equation for y when $y = 0$, $t = t_2$, $v_0 = 0$, and $y_0 = H$.

$$y = -\frac{gt^2}{2} + (v_0 \sin\theta)t + y_0$$

$$0 = -\frac{gt_2^2}{2} + H$$

$$t_2^2 = \sqrt{\frac{2H}{g}}$$

Customary U.S. Solution

The maximum height is

$$H = z + \frac{v_0^2 \sin^2\theta}{2g}$$

$$= 12{,}000 \text{ ft} + \frac{\left(600 \frac{\text{ft}}{\text{sec}}\right)^2 (\sin 30°)^2}{(2)\left(32.2 \frac{\text{ft}}{\text{sec}^2}\right)}$$

$$= 13{,}398 \text{ ft}$$

The time for the bomb to reach the ground is

$$t_2 = \sqrt{\frac{2H}{g}} = \sqrt{\frac{(2)(13{,}398 \text{ ft})}{32.2 \frac{\text{ft}}{\text{sec}^2}}} = 28.85 \text{ sec}$$

$$t = t_1 + t_2 = 9.32 \text{ sec} + 28.85 \text{ sec}$$

$$= 38.17 \text{ sec} \quad (38 \text{ sec})$$

SI Solution

The maximum height is

$$H = z + \frac{v_0^2 \sin^2\theta}{2g}$$

$$= 3600 \text{ m} + \frac{\left(180 \frac{\text{m}}{\text{s}}\right)^2 (\sin 30°)^2}{(2)\left(9.81 \frac{\text{m}}{\text{s}^2}\right)}$$

$$= 4012.8 \text{ m} \quad (4000 \text{ m})$$

$$t_1 = \frac{1}{2}\left(\frac{2v_0 \sin\theta}{g}\right) = \frac{\left(600 \frac{\text{ft}}{\text{sec}}\right)(\sin 30°)}{32.2 \frac{\text{ft}}{\text{sec}^2}} = 9.32 \text{ sec}$$

The time for the bomb to reach the ground is

$$t_2 = \sqrt{\frac{2H}{g}} = \sqrt{\frac{(2)(4012.8 \text{ m})}{9.81 \frac{\text{m}}{\text{s}^2}}} = 28.60 \text{ s}$$

$$t = t_1 + t_2 = 9.17 \text{ s} + 28.60 \text{ s}$$

$$= 37.77 \text{ s} \quad (38 \text{ s})$$

The answer is (B).

7. The displacement is

$$\theta = \theta(3) - \theta(1) = \int_1^3 \omega(t)\, dt$$

$$= \int_1^3 (6t^2 - 10t)\, dt = 2t^3 - 5t^2 + C\Big|_1^3$$

$$[C = 0 \text{ since } \theta(0) = 0]$$

$$= (2)(3)^3 - (5)(3)^2 - (2)(1)^3 + (5)(1)^2$$

$$= 12 \text{ rad}$$

The point's position at time t is

$$\theta(t) = \int \omega(t)\, dt = \int (6t^2 - 10t)\, dt = 2t^3 - 5t^2 + C$$

Since the point starts from rest at an arbitrary starting position, let $C = 0$.

The point's position at $t = 1$ is

$$\theta(1) = (2)(1)^3 - (5)(1)^2 = -3$$

Since the point travels in a circle, and since a circle has 2π radians, the position can be expressed as

$$\theta(1) = -3 + 2\pi = 3.28 \text{ rad}$$

The point's position at $t = 3$ is

$$\theta(3) = (2)(3)^3 - (5)(3)^2 = 9$$

Since the point travels in a circle, it returns to its starting position after 2π radians. The position can be expressed as

$$\theta(3) = 9 - 2\pi = 2.72 \text{ rad}$$

The minimum displacement (shortest distance between the particles) is

$$\delta = |\theta_2 - \theta_1| = |2.72 \text{ rad} - 3.28 \text{ rad}|$$
$$= 0.566 \text{ rad} \quad (0.57 \text{ rad})$$

The answer is (A).

8. To find the total distance traveled, check for sign reversals in $\omega(t)$ over the interval $t = 1$ to $t = 3$.

$$6t^2 - 10t = 0$$
$$\text{sign reversal at } t = 5/3$$

The total angle turned is

$$\int_1^3 |\omega(t)|\, dt = \int_1^{5/3} (10t - 6t^2)\, dt + \int_{5/3}^3 (6t^2 - 10t)\, dt$$
$$= 15.26 \text{ rad} \quad (15 \text{ rad})$$

The answer is (D).

9. *Customary U.S. Solution*

The velocity of the car in ft/sec is

$$v_{car} = \frac{\left(45\,\frac{\text{mi}}{\text{hr}}\right)\left(5280\,\frac{\text{ft}}{\text{mi}}\right)}{\left(60\,\frac{\text{sec}}{\text{min}}\right)\left(60\,\frac{\text{min}}{\text{hr}}\right)} = 66 \text{ ft/sec}$$

The separation distance after 1 sec is

$$s(1) = \sqrt{\left(200 \text{ ft} + \left(15\,\frac{\text{ft}}{\text{sec}}\right)(1 \text{ sec})\right)^2 + \left(\left(66\,\frac{\text{ft}}{\text{sec}}\right)(1 \text{ sec})\right)^2} = 224.9 \text{ ft}$$

The separation velocity is the difference in components of the car's and balloon's velocities along a mutually parallel line. Use the separation vector as this line.

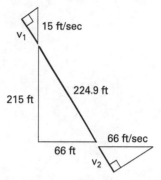

(not to scale)

$$v_1 = \left(15\,\frac{\text{ft}}{\text{sec}}\right)\left(\frac{215 \text{ ft}}{224.9 \text{ ft}}\right) = 14.34 \text{ ft/sec}$$

$$v_2 = \left(66\,\frac{\text{ft}}{\text{sec}}\right)\left(\frac{66 \text{ ft}}{224.9 \text{ ft}}\right) = 19.37 \text{ ft/sec}$$

$$\Delta v = v_1 + v_2 = 14.34\,\frac{\text{ft}}{\text{sec}} + 19.37\,\frac{\text{ft}}{\text{sec}}$$
$$= 33.71 \text{ ft/sec} \quad (34 \text{ ft/sec})$$

SI Solution

The velocity of the car in m/s is

$$v_{car} = \frac{\left(72\,\frac{\text{km}}{\text{h}}\right)\left(1000\,\frac{\text{m}}{\text{km}}\right)}{\left(60\,\frac{\text{s}}{\text{min}}\right)\left(60\,\frac{\text{min}}{\text{h}}\right)} = 20 \text{ m/s}$$

The separation distance after 1 s is

$$s(1) = \sqrt{\left(60 \text{ m} + \left(4.5\,\frac{\text{m}}{\text{s}}\right)(1 \text{ s})\right)^2 + \left(\left(20\,\frac{\text{m}}{\text{s}}\right)(1 \text{ s})\right)^2}$$
$$= 67.53 \text{ m}$$

The separation velocity is the difference in components of the car's and balloon's velocities along a mutually parallel line. Use the separation vector as this line.

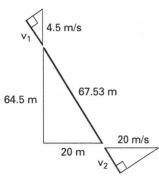

(not to scale)

$$v_1 = \left(4.5 \ \frac{m}{s}\right)\left(\frac{64.5 \ m}{67.53 \ m}\right) = 4.298 \ m/s$$

$$v_2 = \left(20 \ \frac{m}{s}\right)\left(\frac{20 \ m}{67.53 \ m}\right) = 5.923 \ m/s$$

$$\Delta v = v_1 + v_2 = 4.298 \ \frac{m}{s} + 5.923 \ \frac{m}{s}$$
$$= 10.22 \ m/s \quad (10 \ m/s)$$

The answer is (B).

10. *Customary U.S. Solution*

The angular velocity of the wheel is

$$\omega = \frac{v_C}{r} = \frac{\left(28 \ \frac{mi}{hr}\right)\left(5280 \ \frac{ft}{mi}\right)\left(12 \ \frac{in}{ft}\right)}{(12 \ in)\left(60 \ \frac{sec}{min}\right)\left(60 \ \frac{min}{hr}\right)}$$
$$= 41.07 \ rad/sec$$

The distance from the valve stem to the instant center of the point of contact of the wheel and the surface is determined from the law of cosines for the triangle defined by the valve stem, instant center, and center of wheel.

$$r^2 = r_{wheel}^2 + r_{stem}^2 - 2r_{wheel}r_{stem}\cos\phi$$
$$= (12 \ in)^2 + (6 \ in)^2 - (2)(12 \ in)(6 \ in)(\cos 135°)$$
$$r = 16.79 \ in$$

The tangential velocity is

Plane Circular Motion

$$v_t = r\omega$$
$$= \frac{(16.79 \ in)\left(41.07 \ \frac{rad}{sec}\right)}{12 \ \frac{in}{ft}}$$
$$= 57.46 \ ft/sec \quad (57 \ ft/sec)$$

SI Solution
The angular velocity of the wheel is

$$\omega = \frac{v_C}{r} = \frac{12.5 \ \frac{m}{s}}{0.305 \ m} = 40.98 \ rad/s$$

The distance from the valve stem to the instant center at the point of contact of the wheel and the surface is determined from the law of cosines for the triangle defined by the valve stem, instant center, and center of wheel.

$$r^2 = r_{wheel}^2 + r_{stem}^2 - 2r_{wheel}r_{stem}\cos\phi$$
$$= (0.305 \ m)^2 + (0.150 \ m)^2$$
$$\quad - (2)(0.305 \ m)(0.150 \ m)(\cos 135°)$$
$$r = 0.425 \ m$$

The tangential velocity is

Plane Circular Motion

$$v_t = r\omega$$
$$= (0.425 \ m)\left(40.98 \ \frac{rad}{s}\right)$$
$$= 17.4 \ m/s \quad (17 \ m/s)$$

The answer is (B).

11. *Customary U.S. Solution*

The velocity of point A is

Plane Circular Motion

$$v_{A/O} = r_{outer}\omega = \left(3 \ \frac{ft}{rad}\right)\left(5 \ \frac{rad}{sec}\right)$$
$$= 15 \ ft/sec, \ 45° \ below \ horizontal, \ to \ the \ right$$

The velocity of point B is

Plane Circular Motion

$$v_{B/O} = r_{outer}\omega = 15 \ ft/sec, \ horizontal, \ to \ the \ left$$

SI Solution

The velocity of point A is

Plane Circular Motion
$$v_{A/O} = r_{outer}\omega = \left(0.9 \ \frac{m}{rad}\right)\left(5 \ \frac{rad}{s}\right)$$
$$= 4.5 \ m/s, \ 45° \ \text{below horizontal,} \ \text{to the right}$$

The velocity of point B is

Plane Circular Motion
$$v_{B/O} = r_{outer}\omega = 4.5 \ m/s, \ \text{horizontal, to the left}$$

The answer is (D).

12. *Customary U.S. Solution*

From the equation for tangential velocity, the angular velocity is

Plane Circular Motion
$$v_t = r\omega$$
$$\omega = \frac{v_0}{r_{inner}}$$
$$= \frac{10 \ \frac{ft}{sec}}{2 \ \frac{ft}{rad}}$$
$$= 5 \ rad/sec$$

From the law of cosines,

$$|AB|^2 = r_A^2 + r_B^2 - 2r_A r_B \cos\phi$$
$$= (3 \ ft)^2 + (3 \ ft)^2 - (2)(3 \ ft)^2(\cos 135°)$$
$$= 30.728 \ ft^2$$
$$|AB| = 5.543 \ ft$$

From the equation for tangential velocity, using the absolute value of AB as the radius,

Plane Circular Motion
$$v_{B/A} = r_{outer}\omega = |AB|\omega = (5.543 \ ft)\left(5 \ \frac{rad}{sec}\right)$$
$$= 27.72 \ ft/sec \ (28 \ ft/sec), \ 22.5° \ (23°)$$
above horizontal, to the left

SI Solution

From the equation for tangential velocity, the angular velocity is

Plane Circular Motion
$$v_t = r\omega$$
$$\omega = \frac{v_0}{r_{inner}}$$
$$= \frac{3 \ \frac{m}{s}}{0.6 \ \frac{m}{rad}}$$
$$= 5 \ rad/s$$

From the law of cosines,

$$|AB|^2 = r_A^2 + r_B^2 - 2r_A r_B \cos\phi$$
$$= (0.9 \ m)^2 + (0.9 \ m)^2 - (2)(0.9 \ m)^2(\cos 135°)$$
$$= 2.7655 \ m^2$$
$$|AB| = 1.663 \ m$$

From the equation for tangential velocity, using the absolute value of AB as the radius,

Plane Circular Motion
$$v_{B/A} = r_{outer}\omega = |AB|\omega = (1.663 \ m)\left(5 \ \frac{rad}{s}\right)$$
$$= 8.315 \ m/s \ (8.3 \ m/s), \ 22.5° \ (23°)$$
above horizontal, to the left

The answer is (A).

57 Kinetics

Content in blue refers to the *NCEES Handbook*.

PRACTICE PROBLEMS

1. A 10 lbm (5 kg) mass is tied to a 2 ft (50 cm) string and whirled at 5 rev/sec horizontally to the ground. The centripetal acceleration is most nearly

(A) 1400 ft/sec^2 (350 m/s^2)
(B) 1600 ft/sec^2 (400 m/s^2)
(C) 1800 ft/sec^2 (450 m/s^2)
(D) 2000 ft/sec^2 (490 m/s^2)

2. A 10 lbm (5 kg) mass is tied to a 2 ft (50 cm) string and whirled at 5 rev/sec horizontally to the ground. The centrifugal force is most nearly

(A) 490 lbf (1900 N)
(B) 610 lbf (2500 N)
(C) 750 lbf (2900 N)
(D) 820 lbf (3100 N)

3. A 10 lbm (5 kg) mass is tied to a 2 ft (50 cm) string and whirled at 5 rev/sec horizontally to the ground. The angular momentum is most nearly

(A) 39 ft-lbf-sec (39 J·s)
(B) 44 ft-lbf-sec (44 J·s)
(C) 53 ft-lbf-sec (53 J·s)
(D) 71 ft-lbf-sec (71 J·s)

4. Sand is dropping at the rate of 560 lbm/min (250 kg/min) onto a conveyor belt moving with a velocity of 3.2 ft/sec (0.98 m/s). The force required to keep the belt moving is most nearly

(A) 0.93 lbf (4.1 N)
(B) 1.2 lbf (5.3 N)
(C) 2.3 lbf (10 N)
(D) 4.8 lbf (14 N)

5. The impulse imparted to a 0.4 lbm (0.2 kg) baseball that approaches the batter at 90 ft/sec (30 m/s) and leaves at 130 ft/sec (40 m/s) is most nearly

(A) 1.8 lbf-sec (9.4 N·s)
(B) 2.7 lbf-sec (14 N·s)
(C) 4.1 lbf-sec (21 N·s)
(D) 8.9 lbf-sec (46 N·s)

6. At what approximate velocity will a 1000 lbm (500 kg) gun mounted on wheels recoil if a 2.6 lbm (1.2 kg) projectile is propelled from it at a velocity of 2100 ft/sec (650 m/s)?

(A) 3.7 ft/sec (1.1 m/s)
(B) 4.8 ft/sec (1.4 m/s)
(C) 5.5 ft/sec (1.6 m/s)
(D) 15 ft/sec (4.5 m/s)

7. A 0.15 lbm (60 g) bullet traveling 2300 ft/sec (700 m/s) embeds itself in a 9 lbm (4.5 kg) wooden block initially at rest. The block's velocity immediately after impact is most nearly

(A) 24 ft/sec (7.2 m/s)
(B) 38 ft/sec (9.2 m/s)
(C) 66 ft/sec (20 m/s)
(D) 110 ft/sec (33 m/s)

8. A nozzle discharges 40 gal/min (0.15 m^3/min) of water at 60 ft/sec (20 m/s). There is no splashback. The total force required to hold a flat vertical plate in front of the nozzle is most nearly

(A) 10 lbf (50 N)
(B) 20 lbf (100 N)
(C) 40 lbf (200 N)
(D) 80 lbf (400 N)

SOLUTIONS

1. *Customary U.S. Solution*

The equation for tangential velocity is

Plane Circular Motion
$$v_t = r\omega$$

For circular motion, the equation for angular velocity is

$$\omega = 2\pi N$$

Calculate the tangential velocity.

$$\begin{aligned}v_t &= r\omega \\ &= r(2\pi N) \\ &= (2 \text{ ft})\left(2\pi \ \frac{\text{rad}}{\text{rev}}\right)\left(5 \ \frac{\text{rev}}{\text{sec}}\right) \\ &= 62.83 \text{ ft/sec}\end{aligned}$$

Calculate the centripetal acceleration (i.e., the acceleration normal to the path of motion).

Centripetal Acceleration
$$\begin{aligned}a &= \frac{dv}{dt} = \frac{v^2}{r} \\ &= \frac{v_t^2}{r} \\ &= \frac{\left(62.83 \ \frac{\text{ft}}{\text{sec}}\right)^2}{2 \text{ ft}} \\ &= 1974 \text{ ft/sec}^2\end{aligned}$$

SI Solution

The equation for tangential velocity is

Plane Circular Motion
$$v_t = r\omega$$

For circular motion, the equation for angular velocity is

$$\omega = 2\pi N$$

Calculate the tangential velocity.

$$\begin{aligned}v_t &= r\omega \\ &= r(2\pi N) \\ &= (0.5 \text{ m})\left(2\pi \ \frac{\text{rad}}{\text{rev}}\right)\left(5 \ \frac{\text{rev}}{\text{s}}\right) \\ &= 15.708 \text{ m/s}\end{aligned}$$

Calculate the centripetal acceleration (i.e., the acceleration normal to the path of motion).

Centripetal Acceleration
$$\begin{aligned}a &= \frac{dv}{dt} = \frac{v^2}{r} \\ &= \frac{v_t^2}{r} \\ &= \frac{\left(15.708 \ \frac{\text{m}}{\text{sec}}\right)^2}{0.5 \text{ m}} \\ &= 493.5 \text{ m/s}^2\end{aligned}$$

The answer is (D).

2. *Customary U.S. Solution*

In order to calculate the centrifugal force, first calculate the centripetal acceleration.

The equation for tangential velocity is

Plane Circular Motion
$$v_t = r\omega$$

For circular motion, the equation for angular velocity is

$$\omega = 2\pi N$$

Calculate the tangential velocity.

$$\begin{aligned}v_t &= r\omega \\ &= r(2\pi N) \\ &= (2 \text{ ft})\left(2\pi \ \frac{\text{rad}}{\text{rev}}\right)\left(5 \ \frac{\text{rev}}{\text{sec}}\right) \\ &= 62.83 \text{ ft/sec}\end{aligned}$$

Calculate the centripetal acceleration (i.e., the acceleration normal to the path of motion).

Centripetal Acceleration
$$\begin{aligned}a &= \frac{dv}{dt} = \frac{v^2}{r} \\ &= \frac{v_t^2}{r} \\ &= \frac{\left(62.83 \ \frac{\text{ft}}{\text{sec}}\right)^2}{2 \text{ ft}} \\ &= 1974 \text{ ft/sec}^2\end{aligned}$$

Calculate the centrifugal force.

$$F = \frac{ma}{g_c}$$

$$F = \frac{ma}{g_c} = \frac{(10\text{ lbm})\left(1974\ \dfrac{\text{ft}}{\text{sec}^2}\right)}{32.174\ \dfrac{\text{lbm-ft}}{\text{lbf-sec}^2}}$$

$$= 613.5\text{ lbf} \quad (610\text{ lbf}) \quad [\text{directed outward}]$$

SI Solution

In order to calculate the centrifugal force, first calculate the centripetal acceleration.

The equation for tangential velocity is

Plane Circular Motion

$$v_t = r\omega$$

For circular motion, the equation for angular velocity is

$$\omega = 2\pi N$$

Calculate the tangential velocity.

$$v_t = r\omega$$
$$= r(2\pi N)$$
$$= (0.5\text{ m})\left(2\pi\ \dfrac{\text{rad}}{\text{rev}}\right)\left(5\ \dfrac{\text{rev}}{\text{s}}\right)$$
$$= 15.708\text{ m/s}$$

Calculate the centripetal acceleration (i.e., the acceleration normal to the path of motion).

Centripetal Acceleration

$$a = \frac{dv}{dt} = \frac{v^2}{r}$$
$$= \frac{v_t^2}{r}$$
$$= \frac{\left(15.708\ \dfrac{\text{m}}{\text{sec}}\right)^2}{0.5\text{ m}}$$
$$= 493.5\text{ m/s}^2$$

Calculate the centrifugal force.

Units

$$F = \frac{ma}{g_c}$$

$$F = ma$$
$$= (5\text{ kg})\left(493.5\ \dfrac{\text{m}}{\text{s}^2}\right)$$
$$= 2467.5\text{ N} \quad (2500\text{ N}) \quad [\text{directed outward}]$$

The answer is (B).

3. *Customary U.S. Solution*

The equation for the angular momentum is

Angular Momentum

$$\mathbf{H}_0 = \mathbf{r} \times m\mathbf{v}$$

Since the tangential velocity is perpendicular to the radius,

$$H = \frac{|r||mv_t|\sin 90°}{g_c} = \frac{mrv_t}{g_c}$$

Calculate the tangential velocity.

Plane Circular Motion

$$v_t = r\omega$$

For circular motion, the angular velocity is

$$\omega = 2\pi N$$

Calculate the tangential velocity.

$$v_t = r\omega$$
$$= r(2\pi N)$$
$$= (2\text{ ft})\left(2\pi\ \dfrac{\text{rad}}{\text{rev}}\right)\left(5\ \dfrac{\text{rev}}{\text{sec}}\right)$$
$$= 62.83\text{ ft/sec}$$

Substitute the known values into the equation for the angular momentum.

$$H = \frac{mrv_t}{g_c}$$

$$= \frac{(10\text{ lbm})(2\text{ ft})\left(62.83\ \dfrac{\text{ft}}{\text{sec}}\right)}{32.174\ \dfrac{\text{lbm-ft}}{\text{lbf-sec}^2}}$$

$$= 39.05\text{ ft-lbf-sec} \quad (39\text{ ft-lbf-sec})$$

SI Solution

The equation for the angular momentum is

Angular Momentum

$$\mathbf{H}_0 = \mathbf{r} \times m\mathbf{v}$$

Since the tangential velocity is perpendicular to the radius,

$$H = \frac{|r||m v_t| \sin 90°}{g_c} = \frac{mr v_t}{g_c}$$

Calculate the tangential velocity.

Plane Circular Motion

$$v_t = r\omega$$

For circular motion, the angular velocity is

$$\omega = 2\pi N$$

Calculate the tangential velocity.

$$\begin{aligned} v_t &= r\omega \\ &= r(2\pi N) \\ &= (0.5 \text{ m})\left(2\pi \frac{\text{rad}}{\text{rev}}\right)\left(5 \frac{\text{rev}}{\text{sec}}\right) \\ &= 15.708 \text{ m/s} \end{aligned}$$

Substitute the known values into the equation for angular momentum.

$$\begin{aligned} H &= mr v_t \\ &= (0.5 \text{ m})(5 \text{ kg})\left(15.708 \frac{\text{m}}{\text{s}}\right) \\ &= 39.27 \text{ J·s} \quad (39 \text{ J·s}) \end{aligned}$$

The answer is (A).

4. *Customary U.S. Solution*

The mass flow rate of the sand is constant, but the velocity changes. Apply Newton's second law of motion for constant mass and calculate the force required to keep the belt moving.

Newton's Second Law (Equations of Motion)

$$\sum F = m\frac{dv}{dt} = m\frac{dv}{g_c dt} = \frac{\left(560 \frac{\text{lbm}}{\text{min}}\right)\left(3.2 \frac{\text{ft}}{\text{sec}} - 0 \frac{\text{ft}}{\text{sec}}\right)}{\left(32.174 \frac{\text{lbm-ft}}{\text{lbf-sec}^2}\right)\left(60 \frac{\text{sec}}{\text{min}}\right)}$$

$$= 0.9282 \text{ lbf} \quad (0.93 \text{ lbf})$$

SI Solution

The mass flow rate of the sand is constant, but the velocity changes. Apply Newton's second law of motion for constant mass and calculate the force required to keep the belt moving.

Newton's Second Law (Equations of Motion)

$$\sum F = m\frac{dv}{dt} = \frac{\left(250 \frac{\text{kg}}{\text{min}}\right)\left(0.98 \frac{\text{m}}{\text{s}} - 0 \frac{\text{m}}{\text{s}}\right)}{60 \frac{\text{s}}{\text{min}}}$$

$$= 4.083 \text{ N} \quad (4.1 \text{ N})$$

The answer is (A).

5. *Customary U.S. Solution*

Apply Newton's second law of motion for constant mass.

Newton's Second Law (Equations of Motion)

$$\sum F = m\frac{dv}{dt}$$

Rearranging, the equation for the impulse is

$$Fdt = \frac{m\Delta v}{g_c} = \frac{m(v_2 - v_1)}{g_c}$$

In this case, the final and initial velocities are in opposite directions.

$$v_2 = 130 \text{ ft/sec}$$
$$v_1 = -90 \text{ ft/sec}$$

The impulse is

$$\frac{m(v_2 - v_1)}{g_c} = \frac{(0.4 \text{ lbm})\left(130 \frac{\text{ft}}{\text{sec}} - \left(-90 \frac{\text{ft}}{\text{sec}}\right)\right)}{32.174 \frac{\text{lbm-ft}}{\text{lbf-sec}^2}}$$

$$= 2.73 \text{ lbf-sec} \quad (2.7 \text{ lbf-sec})$$

SI Solution

Apply Newton's second law of motion for constant mass.

Newton's Second Law (Equations of Motion)

$$\sum F = m\frac{dv}{dt}$$

Rearranging, the equation for the impulse is

$$Fdt = m\Delta v = m(v_2 - v_1)$$

In this case, the final and initial velocities are in opposite directions.

$$v_2 = 40 \text{ m/s}$$
$$v_1 = -30 \text{ m/s}$$

The impulse is

$$m(v_2 - v_1) = (0.2 \text{ kg})\left(40 \text{ } \frac{\text{m}}{\text{s}} - \left(-30 \text{ } \frac{\text{m}}{\text{s}}\right)\right)$$
$$= 14 \text{ N·s}$$

The answer is (B).

6. *Customary U.S. Solution*

Since momentum is always conserved, the momentum of the projectile is transferred to the momentum of the gun, but in the opposite direction.

$$\frac{m_{\text{gun}} v_{\text{gun}}}{g_c} = \frac{-m_{\text{proj}} v_{\text{proj}}}{g_c}$$

$$v_{\text{gun}} = \frac{-m_{\text{proj}} v_{\text{proj}}}{m_{\text{gun}}} = \frac{-(2.6 \text{ lbm})\left(2100 \text{ } \frac{\text{ft}}{\text{sec}}\right)}{1000 \text{ lbm}}$$
$$= -5.46 \text{ ft/sec} \quad (5.5 \text{ ft/sec})$$

SI Solution

Since momentum is always conserved, the momentum of the projectile is transferred to the momentum of the gun but in the opposite direction.

$$m_{\text{gun}} v_{\text{gun}} = -m_{\text{proj}} v_{\text{proj}}$$

$$v_{\text{gun}} = -\frac{m_{\text{proj}} v_{\text{proj}}}{m_{\text{gun}}} = \frac{-(1.2 \text{ kg})\left(650 \text{ } \frac{\text{m}}{\text{s}}\right)}{500 \text{ kg}}$$
$$= -1.56 \text{ m/s} \quad (1.6 \text{ m/s})$$

The answer is (C).

7. *Customary U.S. Solution*

Due to conservation of momentum, the momentum before impact will be equal to the momentum after impact.

Coefficient of Restitution
$$m_1 v_1 + m_2 v_2 = m_1 v_1' + m_2 v_2'$$

Since the bullet is embedded in the block, the bullet and the block travel together with the same velocity. The equation becomes

$$\frac{m_{\text{bullet}} v_{\text{bullet}}}{g_c} = \frac{m_{(\text{bullet+block})} v_{(\text{bullet+block})}}{g_c}$$

Calculate the velocity after the impact.

$$v_{(\text{bullet+block})} = \frac{m_{\text{bullet}} v_{\text{bullet}}}{m_{(\text{bullet+block})}} = \frac{(0.15 \text{ lbm})\left(2300 \text{ } \frac{\text{ft}}{\text{sec}}\right)}{0.15 \text{ lbm} + 9 \text{ lbm}}$$
$$= 37.7 \text{ ft/sec} \quad (38 \text{ ft/sec})$$

SI Solution

Due to conservation of momentum, the momentum before impact will be equal to the momentum after impact.

Coefficient of Restitution
$$m_1 v_1 + m_2 v_2 = m_1 v_1' + m_2 v_2'$$

Since the bullet is embedded in the block, the bullet and the block travel together with the same velocity. The equation becomes

$$m_{\text{bullet}} v_{\text{bullet}} = m_{(\text{bullet+block})} v_{(\text{bullet+block})}$$

Calculate the velocity after the impact.

$$v_{(\text{bullet+block})} = \frac{m_{\text{bullet}} v_{\text{bullet}}}{m_{(\text{bullet+block})}} = \frac{(0.06 \text{ kg})\left(700 \text{ } \frac{\text{m}}{\text{s}}\right)}{0.06 \text{ kg} + 4.5 \text{ kg}}$$
$$= 9.21 \text{ m/s} \quad (9.2 \text{ m/s})$$

The answer is (B).

8. *Customary U.S. Solution*

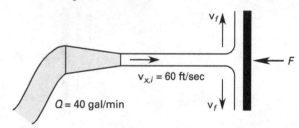

From the impulse-momentum principle, the external force on the water jet is equal to the rate of change of momentum of the jet.

Impulse-Momentum Principle
$$\sum F = \sum Q_2 \rho_2 v_2 - \sum Q_1 \rho_1 v_1$$

From the conservation of mass principle, the mass flow rate of the jet is constant.

Continuity Equation
$$\dot{m} = \rho Q = \rho_1 Q_1 = \rho_2 Q_2 = \text{constant}$$

Apply the impulse-momentum principle in the direction of the jet (the x-direction)

$$F_x = \frac{\dot{m}(v_{2x} - v_{1x})}{g_c}$$

Calculate the mass flow rate. The density of water at standard conditions is 62.4 lbm/ft³. [Properties of Water at Standard Conditions]

$$\dot{m} = \rho Q = \frac{\left(62.4 \ \frac{\text{lbm}}{\text{ft}^3}\right)\left(40 \ \frac{\text{gal}}{\text{min}}\right)}{\left(60 \ \frac{\text{sec}}{\text{min}}\right)\left(7.48 \ \frac{\text{gal}}{\text{ft}^3}\right)}$$

$$= 5.56 \ \text{lbm/sec}$$

Since the jet comes to a standstill after impact, the x-component of the velocity after the jet hits the wall is 0 ft/sec. The x-component of the velocity before the jet hits the wall is 60 ft/sec.

Substitute the known values in the equation for the impulse-momentum principle.

$$F_x = \frac{\left(5.56 \ \frac{\text{lbm}}{\text{sec}}\right)\left(0 \ \frac{\text{ft}}{\text{sec}} - 60 \ \frac{\text{ft}}{\text{sec}}\right)}{32.174 \ \frac{\text{lbm-ft}}{\text{lbf-sec}^2}}$$

$$= -10.36 \ \text{lbf} \quad (10 \ \text{lbf})$$

SI Solution

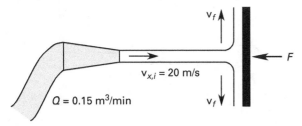

From the impulse-momentum principle, the external force on the water jet is equal to the rate of change of momentum of the jet.

Impulse-Momentum Principle

$$\sum F = \sum Q_2 \rho_2 v_2 - \sum Q_1 \rho_1 v_1$$

From the conservation of mass principle, the mass flow rate of the jet is constant.

Continuity Equation

$$\dot{m} = \rho Q = \rho_1 Q_1 = \rho_2 Q_2 = \text{constant}$$

Apply the impulse-momentum principle in the direction of the jet (the x-direction).

$$F_x = \dot{m}(v_{2x} - v_{1x})$$

Calculate the mass flow rate. The density of water at standard conditions is 1000 kg/m³. [Properties of Water at Standard Conditions]

$$\dot{m} = \rho Q = \frac{\left(1000 \ \frac{\text{kg}}{\text{m}^3}\right)\left(0.15 \ \frac{\text{m}^3}{\text{min}}\right)}{60 \ \frac{\text{s}}{\text{min}}}$$

$$= 2.5 \ \text{kg/s}$$

Since the jet comes to a standstill after impact, the x-component of the velocity after the jet hits the wall is 0 m/s. The x-component of the velocity before the jet hits the wall is 20 m/s.

Substitute the known values in the equation.

$$F_x = \left(2.5 \ \frac{\text{kg}}{\text{s}}\right)\left(0 \ \frac{\text{m}}{\text{s}} - 20 \ \frac{\text{m}}{\text{s}}\right)$$

$$= -50 \ \text{N} \quad (50 \ \text{N})$$

The answer is (A).

58 Mechanisms and Power Transmission Systems

Content in blue refers to the NCEES Handbook.

PRACTICE PROBLEMS

1. The lever shown rotates about point O. Arms OA and OB are fixed in relation to one another. The length of OA is 3 ft. About point O, the angular velocity of the lever, ω, is 10 rad/sec, and the angular acceleration of the lever, α, is -20 rad/sec^2. What is most nearly the magnitude of the total acceleration of point A?

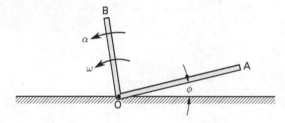

(A) 60 ft/sec^2
(B) 310 ft/sec^2
(C) 350 ft/sec^2
(D) 410 ft/sec^2

2. The lever shown rotates freely about point O. Arms OA and OB are fixed in relation to one another. The length of OA is 4 ft. The angular velocity of the lever, ω, is 100 rad/sec, and the angular acceleration of the lever, α, is constant at -5 rad/sec^2. What is most nearly the distance traveled by point A before the rotational speed is reduced to zero?

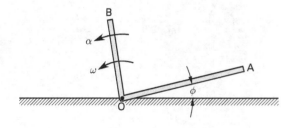

(A) 2000 ft
(B) 3000 ft
(C) 4000 ft
(D) 5000 ft

3. The lever shown is free to rotate about point O. Arms OA and OB are fixed in relation to one another. The length of OA is 1.5 ft, and the length of OB is 2.5 ft. At a particular moment, the velocity of point A, is $-70.5\mathbf{i} - 25.5\mathbf{j}$ ft/sec. Considering clockwise rotation as positive, what is most nearly the velocity of point B at that moment?

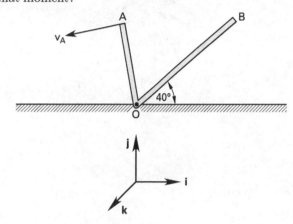

(A) $-77.4\mathbf{i} - 111\mathbf{j}$ ft/sec
(B) $-80.5\mathbf{i} + 96.0\mathbf{j}$ ft/sec
(C) $-98.4\mathbf{i} - 42.5\mathbf{j}$ ft/sec
(D) $-105\mathbf{i} + 48.5\mathbf{j}$ ft/sec

4. A flywheel is designed as a solid disk 1 in thick and 20 in in diameter, with a concentric 4 in diameter mounting hole. It is manufactured from cast iron with an ultimate tensile strength of 30,000 lbf/in^2 and a Poisson's ratio of 0.27. The density of the cast iron is 0.26 lbm/in^3. Using a factor of safety of 10, the maximum safe speed for the flywheel is most nearly

(A) 1700 rpm
(B) 2200 rpm
(C) 3500 rpm
(D) 8700 rpm

5. A simple epicyclic gearbox with one planet has gears with 24, 40, and 104 teeth on the sun, planet, and internal ring gears, respectively. The sun rotates clockwise at

50 rpm. The ring gear is fixed. The rotational velocity of the planet carrier is most nearly

(A) 9.4 rpm
(B) 12 rpm
(C) 17 rpm
(D) 23 rpm

6. Refer to the epicyclic gear set illustrated. Gear A rotates counterclockwise on a fixed center at 100 rpm. The ring gear rotates. The sun gear is fixed. Each gear has the number of teeth indicated. What is most nearly the rotational speed of the planet carrier?

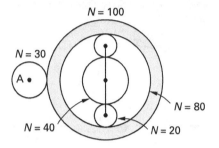

(A) 20 rpm
(B) 50 rpm
(C) 90 rpm
(D) 120 rpm

7. An epicyclic gear train consists of a ring gear, three planets, and a fixed sun gear. 15 hp are transmitted through the input ring gear, which turns clockwise at 1500 rpm. The diametral pitch is 10 per inch with a 20° pressure angle. The pitch diameters are at 5 in, 2½ in, and 10 in for the sun, planet, and ring gear, respectively.

What are most nearly the velocities of the sun, ring, and carrier gears, respectively?

(A) 0 rpm, 1000 rpm, 1500 rpm
(B) 0 rpm, 1500 rpm, 1000 rpm
(C) 1000 rpm, 1500 rpm, 1500 rpm
(D) 1500 rpm, 0 rpm, 1500 rpm

8. An epicyclic gear train consists of a ring gear, three planets, and a fixed sun gear. 15 hp (11 kW) are transmitted through the input ring gear, which turns clockwise at 1500 rpm. The diametral pitch is 10 per inch with a 20° pressure angle. The pitch diameters are at 5 in, 2½ in, and 10 in (127 mm, 63.5 mm, 254 mm) for the sun, planet, and ring gear, respectively. The velocities of the sun, carrier, and ring gears are 0 rpm, 1000 rpm, 1500 rpm, respectively. Approximately what torques are on the input and output shafts, respectively?

(A) 50 ft-lbf (70 N·m); 80 ft-lbf (110 N·m)
(B) 60 ft-lbf (80 N·m); 50 ft-lbf (70 N·m)
(C) 80 ft-lbf (110 N·m); 60 ft-lbf (80 N·m)
(D) 90 ft-lbf (120 N·m); 80 ft-lbf (110 N·m)

9. Bars r_0, r_1, r_2, and r_3 constitute a four-bar linkage as shown. With the origin, O, at the base of bar r_3, and positive directions upward and to the right, what are the coordinates of the rotating end of r_3 (point B) when θ_1 is 45°?

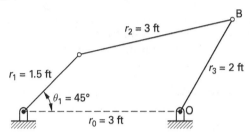

(A) $x = 1.0$ ft; $y = 1.7$ ft
(B) $x = 0.88$ ft; $y = 1.5$ ft
(C) $x = 0.67$ ft; $y = 1.5$ ft
(D) $x = 0.64$ ft; $y = 1.9$ ft

10. Bars r_0, r_1, r_2, and r_3 constitute a four-bar linkage, as shown. At a particular moment, θ_1 is 45°, θ_2 is 13.1°, ω_1 is 150 rad/sec, and ω_2 is -27.2 rad/sec. Using the shown unit vectors $\mathbf{i}$, $\mathbf{j}$, and $\mathbf{k}$, what is most nearly the velocity of point B at that moment?

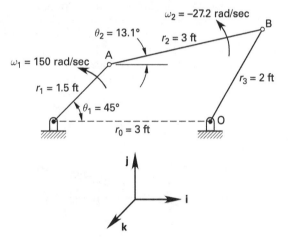

(A) 110 ft/sec∠140°
(B) 130 ft/sec∠150°
(C) 150 ft/sec∠150°
(D) 160 ft/sec∠150°

SOLUTIONS

1. The magnitude of the total acceleration is found by summing the components of normal acceleration, a_n, and tangential acceleration, a_t.

$$a_n = -r\omega^2 = -(3\text{ ft})\left(10\ \frac{\text{rad}}{\text{sec}}\right)^2$$
$$= -300\text{ ft/sec}^2$$

$$a_t = r\alpha = (3\text{ ft})\left(-20\ \frac{\text{rad}}{\text{sec}^2}\right) = -60\text{ ft/sec}^2$$

$$a = \sqrt{\left(-300\ \frac{\text{ft}}{\text{sec}^2}\right)^2 + \left(-60\ \frac{\text{ft}}{\text{sec}^2}\right)^2}$$
$$= 306\text{ ft/sec}^2\ (310\text{ ft/sec}^2)$$

The answer is (B).

2. Use the equation for the angular velocity of the lever, ω. When the rotational speed is zero, $\omega = 0$. Rearranging the equation to solve for time, t,

Mass Radius of Gyration

$$\omega = \omega_0 + \alpha_c t$$

$$t = \frac{-\omega_0}{\alpha_c} = \frac{-100\ \frac{\text{rad}}{\text{sec}}}{-5\ \frac{\text{rad}}{\text{sec}^2}} = 20\text{ sec}$$

The angular distance traveled by the lever, θ, is

Mass Radius of Gyration

$$\theta = \theta_0 + \omega_0 t + \tfrac{1}{2}\alpha_c t^2$$
$$= 0 + \left(100\ \frac{\text{rad}}{\text{sec}}\right)(20\text{ sec}) + \left(\tfrac{1}{2}\right)\left(-5\ \frac{\text{rad}}{\text{sec}^2}\right)(20\text{ sec})^2$$
$$= 1000\text{ rad}$$

The distance traveled by point A, L_A, is

$$L_A = \theta L = (1000\text{ rad})(4\text{ ft}) = 4000\text{ ft}$$

The answer is (C).

3. Use the traditional definitions of unit vectors **i**, **j**, and **k**. The rotation is counterclockwise since both components of the velocity are negative.

The magnitude of the angular velocity of the lever, ω, is found from the equation for the tangential velocity.

Plane Circular Motion

$$v_t = r\omega$$

$$\omega = \frac{v}{L}$$
$$= \frac{\sqrt{v_x^2 + v_y^2}}{L}$$
$$= \frac{\sqrt{\left(-70.5\ \frac{\text{ft}}{\text{sec}}\right)^2 + \left(-25.5\ \frac{\text{ft}}{\text{sec}}\right)^2}}{1.5\text{ ft}}$$
$$= 50\text{ rad/sec}$$

The position coordinates of OB_x and OB_y are

$$OB_x = OB\cos\theta = (2.5\text{ ft})\cos 40° = 1.92\text{ ft}$$
$$OB_y = OB\sin\theta = (2.5\text{ ft})\sin 40° = 1.61\text{ ft}$$

Unit vector **k** is oriented along the rotational axis. The velocity of point B, v_B, is found from the cross product of the angular velocity and the link length.

$$\mathbf{v}_B = \omega \times \mathbf{OB}$$

Calculate the vector cross product.

$$\mathbf{v}_B = \begin{vmatrix} \mathbf{i} & \mathbf{j} & \mathbf{k} \\ 0 & 0 & 50\text{ rad/sec} \\ 1.92\text{ ft} & 1.61\text{ ft} & 0 \end{vmatrix}$$
$$= -80.5\mathbf{i} + 96.0\mathbf{j}\text{ ft/sec}$$

The answer is (B).

4. Treat the flywheel as a rotating ring ($r_o < = 10\ t$, which is satisfied in this case). First, consider the maximum tangential stress.

For a factor of safety of 10,

$$\sigma_{t,\max} = \frac{S_{ut}}{\text{FS}} = \frac{30{,}000\ \frac{\text{lbf}}{\text{in}^2}}{10} = 3000\text{ lbf/in}^2$$

Use the equation for stress in a rotating ring. Set $r = r_i$ for the maximum tangential stress, then solve for the angular velocity.

Rotating Rings

$$\sigma_t = \rho\omega^2\left(\frac{3+v}{8}\right)\left(r_i^2 + r_o^2 + \frac{r_i^2 r_o^2}{r^2} - \frac{1+3v}{3+v}r^2\right)$$

$$= \rho\omega^2\left(\frac{3+v}{8}\right)\left(r_i^2 + 2r_o^2 - \frac{1+3v}{3+v}r_i^2\right)$$

$$\omega = \sqrt{\frac{4\sigma_t}{\rho\left((3+v)r_o^2 + (1-v)r_i^2\right)}}$$

$$= \sqrt{\frac{(4)\left(3000\,\frac{\text{lbf}}{\text{in}^2}\right)\left(32.2\,\frac{\text{lbm-ft}}{\text{lbf-sec}^2}\right)\left(\frac{12\,\text{in}}{\text{ft}}\right)}{\left(0.26\,\frac{\text{lbm}}{\text{in}^3}\right)\left((3+0.27)\left(\frac{20\,\text{in}}{2}\right)^2 + (1-0.27)r\left(\frac{4\,\text{in}}{2}\right)^2\right)}}$$

$$= 232.5 \text{ rad/sec}$$

Convert to revolutions per minute to find the maximum safe speed based on the tangential stress.

$$n = \frac{\omega}{2\pi\,\frac{\text{rad}}{\text{rev}}} = \frac{\left(232.5\,\frac{\text{rad}}{\text{sec}}\right)\left(60\,\frac{\text{sec}}{\text{min}}\right)}{2\pi\,\frac{\text{rad}}{\text{rev}}} = 2220 \text{ rpm}$$

Use the equation for stress in a rotating ring and set $r = \sqrt{r_o r_i}$ for the maximum radial stress, then solve for the angular velocity.

Rotating Rings

$$\sigma_r = \rho\omega^2\left(\frac{3+v}{8}\right)\left(r_i^2 + r_o^2 - \frac{r_i^2 r_o^2}{r^2} - r^2\right)$$

$$= \rho\omega^2\left(\frac{3+v}{8}\right)\left(r_i^2 + r_o^2 - \frac{r_i^2 r_o^2}{r_o r_i} - r_o r_i\right)$$

$$= \rho\omega^2\left(\frac{3+v}{8}\right)(r_o - r_i)^2$$

$$\omega = \sqrt{\frac{\sigma_r}{\rho\left((3+v)(r_0-r_i)^2\right)}}$$

$$= \sqrt{\frac{(8)\left(3000\,\frac{\text{lbf}}{\text{in}^2}\right)\left(12\,\frac{\text{in}}{\text{ft}}\right)\left(32.2\,\frac{\text{lbm-ft}}{\text{lbf-sec}^2}\right)}{\left(0.26\,\frac{\text{lbm}}{\text{in}^3}\right)\left((3+0.27)\left(\frac{20\,\text{in}}{2} - \frac{4\,\text{in}}{2}\right)^2\right)}}$$

$$= 412.8 \text{ rad/sec}$$

Convert to revolutions per minute to find the maximum safe speed due to the radial stress.

$$n = \frac{\omega}{2\pi} = \frac{\left(412.8\,\frac{\text{rad}}{\text{sec}}\right)\left(60\,\frac{\text{sec}}{\text{min}}\right)}{2\pi\,\frac{\text{rad}}{\text{rev}}} = 3942 \text{ rpm}$$

The flywheel's maximum safe speed is limited by tangential stress and is 2220 rpm (2200 rpm).

The answer is (B).

5. Find the rotation of the driver. The sun is the driving gear, therefore $A = 24$. The planet is the driven gear, therefore $B = 40$. The internal ring is fixed, therefore $C = 104$.

Types of Planetary Gears

$$D = 1 + \frac{C}{A} = 1 + \frac{104}{24} = 5.333$$

Rearrange the equation for D as a function of rotational motion to find the speed of the carrier.

$$D = \frac{\omega_{\text{driver}}}{\omega_{\text{follower}}} = \frac{\omega_{\text{sun}}}{\omega_{\text{carrier}}}$$

$$\omega_{\text{carrier}} = \frac{\omega_{\text{sun}}}{D}$$

$$= \frac{50\,\frac{\text{rev}}{\text{min}}}{5.333}$$

$$= 9.375 \text{ rpm} \quad (9.4 \text{ rpm})$$

The answer is (A).

6. Since the ring gear rotates in a different direction from gear A, the rotational velocity of the ring gear is

$$\omega_{\text{ring}} = \omega_A\left(-\frac{N_A}{N_{\text{ring}}}\right) = \left(-100\,\frac{\text{rev}}{\text{min}}\right)\left(-\frac{30\,\text{teeth}}{100\,\text{teeth}}\right)$$

$$= 30 \text{ rev/min} \quad \text{[clockwise]}$$

Find the rotation of the driver. The internal ring is the driving gear, therefore $A = 80$. The planet is the driven gear, therefore $B = 20$. The internal ring is fixed, therefore $C = 40$.

Types of Planetary Gears

$$D = 1 + \frac{C}{A} = 1 + \frac{40}{80} = 1.5$$

Rearrange the equation for D as a function of rotational motion to find the speed of the carrier.

$$D = \frac{\omega_{\text{driver}}}{\omega_{\text{follower}}} = \frac{\omega_{\text{ring}}}{\omega_{\text{carrier}}}$$

$$\omega_{\text{carrier}} = \frac{\omega_{\text{ring}}}{D}$$

$$= \frac{30 \, \frac{\text{rev}}{\text{min}}}{1.5}$$

$$= 20 \text{ rpm}$$

The answer is (A).

7. The diametral pitch equation can be modified to calculate the number of teeth.

Involute Gear Tooth Nomenclature

$$p_d = \frac{N}{D}$$

$$N = p_d D$$

For the sun gear,

$$N_{\text{sun}} = p_d = \left(10 \, \frac{\text{teeth}}{\text{in}}\right)(5 \text{ in}) = 50 \text{ teeth}$$

For the planet gears,

$$N_{\text{planet}} = p_d = \left(10 \, \frac{\text{teeth}}{\text{in}}\right)(2.5 \text{ in}) = 25 \text{ teeth}$$

For the ring gear,

$$N_{\text{ring}} = p_d = \left(10 \, \frac{\text{teeth}}{\text{in}}\right)(10 \text{ in}) = 100 \text{ teeth}$$

Since the sun gear is fixed, its rotational speed is zero ($\omega_{\text{sun}} = 0$ rpm).

The rotational speed of the ring gear is $\omega_{\text{ring}} = 1500$ rpm.

Since the ring and sun gears rotate in different directions, the train value is negative and can be found by calculating the ratio between N_{ring} and N_{sun}.

$$\text{TV} = -\frac{N_{\text{ring}}}{N_{\text{sun}}} = -\frac{100 \text{ teeth}}{50 \text{ teeth}} = -2$$

The rotational speed of the sun gear can be found by solving the equation for the angular velocity for speed of the sun gear,

$$\omega_{\text{sun}} = (\text{TV})\omega_{\text{ring}} + (1 - \text{TV})\omega_{\text{carrier}}$$

$$0 \, \frac{\text{rev}}{\text{min}} = (-2)\left(1500 \, \frac{\text{rev}}{\text{min}}\right) + (1 - (-2))\omega_{\text{carrier}}$$

Solve for the rotational speed of the carrier gear.

$$\omega_{\text{carrier}} = \frac{(2)\left(1500 \, \frac{\text{rev}}{\text{min}}\right)}{3} = 1000 \text{ rpm}$$

The answer is (B).

8. *Customary U.S. Solution*

The torque on the input shaft can be found using shaft horsepower equation.

Shaft-Horsepower Relationship and Force-Horsepower Relationship

$$\text{HP} = \frac{Tn}{63{,}025}$$

$$T_{\text{in,ft-lbf}} = \frac{(\text{HP})\left(63{,}025 \, \frac{\text{in-lbf}}{\text{hp-min}}\right)}{n_{\text{rpm,ring}}}$$

$$= \frac{(15 \text{ hp})\left(63{,}025 \, \frac{\text{in-lbf}}{\text{hp-min}}\right)}{\left(1500 \, \frac{\text{rev}}{\text{min}}\right)\left(12 \, \frac{\text{in}}{\text{ft}}\right)}$$

$$= 52.52 \text{ ft-lbf} \quad (50 \text{ ft-lbf})$$

The torque on the output shaft is

Shaft-Horsepower Relationship and Force-Horsepower Relationship

$$\text{HP} = \frac{Tn}{63{,}025}$$

$$T_{\text{out,ft-lbf}} = \frac{(\text{HP})\left(63{,}025 \, \frac{\text{in-lbf}}{\text{hp-min}}\right)}{n_{\text{carrier}}}$$

$$= \frac{(15 \text{ hp})\left(63{,}025 \, \frac{\text{in-lbf}}{\text{hp-min}}\right)}{\left(1000 \, \frac{\text{rev}}{\text{min}}\right)\left(12 \, \frac{\text{in}}{\text{ft}}\right)}$$

$$= 78.78 \text{ ft-lbf} \quad (80 \text{ ft-lbf})$$

SI Solution

The equation for the torque on the input shaft is

Shaft-Horsepower Relationship and Force-Horsepower Relationship

$$\text{HP} = \frac{Tn}{63{,}025}$$

For SI units, this can be rewritten as

$$\text{HP} = \frac{Tn}{9549 \dfrac{\text{N·m}}{\text{kW·min}}}$$

The torque on the input shaft is

$$T_{\text{in,N·m}} = \frac{(\text{HP})\left(9549 \dfrac{\text{N·m}}{\text{kW·min}}\right)}{n_{\text{rpm,ring}}}$$

$$= \frac{(11 \text{ kW})\left(9549 \dfrac{\text{N·m}}{\text{kW·min}}\right)}{1500 \dfrac{\text{rev}}{\text{min}}}$$

$$= 70.03 \text{ N·m} \quad (70 \text{ N·m})$$

The torque on the output shaft is

$$T_{\text{out,N·m}} = \frac{(\text{HP})\left(9549 \dfrac{\text{N·m}}{\text{kW·min}}\right)}{n_{\text{carrier}}}$$

$$= \frac{(11 \text{ kW})\left(9549 \dfrac{\text{N·m}}{\text{kW·min}}\right)}{1000 \dfrac{\text{rev}}{\text{min}}}$$

$$= 105 \text{ N·m} \quad (110 \text{ N·m})$$

The answer is (A).

9. Use a diagonal line to divide the four-bar linkage into two triangles as shown.

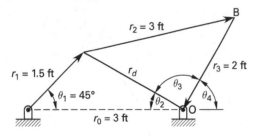

From the law of cosines, the length of the diagonal, r_d, is

$$r_d = \sqrt{r_0^2 + r_1^2 - 2r_0 r_1 \cos\theta_1}$$

$$= \sqrt{(3 \text{ ft})^2 + (1.5 \text{ ft})^2 - (2)(3 \text{ ft})(1.5 \text{ ft})\cos 45°}$$

$$= 2.2 \text{ ft}$$

Use the law of sines to solve for θ_2.

$$\frac{\sin\theta_2}{r_1} = \frac{\sin\theta_1}{r_d}$$

$$\sin\theta_2 = \frac{r_1 \sin\theta_1}{r_d} = \frac{(1.5 \text{ ft})\sin 45°}{2.2 \text{ ft}} = 0.482$$

$$\theta_2 = 28.82°$$

Use the law of cosines to solve for θ_3.

$$\cos\theta_3 = \frac{r_2^2 - r_d^2 - r_3^2}{-2r_d r_3} = \frac{(3 \text{ ft})^2 - (2.2 \text{ ft})^2 - (2 \text{ ft})^2}{(-2)(2.2 \text{ ft})(2 \text{ ft})}$$

$$= -0.018$$

$$\theta_3 = 91.03°$$

Angle θ_4 is

$$\theta_4 = 180° - \theta_3 - \theta_2 = 180° - 91.03° - 28.82°$$

$$= 60.15°$$

The coordinates for the rotating end of bar r_3 are

$$x = r_3 \cos\theta_4 = (2 \text{ ft})\cos 60.15° = 1.00 \text{ ft} \quad (1.0 \text{ ft})$$
$$y = r_3 \sin\theta_4 = (2 \text{ ft})\sin 60.15° = 1.73 \text{ ft} \quad (1.7 \text{ ft})$$

The answer is (A).

10. With respect to the base of bar r_1, the coordinates of point A are

$$x = r_1 \cos\theta_1 = (1.5 \text{ ft})\cos 45° = 1.06 \text{ ft}$$
$$y = r_1 \sin\theta_1 = (1.5 \text{ ft})\sin 45° = 1.06 \text{ ft}$$

With respect to the base of bar r_2, the coordinates of point B are

$$x = r_2 \cos\theta_2 = (3 \text{ ft})\cos 13.1° = 2.92 \text{ ft}$$
$$y = r_2 \cos\theta_2 = (3 \text{ ft})\sin 13.1° = 0.68 \text{ ft}$$

Expressing both velocities in terms of the unit vectors, the velocity of point B, $\mathbf{v}_B$, is the simple sum of the two vector velocities.

$$\mathbf{v}_B = \mathbf{v}_1 + \mathbf{v}_2 = (\boldsymbol{\omega}_1 \times \mathbf{r}_1) + (\boldsymbol{\omega}_2 \times \mathbf{r}_2)$$

Take the vector cross products.

$$\mathbf{v}_B = \begin{vmatrix} \mathbf{i} & \mathbf{j} & \mathbf{k} \\ 0 & 0 & 150 \text{ rad/sec} \\ 1.06 \text{ ft} & 1.06 \text{ ft} & 0 \end{vmatrix}$$
$$+ \begin{vmatrix} \mathbf{i} & \mathbf{j} & \mathbf{k} \\ 0 & 0 & -27.2 \text{ rad/sec} \\ 2.92 \text{ ft} & 0.68 \text{ ft} & 0 \end{vmatrix}$$
$$= -140.5\mathbf{i} \text{ ft/sec} + 79.6\mathbf{j} \text{ ft/sec}$$
$$= 161 \text{ ft/sec} \angle 150.5° \quad (160 \text{ ft/sec} \angle 150°)$$

The answer is (D).

59 Vibrating Systems

Content in blue refers to the NCEES Handbook.

PRACTICE PROBLEMS

1. When an 8 lbm (3.6 kg) mass is attached on the end of a spring, the spring stretches 5.9 in (150 mm). A dashpot with a damping coefficient of 0.50 lbf-sec/ft (7.3 N·s/m) opposes movement of the mass. The mass is initially at rest. The natural frequency is most nearly

(A) 0.4 rad/sec (0.4 rad/s)
(B) 2.0 rad/sec (2.0 rad/s)
(C) 8.0 rad/sec (8.0 rad/s)
(D) 65 rad/sec (65 rad/s)

2. When an 8 lbm (3.6 kg) mass is attached on the end of a spring, the spring stretches 5.9 in (150 mm). A dashpot with a damping coefficient of 0.50 lbf-sec/ft (7.3 N·s/m) opposes movement of the mass. The mass is initially at rest. The damping ratio is most nearly

(A) 0.02
(B) 0.08
(C) 0.1
(D) 0.4

3. A uniform bar with a mass of 5 lbm (2.3 kg) carries a concentrated mass of 3 lbm (1.4 kg) at its free end. The bar is hinged at one end and supported by an outboard spring as shown.

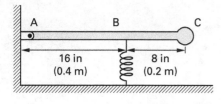

The deflection of the spring from its unstressed position is 0.55 in (1.4 cm). The angle of the defection is 0.034 radian. There is no damping. The natural linear frequency of vibration is most nearly

(A) 4 Hz
(B) 7 Hz
(C) 20 Hz
(D) 50 Hz

4. A 300 lbm (140 kg) electromagnet at the end of a cable holds 200 lbm (90 kg) of scrap metal. The total equivalent stiffness of the cable and crane boom is 1000 lbf/in (175 kN/m). The current to the electromagnet is cut off suddenly, and the scrap falls away. Neglect damping. The frequency of oscillation of the electromagnet is most nearly

(A) 5.7 Hz
(B) 10 Hz
(C) 23 Hz
(D) 84 Hz

5. A simple steel beam of thickness 1.5 in (40 mm) supports two equal concentrated loads of 15,000 lbf (70 kN) each as shown in the figure. The width of the beam is 10 in (250 mm).

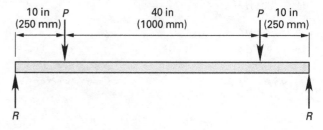

The modulus of elasticity of the steel is 2.9×10^7 lbf/in² (200 GPa). The natural linear frequency of vibration in the vertical direction is most nearly

(A) 1.6 Hz
(B) 2.7 Hz
(C) 3.5 Hz
(D) 5.6 Hz

6. A 175 lbm (80 kg), single-cylinder air compressor is mounted on four identical, equally loaded corner springs. The motor turns at 1200 rpm. During each cycle, a disturbing force is generated by a 3.6 lbm (1.6 kg) imbalance acting at a radius of 3 in (75 mm). Damping is insignificant. It is desired that the dynamic force transmitted to the base be limited to 5% of the disturbing force. The individual spring stiffness needed is most nearly

(A) 44 lbf/in (7.5 kN/m)

(B) 68 lbf/in (12 kN/m)

(C) 85 lbf/in (15 kN/m)

(D) 170 lbf/in (29 kN/m)

7. A hydraulic oil pump compresses a cork mounting pad by 0.02 in (0.51 mm) when it is not running. The pump is turned at 1725 rpm. Neglect damping. The magnification factor of the mounting pad is increased by most nearly

(A) 25%

(B) 45%

(C) 75%

(D) 100%

8. A length of square structural steel tubing (modulus of elasticity is 29×10^6 lbf/in^2) is used as a flagpole. The tube measures 4 in × 4 in externally, and the wall thickness is 0.125 in. The flagpole is fixed at the bottom and extends vertically 22 ft. The total flagpole mass is 150 lbm. The flagpole has a damping ratio of 0.119. The damped linear natural frequency of the flagpole is most nearly

(A) 1.2 Hz

(B) 2.0 Hz

(C) 5.0 Hz

(D) 7.0 Hz

9. A mass of 100 lbm (45 kg) is supported uniformly by a spring system. The spring system has a combined stiffness of 1200 lbf/ft (17.5 kN/m). A dashpot has been installed.

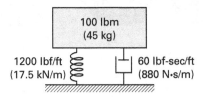

The undamped natural frequency is most nearly

(A) 3.0 rad/sec (3.0 rad/s)

(B) 5.0 rad/sec (5.0 rad/s)

(C) 13 rad/sec (13 rad/s)

(D) 20 rad/sec (20 rad/s)

SOLUTIONS

1. *Customary U.S. Solution*

The natural frequency is

Free Vibration
$$\omega_n = \sqrt{\frac{g}{\delta_{st}}}$$
$$= \sqrt{\frac{\left(32.17 \frac{\text{ft}}{\text{sec}^2}\right)\left(12 \frac{\text{in}}{\text{ft}}\right)}{5.9 \text{ in}}}$$
$$= 8.08 \text{ rad/s} \quad (8.0 \text{ rad/s})$$

SI Solution

The natural frequency is

Free Vibration
$$\omega_n = \sqrt{\frac{g}{\delta_{st}}}$$
$$= \sqrt{\frac{9.807 \frac{\text{m}}{\text{s}^2}}{0.15 \text{ m}}}$$
$$= 8.08 \text{ rad/s} \quad (8.0 \text{ rad/s})$$

The answer is (C).

2. *Customary U.S. Solution*

Calculate the spring constant by using the equation for free vibration and the static deflection.

Free Vibration
$$mg = k\delta_{st}$$
$$k = \frac{m\left(\frac{g}{g_c}\right)}{\delta_{st}}$$
$$= \left(\frac{8 \text{ lbm}}{5.9 \text{ in}}\right)\left(\frac{32.17 \frac{\text{ft}}{\text{sec}^2}}{32.17 \frac{\text{lbm-ft}}{\text{lbf-sec}^2}}\right)$$
$$= 1.356 \text{ lbf/in}$$

Use the equation for the critical damping coefficient to find the damping ratio.

Torsional Vibration
$$c_{\text{critical}} = 2m\omega_n = 2m\sqrt{\frac{k}{\frac{m}{g_c}}} = 2\sqrt{\frac{km}{g_c}}$$

The damping ratio is

Vibration Transmissibility, Base Motion
$$\zeta = \frac{c}{c_{\text{critical}}}$$
$$= \frac{c}{2\sqrt{\frac{km}{g_c}}}$$
$$= \frac{0.50 \frac{\text{lbf-sec}}{\text{ft}}}{(2)\sqrt{\frac{\left(1.356 \frac{\text{lbf}}{\text{in}}\right)\left(12 \frac{\text{in}}{\text{ft}}\right)(8 \text{ lbm})}{32.17 \frac{\text{lbm-ft}}{\text{lbf-sec}^2}}}}$$
$$= 0.124 \quad (0.1)$$

SI Solution

Calculate the spring constant by using the equation for free vibration and the static deflection.

Free Vibration
$$mg = k\delta_{st}$$
$$k = \frac{mg}{\delta_{st}}$$
$$= \frac{(3.6 \text{ kg})\left(9.807 \frac{\text{m}}{\text{s}^2}\right)}{0.15 \text{ m}}$$
$$= 235.4 \text{ N/m}$$

Use the equation for the critical damping coefficient to find the damping ratio.

Torsional Vibration
$$c_{\text{critical}} = 2m\omega_n = (2m)\sqrt{\frac{k}{m}} = 2\sqrt{km}$$

The damping ratio is

Vibration Transmissibility, Base Motion
$$\zeta = \frac{c}{c_{\text{critical}}}$$
$$= \frac{c}{2\sqrt{km}}$$
$$= \frac{7.3 \frac{\text{N s}}{\text{m}}}{2\sqrt{\left(235.4 \frac{\text{N}}{\text{m}}\right)(3.6 \text{ kg})}}$$
$$= 0.125 \quad (0.1)$$

The answer is (C).

3. *Customary U.S. Solution*

First, consider static equilibrium. The bar mass is considered concentrated at 12 in (24 in/2) from the hinge.

$$\sum M_A = 0$$

$$(3 \text{ lbm}) \left(\frac{32.17 \frac{\text{ft}}{\text{sec}^2}}{32.17 \frac{\text{lbm-ft}}{\text{lbf-sec}^2}} \right) (24 \text{ in})$$

$$+ (5 \text{ lbm}) \left(\frac{32.17 \frac{\text{ft}}{\text{sec}^2}}{32.17 \frac{\text{lbm-ft}}{\text{lbf-sec}^2}} \right) (12 \text{ in}) - M_{\text{spring}} = 0$$

$$M_{\text{spring}} = 132 \text{ in-lbf}$$

Find the equivalent torsional spring constant using the relation between the deflection and the moment. [Mechanical Springs]

$$k_t = \frac{M_{\text{spring}}}{\theta} = \frac{132 \text{ in-lbf}}{0.034 \text{ rad}} = 3837 \text{ in-lbf}$$

Find the mass moment of inertia of the bar rotating about its end.

Mass and Mass Moments of Inertia of Geometric Shapes

$$I_{\text{bar}} = \frac{mL^2}{3} = \frac{(5 \text{ lbm})(24 \text{ in})^2}{3} = 960 \text{ lbm-in}^2$$

The mass moment of inertia of the concentrated mass is

$$I_{\text{mass}} = mL^2 = (3 \text{ lbm})(24 \text{ in})^2 = 1728 \text{ lbm-in}^2$$

The total mass moment of inertia of the system is

$$I = I_{\text{bar}} + I_{\text{mass}} = 960 \text{ lbm-in}^2 + 1728 \text{ lbm-in}^2$$
$$= 2688 \text{ lbm-in}^2$$

The natural frequency is

Torsional Vibration

$$\omega_n = \sqrt{\frac{k_t g_c}{I}}$$

$$= \sqrt{\frac{(3837 \text{ in-lbf}) \left(32.17 \frac{\text{lbm-ft}}{\text{lbf-sec}^2} \right) \left(12 \frac{\text{in}}{\text{ft}} \right)}{2688 \text{ lbm-in}^2}}$$

$$= 23.49 \text{ rad/sec}$$

The undamped natural period of vibration is expressed by

Free Vibration

$$\tau_n = \frac{2\pi}{\omega_n} = \frac{1}{f}$$

Use the equation for the undamped natural period of vibration to find the natural linear frequency.

$$\frac{1}{f} = \frac{2\pi}{\omega_n}$$

$$f = \frac{\omega_n}{2\pi}$$

$$= \frac{23.49 \frac{\text{rad}}{\text{sec}}}{2\pi \frac{\text{rad}}{\text{rev}}}$$

$$= 3.74 \text{ Hz} \quad (4 \text{ Hz})$$

SI Solution

First, consider static equilibrium. The bar mass is considered concentrated at 0.3 m (0.6 m/2) from the hinge.

$$\sum M_A = 0$$

$$(1.4 \text{ kg}) \left(9.807 \frac{\text{m}}{\text{s}^2} \right) (0.6 \text{ m})$$

$$+ (2.3 \text{ kg}) \left(9.807 \frac{\text{m}}{\text{s}^2} \right) (0.3 \text{ m}) - M_{\text{spring}} = 0$$

$$M_{\text{spring}} = 15.01 \text{ N·m}$$

The equivalent torsional spring constant is found using the relation between the deflection and moment. [Mechanical Springs]

$$k_t = \frac{M_{\text{spring}}}{\theta} = \frac{15.01 \text{ N·m}}{0.034 \text{ rad}} = 428.9 \text{ N·m}$$

Find the mass moment of inertia of the bar rotating about its end.

Mass and Mass Moments of Inertia of Geometric Shapes

$$I_{\text{bar}} = \frac{mL^2}{3} = \frac{(2.3 \text{ kg})(0.60 \text{ m})^2}{3}$$
$$= 0.276 \text{ kg·m}^2$$

The mass moment of inertia of the concentrated mass is

$$I_{\text{mass}} = mL^2 = (1.4 \text{ kg})(0.60 \text{ m})^2 = 0.504 \text{ kg·m}^2$$

The total mass moment of inertia of the system is

$$I = I_{bar} + I_{mass} = 0.276 \text{ kg·m}^2 + 0.504 \text{ kg·m}^2$$
$$= 0.78 \text{ kg·m}^2$$

The natural frequency is

Torsional Vibration

$$\omega_n = \sqrt{\frac{k_t}{I}}$$
$$= \sqrt{\frac{428.9 \text{ N·m}}{0.78 \text{ kg·m}^2}} = 23.45 \text{ rad/s}$$

The undamped natural period of vibration is expressed by

Free Vibration

$$\tau_n = \frac{2\pi}{\omega_n} = \frac{1}{f}$$

Use the equation for the undamped natural period of vibration to find the natural linear frequency.

$$\frac{1}{f} = \frac{2\pi}{\omega_n}$$
$$f = \frac{\omega_n}{2\pi}$$
$$= \frac{23.45 \frac{\text{rad}}{\text{s}}}{2\pi \frac{\text{rad}}{\text{rev}}}$$
$$= 3.73 \text{ Hz} \quad (4 \text{ Hz})$$

The answer is (A).

4. *Customary U.S. Solution*

The static deflection caused by the electromagnet is

Free Vibration

$$mg = k\delta_{st}$$
$$\delta_{st} = \frac{m\left(\frac{g}{g_c}\right)}{k}$$
$$= \left(\frac{300 \text{ lbm}}{1000 \frac{\text{lbf}}{\text{in}}}\right)\left(\frac{32.17 \frac{\text{ft}}{\text{sec}^2}}{32.17 \frac{\text{lbm-ft}}{\text{lbf-sec}^2}}\right)$$
$$= 0.3 \text{ in}$$

The undamped natural period of vibration is expressed by

Free Vibration

$$\tau_n = \frac{2\pi}{\omega_n} = \frac{1}{f}$$

The natural linear frequency is

$$\frac{1}{f} = \frac{2\pi}{\omega_n}$$
$$f = \left(\frac{1}{2\pi}\right)\omega_n$$
$$= \left(\frac{1}{2\pi}\right)\sqrt{\frac{g}{\delta_{st}}}$$
$$= \left(\frac{1}{2\pi}\right)\sqrt{\frac{\left(32.17 \frac{\text{ft}}{\text{sec}^2}\right)\left(12 \frac{\text{in}}{\text{ft}}\right)}{0.3 \text{ in}}}$$
$$= 5.71 \text{ Hz} \quad (5.7 \text{ Hz})$$

SI Solution

The static deflection caused by the electromagnet is

Free Vibration

$$mg = k\delta_{st}$$
$$\delta_{st} = \frac{mg}{k} = \frac{(140 \text{ kg})\left(9.807 \frac{\text{m}}{\text{s}^2}\right)}{\left(175 \frac{\text{kN}}{\text{m}}\right)\left(1000 \frac{\text{N}}{\text{kN}}\right)}$$
$$= 0.00785 \text{ m}$$

The undamped natural period of vibration is expressed by

Free Vibration

$$\tau_n = \frac{2\pi}{\omega_n} = \frac{1}{f}$$

The natural linear frequency is

$$\frac{1}{f} = \frac{2\pi}{\omega_n}$$
$$f = \left(\frac{1}{2\pi}\right)\omega_n$$
$$= \left(\frac{1}{2\pi}\right)\sqrt{\frac{g}{\delta_{st}}}$$
$$= \left(\frac{1}{2\pi}\right)\sqrt{\frac{9.807 \frac{\text{m}}{\text{s}^2}}{0.00785 \text{ m}}}$$
$$= 5.63 \text{ Hz} \quad (5.7 \text{ Hz})$$

The answer is (A).

5. *Customary U.S Solution*

Calculate the moment of inertia of the beam.

$$I_x = \frac{bh^3}{12} = \frac{(10 \text{ in})(1.5 \text{ in})^3}{12}$$
$$= 2.8125 \text{ in}^4$$

Calculate the length of the beam.

$$L = 40 \text{ in} + 2(10 \text{ in}) = 60 \text{ in}$$

Calculate the deflection at the center of the span [Bending Moment, Vertical Shear, and Deflection of Beams of Uniform Cross Section, Under Various Conditions of Loading]

$$y = \frac{Pd}{24EI}(3l^2 - 4d^2)$$
$$= \frac{(15000 \text{ lbf})(10 \text{ in})}{(24)\left(2.9 \times 10^7 \frac{\text{lbf}}{\text{in}^2}\right)(2.8125 \text{ in}^4)}(3(60 \text{ in})^2 - 4(10 \text{ in})^2)$$
$$= 0.7969 \text{ in}$$

The static deflection is

$$\delta_{st} = y = 0.7969 \text{ in} = (0.7969 \text{ in})\left(\frac{1 \text{ ft}}{12 \text{ in}}\right)$$
$$= 0.0664 \text{ ft}$$

Calculate the undamped natural circular frequency of the beam [Free Vibration].

$$\omega_n = \sqrt{\frac{g}{\delta_{st}}} = \sqrt{\frac{32.2 \frac{\text{ft}}{\text{sec}^2}}{0.0664 \text{ ft}}} = 22.02 \text{ rad/sec}$$

Calculate the natural linear frequency of vibration.

$$f = \frac{\omega_n}{2\pi} = \frac{22.02 \frac{\text{rad}}{\text{sec}}}{2\pi} = 3.504 \text{ Hz} \quad (3.5 \text{ Hz})$$

SI Solution

Content in blue refers to the *NCEES Handbook*.

Calculate the moment of inertia of the beam.

$$I_x = \frac{bh^3}{12} = \left(\frac{(250 \text{ mm})(40 \text{ mm})^3}{12}\right)\left(\frac{1 \text{ m}}{1000 \text{ mm}}\right)^4$$
$$= 1.33 \times 10^{-6} \text{ m}^4$$

Calculate the length of the beam.

$$L = 1000 \text{ mm} + 2(250 \text{ mm}) = 1500 \text{ mm} = 1.5 \text{ m}$$

Calculate the deflection at the center of the span [Bending Moment, Vertical Shear, and Deflection of Beams of Uniform Cross Section, Under Various Conditions of Loading]

$$y = \frac{Pd}{24EI}(3l^2 - 4d^2)$$
$$= \frac{(70 \text{ kN})(0.25 \text{ m})}{(24)(200 \times 10^6 \text{ kPa})(1.33 \times 10^{-6} \text{ m}^4)}(3(1.5 \text{ m})^2 - 4(0.25 \text{ m})^2)$$
$$= 0.0178 \text{ m}$$

The static deflection is, $\delta_{st} = y = 0.0178 \text{ m}$

Calculate the undamped natural circular frequency of the beam [Free Vibration].

$$\omega_n = \sqrt{\frac{g}{\delta_{st}}} = \sqrt{\frac{9.81 \frac{\text{m}}{\text{sec}^2}}{0.0178 \text{ m}}} = 23.48 \text{ rad/sec}$$

Calculate the natural linear frequency of vibration.

$$f = \frac{\omega_n}{2\pi} = \frac{23.48 \frac{\text{rad}}{\text{sec}}}{2\pi} = 3.74 \text{ Hz} \quad (3.5 \text{ Hz})$$

The answer is (C).

6. *Customary U.S. Solution*

The forcing frequency is

$$\omega_f = \frac{\left(1200 \frac{\text{rev}}{\text{min}}\right)\left(2\pi \frac{\text{rad}}{\text{rev}}\right)}{60 \frac{\text{sec}}{\text{min}}} = 125.7 \text{ rad/sec}$$

Find the out-of-balance force caused by the rotating imbalance. [Newton's Second Law (Equations of Motion)] [Plane Circular Motion]

$$F_o = \frac{m_o \omega^2 r}{g_c} = \frac{(3.6 \text{ lbm})\left(125.7 \frac{\text{rad}}{\text{sec}}\right)^2 (3 \text{ in})}{\left(32.17 \frac{\text{lbm-ft}}{\text{lbf-sec}^2}\right)\left(12 \frac{\text{in}}{\text{ft}}\right)}$$
$$= 441.6 \text{ lbf} \quad (440 \text{ lbf})$$

The transmissibility is [Vibration Transmissibility, Base Motion]

$$TR = \frac{|F_{\text{transmitted}}|}{F_{\text{applied}}} = 0.05$$

Find the principal (positive) transmissibility for negligible damping.

Forced Vibration Under Harmonic Force

$$TR = \frac{1}{1 - \left(\frac{\omega_f}{\omega_n}\right)^2}$$

$$\frac{\omega_f}{\omega_n} = \sqrt{\frac{1}{TR} + 1} = \sqrt{\frac{1}{0.05} + 1} = 4.5826$$

The required natural frequency is

$$\omega_n = \frac{\omega_f}{4.5826} = \frac{125.7 \frac{\text{rad}}{\text{sec}}}{4.5826}$$
$$= 27.43 \text{ rad/sec}$$

The required stiffness of the system is

Free Vibration

$$\omega_n = \sqrt{\frac{k_{eq}}{m}}$$

$$k_{eq} = \frac{m\omega_n^2}{g_c}$$

$$= \frac{(175 \text{ lbm})\left(27.43 \frac{\text{rad}}{\text{sec}}\right)^2}{\left(32.17 \frac{\text{lbm-ft}}{\text{lbf-sec}^2}\right)\left(12 \frac{\text{in}}{\text{ft}}\right)}$$

$$= 340.8 \text{ lbf/in}$$

For four identical springs in parallel, the required stiffness for an individual spring is

$$k = \frac{k_{eq}}{4} = \frac{340.8 \frac{\text{lbf}}{\text{in}}}{4}$$
$$= 85.2 \text{ lbf/in} \quad (85 \text{ lbf/in})$$

SI Solution

The forcing frequency is

$$\omega_f = \frac{\left(1200 \frac{\text{rev}}{\text{min}}\right)\left(2\pi \frac{\text{rad}}{\text{rev}}\right)}{60 \frac{\text{s}}{\text{min}}} = 125.7 \text{ rad/s}$$

Find the out-of-balance force caused by the rotating imbalance. [Newton's Second Law (Equations of Motion)] [Plane Circular Motion]

$$F_o = m_o\omega^2 r = (1.6 \text{ kg})\left(125.7 \frac{\text{rad}}{\text{s}}\right)^2 (0.075 \text{ m})$$
$$= 1896 \text{ N} \quad (1900 \text{ N})$$

The transmissibility is [Vibration Transmissibility, Base Motion]

$$TR = \frac{|F_{\text{transmitted}}|}{F_{\text{applied}}} = 0.05$$

Find the principal (positive) transmissibility for negligible damping.

Forced Vibration Under Harmonic Force

$$TR = \frac{1}{1 - \left(\frac{\omega_f}{\omega_n}\right)^2}$$

$$\frac{\omega_f}{\omega_n} = \sqrt{\frac{1}{TR} + 1} = \sqrt{\frac{1}{0.05} + 1} = 4.5826$$

The required natural frequency is

$$\omega_n = \frac{\omega_f}{4.5826} = \frac{125.7 \frac{\text{rad}}{\text{s}}}{4.5826}$$
$$= 27.43 \text{ rad/s}$$

The required stiffness of the system is

Free Vibration

$$\omega_n = \sqrt{\frac{k}{m}}$$

$$k_{eq} = m\omega_n^2$$

$$= (80 \text{ kg})\left(27.43 \frac{\text{rad}}{\text{s}}\right)^2$$

$$= 60\,192 \text{ N/m}$$

For four identical springs in parallel, the required stiffness for an individual spring is

$$k = \frac{k_{eq}}{4} = \frac{60\,192 \frac{\text{N}}{\text{m}}}{4}$$
$$= 15\,048 \text{ N/m} \quad (15 \text{ kN/m})$$

The answer is (C).

7. *Customary U.S Solution*

The forcing frequency is

$$\omega = (2\pi)\left(\frac{1725\,\frac{\text{rev}}{\text{min}}}{\frac{60\text{ s}}{\text{min}}}\right) = 181\text{ rad/s}$$

The natural frequency is given by

Free Vibration

$$\omega_n = \sqrt{\frac{g}{\delta_{st}}} = \sqrt{\frac{32.2\,\frac{\text{ft}}{\text{sec}^2}}{\frac{0.02\text{ in}}{\frac{12\text{ in}}{1\text{ ft}}}}} = 139\text{ rad/s}$$

Calculate the frequency ratio.

Forced Vibration Under Harmonic Force

$$r = \frac{\omega}{\omega_n} = \frac{181\,\frac{\text{rad}}{\text{s}}}{139\,\frac{\text{rad}}{\text{s}}} = 1.302$$

Find the magnification factor, with negligible damping.

Forced Vibration Under Harmonic Force

$$M = \frac{1}{\sqrt{(1-r^2)^2}} = \frac{1}{\sqrt{(1-1.302^2)^2}} = 1.44$$

This is a 44% increase in force. The closest answer option is 45%.

SI Solution

The forcing frequency is

$$\omega = (2\pi)\left(\frac{1725\,\frac{\text{rev}}{\text{min}}}{\frac{60\text{ s}}{\text{min}}}\right) = 181\text{ rad/s}$$

The natural frequency is given by

Free Vibration

$$\omega_n = \sqrt{\frac{g}{\delta_{st}}} = \sqrt{\frac{9.81\,\frac{\text{m}}{\text{s}^2}}{\frac{0.51\text{ mm}}{\frac{1000\text{ mm}}{1\text{ m}}}}} = 139\text{ rad/s}$$

Calculate the frequency ratio.

Forced Vibration Under Harmonic Force

$$r = \frac{\omega}{\omega_n} = \frac{181\,\frac{\text{rad}}{\text{s}}}{139\,\frac{\text{rad}}{\text{s}}} = 1.302$$

Find the magnification factor, with negligible damping.

Forced Vibration Under Harmonic Force

$$M = \frac{1}{\sqrt{(1-r^2)^2}} = \frac{1}{\sqrt{(1-1.302^2)^2}} = 1.44$$

This is a 44% (45%) increase in force.

The answer is (B).

8. Find the moment of inertia of the flagpole.

Properties of Various Shapes

$$I = \frac{bh^3}{12}$$

$$= \frac{(b_o h_o^3 - b_i h_i^3)}{12}$$

$$= \left(\frac{1}{12}\right)\begin{pmatrix}(4\text{ in})(4\text{ in})^3 - (4\text{ in} - (2)(0.125\text{ in}))\\ \times (4\text{ in} - (2)(0.125\text{ in}))^3\end{pmatrix}$$

$$= 4.854\text{ in}^4$$

The flagpole will deflect due to its own weight since there is no external load. The stiffness of the flagpole will be the weight of the flagpole (simulated as a cantilever beam with uniform load) divided by the deflection.

For a simple cantilever with distributed weight, $w = W/L$, the stiffness can be calculated as W/δ. [Bending Moment, Vertical Shear, and Deflection of Beams of Uniform Cross Section, Under Various Conditions of Loading]

$$k = \frac{W}{\delta} = \frac{W}{\frac{wL^4}{8EI}} = \frac{W}{\frac{WL^3}{8EI}}$$

$$= \frac{8EI}{L^3}$$

$$= \frac{(8)\left(29 \times 10^6\,\frac{\text{lbf}}{\text{in}^2}\right)(4.854\text{ in}^4)}{\left((22\text{ ft})\left(12\,\frac{\text{in}}{\text{ft}}\right)\right)^3}$$

$$= 61.20\text{ lbf/in}$$

Only the linear natural frequency, f, has units of Hz (not rad/sec). The linear natural frequency is

Free Vibration
$$\frac{1}{f} = \frac{2\pi}{\omega_n}$$

$$f = \frac{\omega_n}{2\pi} = \frac{1}{2\pi}\sqrt{\frac{kg_c}{m}}$$

$$= \frac{1}{2\pi}\sqrt{\frac{\left(61.20\ \frac{\text{lbf}}{\text{in}}\right)\left(32.17\ \frac{\text{lbm-ft}}{\text{lbf-sec}^2}\right)\left(12\ \frac{\text{in}}{\text{ft}}\right)}{150\ \text{lbm}}}$$

$$= 1.998\ \text{Hz}$$

The damped linear natural frequency is

Torsional Vibration
$$\omega_d = \omega_n\sqrt{1 - \zeta^2}$$
$$2\pi f_d = 2\pi f\sqrt{1 - \zeta^2}$$
$$f_d = f\sqrt{1 - \zeta^2}$$
$$= (1.998\ \text{Hz})\sqrt{1 - (0.119)^2}$$
$$= 1.984\ \text{Hz}\quad(2.0\ \text{Hz})$$

The answer is (B).

9. *Customary U.S. Solution*

Write the system differential equation for a force input, f.

Forced Vibration Under Harmonic Force
$$m\ddot{x} + c\dot{x} + kx = F_0\cos\omega t = f$$

Divide by m to write the equation in terms of natural frequency and damping factor.

$$x'' + \left(\frac{c}{m}\right)x' + \left(\frac{k}{m}\right)x = \frac{f}{m}$$

$$x'' + 2\zeta\omega_n x' + \omega_n^2 x = \frac{f}{\frac{k}{\omega_n^2}} = \omega_n^2\left(\frac{f}{k}\right)$$

Define the forcing function as $h = f/k$.

The system equation is a second-order linear differential equation with constant coefficients, and the forcing function is a step of height h.

$$x'' + 2\zeta\omega_n x' + \omega_n^2 x = \omega_n^2 h$$

The undamped natural frequency is

Free Vibration
$$\omega_n = \sqrt{\frac{kg_c}{m}} = \sqrt{\frac{\left(1200\ \frac{\text{lbf}}{\text{ft}}\right)\left(32.2\ \frac{\text{lbm-ft}}{\text{lbf-sec}^2}\right)}{100\ \text{lbm}}}$$

$$= 19.66\ \text{rad/sec}\quad(20\ \text{rad/sec})$$

SI Solution

The steps to derive the system equation are the same as in the customary U.S. solution.

The undamped natural frequency is

Free Vibration
$$\omega_n = \sqrt{\frac{k}{m}} = \sqrt{\frac{17\,500\ \frac{\text{N}}{\text{m}}}{45\ \text{kg}}} = 19.7\ \text{rad/s}\quad(20\ \text{rad/s})$$

The answer is (D).

60 Modeling of Engineering Systems

Content in blue refers to the NCEES Handbook.

PRACTICE PROBLEMS

1. A system diagram for a system of ideal elements is shown.

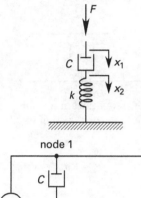

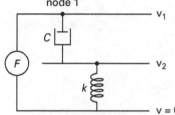

What is the differential equation for node 1 and node 2, respectively?

(A) $F = (x_1' - x_2');\ F = kx_1$

(B) $F = (x_1 - x_2);\ F = kx_1 + kx_2$

(C) $F = C(x_1' - x_2');\ F = kx_2$

(D) $F = C(x_2' - x_1');\ F = kx_2 - kx_1$

2. A system diagram and a system of ideal elements are shown.

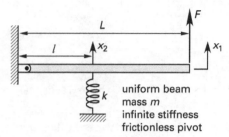

What is the differential equation for the system?

(A) $FL = \left(\frac{1}{5}mL^2\right)\theta - kl^2\theta$

(B) $FL = \left(\frac{1}{5}mL^2\right)\theta'' + kl^2\theta$

(C) $FL = \left(\frac{1}{3}mL^2\right)\theta'' + kl^2\theta$

(D) $FL = \left(\frac{1}{3}mL^2\right)\theta' - kl^2\theta$

3. The coupling of a railroad car is modeled as the mechanical system shown. Assume all elements are linear. What are the system equations that describe the positions x_1 and x_2 as functions of time?

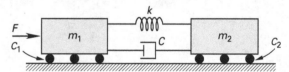

(A) $F = m_1 x_1' + C_1 x_1' + C(x_1' + x_2') + k(x_1 - x_2);$
$0 = C_2 x_2' + m_2 x_2'' + C(x_1' - x_2') + k(x_2 + x_1)$

(B) $F = m_1 x_1'' + C_1 x_1'' + C(x_2' + x_1') + k(x_1 + x_2);$
$0 = C_2 x_2' + m_2 x_2'' + C(x_1' - x_2') + k(x_2 + x_1)$

(C) $F = m_1 x_1'' + C_1 x_1'' + C(x_1' - x_2') + k(x_1 - x_2);$
$0 = C_2 x_2'' + m_2 x_2' + C(x_1' + x_2') + k(x_2 - x_1)$

(D) $F = m_1 x_1'' + C_1 x_1' + C(x_1' - x_2') + k(x_1 - x_2);$
$0 = C_2 x_2' + m_2 x_2'' + C(x_2' - x_1') + k(x_2 - x_1)$

4. Water is discharged freely at a constant rate into an open tank. Water flows out of the tank through a drain with a resistance to flow.

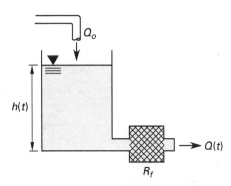

Using idealized elements, the system can be diagrammed as shown.

The fluid capacitance of the water in the tank is C_f. One end of each of the two energy sources, Q_1 and Q_2, connects to $p = 0$. What is the differential equation for node 1 and node 2, respectively, that describes the response of the system?

(A) $Q_1 = C_{f_1}\left(\dfrac{dp_1}{dt}\right) + \left(\dfrac{1}{R_f}\right)(p_1 - p_2);$

$Q_2 = \left(\dfrac{1}{R_f}\right)(p_1 - p_2)$

(B) $Q_1 = C_{f_1}\left(\dfrac{dp_1}{dt}\right) - \left(\dfrac{1}{R_f}\right)(p_2 + p_1);$

$Q_2 = \left(\dfrac{1}{C_{f_1}}\right)(p_2 + p_1)$

(C) $Q_1 = C_{f_1}\left(\dfrac{dp_2}{dt}\right) + \left(\dfrac{1}{R_f}\right)(p_2 - p_1);$

$Q_2 = \left(\dfrac{1}{R_f}\right)(p_1 + p_2)$

(D) $Q_1 = C_{f_1}\left(\dfrac{dp_1}{dt}\right) - \left(\dfrac{1}{R_f}\right)(p_1 - p_2);$

$Q_2 = \left(\dfrac{1}{C_{f_1}}\right)(p_1 - p_2)$

5. Water is pumped into the bottom of an open tank.

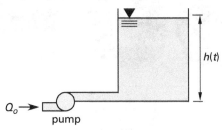

The fluid resistance in the entrance pipe is R_f. The pressure at the entrance is p_1, and the pressure in the tank is p_2. The pressure at the top of the tank is $p = 0$. Using idealized elements, the system can be diagrammed as shown.

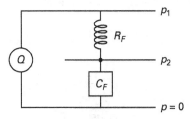

What is the differential equation for the resistor and the capacitor, respectively, that describes the response of the system?

(A) $Q = \dfrac{p_1 + p_2}{C_f}; Q = R_f\left(\dfrac{dp_2 - dp_1}{dt}\right)$

(B) $Q = \dfrac{p_2 - p_1}{R_f}; Q = C_f\left(\dfrac{dp_1 + dp_2}{dt}\right)$

(C) $Q = \dfrac{p_1 - p_2}{C_f}; Q = C_f\left(\dfrac{dp_1 + dp_2}{dt}\right)$

(D) $Q = \dfrac{p_1 - p_2}{R_f}; Q = C_f\left(\dfrac{dp_2}{dt}\right)$

SOLUTIONS

1. The velocity of the plunger is v_1. The velocity of the body of the damper is the same as the upper part of the spring, v_2. The other end of the force and the spring is attached to the stationary wall at $v = 0$. The force from the source is the same force experienced by the dashpot. One of the system equations is based on node 1. Expanding with Eq. 60.5,

$$F = F_C = C(v_1 - v_2) = C(x_1' - x_2')$$

The force from the source is the same force experienced by the spring. A second system equation is based on node 2. Expanding with Eq. 60.4,

$$F = F_k = k(x_2 - 0) = kx_2$$

The answer is (C).

2. Treat this as a rotational system. The applied rotational torque is

$$T = FL$$

The equivalent torsional spring constant is

$$k_r = \frac{M_{\text{resisting}}}{\theta} = \frac{F_k l}{\theta} = \frac{kx_2 l}{\theta}$$

However, $x_2 = l \sin\theta$ and $\theta \approx \sin\theta$ for small angles.

$$k_r = kl^2$$

The moment of inertia of the beam about the hinge point is

$$I = \tfrac{1}{3}mL^2$$

The equivalent rotational system is

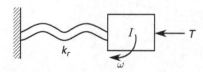

The angular velocity of the end of the spring connected to the inertial element is ω. The other end of the spring is attached to the stationary wall at $\omega = 0$.

The torque leaving the source splits: some of it goes through I and some of it goes through k_r. The conservation law is written to conserve torque in the ω line.

$$T = T_I + T_{k_r}$$

Expanding with Eq. 60.6 and Eq. 60.7,

$$T = I\alpha + k_r(\theta - 0)$$
$$FL = \left(\tfrac{1}{3}mL^2\right)\theta'' + kl^2\theta$$

The answer is (C).

3. The velocity of the end of the spring connected to m_1 is v_1. This is also the velocity of the plunger and the velocity of the viscous damper, C_1. The velocity of the end of the spring connected to m_2 is v_2. This is also the velocity of the body of the damper and the velocity of the viscous damper, C_2. The other end of each mass connects to $v = 0$. The system diagram is

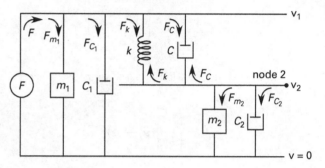

The force leaving the source splits: some of it goes through m_1, C_1, k, and C. The conservation law is written to conserve force in the v_1 line. The equation is based on node 1.

$$F = F_{m_1} + F_{C_1} + F_k + F_C$$

Expanding with Eq. 60.3, Eq. 60.4, and Eq. 60.5,

$$F = m_1 a_1 + C_1(v_1 - 0) + C(v_1 - v_2) + k(x_1 - x_2)$$
$$= m_1 x_1'' + C_1 x_1' + C(x_1' - x_2') + k(x_1 - x_2)$$

The same conservation principle based on node 2 is used to conserve force in the v_2 line. Expanding with Eq. 60.3, Eq. 60.4, and Eq. 60.5,

$$0 = F_{C_2} + F_{m_2} + F_C + F_k$$
$$= C_2(v_2 - 0) + m_2 a_2 + C(v_2 - v_1) + k(x_2 - x_1)$$
$$= C_2 x_2' + m_2 x_2'' + C(x_2' - x_1') + k(x_2 - x_1)$$

The answer is (D).

4. From Eq. 60.10, the flow through the capacitor is

$$Q_{C_{f_1}} = C_{f_1}\left(\frac{dp_1}{dt}\right)$$

From Eq. 60.12, the flow through the resistor is

$$Q_{R_f} = \frac{p_1 - p_2}{R_f}$$

One of the system equations is based on conservation of flow at node 1.

$$\begin{aligned}Q_1 &= Q_{C_{f_1}} + Q_{R_f} \\ &= C_{f_1}\left(\frac{dp_1}{dt}\right) + \left(\frac{1}{R_f}\right)(p_1 - p_2)\end{aligned}$$

The second system equation is based on conservation of flow at node 2.

$$Q_2 = \left(\frac{1}{R_f}\right)(p_1 - p_2)$$

The answer is (A).

5. The source flow, Q, is the same flow through the resistor and the capacitor. Use Eq. 60.12 for the resistor.

$$Q = \frac{p_1 - p_2}{R_f}$$

Use Eq. 60.10 for the capacitor.

$$Q = C_f\left(\frac{dp_2}{dt}\right)$$

The answer is (D).

61 Analysis of Engineering Systems

Content in blue refers to the *NCEES Handbook*.

PRACTICE PROBLEMS

1. A block diagram is shown.

Simplifying, the overall system gain is

(A) $\dfrac{ABC}{1+BC-ABD}$

(B) $\dfrac{AB}{1-BC-AD}$

(C) $\dfrac{BC}{1+AC+ABD}$

(D) $\dfrac{ABD}{1-ABD}$

2. A block diagram is shown.

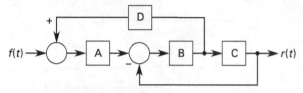

Simplifying, the overall system gain is

(A) $\dfrac{ABC}{BC-ABD}$

(B) $\dfrac{AEC}{1+AD-ABD}$

(C) $\dfrac{BC}{1+BC-ABD}$

(D) $\dfrac{ABC}{1-BE+ABD}$

3. A constant-speed motor/magnetic clutch drive train is monitored and controlled by a speed-sensing tachometer. The entire system is modeled as a control system block diagram, as shown. (The lowercase letters represent small-signal increments from the reference values.) When the control system is operating, the desired motor speed, n (in rpm), is set with a speed-setting potentiometer. The setting is compared to the tachometer output. The comparator output error (in volts), controls the clutch. A current, i (in amps), passes through the clutch coil. The external load torque, $t_L m$ (in in-lbf), is seen by the clutch and is countered by the clutch output torque, t (in in-lbf). See *Illustration for problem 3, 4, and 5.*

4. A constant-speed motor/magnetic clutch drive train is monitored and controlled by a speed-sensing tachometer. The entire system is modeled as a control system block diagram, as shown. (The lowercase letters represent small-signal increments from the reference values.) When the control system is operating, the desired motor speed, n (in rpm), is set with a speed-setting potentiometer. The setting is compared to the tachometer output. The comparator output error (in volts), controls the clutch. A current, i (in amps), passes through the clutch coil. The external load torque, $t_L m$ (in in-lbf), is seen by the clutch and is countered by the clutch output torque, t (in in-lbf). See *Illustration for problem 3, 4, and 5*

Assume that you have to select the comparator gain and that it doesn't have to be 0.1. Using the root-locus method or either the Routh or Nyquist stability criterion, the limits of the comparator gain that cause the closed-loop system to be unstable are most nearly

(A) $1+250K > 0, K > -0.004$

(B) $1+250K < 1, K > -0.002$

(C) $1+500K > 0, K > -0.002$

(D) $1+500K < 0, K > -0.004$

Illustration for problem 3, 4, and 5

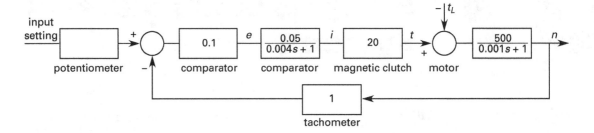

5. A constant-speed motor/magnetic clutch drive train is monitored and controlled by a speed-sensing tachometer. The entire system is modeled as a control system block diagram, as shown. (The lowercase letters represent small-signal increments from the reference values.) When the control system is operating, the desired motor speed, n (in rpm), is set with a speed-setting potentiometer. The setting is compared to the tachometer output. The comparator output error (in volts), controls the clutch. A current, i (in amps), passes through the clutch coil. The external load torque, $t_L m$ (in in-lbf), is seen by the clutch and is countered by the clutch output torque, t (in in-lbf). See *Illustration for Problems 3 and 4*.

What kind of control is necessary to improve the steady-state response of the closed-loop system to constant disturbances in the load torque?

(A) integral

(B) logic

(C) on-off

(D) proportional

SOLUTIONS

1. Draw the block diagram.

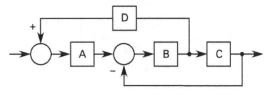

From Fig. 61.5, use case 7 to move the extreme right pick-off point to the left of C.

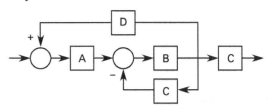

Use case 6 to combine the two summing points on the left.

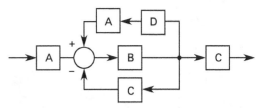

Use case 1 to combine boxes in series in the upper feedback loop.

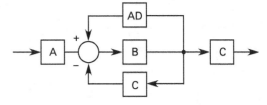

Use case 2 to combine the two feedback loops.

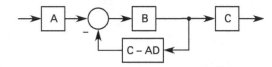

Use case 3 to simplify the remaining feedback loop.

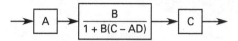

Use case 1 to combine boxes in series to determine the system gain.

$$G_{\text{loop}} = \frac{ABC}{1 + BC - ABD}$$

The answer is (A).

2. Draw the block diagram.

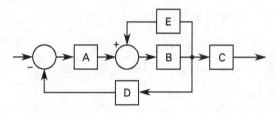

From Fig. 61.5, use case 6 to combine the two summing points on the left.

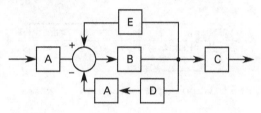

Use case 1 to combine boxes in series in the lower feedback loop.

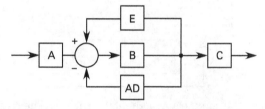

Use case 2 to combine the two feedback loops.

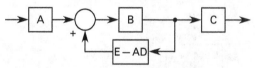

Use case 3 to simplify the remaining feedback loop.

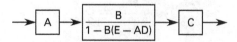

Use case 1 to combine boxes in series to determine the system gain.

$$G_{\text{loop}} = \frac{ABC}{1 - BE + ABD}$$

The answer is (D).

3. First, redraw the system in more traditional form.

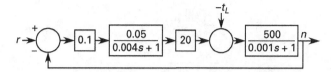

From Fig. 61.5, use case 1 to combine boxes in series.

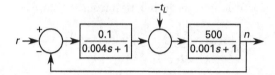

Use case 5 to combine the two summing points.

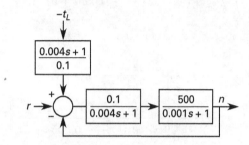

Use case 1 to combine boxes in series.

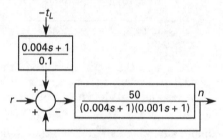

From Fig. 61.5, use case 3 to simplify the rules for simplifying block diagrams, simplify the feedback loop.

$$G_c(s) = \frac{\dfrac{50}{(0.004s+1)(0.001s+1)}}{1 + \dfrac{50}{(0.004s+1)(0.001s+1)}}$$

$$= \frac{50}{(0.004s+1)(0.001s+1) + 50}$$

$$= \frac{50}{(4 \times 10^{-6})s^2 + 0.005s + 51}$$

The closed-loop transfer function from r to n is

$$T_1(s) = \frac{N(s)}{R(s)} = G_c(s)$$

$$= \frac{50}{(4 \times 10^{-6})s^2 + 0.005s + 51}$$

The closed-loop transfer function from $-t_L$ to n is

$$T_2(s) = \frac{N(s)}{-T_L(s)} = \left(\frac{0.004s+1}{0.1}\right) G_c(s)$$

$$= \frac{(500)(0.004s+1)}{(4 \times 10^{-6})s^2 + 0.005s + 51}$$

The response to a step change in input t_L is

$$N(s) = -T_L(s) \left(\frac{0.004s+1}{0.1}\right) G_c(s)$$

$$= \left(\frac{-T_L}{s}\right) \left(\frac{(500)(0.004s+1)}{(4 \times 10^{-6})s^2 + 0.005s + 51}\right)$$

$$= \frac{-(1.25 \times 10^8)\left(\frac{T_L}{s}\right)(0.004s+1)}{s^2 + 1250s + (1.275 \times 10^7)}$$

The response to a step change in t_L will be similar to a step change in r. However, the numerator, $0.004s + 1$, will cause the response to deviate from second order. The response will still be oscillatory and very fast.

Use the final value theorem to find the steady-state error.

$$n = \lim_{s \to 0} sN(s)$$

$$= \lim_{s \to 0} \left(\frac{-s(1.25 \times 10^8)\left(\frac{T_L}{s}\right)(0.004s+1)}{s^2 + 1250s + (1.27 \times 10^7)}\right)$$

$$= -9.84 T_L$$

The new steady-state error for both r and t_L inputs is

$$e = R - n = R - (0.98R - 9.84 T_L) = 0.02 R + 9.84 T_L$$

Thus, the load torque, $-t_L$, contributes to the steady-state error.

The response is oscillatory. The plot of the response due to a step change in $-t_L$ is

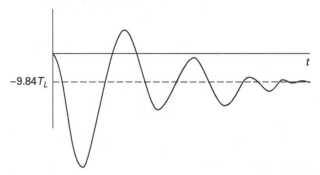

4. First, redraw the system in more traditional form.

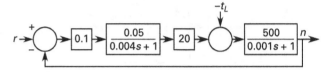

From Fig. 61.5, using the rules for simplifying block diagrams, combine boxes in series.

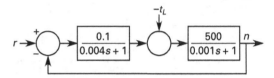

Use case 5 to combine the two summing points.

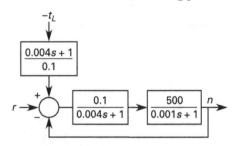

Replace the comparator gain of 0.1 with K. The reduced system using case 1 is

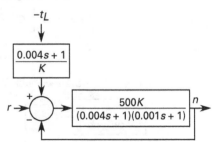

The closed-loop transfer function from r to n is

$$T_1(s) = \frac{N(s)}{R(s)} = \frac{\dfrac{500K}{(0.004s+1)(0.001s+1)}}{1 + \dfrac{500K}{(0.004s+1)(0.001s+1)}}$$

$$= \frac{500K}{(4 \times 10^{-6})s^2 + 0.005s + 1 + 500K}$$

The Routh-Hurwitz table is

$$\begin{bmatrix} a_0 & a_2 \\ a_1 & a_3 \\ b_1 & b_2 \end{bmatrix} = \begin{bmatrix} 4 \times 10^{-6} & 1 + 500K \\ 0.005 & 0 \\ b_1 & b_2 \end{bmatrix}$$

Calculate the remaining coefficients.

$$b_1 = \frac{a_1 a_2 - a_0 a_3}{a_1}$$

$$= \frac{(0.005)(1 + 500K) - (4 \times 10^{-6})(0)}{0.005}$$

$$= 1 + 500K$$

For a stable system, there can be no sign changes in the first column of the table.

Thus,

$$1 + 500K > 0$$
$$K > -0.002$$

The answer is (C).

5. The closed-loop system steady-state response can be improved by adding integral control. This will effectively compensate for any steady-state disturbances due to clutch output torque and will provide a zero steady-state error for a step input for the time response, $r(t)$. This addition has a side effect of reducing the stability margin of the system. However, if properly designed, the system will still be stable.

The answer is (A).

62 Management Science

Content in blue refers to the *NCEES Handbook*.

PRACTICE PROBLEMS

1. Printed circuit boards are manufactured in four consecutive departmental operations. Units move sequentially through departments 1, 2, 3, and 4. Each operation occurs at a station, each of which has a single employee overseeing a specific machine. Employees in all departments work from 8:00 a.m. to 5:00 p.m. and have one hour total for lunch and personal breaks. Defects created within a department are found before the units are passed to the subsequent department. Defective units are discarded. There are 52 work weeks per year, each of which has 5 work days. No units are produced during set-up, downtime, maintenance, or record-keeping periods. Units left incomplete at the end of one day are completed on the following day. The amount of time required for each department's tasks and the percentage of defective units found in each department are shown.

	department			
	1	2	3	4
production time (sec/unit)	6	10	11	45
set-up time (min/day)	16	8	20	5
downtime (min/day)	12	10	15	0
maintenance time (min/day)	8	12	8	0
record-keeping (min/day)	6	6	6	30
percentage defects	4%	6%	3%	2%

What is the maximum number of units that can be produced in one year if each department has only a single station (machine)?

(A) 72,000 units/yr
(B) 110,000 units/yr
(C) 140,000 units/yr
(D) 150,000 units/yr

2. Printed circuit boards are manufactured in four consecutive departmental operations. Units move sequentially through departments 1, 2, 3, and 4. Each operation occurs at a station, each of which has a single employee overseeing a specific machine. Employees in all departments work from 8:00 a.m. to 5:00 p.m. and have one hour total for lunch and personal breaks. Defects created within a department are found before the units are passed to the subsequent department. Defective units are discarded. There are 52 work weeks per year, each of which has 5 work days. No units are produced during set-up, downtime, maintenance, or record-keeping periods. Units left incomplete at the end of one day are completed on the following day.

	department			
	1	2	3	4
production time (sec/unit)	6	10	11	45
set-up time (min/day)	16	8	20	5
downtime (min/day)	12	10	15	0
maintenance time (min/day)	8	12	8	0
record-keeping (min/day)	6	6	6	30
percentage defects	4%	6%	3%	2%

If the production goal is 900,000 completed units per year, what is the minimum number (total) of machines needed in departments 1, 2, 3, and 4, respectively?

(A) 1, 2, 2, 6
(B) 1, 2, 3, 7
(C) 2, 2, 2, 6
(D) 2, 2, 3, 7

3. Four workers perform operations 1, 2, 3, and 4 in sequence on a manual assembly line. Each station performs its operation only once on the product before sending the product on to the next operation. Operation times at the stations are shown. (Travel times are included in the operation times.)

station	time (min/unit)
1	0.6
2	0.6
3	0.9
4	0.8

A fifth "floating" station has the ability to assist any of the four stations. The fifth station works with the same efficiencies and times as the four stations. There is no fixed assignment for this fifth station. The fifth station is allowed to help any station that needs it.

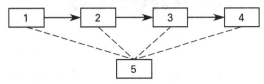

The operators of all five stations are permitted a 10-minute break each hour.

What is the fraction of station 5's time allocated to station 4?

(A) 0.04

(B) 0.55

(C) 0.38

(D) 0.25

4. Four workers perform operations 1, 2, 3, and 4 in sequence on a manual assembly line. Each station performs its operation only once on the product before sending the product on to the next operation. Operation times at the stations are as shown. (Travel times are included in the operation times.)

station	time (min/unit)
1	0.6
2	0.6
3	0.9
4	0.8

A fifth "floating" station has the ability to assist any of the four stations. The fifth station works with the same efficiencies and times as the four stations. There is no fixed assignment for this fifth station. The fifth station is allowed to help any station that needs it. The operators of all five stations are permitted a 10-minute break each hour.

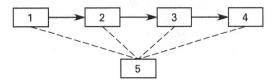

Neglecting the initial (transient) performance, what is the maximum production rate that can be achieved assuming that the fifth station is assigned to work optimally?

(A) 66 units/hr

(B) 71 units/hr

(C) 86 units/hr

(D) 277 units/hr

5. The activities that constitute a project are listed. The project starts at $t=0$. There is no expected finish time for the project.

activity	predecessors	successors	duration
start	–	A	0
A	start	B, C, D	7
B	A	G	6
C	A	E, F	5
D	A	G	2
E	C	H	13
F	C	H, I	4
G	D, B	I	18
H	E, F	finish	7
I	F, G	finish	5
finish	H, I	–	0

The critical path network is shown.

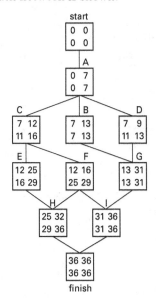

What is the critical path?

(A) A-B-G-I
(B) A-C-F-I
(C) A-C-E-H
(D) A-D-G-I

6. The activities that constitute a project are listed. The project starts at $t = 0$.

activity	predecessors	successors	duration
start	–	A	0
A	start	B, C, D	7
B	A	G	6
C	A	E, F	5
D	A	G	2
E	C	H	13
F	C	H, I	4
G	D, B	I	18
H	E, F	finish	7
I	F, G	finish	5
finish	H, I	–	0

The critical path network is shown.

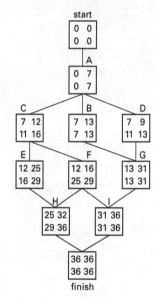

The earliest finish is most nearly

(A) 25
(B) 29
(C) 32
(D) 36

7. The activities that constitute a project are listed. The project starts at $t = 0$.

activity	predecessors	successors	duration
start	–	A	0
A	start	B, C, D	7
B	A	G	6
C	A	E, F	5
D	A	G	2
E	C	H	13
F	C	H, I	4
G	D, B	I	18
H	E, F	finish	7
I	F, G	finish	5
finish	H, I	–	0

The critical path network is shown.

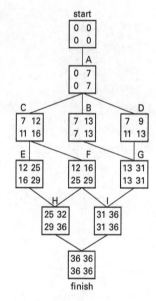

The latest finish is most nearly

(A) 25
(B) 29
(C) 32
(D) 36

8. The activities that constitute a project are listed. The project starts at $t=0$.

activity	predecessors	successors	duration
start	–	A	0
A	start	B, C, D	7
B	A	G	6
C	A	E, F	5
D	A	G	2
E	C	H	13
F	C	H, I	4
G	D, B	I	18
H	E, F	finish	7
I	F, G	finish	5
finish	H, I	–	0

The critical path network is shown.

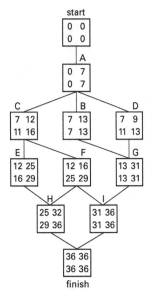

The slack along the critical path is most nearly

(A) 0
(B) 1
(C) 4
(D) 5

9. The activities that constitute a project are listed. The project starts at $t=0$.

activity	predecessors	successors	duration
start	–	A	0
A	start	B, C, D	7
B	A	G	6
C	A	E, F	5
D	A	G	2
E	C	H	13
F	C	H, I	4
G	D, B	I	18
H	E, F	finish	7
I	F, G	finish	5
finish	H, I	–	0

The critical path network is shown.

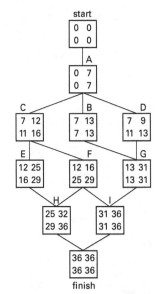

The float along the critical path is most nearly

(A) 0
(B) 1
(C) 4
(D) 5

10. A manufacturing company would like to determine if a new process for manufacturing bolts is within control limits. Over the course of a workday, 10 sample sets are taken that measure the diameter of three bolts. Those measurements are shown.

bolt diameter measurements

sample set number	sample number		
	1	2	3
1	0.247	0.252	0.249
2	0.249	0.253	0.250
3	0.248	0.251	0.249
4	0.249	0.247	0.252
5	0.253	0.248	0.251
6	0.255	0.254	0.251
7	0.248	0.247	0.251
8	0.246	0.249	0.247
9	0.252	0.246	0.247
10	0.250	0.255	0.253

The centerline, upper control limit, and lower control limit of this process are most nearly

(A) 0.248, 0.2545, 0.2425

(B) 0.250, 0.2569, 0.2431

(C) 0.250, 0.2545, 0.2455

(D) 0.252, 0.2589, 0.2451

11. A manufacturing company needs to determine if a new process for manufacturing bolts is within control limits. A study is conducted on a new process, for which the $\bar{X}$ chart is shown.

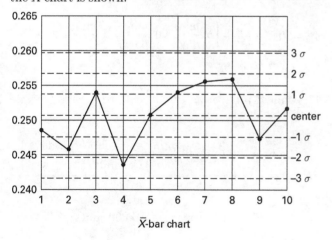

$\bar{X}$-bar chart

Is the new process within control limits, and if not, why is it out of control?

(A) The process is within control limits.

(B) The process is out of control due to points falling more than 2σ from the center line.

(C) The process is out of control due to two out of three consecutive points falling on the same side of and more than 1σ from the center line.

(D) The process is out of control due to three out of four consecutive points falling on the same side of and more than 1σ unit from the center line.

SOLUTIONS

1. Calculate the annual production rate for each of the departments. Assume each department has an infinite backlog of units processed by the previous department.

The number of work days per year is

$$\text{WDY} = \left(5 \ \frac{\text{days}}{\text{wk}}\right)\left(52 \ \frac{\text{wk}}{\text{yr}}\right) = 260 \ \text{days/yr}$$

Each employee works 8 hours per day, so the number of minutes worked per employee per day is

$$\left(8 \ \frac{\text{hr}}{\text{day}}\right)\left(60 \ \frac{\text{min}}{\text{hr}}\right) = 480 \ \text{min/day}$$

In general, the productive time available per machine is

$$M = 480 \ \frac{\text{min}}{\text{day}} - \left(\begin{array}{c}\frac{\text{set-up}}{\text{day}} + \frac{\text{downtime}}{\text{day}} \\ + \frac{\text{maintenance}}{\text{day}} \\ + \frac{\text{record-keeping}}{\text{day}}\end{array}\right)$$

The equation for the production from a single department is

$$D_{\text{out}} = \frac{(1 - \text{defect fraction})(\text{time available})}{\text{production time}}$$

Department 1:

The productive time available is

$$M_1 = 480 \ \frac{\text{min}}{\text{day}} - \left(16 \ \frac{\text{min}}{\text{day}} + 12 \ \frac{\text{min}}{\text{day}} + 8 \ \frac{\text{min}}{\text{day}} + 6 \ \frac{\text{min}}{\text{day}}\right)$$
$$= 438 \ \text{min/day}$$

The annual production rate of defect-free units from a single machine is

$$D_{1,\text{out}} = \frac{(1 - 0.04)\left(438 \ \frac{\text{min}}{\text{day}}\right)\left(60 \ \frac{\text{sec}}{\text{min}}\right)\left(260 \ \frac{\text{days}}{\text{yr}}\right)}{6 \ \frac{\text{sec}}{\text{unit}}}$$
$$= 1{,}093{,}248 \ \text{units/yr}$$

Department 2:

The productive time available is

$$M_2 = 480 \ \frac{\text{min}}{\text{day}} - \left(8 \ \frac{\text{min}}{\text{day}} + 10 \ \frac{\text{min}}{\text{day}} + 12 \ \frac{\text{min}}{\text{day}} + 6 \ \frac{\text{min}}{\text{day}}\right)$$
$$= 444 \ \text{min/day}$$

$$D_{2,\text{out}} = \frac{(1 - 0.06)\left(444 \ \frac{\text{min}}{\text{day}}\right)\left(60 \ \frac{\text{sec}}{\text{min}}\right)\left(260 \ \frac{\text{days}}{\text{yr}}\right)}{10 \ \frac{\text{sec}}{\text{unit}}}$$
$$= 651{,}082 \ \text{units/yr}$$

Department 3:

The productive time available is

$$M_3 = 480 \ \frac{\text{min}}{\text{day}} - \left(20 \ \frac{\text{min}}{\text{day}} + 15 \ \frac{\text{min}}{\text{day}} + 8 \ \frac{\text{min}}{\text{day}} + 6 \ \frac{\text{min}}{\text{day}}\right)$$
$$= 431 \ \text{min/day}$$

The annual production rate of defect-free units from a single machine is

$$D_{3,\text{out}} = \frac{(1 - 0.03)\left(431 \ \frac{\text{min}}{\text{day}}\right)\left(60 \ \frac{\text{sec}}{\text{min}}\right)\left(260 \ \frac{\text{days}}{\text{yr}}\right)}{11 \ \frac{\text{sec}}{\text{unit}}}$$
$$= 592{,}899 \ \text{units/yr}$$

Department 4:

The productive time available is

$$M_4 = 480 \ \frac{\text{min}}{\text{day}} - \left(5 \ \frac{\text{min}}{\text{day}} + 0 \ \frac{\text{min}}{\text{day}} + 0 \ \frac{\text{min}}{\text{day}} + 30 \ \frac{\text{min}}{\text{day}}\right)$$
$$= 445 \ \text{min/day}$$

The annual production rate of defect-free units from a single machine is

$$D_{4,\text{out}} = \frac{(1 - 0.02)\left(445 \ \frac{\text{min}}{\text{day}}\right)\left(60 \ \frac{\text{sec}}{\text{min}}\right)\left(260 \ \frac{\text{days}}{\text{yr}}\right)}{45 \ \frac{\text{sec}}{\text{unit}}}$$
$$= 151{,}181 \ \text{units/yr} \quad (150{,}000 \ \text{units/yr})$$

Department 4 is the bottleneck. The annual production rate is limited to 151,181 units/yr (150,000 units/yr).

The answer is (D).

2. Calculate the annual production rate for each of the departments. Then work backward from the desired output. Assume each department has an infinite backlog of units processed by the previous department.

The number of work days per year is

$$\text{WDY} = \left(5 \; \frac{\text{days}}{\text{wk}}\right)\left(52 \; \frac{\text{wk}}{\text{yr}}\right) = 260 \; \text{days/yr}$$

Each employee works 8 hours per day, so the number of minutes worked per employee is

$$\left(8 \; \frac{\text{hr}}{\text{day}}\right)\left(60 \; \frac{\text{min}}{\text{hr}}\right) = 480 \; \text{min/day}$$

In general, the productive time available per machine is

$$M = 480 \; \frac{\text{min}}{\text{day}} - \left(\begin{array}{l}\frac{\text{set-up}}{\text{day}} + \frac{\text{downtime}}{\text{day}} \\ + \frac{\text{maintenance}}{\text{day}} \\ + \frac{\text{record-keeping}}{\text{day}}\end{array}\right)$$

The production from a single department is

$$D_{\text{out}} = \frac{(1 - \text{defect fraction})(\text{time available})}{\text{production time}}$$

Department 1:

The productive time available is

$$M_1 = 480 \; \frac{\text{min}}{\text{day}} - \left(16 \; \frac{\text{min}}{\text{day}} + 12 \; \frac{\text{min}}{\text{day}} + 8 \; \frac{\text{min}}{\text{day}} + 6 \; \frac{\text{min}}{\text{day}}\right)$$
$$= 438 \; \text{min/day}$$

The annual production rate of defect-free units from a single machine is

$$D_{1,\text{out}} = \frac{(1 - 0.04)\left(438 \; \frac{\text{min}}{\text{day}}\right)\left(60 \; \frac{\text{sec}}{\text{min}}\right)\left(260 \; \frac{\text{days}}{\text{yr}}\right)}{6 \; \frac{\text{sec}}{\text{unit}}}$$
$$= 1{,}093{,}248 \; \text{units/yr}$$

Department 2:

The productive time available is

$$M_2 = 480 \; \frac{\text{min}}{\text{day}} - \left(8 \; \frac{\text{min}}{\text{day}} + 10 \; \frac{\text{min}}{\text{day}} + 12 \; \frac{\text{min}}{\text{day}} + 6 \; \frac{\text{min}}{\text{day}}\right)$$
$$= 444 \; \text{min/day}$$

The annual production rate of defect-free units from a single machine is

$$D_{2,\text{out}} = \frac{(1 - 0.06)\left(444 \; \frac{\text{min}}{\text{day}}\right)\left(60 \; \frac{\text{sec}}{\text{min}}\right)\left(260 \; \frac{\text{days}}{\text{yr}}\right)}{10 \; \frac{\text{sec}}{\text{unit}}}$$
$$= 651{,}082 \; \text{units/yr}$$

Department 3:

The productive time available is

$$M_3 = 480 \; \frac{\text{min}}{\text{day}} - \left(20 \; \frac{\text{min}}{\text{day}} + 15 \; \frac{\text{min}}{\text{day}} + 8 \; \frac{\text{min}}{\text{day}} + 6 \; \frac{\text{min}}{\text{day}}\right)$$
$$= 431 \; \text{min/day}$$

The annual production rate of defect-free units from a single machine is

$$D_{3,\text{out}} = \frac{(1 - 0.03)\left(431 \; \frac{\text{min}}{\text{day}}\right)\left(60 \; \frac{\text{sec}}{\text{min}}\right)\left(260 \; \frac{\text{days}}{\text{yr}}\right)}{11 \; \frac{\text{sec}}{\text{unit}}}$$
$$= 592{,}899 \; \text{units/yr}$$

Department 4:

The productive time available is

$$M_4 = 480 \; \frac{\text{min}}{\text{day}} - \left(5 \; \frac{\text{min}}{\text{day}} + 0 \; \frac{\text{min}}{\text{day}} + 0 \; \frac{\text{min}}{\text{day}} + 30 \; \frac{\text{min}}{\text{day}}\right)$$
$$= 445 \; \text{min/day}$$

The annual production rate of defect-free units from a single machine is

$$D_{4,\text{out}} = \frac{(1 - 0.02)\left(445 \; \frac{\text{min}}{\text{day}}\right)\left(60 \; \frac{\text{sec}}{\text{min}}\right)\left(260 \; \frac{\text{days}}{\text{yr}}\right)}{45 \; \frac{\text{sec}}{\text{unit}}}$$
$$= 151{,}181 \; \text{units/yr} \quad (150{,}000 \; \text{units/yr})$$

Department 4:

Each machine in department 4 produces 151,181 defect-free units per year. In order to produce 900,000 units per year, the number of machines needed is

$$n_4 = \frac{900{,}000 \; \frac{\text{units}}{\text{yr}}}{151{,}181 \; \frac{\text{units}}{\text{machine-yr}}}$$
$$= 5.95 \; \text{machines} \quad (6 \; \text{machines})$$

Department 3:

In order to produce 900,000 units per year and supply the extra that will become defective in the subsequent processing, the number of machines needed is

$$n_3 = \frac{900{,}000 \, \frac{\text{units}}{\text{yr}}}{(1-0.02)\left(592{,}899 \, \frac{\text{units}}{\text{machine-yr}}\right)}$$

$$= 1.55 \text{ machines} \quad (2 \text{ machines})$$

Department 2:

In order to produce 900,000 units per year and supply the extra that will become defective in the subsequent processing, the number of machines needed is

$$n_2 = \frac{900{,}000 \, \frac{\text{units}}{\text{yr}}}{(1-0.03)(1-0.02)\left(651{,}082 \, \frac{\text{units}}{\text{machine-yr}}\right)}$$

$$= 1.45 \text{ machines} \quad (2 \text{ machines})$$

Department 1:

In order to produce 900,000 units per year and supply the extra that will become defective in the subsequent processing, the number of machines needed is

$$n_4 = \frac{900{,}000 \, \frac{\text{units}}{\text{yr}}}{(1-0.06)(1-0.03)(1-0.02)} \times \left(1{,}093{,}248 \, \frac{\text{units}}{\text{machine-yr}}\right)$$

$$= 0.92 \text{ machine} \quad (1 \text{ machine})$$

The answer is (A).

3. Disregard the 10 min per hour shift break, since this break reduces the capacity of all stations by the same percentage.

Determine the maximum output per hour for each station.

station	output
1	$\dfrac{60 \, \frac{\text{min}}{\text{hr}}}{0.6 \, \frac{\text{min}}{\text{unit}}} = 100 \text{ units/hr}$
2	$\dfrac{60 \, \frac{\text{min}}{\text{hr}}}{0.6 \, \frac{\text{min}}{\text{unit}}} = 100 \text{ units/hr}$
3	$\dfrac{60 \, \frac{\text{min}}{\text{hr}}}{0.9 \, \frac{\text{min}}{\text{unit}}} = 66.67 \text{ units/hr}$
4	$\dfrac{60 \, \frac{\text{min}}{\text{hr}}}{0.8 \, \frac{\text{min}}{\text{unit}}} = 75 \text{ units/hr}$

Stations 3 and 4 are the bottleneck operations. Intuitively, operation 3 needs help the most, followed by operation 4.

Start by allocating station 5 capacity to the slowest operation, operation 3. Try to bring station 3 up to the same capacity as stations 1 and 2. To do so requires station 5 to produce $100 - 66.67 = 33.33$ units per hour. Since station 5 works at the same speed as station 3, the fraction of time station 5 needs to assist station 3 is $33.33/66.67 = 0.5$ (50%). This leaves 50% of station 5's time available to assist other stations.

Next, allocate the remaining station 5 time to station 4. To bring station 4 up to 100 units per hour will require station 5 to produce $100 - 75 = 25$ units per hour. Since station 5 works at the same rate as station 4, the fraction of time station 5 needs to assist station 4 is $25/75 = 0.3333$ (33.33%).

So, all of the stations have been brought up to 100 units per hour. Station 5 still has some remaining capacity: $100\% - 50\% - 33.33\% = 16.67\%$. This remaining capacity needs to be allocated to all of the remaining stations to bring them all up to the same output rate.

All of the extra time must come from the remaining capacity of station 5, since all other stations are working at their individual capacities. Station 5 has 17% of its time left, and it must allocate its time in the same fraction (ratio) as the assembly times.

operation	time	fraction of total	ratio × 17%
1	0.6	0.2069	3.52%
2	0.6	0.2069	3.52%
3	0.9	0.3103	5.27%
4	0.8	0.2759	4.69%
totals	2.9	1.0000	17.00%

Therefore, 3.52% (0.04) of station 5's time will be given to operations 1 and 2. Operation 3 will receive 50% + 5.27% = 55.27% (0.55) of station 5's time. Operation 4 will receive 33.33% + 4.69% = 38.02% (0.38).

The answer is (C).

4. If station 5 is assigned to assist each station optimally, the available production time is distributed equally across all stations. The total production time for a unit is

$$0.6\,\frac{\min}{\text{unit}} + 0.6\,\frac{\min}{\text{unit}} + 0.9\,\frac{\min}{\text{unit}} + 0.8\,\frac{\min}{\text{unit}} = 2.9\ \min/\text{unit}$$

The actual time available to work per hour across all five stations, including the break time, is

$$(5\ \text{stations})\frac{(60\ \min - 10\ \min)}{\text{hr}} = 250\,\frac{\min}{\text{hr}}$$

The total number of units that can be produced per hour is

$$\frac{250\,\frac{\min}{\text{hr}}}{2.9\,\frac{\min}{\text{unit}}} = 86.2\,\frac{\text{units}}{\text{hr}} \quad \left(86\,\frac{\text{units}}{\text{hr}}\right)$$

The answer is (C).

5. The activity is critical if the earliest start equals the latest start.

The critical path is A-B-G-I.

The answer is (A).

6. By the earliest start (ES) rule, the earliest start time for an activity leaving a particular node is equal to the largest of the earliest finish times for all activities entering the node.

The earliest finish is 36.

The answer is (D).

7. By the latest finish (LF) rule, the latest finish time for an activity entering a particular node is equal to the smallest of the latest start times for all activities leaving the node.

The latest finish is 36.

The answer is (D).

8. There can be no slack along the critical path.

The answer is (A).

9. There can be no float along a critical path.

The answer is (A).

10. Calculate the mean and range for each sample set.

$$\overline{X}_1 = \left(\frac{1}{n}\right)\sum_{i=1}^{n} X_i$$

$$= \left(\frac{1}{3}\right)(0.247 + 0.252 + 0.249)$$

$$= 0.249$$

$\overline{R}$ is calculated by subtracting the smallest number from the largest number in the sample set. The means and ranges are as tabulated.

bolt diameter measurements

sample set number	sample number 1	2	3	mean $\overline{X}$	range $\overline{R}$
1	0.247	0.252	0.249	0.249	0.005
2	0.249	0.253	0.250	0.251	0.004
3	0.248	0.251	0.249	0.249	0.003
4	0.249	0.247	0.252	0.249	0.005
5	0.253	0.248	0.251	0.251	0.005
6	0.255	0.254	0.251	0.253	0.004
7	0.248	0.247	0.251	0.249	0.004
8	0.246	0.249	0.247	0.247	0.003
9	0.252	0.246	0.247	0.248	0.006
10	0.250	0.255	0.253	0.253	0.005
			$\overline{\overline{X}}$	0.250	
			$\overline{\overline{R}}$	0.004	

The overall mean and overall range for all of the sample sets are

Dispersion, Mean, Median, and Mode Values

$$\overline{\overline{X}} = \left(\frac{1}{n}\right)\sum_{i=1}^{n} X_i$$

$$= \left(\frac{1}{10}\right)\!\left(\!\begin{array}{l}0.249 + 0.251 + 0.249 + 0.249 + 0.251 \\ +0.253 + 0.249 + 0.247 + 0.248 + 0.253\end{array}\!\right)$$

$$= 0.250$$

$$\overline{R} = \left(\frac{1}{n}\right)\sum_{i=1}^{n} R_i$$

$$= \left(\frac{1}{10}\right)\!\left(\!\begin{array}{l}0.005 + 0.004 + 0.003 + 0.005 + 0.005 \\ +0.004 + 0.004 + 0.003 + 0.006 + 0.005\end{array}\!\right)$$

$$= 0.004$$

The control-chart limit factor A_2 for a sample size of 3 is 1.02.[Statistical Quality Control]

The upper and lower control limits are

Control-Limit Calculations

$$\begin{aligned}\text{UCL} &= \overline{X} + A_2\overline{R} \\ &= 0.25 + (1.02)(0.004) \\ &= 0.254455 \quad (0.2545) \\ \text{LCL} &= \overline{X} - A_2\overline{R} \\ &= 0.25 - (1.02)(0.004) \\ &= 0.245479 \quad (0.2455)\end{aligned}$$

The answer is (C).

11. This process is in control as the process agrees with all 8 criteria from 2.4.1 Tests for Out of Control, for Three-Sigma Control Units. While answers B, C, and D all occur in the data presented, they are not in agreement with any of the 8 criteria.

The answer is (A).

Instrumentation and Measurements

Content in blue refers to the *NCEES Handbook*.

PRACTICE PROBLEMS

1. Diesel engines in a waste-to-energy plant are powered by a gaseous mixture of methane and other digestion gases. The combustion conditions in each cylinder are continuously monitored by pressure and temperature transducers. A particular cylinder's indicator diagram from top dead center to bottom dead center is shown. The scale selected in the monitoring software is set to 111 kPa/mm.

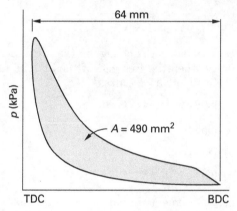

Most nearly, what is the mean effective pressure in the cylinder?

(A) 7.7 kPa

(B) 15 kPa

(C) 360 kPa

(D) 850 kPa

2. The temperature of a steel girder during a fire test is measured with a type 404 platinum resistance temperature detector (RTD) and a simple two-wire bridge, as shown.

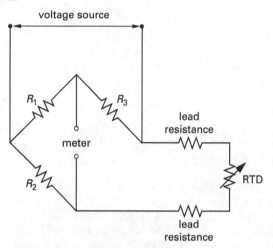

The RTD has a resistance, R, of 100 Ω at 0°C, and a temperature coefficient, α, of 0.00385 1/°C. The values of the bridge resistances are $R_1 = 1000$ Ω and $R_3 = 1000$ Ω. When the meter is nulled out during a test, $R_2 = 376$ Ω. The lead resistances are each 100 Ω. The resistance of a platinum RTD is found from the equation shown.

$$R_T = R_0(1 + AT + BT^2)$$

Assume the coefficient B for the resistance of the RTD is zero. Most nearly, what is the temperature of the beam during the test?

(A) 130°C

(B) 175°C

(C) 200°C

(D) 450°C

3. A load cell measures tensile force with a steel bar and a bonded strain gauge, as shown. The modulus of elasticity of the steel is 30×10^6 lbf/in^2. The bar's tensile area is 1.0 in^2. The gage factor of the strain gauge is 6.0. The unloaded strain gauge resistance is 100 Ω, and the loaded resistance is 100.1 Ω.

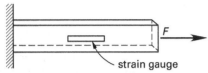

The equation relating strain to changes in resistance of the strain gage is

$$\epsilon = \frac{\Delta R_g}{(\text{GF})R_g}$$

The applied load is most nearly

(A) 2500 lbf

(B) 5000 lbf

(C) 17,000 lbf

(D) 20,000 lbf

4. A strain gauge is bonded to each of the four vertical bars of a grain hopper scale, as shown. The hopper has a tare weight of 1500 lbf. The support bars have a modulus of elasticity of 20×10^6 lbf/in^2, and each bar has a cross-sectional area of 2 in^2. Each strain gauge has a gage factor of 3 and has an initial resistance of 300 Ω when originally applied to the unstressed bars. After the grain is loaded into the hopper, the strain gauge resistances are 300.05 Ω, 300.09 Ω, 300.11 Ω, and 300.03 Ω, respectively.

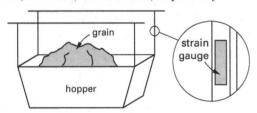

The average strain found by the gauge is a function of the change in the resistance of the strain gage and the gage factor, as shown.

$$\epsilon = \frac{\Delta R_g}{(\text{GF})R_g}$$

What is most nearly the weight of the grain loaded into the hopper?

(A) 9500 lbf

(B) 11,000 lbf

(C) 12,000 lbf

(D) 14,000 lbf

5. A strain gauge is bonded to the top of a rectangular aluminum bar 12 cm long, 2 cm high, and 4 cm wide, as shown. The gauge has a gage factor of 3.0 and a nominal (strain-free) resistance of 350 Ω. The aluminum has a modulus of elasticity of 73 GPa. The bar is loaded as a cantilever beam. The distance from the applied load to the center of the strain gauge is 10 cm. The instrumentation consists of a simple Wheatstone bridge consisting of two 1000 Ω resistors, a variable resistor, and the strain gauge. Lead resistance is negligible.

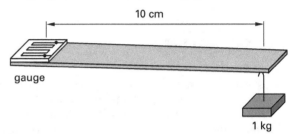

The average strain found by the gauge is a function of the change in the resistance of the strain gage and the gage factor, as shown.

$$\epsilon = \frac{\Delta R_g}{(\text{GF})R_g}$$

Most nearly, what is the change in the resistance of the strain gauge when the free end of the aluminum bar is loaded vertically by a 1 kg mass?

(A) 0.00054 Ω

(B) 0.0053 Ω

(C) 0.048 Ω

(D) 0.50 Ω

6. Which statement about a proportional-integral-derivative (PID) controller is correct?

(A) The integral term ensures the controller controls at setpoint, and the derivative term increases controller damping and enhances controller stability.

(B) The derivative term ensures the PID controls at setpoint. The integral term increases controller damping and enhances controller stability.

(C) Adjusting the controller gain to the optimum setting will force the controller to control at setpoint.

(D) Controller stability is independent of the controller gain.

SOLUTIONS

1. The average height of the pressure plot is

$$\bar{h} = \frac{A}{w} = \frac{490 \text{ mm}^2}{64 \text{ mm}} = 7.656 \text{ mm}$$

The mean effective pressure is the product of the average height of the pressure plot, $\bar{h}$ and the scale, k.

$$\bar{p} = k\bar{h} = \left(111 \frac{\text{kPa}}{\text{mm}}\right)(7.656 \text{ mm})$$
$$= 849.8 \text{ kPa} \quad (850 \text{ kPa})$$

The answer is (D).

2. When the meter is nulled out, the resistance of the leg containing the resistance temperature detector (RTD) is

$$R_4 = R_{\text{RTD}} + 2R_{\text{lead}} = \frac{R_2 R_3}{R_1} = \frac{(376 \text{ }\Omega)(1000 \text{ }\Omega)}{1000 \text{ }\Omega}$$
$$= 376 \text{ }\Omega$$

The resistance of the RTD is

$$R_{\text{RTD}} = R_4 - 2R_{\text{lead}} = 376 \text{ }\Omega - (2)(100 \text{ }\Omega) = 176 \text{ }\Omega$$

For a platinum RTD with $\alpha = 0.00385 \text{ 1/°C}$, the resistance is

$$R_T = R_0(1 + AT + BT^2)$$
$$= R_0 \begin{pmatrix} 1 + (3.9083 \times 10^{-3})T \\ + BT^2 \end{pmatrix}$$
$$\approx R_0(1 + (3.9083 \times 10^{-3})T)$$
$$176 \text{ }\Omega \approx (100 \text{ }\Omega)(1 + (3.9083 \times 10^{-3})T)$$
$$T \approx 194.5°C \quad (200°C)$$

(If the squared term is kept, the temperature is 200.3°C.)

The answer is (C).

3. The strain is

$$\epsilon = \frac{\Delta R_g}{(\text{GF})R_g} = \frac{100.1 \text{ }\Omega - 100 \text{ }\Omega}{(6.0)(100 \text{ }\Omega)} = 1.67 \times 10^{-4}$$

From the equation for the modulus of elasticity, the stress is

Uniaxial Loading and Deformation

$$E = \frac{\sigma}{\epsilon}$$
$$\sigma = E\epsilon$$
$$= \left(30 \times 10^6 \frac{\text{lbf}}{\text{in}^2}\right)(1.67 \times 10^{-4})$$
$$= 5.0 \times 10^3 \text{ lbf/in}^2$$

From the equation for stress, the applied load is

Uniaxial Loading and Deformation

$$\sigma = \frac{P}{A}$$
$$F = \sigma A$$
$$= \left(5.0 \times 10^3 \frac{\text{lbf}}{\text{in}^2}\right)(1.0 \text{ in}^2)$$
$$= 5.0 \times 10^3 \text{ lbf} \quad (5000 \text{ lbf})$$

The answer is (B).

4. Work with average values. The average strain gauge resistance is

$$R_{\text{ave}} = \frac{300.05 \text{ }\Omega + 300.09 \text{ }\Omega + 300.11 \text{ }\Omega + 300.03 \text{ }\Omega}{4}$$
$$= 300.07 \text{ }\Omega$$

The average strain is

$$\epsilon_{\text{ave}} = \frac{\Delta R_g}{(\text{GF})R_g} = \frac{300.07 \text{ }\Omega - 300 \text{ }\Omega}{(3)(300 \text{ }\Omega)}$$
$$= 7.777 \times 10^{-5} \text{ in/in}$$

From the equation for the modulus of elasticity, the stress is

Uniaxial Loading and Deformation

$$E = \frac{\sigma}{\epsilon}$$
$$\sigma_{\text{ave}} = E\epsilon_{\text{ave}}$$
$$= \left(20 \times 10^6 \frac{\text{lbf}}{\text{in}^2}\right)\left(7.777 \times 10^{-5} \frac{\text{in}}{\text{in}}\right)$$
$$= 1555 \text{ lbf/in}^2$$

From the equation for stress, the average load is

Uniaxial Loading and Deformation

$$\sigma = \frac{P}{A}$$

$$\begin{aligned} P_{ave} &= \sigma_{ave} A \\ &= \left(1555 \; \frac{\text{lbf}}{\text{in}^2}\right)(2 \text{ in}^2) \\ &= 3110 \text{ lbf} \end{aligned}$$

The total weight of the hopper and grain is

$$\begin{aligned} W_t &= 4 P_{ave} = (4)(3110 \text{ lbf}) \\ &= 12{,}440 \text{ lbf} \end{aligned}$$

Therefore, the weight of the grain is

$$\begin{aligned} W_g &= W_t - W_h = 12{,}440 \text{ lbf} - 1500 \text{ lbf} \\ &= 10{,}940 \text{ lbf} \quad (11{,}000 \text{ lbf}) \end{aligned}$$

The answer is (B).

5. The moment of inertia of the bar in bending is

Properties of Various Shapes

$$I = \frac{bh^3}{12} = \frac{(4 \text{ cm})(2 \text{ cm})^3}{12}$$

$$= 2.667 \text{ cm}^4$$

The stress in the bar is

Stresses in Beams

$$\sigma = \pm \frac{Mc}{I} = \frac{mgLc}{I}$$

$$= \frac{(1 \text{ kg})\left(9.81 \; \frac{\text{m}}{\text{s}^2}\right)(10 \text{ cm})\left(\frac{2 \text{ cm}}{2}\right)\left(100 \; \frac{\text{cm}}{\text{m}}\right)^2}{2.667 \text{ cm}^4}$$

$$= 367{,}829 \text{ Pa}$$

From the equation for the modulus of elasticity, the strain is

Uniaxial Loading and Deformation

$$E = \frac{\sigma}{\epsilon}$$

$$\begin{aligned} \epsilon &= \frac{\sigma}{E} \\ &= \frac{367{,}829 \text{ Pa}}{(73 \text{ GPa})\left(10^9 \; \frac{\text{Pa}}{\text{GPa}}\right)} \\ &= 5.04 \times 10^{-6} \end{aligned}$$

From the equation for the average strain, the change in resistance is

$$\epsilon = \frac{\Delta R_g}{(\text{GF})R_g}$$

$$\begin{aligned} \Delta R_g &= \epsilon(\text{GF})R_g \\ &= (5.04 \times 10^{-6})(3.0)(350 \; \Omega) \\ &= 0.00529 \; \Omega \quad (0.0053 \; \Omega) \end{aligned}$$

The answer is (B).

6. Controller stability is affected by controller gain. If the gain is too high, the controller may become unstable; if the gain is too low, the controller response time may be too slow. The controller gain cannot be forced to operate at setpoint by adjusting the controller gain to an optimum setting. The integral term compensates for offset and ensures the controller controls at setpoint. The derivative term increases controller damping and enhances controller stability.

The answer is (A).

64 Manufacturing Processes

Content in blue refers to the *NCEES Handbook*.

PRACTICE PROBLEMS

1. A 10 mil (250 μm) adhesive with a shear strength of 1500 lbf/in² (10 MPa) is used in a lap joint between two 0.20 in (5 mm) aluminum sheets. The aluminum has a yield strength of 15,000 lbf/in² (100 MPa).

The adhesive is loaded in pure shear. Use a stress concentration factor of 2. The joint is to be as strong as the aluminum. The width (overlap) of adhesive joint that is required is most nearly

(A) 2.3 in (0.058 m)

(B) 4.0 in (0.10 m)

(C) 5.3 in (0.13 m)

(D) 6.9 in (0.17 m)

2. The welding symbol shown calls for a

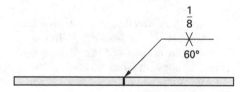

(A) double-welded butt joint

(B) single-welded butt joint

(C) single-full fillet lap joint

(D) double-welded fillet lap joint

3. The tool life equation for tool bit A is $v_A T_A^{0.3} = 605$, and the equation for tool bit B is $v_B T_B^{0.15} = 386$. v is the cutting speed in feet per minute, and T is the tool life in minutes. At a particular cutting speed, tool bit B will last twice as long as tool bit A before developing wear. This cutting speed is most nearly

(A) 120 ft/min

(B) 200 ft/min

(C) 340 ft/min

(D) 480 ft/min

4. For an orthogonal metal cutting operation, the cutting speed is 200 ft/min, the thrust force, normal to the cutting velocity vector, is 60 lbf, and the total resultant force vector is 100 lbf. The amount of power input needed for the cutting operation is most nearly

(A) 0.2 hp

(B) 0.3 hp

(C) 0.4 hp

(D) 0.5 hp

5. Four proposed designs are submitted to a company to attach a rubber foot to a steel bracket. Due to design constraints, the mounting bracket must be bonded to the rubber pad to prevent tearing during shear loads. Which of the four proposed designs provides the greatest bond strength?

(A)

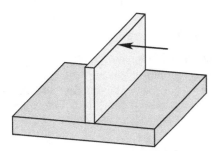

(B)

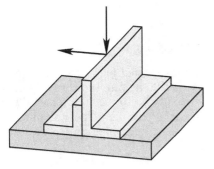

(C)

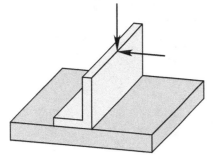

(D)

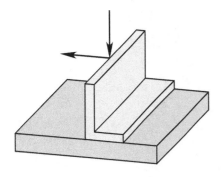

6. What are some typical methods to reduce peel stresses in an adhesively bonded lap joint?

(A) Increase the area of the materials being bonded.

(B) Include mechanical interferences or stops in the design.

(C) Change the type of materials being bonded.

(D) both A and B

7. Select the correct welding symbol for the assembly shown.

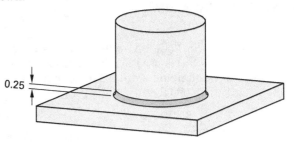

(A)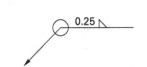

(B)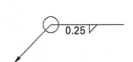

(C)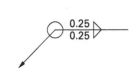

(D)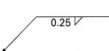

8. A 300 ft length of insulated steel pipe carries fluid that ranges from 0°F to 500°F. A linear expansion device is installed at one end of the pipe to keep stresses due to thermal expansion from exceeding 18.9 ksi, which is the maximum the pipe can safely withstand. The minimum length of expansion that the device must absorb is most nearly

(A) 6.7 in

(B) 7.7 in

(C) 8.5 in

(D) 9.4 in

SOLUTIONS

1. *Customary U.S. Solution*

Assume the unit length ($L = 1$ in) for the joint. The area of the adhesive and the area of the aluminum are

$$A_{\text{adhesive}} = wL = w_{\text{in}}(1 \text{ in}) = w$$
$$A_{\text{aluminum}} = Lt = (1 \text{ in})(0.2 \text{ in}) = 0.2 \text{ in}^2$$

Find the loading on the cross section from the equation for the stress on the cross section.

Uniaxial Loading and Deformation

$$\sigma = \frac{P}{A_{\text{aluminum}}}$$

$$P = \sigma A_{\text{aluminum}} = \left(15{,}000 \, \frac{\text{lbf}}{\text{in}^2}\right)(0.2 \text{ in}^2)$$
$$= 3000 \text{ lbf}$$

Given a shear stress of 1500 lbf/in² and a stress concentration factor of 2, the required width of the adhesive is

$$\sigma = \frac{2P}{A_{\text{adhesive}}}$$

$$1500 \, \frac{\text{lbf}}{\text{in}^2} = \frac{(2)(3000 \text{ lbf})}{w}$$

$$w = 4.0 \text{ in}$$

SI Solution

Assume the unit length ($L = 1$ cm) for the joint. The area of the adhesive and the area of the aluminum are

$$A_{\text{adhesive}} = wL = w_{\text{in}}(0.01 \text{ m}) = 0.01w$$
$$A_{\text{aluminum}} = Lt = (0.01 \text{ m})(0.005 \text{ m})$$
$$= 0.00005 \text{ m}^2$$

Find the loading on the cross section from the equation for the stress on the cross section.

Uniaxial Loading and Deformation

$$\sigma = \frac{P}{A_{\text{aluminum}}}$$

$$P = \sigma A_{\text{aluminum}} = (100 \times 10^6 \text{ Pa})(0.00005 \text{ m}^2)$$
$$= 5000 \text{ N}$$

Given a shear stress of 1500 lbf/in² and a stress concentration factor of 2, the required width of the adhesive is

$$\sigma = \frac{2P}{A_{\text{adhesive}}}$$

$$10 \times 10^6 \text{ Pa} = \frac{(2)(5000 \text{ N})}{0.01w}$$

$$w = 0.10 \text{ m}$$

The answer is (B).

2. The two pieces are joined end to end, so this is a butt joint. The V shapes on both sides of the arrow indicate welding on both sides of the joint. [Basic Weld Symbols]

The symbol indicates a double-welded butt joint.

The answer is (A).

3. Rearrange the two tool life equations in terms of the cutting speed.

$$v_A = \frac{605}{T_A^{0.3}}$$

$$v_B = \frac{386}{T_B^{0.15}}$$

If the same cutting speed is used for both tool bits,

$$\frac{605}{T_A^{0.3}} = \frac{386}{T_B^{0.15}}$$

Since $T_B = 2T_A$, substitute $2T_A$ for T_B and solve for T_A.

$$\frac{605}{T_A^{0.3}} = \frac{386}{(2T_A)^{0.15}}$$

$$\frac{T_A^{0.3}}{T_A^{0.15}} = T_A^{0.15} = \frac{(605)(2)^{0.15}}{386} = 1.739$$

$$T_A = 39.99$$

The cutting speed is

$$v_A = \frac{605}{T_A^{0.3}} = \frac{605}{(39.99 \text{ min})^{0.3}}$$
$$= 200.06 \text{ ft/min} \quad (200 \text{ ft/min})$$

The answer is (B).

4. The resultant force F_R between the tool and the chip can be resolved into two components: a horizontal cutting force, F_H, and a vertical thrust force, F_V.

Resultant (Two Dimensions)

$$F = \left[\left(\sum_{i=1}^{n} F_{x,i}\right)^2 + \left(\sum_{i=1}^{n} F_{y,i}\right)^2\right]^{1/2}$$

Therefore,

$$F_R = \sqrt{F_H^2 + F_v^2}$$

Calculate the horizontal cutting force.

$$F_H = \sqrt{F_R^2 - F_v^2} = \sqrt{(100 \text{ lbf})^2 - (60 \text{ lbf})^2}$$
$$= 80 \text{ lbf}$$

The power needed for the orthogonal cutting operation is

Power and Efficiency

$$P = F_H \text{v}$$

$$= \frac{(80 \text{ lbf})\left(200 \dfrac{\text{ft}}{\text{min}}\right)}{33{,}000 \dfrac{\text{ft-min}}{\text{lbf-hp}}}$$

$$= 0.4848 \text{ hp} \quad (0.5 \text{ hp})$$

The answer is (D).

5. Options A, C, and D all have load vectors that negatively affect the strength of the adhesive. [Design Practices That Improve Adhesive Bonding]

The answer is (B).

6. Increasing the surface area for bonding and adding mechanical stops would reduce the peel stresses. [Design Practices That Improve Adhesive Bonding]

The answer is (D).

7. This is an all-around fillet weld with a width of 0.25. The fillet symbol is on the bottom side of the line indicating that the weld is on the arrow side. The circle indicates that it is a weld all the way around the object. Option B shows the welding symbol that represents this weld.

The answer is (B).

8. The modulus of elasticity, E, for steel is 29,000,000 psi, or 29,000 ksi. The coefficient of thermal expansion, α, for steel is 6.5×10^{-6} °F^{-1}. [Typical Material Properties]

The modulus of elasticity is stress, σ, divided by strain, ϵ. The specified stress limit is 18.9 ksi. Use the equation for the modulus of elasticity, and solve for the allowable strain.

Uniaxial Loading and Deformation

$$E = \frac{\sigma}{\epsilon}$$

$$\epsilon = \frac{\sigma}{E} = \frac{18.9 \dfrac{\text{kips}}{\text{in}^2}}{29{,}000 \dfrac{\text{kips}}{\text{in}^2}} = 0.000652$$

Use the formula for the coefficient of thermal expansion, and solve for the change in temperature, ΔT, that will produce the allowable strain.

Thermal Properties

$$\alpha = \frac{\epsilon}{\Delta T}$$

$$\Delta T = \frac{\epsilon}{\alpha} = \frac{0.000652}{6.5 \times 10^{-6} \dfrac{1}{°\text{F}}} = 100°\text{F}$$

The maximum temperature of the pipe is 500°F, so the expansion device must be able to absorb the change in length caused by another 400°F. Find the strain caused by a change in temperature of 400°F.

Thermal Properties

$$\alpha = \frac{\epsilon}{\Delta T}$$

$$\epsilon = \alpha \Delta T = \left(6.5 \times 10^{-6} \dfrac{1}{°\text{F}}\right)(400°\text{F})$$

$$= 0.00260$$

Use the formula for engineering strain, and solve for the change in length corresponding to this strain.

Engineering Strain

$$\epsilon = \frac{\Delta L}{L_o}$$

$$\Delta L = \epsilon L_o = (0.00260)\left((300 \text{ ft})\left(12 \dfrac{\text{in}}{\text{ft}}\right)\right)$$

$$= 9.36 \text{ in} \quad (9.4 \text{ in})$$

The answer is (D).

65 Materials Handling and Processing

Content in blue refers to the *NCEES Handbook*.

PRACTICE PROBLEMS

1. The work performed in raising a 500 kg elevator 30 m is most nearly

(A) 75 kJ
(B) 150 kJ
(C) 15,000 kJ
(D) 150,000 kJ

2. A baghouse collector captures and stores chips, dust, and shavings from a woodworking plant. The particulate matter gathers in the funnel-shaped base of the collector and then falls through an orifice at the bottom. If the particulate matter is cohesive enough, it may form an arched cap over this orifice and block the flow. What is this behavior called?

(A) bridging
(B) choking
(C) corking
(D) piping

3. A large, flat-bottom steel silo has a diameter, D, of 12 ft and a height, h, of 60 ft. The silo is filled with polyethylene pellets (bulk specific weight, ρ, of 35 lbf/ft³ and angle of internal friction, ϕ, of 20°). The vertical pressure can be found from the Janssen equation, as shown.

$$p_h = \frac{\rho g D}{4 g_c \mu k}\left(1 - \exp\left(\frac{-4h\mu k}{D}\right)\right)$$

The coefficient of internal friction is calculated by using,

$$\mu = \tan(\phi)$$

The pressure ratio is found from the Rankine model of earth pressure, as shown.

$$k = \tan^2\left(45° - \frac{\phi}{2}\right)$$

What is most nearly the pressure on the vertical wall at the bottom of the silo?

(A) 280 lbf/ft²
(B) 340 lbf/ft²
(C) 390 lbf/ft²
(D) 570 lbf/ft²

4. A continuous rubber belt conveyor operates at 120 ft/min. The coefficient of friction between the belt and its pulleys is 0.35. The belt wrap angle on the driving pulley is 240°. The maximum belt tension is 750 lbf. The power required to drive the belt is most nearly

(A) 0.6 hp
(B) 1.4 hp
(C) 2.1 hp
(D) 2.9 hp

5. 1800 lbm/hr of sawdust with a bulk density of 11 lbm/ft³ is conveyed by air at standard conditions through a 6 in schedule-40 pipe. The material loading ratio is 1:2. What is most nearly the air velocity?

(A) 1700 ft/min
(B) 2200 ft/min
(C) 2600 ft/min
(D) 4000 ft/min

SOLUTIONS

1. The work due to weight is

$$W_w = -w\Delta s = -(mg)(s_2 - s_1)$$ Work

$$= \frac{-(500 \text{ kg})\left(9.81 \frac{\text{m}}{\text{s}^2}\right)(0 \text{ m} - 30 \text{ m})\left(1 \frac{\text{J}}{\text{kg} \cdot \frac{\text{m}^2}{\text{s}^2}}\right)}{1000 \frac{\text{J}}{\text{kJ}}}$$

$$= 147.15 \text{ kJ} \quad (150 \text{ kJ})$$

The answer is (B).

2. Bridging, also called *arching*, occurs when the contents of a bin or hopper settle into an arched cap that covers the opening. Bridging can be eliminated by reducing cohesion in the material or by enlarging the opening.

The answer is (A).

3. Calculate the coefficient of internal friction.

$$\mu = \tan(\phi) = \tan(20) = 0.3640$$

The pressure ratio is

$$k = \tan^2\left(45° - \frac{\phi}{2}\right)$$
$$= \tan^2\left(45° - \frac{20°}{2}\right)$$
$$= 0.490$$

The vertical pressure on the silo's flat bottom is

$$p_h = \frac{\rho g D}{4 g_c \mu k}\left(1 - \exp\left(\frac{-4 h \mu k}{D}\right)\right)$$

$$= \frac{\left(35 \frac{\text{lbf}}{\text{ft}^3}\right)\left(32.2 \frac{\text{ft}}{\text{sec}^2}\right)(12 \text{ ft})}{(4)\left(32.2 \frac{\text{lbm-ft}}{\text{lbf-sec}^2}\right)(0.3640)(0.490)}$$
$$\times \left(1 - \exp\left(\frac{-(4)(60 \text{ ft})(0.3640)(0.490)}{12 \text{ ft}}\right)\right)$$

$$= 572.1 \text{ lbf/ft}^2 \quad (570 \text{ lbf/ft}^2)$$

The answer is (D).

4. Convert the belt wrap angle to radians.

$$\theta = (240°)\left(\frac{2\pi \text{ rad}}{360°}\right) = \frac{4\pi}{3} \text{ rad}$$

The relationship among the tight-side and loose-side tensions, T_1 and T_2, the coefficient of friction, μ, and the belt wrap angle (in radians), θ, is given by the belt friction equation. Rearranging, the loose-side tension is

Belt Friction

$$F_1 = F_2 e^{\mu\theta}$$

$$F_2 = \frac{F_1}{e^{\mu\theta}} = \frac{750 \text{ lbf}}{e^{(0.35)(4\pi/3 \text{ rad})}}$$
$$= 173.1 \text{ lbf}$$

The power required is

Shaft-Horsepower Relationship and Force-Horsepower Relationship

$$\text{HP} = \frac{F\text{v}}{33{,}000} = \frac{(750 \text{ lbf} - 173.1 \text{ lbf})\left(120 \frac{\text{ft}}{\text{min}}\right)}{33{,}000 \frac{\text{ft-lbf}}{\text{hp-min}}}$$

$$= 2.098 \text{ hp} \quad (2.1 \text{ hp})$$

The answer is (C).

5. The rate of material conveyance per minute is

$$\dot{m}_{\text{sawdust}} = \frac{1800 \frac{\text{lbm}}{\text{hr}}}{60 \frac{\text{min}}{\text{hr}}}$$
$$= 30 \text{ lbm/min}$$

The material-air ratio is 1:2, and the density of dry air at standard conditions is 0.075 lbm/ft³. [Standard Dry Air Conditions at Sea Level]

The flow rate of air is

$$Q_{\text{air}} = \frac{\dot{m}_{\text{air}}}{\rho_{\text{air}}} = \frac{(2)\left(30 \frac{\text{lbm}}{\text{min}}\right)}{0.075 \frac{\text{lbm}}{\text{ft}^3}}$$

$$= 800 \text{ ft}^3/\text{min}$$

The internal area of a 6 in schedule-40 pipe is 28.876 in² (0.2006 ft²). [Schedule 40 Steel Pipe]

The air velocity is

$$v_{air} = \frac{Q_{air}}{A_{pipe}} = \frac{800 \ \frac{ft^3}{min}}{0.2006 \ ft^2}$$
$$= 3988 \ ft/min \quad (4000 \ ft/min)$$

The answer is (D).

66 Fire Protection Sprinkler Systems

Content in blue refers to the *NCEES Handbook*.

PRACTICE PROBLEMS

1. The Safety Data Sheet (SDS) for a liquid substance lists the following properties.

flash point (closed cap)	−4–160°F
flammable range (in air)	0.8–16%
VOC (lbm/gal)	0.0–7.5

How should this material be classified?

(A) flammable

(B) combustible

(C) explosive

(D) not flammable, combustible, or explosive

2. A jet of water flows from an open hydrant hose port at a rate of 900 gal/min. The centerline of the port is 30 in above level grade, and the inside diameter of the port is 2.5 in. Wind and air resistance are negligible. Approximately how far does the center of the water jet travel before hitting the ground?

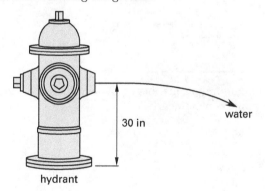

(A) 15 ft

(B) 23 ft

(C) 32 ft

(D) 76 ft

3. The post indicator valve in a fire sprinkler system makes it possible to

  I. restart the system in the event of failure

  II. stop the water supply from outside the building

  III. adjust and optimize flow to the sprinklers

  IV. prevent freezing in winter

  V. maintain and repair the system

(A) I and III only

(B) I and V only

(C) II and IV only

(D) II and V only

4. An unmonitored fire suppression system is being planned for a combined horse barn (moderate amounts of hay), stable, and attached garage/workshop. The building has 11 ft ceilings and a protected area of 2500 ft^2. Water for the system will come from a dedicated tank on the property. A 20% allowance will be added to the sprinkler demand to account for friction loss and multiple fire heads operating simultaneously. What is most nearly the minimum tank size required for the system?

(A) 49,000 gal

(B) 63,000 gal

(C) 70,000 gal

(D) 76,000 gal

5. Pipe intended for a water main is marked "PE3408/PE4710 - PE100." What pipe characteristics are indicated by the designation "PE4710"?

(A) conventional polyethylene pipe with a hydrostatic design basis of 1000 psi

(B) polypropylene pipe with a slow crack growth resistance of more than 7 hr

(C) plastic pipe with a sandbox crush resistance performance evaluation of at least 47 psi

(D) high-performance polyethylene pipe with a hydrostatic design stress of 1000 psi

6. A hydraulically calculated sprinkler design uses a total of 20 identical sprinklers arranged in five identical four-sprinkler runs. The design is based on a minimum water pressure of 15 psi in the supply line. The sprinkler at the end of each run has a discharge coefficient of 6.2 (customary U.S. units) for pure water (SG = 1). According to NFPA 13, what is most nearly the water flow rate that must be supplied for the design to function properly?

(A) 96 gpm

(B) 190 gpm

(C) 240 gpm

(D) 480 gpm

7. A section of sprinkler system piping spans a distance of 100 ft. The pipe run is level and includes two 45° elbows ($K = 0.3$). The pressure at the entrance is 90 psia. The minimum residual pressure head at the end of the run is 4 ft wg. The pipe is standard weight, 2 in B36.10 sprinkler system piping, which has a specific roughness of 0.000005 ft. What is most nearly the maximum flow rate of 60°F water through the pipe run?

(A) 210 gpm

(B) 410 gpm

(C) 620 gpm

(D) 860 gpm

8. When selecting a pump for a fire protection system, which two of the following statements are true?

I. Churn flow head must be no greater than 140% of the rated head.

II. Churn flow head must be at least 140% of the rated head.

III. The capacity at 65% of the rated head must be no greater than 150% of total capacity at the design point.

IV. The capacity at 65% of the rated head must be at least 150% of total capacity at the design point.

(A) I and III only

(B) I and IV only

(C) II and III only

(D) II and IV only

9. A pumper truck draws water from an open reservoir below. The pumper truck and suction piping are 20 ft above the surface of the reservoir. The equivalent length of the suction piping is 45 ft, the diameter of the suction piping is 5 in, and the Hazen-Williams coefficient (C-value) for the suction piping is 90. The truck draws water at a rate of 500 gpm. The atmospheric pressure is 14.7 psia. The temperature in the reservoir is 65°F.

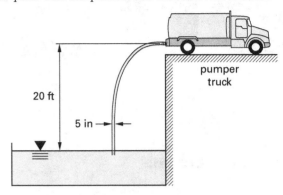

The net positive suction head available (NPSH$_A$) to the pump is most nearly

(A) 5.3 ft

(B) 7.1 ft

(C) 8.8 ft

(D) 9.7 ft

10. A pumper truck draws water from an open reservoir below. The pump suction piping is 14 ft above the surface of the reservoir. The diameter of the suction piping is 4 in, and the Hazen-Williams coefficient (C-value) for the suction piping is 85. The truck draws water from the reservoir at a rate of 600 gpm. The net positive suction head required (NPSH$_R$) for the pump to draw at this rate is 7 ft. The atmospheric pressure is 14.7 psia, and the temperature in the reservoir is 75°F.

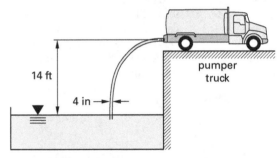

The maximum permissible equivalent length of the suction piping is most nearly

(A) 26 ft

(B) 57 ft

(C) 73 ft

(D) 96 ft

SOLUTIONS

1. The lowest flash point is lower than 100°F, so the substance is flammable. An explosive material does not need an oxidizer. Gasoline requires the oxygen in atmospheric air to burn; therefore, gasoline is not explosive. Flammable range and VOC content are not used in classifying the liquid.

The answer is (A).

2. Use the projectile motion equations. From Table 56.2, the horizontal and vertical distances traveled by the center of the water jet starting at $(0,0)$ in time t are

$$x(t) = v_0 t$$
$$y(t) = H - \frac{1}{2}gt^2$$

When the water hits the ground, the vertical distance traveled, H, is 30 in. Determine the total duration taken to travel this distance.

$$y(t) = H - \frac{1}{2}gt^2$$
$$t = \sqrt{\frac{2H}{g}}$$
$$= \sqrt{\frac{(2)(30 \text{ in})}{\left(32.2 \frac{\text{ft}}{\text{sec}^2}\right)\left(12 \frac{\text{in}}{\text{ft}}\right)}}$$
$$= 0.394 \text{ sec}$$

The flow rate of the water is

$$Q_0 = \frac{900 \frac{\text{gal}}{\text{min}}}{\left(7.48 \frac{\text{gal}}{\text{ft}^3}\right)\left(60 \frac{\text{sec}}{\text{min}}\right)} = 2.005 \text{ ft}^3/\text{sec}$$

The cross-sectional area of the port is

$$A = \frac{\pi d^2}{4} = \frac{\pi \left(\frac{2.5 \text{ in}}{12 \frac{\text{in}}{\text{ft}}}\right)^2}{4}$$
$$= 0.034 \text{ ft}^2$$

The average velocity is

$$v_0 = \frac{Q_0}{A} = \frac{2.005 \frac{\text{ft}^3}{\text{sec}}}{0.034 \text{ ft}^2}$$
$$= 58.97 \text{ ft/sec}$$

The horizontal distance traveled by the water before it hits the ground is

$$x(t) = v_0 t$$
$$= \left(58.97 \frac{\text{ft}}{\text{sec}}\right)(0.394 \text{ sec})$$
$$= 23.23 \text{ ft} \quad (23 \text{ ft})$$

The answer is (B).

3. The post indicator valve (PIV) cannot be used to restart the system in the event of failure, so I is incorrect. The PIV makes it possible to stop the water supply from outside the building in case an emergency situation makes it dangerous to enter, so II is correct. The PIV can only be fully open or fully closed, so adjustment of flow is not possible, making III incorrect. Supply piping for a sprinkler system is buried deep enough to prevent freezing in the winter, so the PIV plays no role in this, and IV is incorrect. The PIV permits the system to be disconnected from the water supply for maintenance and repair, so V is correct.

The answer is (D).

4. Determine the hazard classification for the building. A hazard classification of ordinary hazard group 2 (OH-2) is standard for garages and workshops. Moderate amounts of hay would not change this classification. From Fig. 66.10, the required design density, ρ_S, for an OH-2 occupancy with a protected area, A_c, of 2500 ft² is 0.18 gpm/ft². The sprinkler demand is

$$Q_{\text{sprinkler}} = \rho_S A_c$$
$$= \left(0.18 \frac{\frac{\text{gal}}{\text{min}}}{\text{ft}^2}\right)(2500 \text{ ft}^2)$$
$$= 450 \text{ gpm}$$

From Table 66.5, the sprinkler flow for an unmonitored system in an OH occupancy must be sustained for 90 minutes. The water volume required to support the sprinkler demand, including the 20% allowance, is

$$V_{\text{sprinkler}} = 1.2 Q_{\text{sprinkler}} t = (1.2)\left(450 \frac{\text{gal}}{\text{min}}\right)(90 \text{ min})$$
$$= 48{,}600 \text{ gal}$$

From Table 66.5, the total combined inside and outside hose stream allowance is 250 gpm. The water volume required is

$$V_{\text{hose}} = Q_{\text{hose}} t = \left(250 \frac{\text{gal}}{\text{min}}\right)(90 \text{ min}) = 22{,}500 \text{ gal}$$

The minimum tank size is

$$V_{tank} = V_{sprinkler} + V_{hose} = 48{,}600 \text{ gal} + 22{,}500 \text{ gal}$$
$$= 71{,}100 \text{ gal}$$

Of the options listed, only a tank size of 76,000 gal will hold 71,100 gal.

The answer is (D).

5. It is standard to mark plastic pipe with a designation indicating the type of plastic, followed by four numbers that describe its key properties. From ASTM D3350, "PE" refers to polyethylene; "4" refers to high-performance, high-density polyethylene with a density cell class of 4 (0.947–0.955 g/cc); "7" refers to slow crack growth (SCG) resistance cell class 7 (Pennsylvania notch test (PENT) value > 500 hr); and "10" refers to a 1000 psi hydrostatic design stress with water at 73°F.

The answer is (D).

6. Use Eq. 66.4(b). The system flow rate is

$$Q_{total} = nQ_{single \text{ sprinkler}} = n\left(K\sqrt{\frac{p}{SG}}\right)$$

$$= (20)\left(6.2\sqrt{\frac{15 \frac{\text{lbf}}{\text{in}^2}}{1}}\right)$$

$$= 480 \text{ gpm}$$

The answer is (D).

7. The specific roughness is given. However, sprinkler calculations most commonly use the Hazen-Williams equation, which requires a value for the Hazen-Williams coefficient. An exact conversion from specific roughness to Hazen-Williams coefficient is only possible if the Reynolds number is known. Otherwise, either an iterative solution or an estimate based on sound logic is required. Apparently, the specific roughness is given to help select the Hazen-Williams coefficient.

In this case, per Table 17.2, the specific roughness corresponds to a smooth pipe. In actual practice, this degree of smoothness isn't achievable by steel pipe, so the pipe will either be made of copper or plastic. For smooth copper and plastic pipes, a Hazen-Williams coefficient of 150 is commonly used.

Find the total equivalent length of the pipe. Appendix 66.A lists the equivalent length of a 2 in 45° elbow as 2 ft. Since App. 66.A is based on a Hazen-Williams coefficient of 120, use the modifying factor of 1.51 from Table 66.8.

The total equivalent pipe length of the pipe and two elbows is

$$L_e = L_{pipe} + L_{2 \text{ elbows}} = 100 \text{ ft} + \frac{(2)(2 \text{ ft})}{1.51} = 102.6 \text{ ft}$$

The residual pressure head at the pipe end cannot be less than 4 ft of water, which is equivalent to a pressure of

$$p = \gamma h = \frac{\left(62.4 \frac{\text{lbf}}{\text{ft}^3}\right)(4 \text{ ft})}{\left(12 \frac{\text{in}}{\text{ft}}\right)^2} = 1.73 \text{ lbf/in}^2$$

The maximum friction loss due to the pipe run and elbows is

$$\Delta p_{f,max} = p_1 - p_2 = 90 \frac{\text{lbf}}{\text{in}^2} - 1.73 \frac{\text{lbf}}{\text{in}^2}$$
$$= 88.27 \text{ lbf/in}^2$$

From App. 16.B, the inside diameter, D, of 2 in B36.10 pipe is 2.067 in. Rearrange Eq. 66.10(b) to solve for the flow rate.

$$p_{f,psi} = \frac{4.52 L_{ft} Q_{gpm}^{1.85}}{C^{1.85} D_{in}^{4.87}}$$

$$Q = \left(\frac{p_{f,psi} C^{1.85} D_{in}^{4.87}}{4.52 L_{ft}}\right)^{1/1.85}$$

$$= \left(\frac{\left(88.27 \frac{\text{lbf}}{\text{in}^2}\right)(150)^{1.85}(2.067 \text{ in})^{4.87}}{(4.52)(102.6 \text{ ft})}\right)^{1/1.85}$$

$$= 413.8 \text{ gpm} \quad (410 \text{ gpm})$$

The answer is (B).

8. According to NFPA 20, the head at churn (or shutoff) flow must be no greater than 140% of the rated head. The capacity at 65% of rated head must be no less than 150% of total capacity at the design point.

The answer is (B).

9. From Eq. 18.4(b), the atmospheric head is

$$h_{atm} = \frac{p_{atm}}{\gamma_{water}} = \frac{\left(14.7 \frac{\text{lbf}}{\text{in}^2}\right)\left(12 \frac{\text{in}}{\text{ft}}\right)^2}{62.4 \frac{\text{lbf}}{\text{ft}^3}} = 33.92 \text{ ft}$$

The static suction lift, $h_{z(s)}$, is the distance from the pump inlet to the surface of the reservoir, -20 ft. From the Hazen-Williams equation, Eq. 17.20, the friction loss in feet per foot of pipe is

Pressure Drop of Water Flowing in Circular Pipe (Hazen-Williams)

$$h_{f,\text{ft/ft}} = \frac{10.44 Q_{\text{gpm}}^{1.85}}{C^{1.85} D^{4.87}}$$

$$= \frac{(10.44)\left(500 \dfrac{\text{gal}}{\text{min}}\right)^{1.85}}{(90)^{1.85}(5 \text{ in})^{4.87}}$$

$$= 0.0983 \text{ ft/ft}$$

Multiply the friction loss in feet per foot of pipe by the equivalent length of the suction piping to find the total friction loss.

$$h_f = \left(0.0983 \ \dfrac{\text{ft}}{\text{ft}}\right)(45 \text{ ft})$$

$$= 4.42 \text{ ft}$$

(The Hazen-Williams equation is an empirical formula and not dimensionally consistent.)

By interpolation from App. 23.A, the vapor pressure of water at 65°F is 0.3060 lbf/in², so the vapor pressure head is

$$h_{\text{vp}} = \frac{p_{\text{vapor}}}{\gamma_{\text{water}}} = \frac{\left(0.3060 \ \dfrac{\text{lbf}}{\text{in}^2}\right)\left(12 \ \dfrac{\text{in}}{\text{ft}}\right)^2}{62.4 \ \dfrac{\text{lbf}}{\text{ft}^3}}$$

$$= 0.7062 \text{ ft}$$

From Eq. 18.35, the NPSH$_A$ is

$$\text{NPSH}_A = h_{\text{atm}} + h_{z(s)} - h_{f(s)} - h_{\text{vp}}$$

$$= 33.92 \text{ ft} + (-20 \text{ ft}) - 4.42 \text{ ft} - 0.7062 \text{ ft}$$

$$= 8.7938 \text{ ft} \quad (8.8 \text{ ft})$$

The answer is (C).

10. From Eq. 18.4(b), the atmospheric head is

$$h_{\text{atm}} = \frac{p_{\text{atm}}}{\gamma_{\text{water}}} = \frac{\left(14.7 \ \dfrac{\text{lbf}}{\text{in}^2}\right)\left(12 \ \dfrac{\text{in}}{\text{ft}}\right)^2}{62.4 \ \dfrac{\text{lbf}}{\text{ft}^3}} = 33.92 \text{ ft}$$

The static suction lift, $h_{z(s)}$, is the distance from the pump inlet to the surface of the reservoir, -14 ft.

By interpolation from App. 23.A, the vapor pressure of water at 75°F is 0.4304 lbf/in², so the vapor pressure head is

$$h_{\text{vp}} = \frac{p_{\text{vapor}}}{\gamma_{\text{water}}} = \frac{\left(0.4304 \ \dfrac{\text{lbf}}{\text{in}^2}\right)\left(12 \ \dfrac{\text{in}}{\text{ft}}\right)^2}{62.4 \ \dfrac{\text{lbf}}{\text{ft}^3}} = 0.9932 \text{ ft}$$

The net positive suction head available (NPSH$_A$) must be greater than or equal to NPSH$_R$. Solving Eq. 18.35 for the suction piping loss, and substituting NPSH$_R$ for NPSH$_A$, the maximum allowable suction piping loss is

$$\text{NPSH}_A = h_{\text{atm}} + h_{z(s)} - h_{f(s)} - h_{\text{vp}}$$

$$h_{f(s)} = h_{\text{atm}} + h_{z(s)} - h_{\text{vp}} - \text{NPSH}_R$$

$$= 33.92 \text{ ft} + (-14 \text{ ft}) - 0.9932 \text{ ft} - 7 \text{ ft}$$

$$= 11.93 \text{ ft}$$

Use the Hazen-Williams equation, Eq. 17.20, to find the friction loss in feet of head per foot of suction piping. (The Hazen-Williams equation is an empirical formula and not dimensionally consistent.)

Pressure Drop of Water Flowing in Circular Pipe (Hazen-Williams)

$$h_{f,\text{ft/ft}} = \frac{10.44 Q_{\text{gpm}}^{1.85}}{C^{1.85} D^{4.87}}$$

$$= \frac{\left(600 \ \dfrac{\text{gal}}{\text{min}}\right)^{1.85}}{(85)^{1.85}(4 \text{ in})^{4.87}}$$

$$= 0.4538 \text{ ft/ft}$$

Use the value for the total friction loss (11.93 ft) and the value for the friction loss in feet of head per foot of pipe (0.4536 ft/ft) to find the allowable equivalent length of suction piping.

$$h_f = 0.4538 L$$

$$L = \frac{h_f}{0.4538 \ \dfrac{\text{ft}}{\text{ft}}}$$

$$= \frac{11.93 \text{ ft}}{0.4538 \ \dfrac{\text{ft}}{\text{ft}}}$$

$$= 26.289 \text{ ft} \quad (26 \text{ ft})$$

The answer is (A).

67 Pollutants in the Environment

Content in blue refers to the *NCEES Handbook*.

PRACTICE PROBLEMS

1. A PM-10 sample is collected with a high-volume sampler over a sampling interval of 24 hours and 10 minutes. The sampler draws an average flow of 41 ft^3/min and collects particles on a glass fiber filter. The initial and final masses of the filter are 4.4546 g and 4.4979 g, respectively. The PM-10 concentration is most nearly

(A) 11 μg/m^3

(B) 26 μg/m^3

(C) 42 μg/m^3

(D) 55 μg/m^3

2. An optical discriminator uses light attenuation to size particles. When a particle flows into the sampling chamber, the brightness of a light source at the photodiode is diminished, decreasing the voltage in the detection circuit. When a spherical particle of diameter D enters the sampling chamber, the voltage drop is ΔE_1. Subsequently, when two spherical particles, one with the same diameter, d, and the other with diameter $0.5d$, enter the sampling chamber, the voltage drop is ΔE_2. Most nearly, what is the ratio $\Delta E_2/\Delta E_1$?

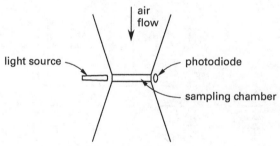

(A) 1.1

(B) 1.3

(C) 1.5

(D) 1.8

3. Which statement related to smog is INCORRECT?

(A) Smog is produced when ozone precursors such as nitrogen dioxide, hydrocarbons, and volatile organic compounds (VOCs) react with sunlight.

(B) Peroxyacyl nitrates contribute to the formation of smog.

(C) Volatile organic compounds (VOCs) contributing to smog are the products of incomplete combustion reactions in automobiles, refineries, and hazardous waste incinerators.

(D) Ground-level ozone is a secondary pollutant that contributes to the formation of smog.

4. Smog generation generally increases with all of the following conditions EXCEPT

(A) temperature inversions

(B) absence of wind

(C) increased commuter traffic

(D) early morning clouds and fog

5. The concentration, C, of nitrogen oxides in a city over a 24-hour period is shown.

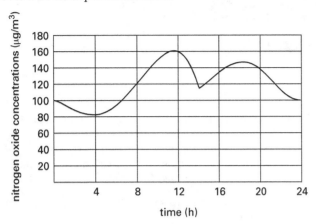

The concentration from $t = 0$ h to $t = 14$ h is described by

$$C_{\mu g/m^3} = 100 - 3.7848t - 2.2609t^2 + 0.6381t^3 - 0.0324t^4$$

The concentration from $t = 14$ h to $t = 24$ h is described by

$$C_{\mu g/m^3} = 5.1885 \times 10^3 - 1.1899 \times 10^3 t + 101.8740t^2 - 3.7638t^3 + 0.0507t^4$$

What is most nearly the average nitrogen oxide concentration over the 24-hour period?

(A) 118 $\mu g/m^3$

(B) 121 $\mu g/m^3$

(C) 244 $\mu g/m^3$

(D) 250 $\mu g/m^3$

SOLUTIONS

1. The duration of sampling is

$$t = (24 \text{ hr})\left(60 \ \frac{\text{min}}{\text{hr}}\right) + 10 \text{ min} = 1450 \text{ min}$$

The volume of air sampled is

$$V = Q_{\text{ave}}t = \left(\frac{41 \ \frac{\text{ft}^3}{\text{min}}}{\left(3.281 \ \frac{\text{ft}}{\text{m}}\right)^3}\right)(1450 \text{ min})$$

$$= 1683 \text{ m}^3$$

The PM-10 concentration is

$$C_{\text{PM-10}} = \frac{m_{\text{particles}}}{V} = \frac{m_{\text{final}} - m_{\text{initial}}}{V}$$

$$= \frac{(4.4979 \text{ g} - 4.4546 \text{ g})\left(10^6 \ \frac{\mu g}{g}\right)}{1683 \text{ m}^3}$$

$$= 25.7 \ \mu g/m^3 \quad (26 \ \mu g/m^3)$$

The answer is (B).

2. From Eq. 69.14, the voltage drop for a single particle is

$$\Delta E = \frac{A_{\text{shadow}} E_0}{A_{\text{detector}}}$$

The ratio of the voltage drop for two particles, E_{2P}, to the voltage drop with the single particle, E_{1P}, is

$$\frac{\Delta E_{2P}}{\Delta E_{1P}} = \frac{\dfrac{A_{2P} E_0}{A_{\text{detector}}}}{\dfrac{A_{1P} E_0}{A_{\text{detector}}}} = \frac{\dfrac{\left(\pi\left(\dfrac{d}{2}\right)^2 + \pi\left(\dfrac{\dfrac{d}{2}}{2}\right)^2\right) E_0}{A_{\text{detector}}}}{\dfrac{\pi\left(\dfrac{d}{2}\right)^2 E_0}{A_{\text{detector}}}}$$

$$= \frac{\left(\dfrac{1}{2}\right)^2 + \left(\dfrac{1}{4}\right)^2}{\left(\dfrac{1}{2}\right)^2}$$

$$= 1.25 \quad (1.3)$$

The answer is (B).

3. Volatile organic compounds (VOCs) are emitted by manufacturing and refining processes, dry cleaners, gasoline stations, print shops, painting operations, municipal wastewater treatment plants, not by combustion sources.

The answer is (C).

4. Temperature inversions are conducive to smog generation because they keep warm air near the ground. An absence of wind causes smog to increase because smog precursors are not dispersed. Smog also increases with traffic, as automobile emissions contain nitrogen oxides and hydrocarbons, which are precursors to smog. Smog is primarily a sunlight-induced reaction, and smog production usually peaks in early afternoon when there is the most sunlight.

The answer is (D).

5. Use integration to calculate the average, C_{ave}.

$$C_{\text{ave},\mu g/m^3} = \frac{\int_{0\,h}^{14\,h} \left(\begin{array}{c} 100 - 3.7848t - 2.2609t^2 \\ +0.6381t^3 - 0.0324t^4 \end{array}\right) dt}{14\,h}$$

$$+ \frac{\int_{14\,h}^{24\,h} \left(\begin{array}{c} 5.1885 \times 10^3 \\ -1.1899 \times 10^3 t \\ +101.8740 t^2 \\ -3.7638 t^3 \\ +0.0507 t^4 \end{array}\right) dt}{10\,h}$$

$$= \frac{\left(\begin{array}{c} 100t - 1.8924t^2 \\ -7.5363 \times 10^{-1} t^3 \\ +0.1595 t^4 - 6.4800 \times 10^{-3} t^5 \end{array}\right)\bigg|_{0\,h}^{14\,h}}{14\,h}$$

$$+ \frac{\left(\begin{array}{c} 5.1885 \times 10^3 t \\ -5.9495 \times 10^2 t^2 \\ +33.9580 t^3 - 0.9410 t^4 \\ +1.0140 \times 10^{-2} t^5 \end{array}\right)\bigg|_{14\,h}^{24\,h}}{10\,h}$$

$$= 244\ \mu g/m^3$$

The answer is (C).

68 Storage and Disposition of Hazardous Materials

Content in blue refers to the *NCEES Handbook*.

PRACTICE PROBLEMS

1. The operator of an underground oil storage tank wishes to permanently convert to above-ground storage. To satisfy federal regulations, what must the operator do?

(A) Drain the existing tank.

(B) Drain, clean, and pressurize the existing tank to 150% of atmospheric pressure.

(C) Drain, clean, and fill the existing tank with sand.

(D) Remove the existing tank, remove and replace soil equal to 100% of the tank volume, and install test wells as per code.

2. According to the Environmental Protection Agency (EPA), leaking tanks and pipes may be repaired as long as the repairs meet "industry codes and standards." Which organization publishes such codes and standards?

(A) American Welding Society (AWS)

(B) National Institute of Standards and Technology (NIST)

(C) International Code Council (ICC)

(D) National Fire Protection Association (NFPA)

3. Contaminants migrate through soils and aquifers at a velocity equal to the

(A) pore velocity

(B) effective velocity

(C) superficial velocity

(D) Darcy velocity

4. Common paint solvents accumulated by a painting contractor are processed in a properly licensed and operating hazardous waste incineration facility. A small amount of mineral ash remains after incineration. Which statement regarding the mineral ash is true?

(A) The ash is considered nonhazardous solid waste and may be disposed of in a municipal solid waste facility that accepts ashes.

(B) The ash is considered an ignitable solid waste and must be buried in a buried (RCRA) municipal solid waste facility.

(C) The waste is considered a "derived from" (D) hazardous waste and must be disposed of in a RCRA hazardous waste facility.

(D) The ash is considered a "K" waste and must be disposed of in a RCRA hazardous waste facility.

5. A large section of town contains vacant factory buildings that, over the years, manufactured numerous unknown products. A developer now wishes to purchase the property from the current owners and convert it to noncommercial use. Public concern centers on possible site contamination that the developer cannot afford to clean up. The property can be referred to as a

(A) Superfund site

(B) designated protection zone

(C) brownfield

(D) undesignated watch site

SOLUTIONS

1. When an underground tank is decommissioned, (1) the regulatory authority must be notified at least 30 days before closing; (2) any contamination must be remediated; (3) the tank must be drained and cleaned by removing all liquids, dangerous vapor levels, and accumulated sludge; and (4) the tank must be removed from the ground or filled with a harmless, chemically inactive solid, such as sand.

The answer is (C).

2. The EPA identifies the following organizations as having relevant codes and standards: API (American Petroleum Institute); ASTM International (formerly American Society for Testing and Materials); KWA (Ken Wilcox Associates, Inc.); NACE International (formerly the National Association of Corrosion Engineers); NFPA (National Fire Protection Association); NLPA (National Leak Prevention Association); PEI (Petroleum Equipment Institute); STI (Steel Tank Institute); and UL (Underwriters Laboratories Inc.).

The answer is (D).

3. Contaminants migrate at the pore velocity, also known as seepage velocity or flow front velocity. Darcy velocity (also known as effective velocity and superficial velocity) does not take the porosity into consideration.

The answer is (A).

4. Although the source material was ignitable, the residual ash is not itself a "specially listed" waste due to its ignitability, corrosivity, reactivity, and toxicity characteristics. Painting is not designated as a specific source industry (such as wood preserving, petroleum refining, and organic chemical manufacturing), so the ash is not a "K" waste. The ash is derived from a hazardous waste, so it is a "derived from" (D) hazardous waste. It must be disposed of in a RCRA hazardous waste facility.

The answer is (C).

5. The current owner, not the developer, would be required to clean up any contamination found. Brownfields are abandoned, idled, or underused industrial and commercial properties where expansion or redevelopment is complicated by actual or suspected environmental contamination.

The answer is (C).

69 Testing and Sampling

Content in blue refers to the *NCEES Handbook*.

PRACTICE PROBLEMS

There are no problems in this book corresponding to Chap. 69 of the *Mechanical Engineering Reference Manual*.

70 Environmental Remediation

Content in blue refers to the *NCEES Handbook*.

PRACTICE PROBLEMS

1. Which statement about petroleum spills and spill remediation is INCORRECT?

(A) As long as the roots of vegetation are kept moist by soil moisture, in situ burning of surface petroleum contaminants has limited long-term environmental effects.

(B) When properly applied in large bodies of water, oil dispersants have little environmental impact.

(C) Washing oil-contaminated sands, soils, and rocks with steam and hot water also removes organisms and nutrients that would otherwise contribute to bioremediation.

(D) Magnetic particle technology helps recover spilled petroleum products for reuse.

2. The treatment of oil-contaminated soil in a large plastic-covered tank is known as

(A) bioremediation
(B) biofiltration
(C) bioreaction
(D) bioventing

3. The performance of a pilot study baghouse follows the filter drag model, with $K_e = 0.6$ in wg-min/ft and $K_s = 0.07$ in wg-ft^2-min/gr. After 20 minutes of operation of the scaled-up baghouse, the dust loading is 27 gr/ft^3, and the air-to-cloth ratio is 3.0 ft/min. Most nearly, what is the pressure drop for the full scale baghouse after 20 minutes?

(A) 2.5 in wg
(B) 4.1 in wg
(C) 5.2 in wg
(D) 7.5 in wg

4. 2500 ft^3/min of air flow into a baghouse. The air has a particulate concentration of 0.8 g/m^3. The instantaneous collection efficiency of the baghouse increases with time as its pores fill with particles, as shown. The collection efficiency is consistent with a rate constant of -1.26 1/hr. Most nearly, what mass of particulate matter is removed in the first five hours of operation?

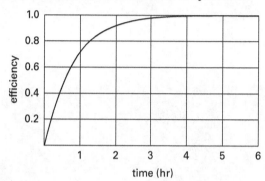

(A) 32 lbm
(B) 36 lbm
(C) 42 lbm
(D) 63 lbm

5. Which statement concerning hazardous waste incinerators is INCORRECT?

(A) The majority of incinerated radioactive waste is high-level waste (HLW) from nuclear power plants.

(B) Open pit incineration results in excessive smoke and high particulate emission.

(C) Single chamber incinerators generally do not meet federal air emission standards.

(D) Multiple chamber incinerators have low particulate emissions and generally meet federal air emission standards without additional air pollution control equipment.

6. A 6 ft high spray tower has a collection efficiency of 80% when removing 10 μm diameter particles. If the tower's height is increased to 8 ft, and if all other characteristics are unchanged, the collection efficiency when removing 7 μm diameter particles will be most nearly

(A) 85%

(B) 91%

(C) 95%

(D) 98%

SOLUTIONS

1. Dispersants are ultimately damaging to the environment. Rather than removing the oil, dispersants break up and distribute the oil, essentially hiding it and making it impossible to remove it from the environment.

The answer is (B).

2. *Bioremediation* is the use of microorganisms to remove pollutants. *Biofiltration* is the use of composting and soil beds to remove pollutants. *Bioreactors* are open or closed tanks that contain dozens or hundreds of slowly rotating disks covered with a biological film of microorganisms used to remove pollutants.

Bioventing is the treatment of contaminated soil in a large plastic-covered tank. Clean air, water, and nutrients are continuously supplied to the tank while off-gases are suctioned off. The off-gases are cleaned with activated carbon adsorption or with thermal or catalytic oxidation prior to discharge.

The answer is (D).

3. The filter drag is

$$S = K_e + K_s W$$
$$= 0.6 \ \frac{\text{in wg-min}}{\text{ft}} + \left(0.07 \ \frac{\text{in wg-ft}^2\text{-min}}{\text{gr}}\right)\left(27 \ \frac{\text{gr}}{\text{ft}^3}\right)$$
$$= 2.49 \ \text{in wg-min/ft}$$

The air-to-cloth ratio is the same as the filtering velocity. The pressure drop is

$$\Delta p = v_{\text{filtering}} S = \left(3.0 \ \frac{\text{ft}}{\text{min}}\right)\left(2.49 \ \frac{\text{in wg-min}}{\text{ft}}\right)$$
$$= 7.47 \ \text{in wg} \quad (7.5 \ \text{in wg})$$

The answer is (D).

4. Instantaneous efficiency (given by the graph) is not average efficiency over time. From Eq. 70.13, the average efficiency of the baghouse, η, over five hours is

$$\eta_{\text{ave}} = \frac{\int_{0 \ \text{hr}}^{5 \ \text{hr}} (1 - e^{-1.26t}) \, dt}{5 \ \text{hr}} = \frac{\left(t + \frac{e^{-1.26t}}{1.26}\right)\Big|_{0 \ \text{hr}}^{5 \ \text{hr}}}{5 \ \text{hr}}$$

$$= \frac{\left(5 \ \text{hr} + \frac{e^{\left(-1.26 \frac{1}{\text{hr}}\right)(5 \ \text{hr})}}{1.26}\right) - \left(0 \ \text{hr} + \frac{e^{\left(-1.26 \frac{1}{\text{hr}}\right)(0 \ \text{hr})}}{1.26}\right)}{5 \ \text{hr}}$$

$$= 0.8416$$

The total mass of particulate matter entering the baghouse in five hours is

$$m_{in} = \rho Q t$$

$$= \frac{\left(0.8 \ \frac{g}{m^3}\right)\left(2500 \ \frac{ft^3}{min}\right)(5 \ hr)\left(60 \ \frac{min}{hr}\right)}{\left(453.6 \ \frac{g}{lbm}\right)\left(3.281 \ \frac{ft}{m}\right)^3}$$

$$= 37.45 \ lbm$$

The mass of particulate matter removed by the filters, m_r, is

$$m_r = \eta_{ave} m_{in} = (0.8416)(37.45 \ lbm)$$
$$= 31.52 \ lbm \quad (32 \ lbm)$$

The answer is (A).

5. The majority of incinerated radioactive waste is low-level waste (LLW).

The answer is (A).

6. Use Eq. 70.41 with the initial conditions to calculate the scrubber constant.

$$K' = \frac{\ln(1-\eta) d_p}{L}$$

$$= \frac{\ln(1-0.8)(10 \ \mu m)\left(10^{-6} \ \frac{m}{\mu m}\right)\left(3.281 \ \frac{ft}{m}\right)}{6 \ ft}$$

$$= -8.80 \times 10^{-6}$$

Using the constant, the scrubber efficiency under the new conditions is

$$\eta = 1 - \exp\left(\frac{K'L}{D_p}\right)$$

$$= 1 - \exp\left(\frac{(-8.80 \times 10^{-6})(8 \ ft)}{(7 \ \mu m)\left(10^{-6} \ \frac{m}{\mu m}\right)\left(3.281 \ \frac{ft}{m}\right)}\right)$$

$$= 0.953 \quad (95\%)$$

The answer is (C).

71 Electricity and Electrical Equipment

Content in blue refers to the NCEES Handbook.

PRACTICE PROBLEMS

1. Most nearly, how much power is dissipated by the circuit shown?

- (A) 1.0 kW
- (B) 3.0 kW
- (C) 4.0 kW
- (D) 48 kW

2. The speed (in rpm) at which an induction motor rotates is

- (A) $\dfrac{120(\text{frequency of the electric source, Hz})}{\text{number of poles}}$
- (B) $\dfrac{120(\text{number of poles})}{\text{frequency of the electric source, Hz}}$
- (C) $\dfrac{120(\text{voltage of the electric source, V})}{\text{frequency of the electric source, Hz}}$
- (D) none of the above

3. A three-phase, four-pole squirrel-cage induction motor operates at full load with an efficiency of 85%, a power factor of 91%, and 3% slip. The operating voltage of the motor is 440 V (rms) at 60 Hz, and the power output of the motor is 200 hp. Use the equation shown to calculate the slip in the motor.

$$N_r = N_s(1-s)$$

The torque developed is most nearly

- (A) 180 ft-lbf
- (B) 450 ft-lbf
- (C) 600 ft-lbf
- (D) 690 ft-lbf

4. A three-phase, four-pole squirrel-cage induction motor operates at full load with an efficiency of 88% and a power factor of 94%. The operating voltage of the motor is 440 V (rms) at 60 Hz, and the power output of the motor is 180 hp. The line current is most nearly

- (A) 130 A
- (B) 150 A
- (C) 210 A
- (D) 280 A

5. The nameplate of an induction motor lists 960 rpm as the full-load speed. Assuming six poles and 4% slip, the frequency the motor was designed for is most nearly

- (A) 24 Hz
- (B) 34 Hz
- (C) 48 Hz
- (D) 50 Hz

6. An insulated air handler unit (AHU) is equipped with a resistive heater. The heater draws 40 kW and is connected to a 460 V (rms) three-phase service. Air at 60°F flows through the AHU at a rate of 5000 ft³/min. Assume operation is adiabatic. The temperature of the air leaving the air handler is most nearly

- (A) 67°F
- (B) 73°F
- (C) 79°F
- (D) 85°F

7. An insulated air handler is equipped with a resistive heater. The heater draws 40 kW and is connected to a 460 V (rms) three-phase service. The rms line current drawn by the heater is most nearly

(A) 29 A

(B) 50 A

(C) 87 A

(D) 150 A

8. A machine shop that operates 14 hours per day, 349 days per year is considering some upgrades to decrease its energy usage. The shop currently uses electric motors generating a total of 600 hp. The upgrade would increase motor electrical-to-mechanical efficiencies from 86% to 95%. The shop is also considering replacing its 50 kW of incandescent lighting with 50 kW of fluorescent lighting. After the upgrade, the overall power factor of the shop will increase from 0.85 to 0.89. The cost for the upgrade is $160,000, and the electricity rate is $0.065/kVA-hr, based on apparent power. The simple payback period for this upgrade is most nearly

(A) 4.3 yr

(B) 5.9 yr

(C) 6.7 yr

(D) 8.1 yr

9. A pump delivers 80 gpm of water to a reservoir at 40 psig located 60 ft above the pump. The water pressure on the suction side of the pump is 10 in Hg (gauge). The total friction head loss in the system is 25 ft of water. Velocity heads are typically small and can be neglected. The pump has an efficiency of 75% and is powered by a three-phase AC motor operating at 120 V with a power factor of 0.82 and an efficiency of 89.5%. The brand of motor is available in 1 hp increments. The maximum amperage needed by the motor is most nearly

(A) 12 A

(B) 24 A

(C) 34 A

(D) 42 A

SOLUTIONS

1. Calculate the equivalent resistance for the two parallel resistors.

$$\frac{1}{R_e} = \frac{1}{R_1} + \frac{1}{R_2} = \frac{1}{0.5 \ \Omega} + \frac{1}{0.5 \ \Omega}$$
$$R_e = 0.25 \ \Omega$$

Although the voltage is AC, the circuit is purely resistive. The power dissipation is real power. The electrical power dissipated is

[Power]

$$P = IV = \frac{V^2}{R_e} = \frac{(110 \ \text{V})^2}{(0.25 \ \Omega)\left(1000 \ \dfrac{\text{W}}{\text{kW}}\right)}$$
$$= 48.4 \ \text{kW} \quad (48 \ \text{kW})$$

The answer is (D).

2. The synchronous speed of an induction motor, N_s, can be calculated by the formula in option A. However, there is always a difference between the synchronous speed, N_s, and the operating speed, N_r. An induction motor always turns a few percent slower than the synchronous speed. This difference (expressed as a percentage of the synchronous speed) is called slip. None of the options mention slip.

The answer is (D).

3. Determine the synchronous speed for a four-pole motor. [Synchronous Speed Motors]

$$N_s = 1800 \ \text{rev/min}$$

Because of slip, the actual speed is

$$N_r = N_s(1-s) = \left(1800 \ \frac{\text{rev}}{\text{min}}\right)(1 - 0.03)$$
$$= 1746 \ \text{rev/min}$$

Use the formula for torque.

[Torques]

$$\text{full load torque in pound-feet} = 5250 \times \frac{\text{horsepower}}{\text{rpm}}$$

$$= (5250)\left(\frac{200 \ \text{hp}}{1746 \ \dfrac{\text{rev}}{\text{min}}}\right)$$
$$= 602 \ \text{ft-lbf} \quad (600 \ \text{ft-lbf})$$

The answer is (C).

4. Use the formula for determining the line current for a three-phase motor.

Power for Different Motor Phases
$$I_{amps} = \frac{P_{hp}(746)}{\sqrt{3}\ V\eta(pf)} = \frac{(180\text{ hp})(746)}{(\sqrt{3})(440\text{ V})(0.88)(0.94)}$$
$$= 213\text{ A} \quad (210\text{ A})$$

The answer is (C).

5. Determine the synchronous speed for a six-pole motor (typically at 60 Hz). [Synchronous Speed Motors]

$$N_s = 1200\text{ rev/min}$$

Calculate the design frequency based on the actual full-load speed, N_a, of 960 rpm. The design frequency is directly proportional to the speed and inversely proportional to the complement of the slip, that is, $(1 - s)$. Therefore, the design frequency is

$$\frac{f_a}{f_s} = \left(\frac{N_a}{N_s}\right)\left(\frac{1-s_s}{1-s_a}\right)$$

$$f_a = f_s\left(\frac{N_a}{N_s}\right)\left(\frac{1-s_s}{1-s_a}\right)$$

$$= (60\text{ Hz})\left(\frac{960\ \frac{\text{rev}}{\text{min}}}{1200\ \frac{\text{rev}}{\text{min}}}\right)\left(\frac{1-0}{1-0.04}\right)$$

$$= 50\text{ Hz}$$

The answer is (D).

6. The temperature of air will increase due to the sensible heat from the heater. Calculate the temperature increase by using the heat gain equation for sensible heating. Use the conversion factor to obtain the sensible heat gain in Btu/hr. [Measurement Relationships]

Heat Gain Calculations Using Standard Air and Water Values
$$q_{s,\text{Btu/hr}} = 1.10\, Q_s\, \Delta T$$

$$\Delta T = \frac{q_s}{1.10\, Q_s} = \frac{(40\text{ kW})\left(3413\ \frac{\text{Btu}}{\text{hr}\cdot\text{kW}}\right)}{\left(1.10\ \frac{\text{Btu-min}}{\text{ft}^3\text{-hr-}°\text{F}}\right)\left(5000\ \frac{\text{ft}^3}{\text{min}}\right)}$$

$$= 24.82°\text{F}$$

Calculate the exit temperature of the air.

$$T_{exit} = T_{inlet} + \Delta T = 60°\text{F} + 24.82°\text{F}$$
$$= 84.82°\text{F} \quad (85°\text{F})$$

The answer is (D).

7. Use the formula for determining the line current for a three-phase motor. Since the load is purely resistive, the power factor is 1.0.

Power for Different Motor Phases
$$I_{amps} = \frac{P_{kW}(1000)}{\sqrt{3}\ V(pf)} = \frac{(40\text{ kW})\left(1000\ \frac{W}{kW}\right)}{(\sqrt{3})(460\text{ V})(1.0)}$$
$$= 50.2\text{ A} \quad (50\text{ A})$$

The answer is (B).

8. Calculate the kilovolt-ampere input to the motors before the upgrade.

Power Factor
$$\text{kilovolt-ampere input (motors)} = \frac{0.746 \times \text{hp}}{E \times \text{power factor}}$$

$$= \frac{\left(0.746\ \frac{\text{kW}}{\text{hp}}\right)(600\text{ hp})}{0.86 \times 0.85}$$

$$= 612.31\text{ kVA}$$

Calculate the kilovolt-ampere input for lighting before the upgrade.

Power Factor
$$\text{kilovolt-ampere input (lighting)} = \frac{\text{kW}}{\text{power factor}}$$

$$= \frac{50\text{ kW}}{0.85}$$

$$= 58.82\text{ kVA}$$

Calculate the total kilovolt-ampere input before upgrade.

$$\text{kVA}_{before} = \text{kVA}_{motor} + \text{kVA}_{lighting}$$
$$= 612.31\text{ kVA} + 58.82\text{ kVA}$$
$$= 671.13\text{ kVA}$$

Similarly, determine the total kilovolt-ampere input after upgrade.

Calculate the kilovolt-ampere input to the motors after the upgrade.

$$\text{kilovolt-ampere input (motors)} = \frac{0.746 \times \text{hp}}{E \times \text{power factor}}^{\text{Power Factor}}$$

$$= \frac{\left(0.746 \, \frac{\text{kW}}{\text{hp}}\right)(600 \text{ hp})}{0.95 \times 0.89}$$

$$= 529.39 \text{ kVA}$$

Calculate the kilovolt-ampere input for lighting after the upgrade.

$$\text{kilovolt-ampere input (lighting)} = \frac{\text{kW}}{\text{power factor}}^{\text{Power Factor}}$$

$$= \frac{50 \text{ kW}}{0.89}$$

$$= 56.18 \text{ kVA}$$

Calculate the total kilovolt-ampere input after upgrade.

$$\text{kVA}_{\text{after}} = \text{kVA}_{\text{motor}} + \text{kVA}_{\text{lighting}}$$
$$= 529.39 \text{ kVA} + 56.18 \text{ kVA}$$
$$= 585.57 \text{ kVA}$$

Calculate the annual cost savings, ΔA, due to the upgrade.

$$\Delta A = (\text{kVA}_{\text{before}} - \text{kVA}_{\text{after}}) \left(\frac{\text{hr of operation}}{\text{yr}}\right) C_{\text{electricity}}$$

$$= (671.13 \text{ kVA} - 585.57 \text{ kVA})$$

$$\times \left(14 \, \frac{\text{hr}}{\text{day}}\right)\left(349 \, \frac{\text{day}}{\text{yr}}\right)\left(0.065 \, \frac{\$}{\text{kVA-hr}}\right)$$

$$= \$27{,}173/\text{yr}$$

Calculate the payback period for the investment in the upgrade.

$$\text{payback period} = \frac{C_{\text{upgrade}}}{\Delta A} = \frac{\$160{,}000}{\frac{\$27{,}173}{\text{yr}}}$$

$$= 5.89 \text{ yr} \quad (5.9 \text{ yr})$$

The answer is (B).

9. Find the volumetric flow rate of water discharged by the pump in cubic feet per second. [Commonly Used Equivalents]

$$Q = \frac{80 \, \frac{\text{gal}}{\text{min}}}{\left(\frac{448.83 \text{ gpm}}{1 \text{ ft}^3/\text{sec}}\right)} = 0.178 \text{ ft}^3/\text{sec}$$

Calculate the suction pressure (reference point 1) (in feet of head) and the reservoir (reference point 2) pressure. [Measurement Relationships]

$$p_1 = \left(\frac{10 \text{ in Hg}}{12 \, \frac{\text{in}}{\text{ft}}}\right)(13.6 \text{ S.G.}) = 11.33 \text{ ft H}_2\text{O}$$

$$p_2 = \left(40 \, \frac{\text{lbf}}{\text{in}^2}\right)(12 \text{ in/ft})^2 = 5760 \text{ lbf/ft}^2$$

Determine the pump head using the mechanical energy equation in terms of head of fluid. [Mechanical Energy Equation in Terms of Energy Per Unit Weight Involving Heads]

Reference point 1 is the pump suction and reference point is the reservoir.

$$\frac{p_1}{\gamma} + \frac{\text{v}_1^2}{2g} + z_1 + h_p = \frac{p_2}{\gamma} + \frac{\text{v}_2^2}{2g} + z_2 + h_{\text{losses}}^{\text{Bernoulli Equation}}$$

Neglect the velocity heads.

$$11.33 \text{ ft H}_2\text{O} + h_p = \frac{5760 \, \frac{\text{lbf}}{\text{ft}^2}}{62.4 \, \frac{\text{lbf}}{\text{ft}^3}}$$

$$+ 60 \text{ ft} + 25 \text{ ft}$$

$$h_p = 165.98 \text{ ft}$$

Calculate the horsepower supplied to the pump.

$$\text{Pump} = \frac{\text{gpm} \times \text{total head}}{3{,}960 \times \text{efficiency}}^{\text{Pump Power Equation}}$$

$$= \frac{\left(80 \, \frac{\text{gal}}{\text{min}}\right)(165.98 \text{ ft})}{(3960)(0.75)}$$

$$= 4.471 \text{ hp}$$

Since the motor is only available in 1 hp increments, round to 5 hp. Using the full load current equation, the amperage of the motor is

Full-Load Current

$$I_{\max} = \frac{746 \times \text{horsepower}}{1.73 \times \text{efficiency} \times \text{voltage} \times \text{power factor}}$$
$$= \frac{(5)(746)}{(1.73)(120 \text{ V})(0.895)(0.82)}$$
$$= 24.48 \text{ A} \quad (24 \text{ A})$$

The answer is (B).

72 Illumination and Sound

Content in blue refers to the *NCEES Handbook*.

PRACTICE PROBLEMS

1. A 60 m × 24 m product assembly area is illuminated by a gridded arrangement of pendant lamps, each producing 18 000 lm. The minimum illumination required at the work surface level is 200 lux (lm/m^2). The lamps are well maintained and have a maintenance factor of 0.8. The utilization factor is 0.4. Approximately how many lamps are required?

(A) 34
(B) 42
(C) 50
(D) 66

2. A gear-driven electric power generation system has a prime mover noise level of 88 dBA, a gear system noise level of 82 dBA, and a generator noise level of 95 dBA. The overall noise level is most nearly

(A) 75 dBA
(B) 86 dBA
(C) 96 dBA
(D) 99 dBA

3. With no machinery operating, the background noise in a room has a sound pressure level of 43 dB. With the machinery operating, the sound pressure level is 45 dB. The sound pressure level due to the machinery alone is most nearly

(A) 2.0 dB
(B) 41 dB
(C) 47 dB
(D) 49 dB

4. 4 ft (1.2 m) from an isotropic sound source, the sound pressure level is 92 dB. The sound pressure level 12 ft (3.6 m) from the source is most nearly

(A) 62 dB
(B) 73 dB
(C) 83 dB
(D) 87 dB

5. If the number of sabins is 50% of the total room area, the maximum possible reduction in sound pressure level is most nearly a

(A) 1.2 dB decrease
(B) 3.0 dB decrease
(C) 4.6 dB decrease
(D) 6.1 dB decrease

6. The sound impact of a machine on its surroundings is being evaluated. The machine is assembled from 18 different parts. The evaluation should consist of the sound rating of the machine as well as the quality and effects of the machine on speech interference. Which of the following methods is best for use in evaluating the machinery?

(A) dBA
(B) NC
(C) RC Mark II
(D) RNC

SOLUTIONS

1. Use the lumen method from Eq. 72.8 to find the number of lamps required.

$$n = \frac{EA}{\Phi(\text{CU})(\text{LLF})} = \frac{\left(200\ \dfrac{\text{lm}}{\text{m}^2}\right)(60\ \text{m})(24\ \text{m})}{(18\,000\ \text{lm})(0.4)(0.8)}$$
$$= 50$$

The answer is (C).

2. From Eq. 72.22, the governing equation for combining multiple noise sources is

$$L_p = 10\log\sum 10^{L_i/10} = 10\log\begin{pmatrix} 10^{88\ \text{dBA}/10} \\ + 10^{82\ \text{dBA}/10} \\ + 10^{95\ \text{dBA}/10}\end{pmatrix}$$
$$= 96\ \text{dBA}$$

The answer is (C).

3. From Eq. 72.22, the combined sound pressure level from multiple noise sources is

$$L_{p,\text{combined}} = 10\log\sum 10^{L_i/10}$$
$$= 10\log\left(10^{43\ \text{dB}/10} + 10^{L_{p,\text{machine}}/10}\right)$$
$$= 45\ \text{dB}$$

Solve for the unknown machinery sound pressure level, $L_{p,\text{machine}}$.

$$10^{L_{p,\text{machine}}/10} = 10^{45\ \text{dB}/10} - 10^{43\ \text{dB}/10}$$
$$L_{p,\text{machine}} = 10\log\left(10^{45\ \text{dB}/10} - 10^{43\ \text{dB}/10}\right)$$
$$= 40.7\ \text{dB} \quad (41\ \text{dB})$$

The answer is (B).

4. The free-field sound pressure is inversely proportional to the distance from the source.

$$\frac{p_2}{p_1} = \frac{r_1}{r_2}$$

From Eq. 72.19, the sound pressure level is

$$L_{p,2} = L_{p,1} + 10\log\left(\frac{p_2}{p_1}\right)^2 = L_{p,1} + 20\log\frac{p_2}{p_1}$$
$$= L_{p,1} + 20\log\left(\frac{r_1}{r_2}\right)$$

Customary U.S. Solution

$$L_{p,2} = L_{p,1} + 20\log\left(\frac{r_1}{r_2}\right) = 92\ \text{dB} + 20\log\left(\frac{4\ \text{ft}}{12\ \text{ft}}\right)$$
$$= 92\ \text{dB} - 9.5\ \text{dB}$$
$$= 82.5\ \text{dB} \quad (83\ \text{dB})$$

SI Solution

$$L_{p,2} = L_{p,1} + 20\log\left(\frac{r_1}{r_2}\right) = 92\ \text{dB} + 20\log\left(\frac{1.2\ \text{m}}{3.6\ \text{m}}\right)$$
$$= 92\ \text{dB} - 9.5\ \text{dB}$$
$$= 82.5\ \text{dB} \quad (83\ \text{dB})$$

The answer is (C).

5. Define A as the total room area.

$$\sum S_1 = 0.50 A$$

The maximum number of sabins is equal to the room area.

$$\sum S_2 = A$$

Use Eq. 72.27 to find the amount of noise reduction.

$$\text{NR} = 10\log\frac{\sum S_1}{\sum S_2} = 10\log\frac{0.50 A}{A} = 10\log 0.50$$
$$= -3.0\ \text{dB} \quad (3.0\ \text{dB decrease})$$

The answer is (B).

6. RC Mark II method is used in evaluating systems with different parts. The RC Mark II method evaluates the quality of the sound of the machine and its effects on speech interference. [Comparison of Sound Rating Methods]

The answer is (C).

73 Engineering Economic Analysis

Content in blue refers to the *NCEES Handbook*.

PRACTICE PROBLEMS

1. At 6% effective annual interest, approximately how much will be accumulated if $1000 is invested for 10 years?

(A) $560
(B) $790
(C) $1600
(D) $1800

2. $2000 is the future worth after four years with 6% effective annual interest. The present worth is most nearly

(A) $520
(B) $580
(C) $1600
(D) $2500

3. At 6% effective annual interest, the amount that should be invested to accumulate $2000 in 20 years is most nearly

(A) $620
(B) $1400
(C) $4400
(D) $6400

4. The effective annual rate of return on an $80 investment that pays back $120 in seven years is most nearly

(A) 4.5%
(B) 5.0%
(C) 5.5%
(D) 6.0%

5. A speculator in land purchases property valued at $15,000 that he expects to hold for 10 years. A monthly rent of $75 is collected from the tenants. (Use the year-end convention.) Taxes are $150 per year, and maintenance costs are $250 per year. In 10 years, the sale price needed to realize a 10% rate of return is most nearly

(A) $26,000
(B) $31,000
(C) $34,000
(D) $36,000

6. An air conditioner costs $2000. The air conditioner can be purchased using a payment plan consisting of 30 equal payments of $89.30 per month. The effective annual interest rate is most nearly

(A) 27%
(B) 35%
(C) 43%
(D) 51%

7. A depreciable item is purchased for $500,000. The salvage value at the end of 25 years is estimated at $100,000. The depreciation in each of the first three years using the straight-line method is

(A) $4000
(B) $16,000
(C) $20,000
(D) $24,000

8. A depreciable item is purchased for $500,000. The salvage value at the end of 25 years is estimated at $100,000. The depreciation in each of the first three years using the sum-of-the-years' digits method is most nearly

(A) $16,000; $16,000; $16,000
(B) $30,000; $28,000; $27,000
(C) $31,000; $30,000; $28,000
(D) $32,000; $31,000; $30,000

9. Two methods are being considered to meet strict air pollution control requirements over the next 10 years. Method A uses equipment with a life of 10 years. Method B uses equipment with a life of five years that will be replaced with new equipment with an additional life of five years. The capacities of the two methods are different, but operating costs do not depend on the throughput. Operation is 24 hours per day, 365 days per year. The effective annual interest rate for this evaluation is 6%.

	method A	method B	
	years 1–10	years 1–5	years 6–10
installation cost	$13,000	$6000	$7000
equipment cost	$10,000	$2000	$2200
operating cost per hour	$10.50	$8.00	$8.00
salvage value	$5000	$2000	$2000
capacity (tons/yr)	50	20	20
life	10 years	5 years	5 years

Which statement accurately depicts the system option for the given range of throughput?

(A) From 0 to 20 tons/yr, method A is the cheapest.

(B) From 20 to 40 tons/yr, method B is the cheapest.

(C) From 50 to 60 tons/yr, method B is the cheapest.

(D) From 60 to 80 tons/yr, method A is the cheapest.

10. A transit district has asked for assistance in determining the proper fare for its bus system. An effective annual interest rate of 6% is to be used. The following additional information was compiled.

cost per bus	$60,000
bus life	20 years
salvage value	$10,000
miles driven per year	37,440
number of passengers per year	80,000
operating cost	$1.00 per mile in the first year, increasing $0.10 per mile each year thereafter

If the transit district decides to set the per-passenger fare at $0.35 for the first year, approximately how much should the passenger fare go up each year thereafter such that the district can break even in 20 years?

(A) $0.022 increase per year

(B) $0.036 increase per year

(C) $0.067 increase per year

(D) $0.070 increase per year

11. A transit district has asked for assistance in determining the proper fare for its bus system. An effective annual interest rate of 6% is to be used. The following additional information was compiled.

cost per bus	$60,000
bus life	20 years
salvage value	$10,000
miles driven per year	37,440
number of passengers per year	80,000
operating cost	$1.00 per mile in the first year, increasing $0.10 per mile each year thereafter

If the transit district decides to set the per-passenger fare at $0.35 for the first year and the per-passenger fare goes up $0.05 each year thereafter, the additional government subsidy (per passenger) needed for the district to break even in 20 years is most nearly

(A) $0.11

(B) $0.12

(C) $0.16

(D) $0.21

12. A chemical pump motor unit is purchased for $14,000. The estimated life is eight years, after which it will be sold for $1800. The depreciation in the first two years by the sum-of-the-years' digits method is most nearly.

(A) $3600

(B) $5080

(C) $6300

(D) $8920

13. A soda ash plant has the water effluent from processing equipment treated in a large settling basin. The settling basin eventually discharges into a river that runs alongside the basin. Recently enacted environmental regulations require all rainfall on the plant to be diverted and treated in the settling basin. A heavy rainfall will cause the entire basin to overflow. An uncontrolled overflow will cause environmental damage and heavy fines. The construction of additional height on the existing basic walls is under consideration.

Data on the costs of construction and expected costs for environmental cleanup and fines are shown. A study has been performed and the probability of a winter with basin overflow in the coming 15 year period by additional basin height has been determined. The soda ash plant management considers 12% to be their minimum rate of return, and it is felt that after 15 years the plant will be closed. The company wants to select the alternative that minimizes its total expected costs.

additional basin height (ft)	probability of a winter having basin overflow	expense for environmental clean up per winter with basin overflow	construction cost
0	15:15	$550,000	$0
5	8:15	$600,000	$600,000
10	3:15	$650,000	$710,000
15	2:15	$900,000	$1,200,000
20	1:15	$800,000	$1,500,000
	50		

The additional height the basin should be built to is most nearly

(A) 5.0 ft

(B) 10 ft

(C) 15 ft

(D) 20 ft

14. A company is insured for $3,500,000 against fire and the insurance rate is $0.69/$1000. The insurance company will decrease the rate to $0.47/$1000 if fire sprinklers are installed. The initial cost of the sprinklers is $7500. Annual costs are $200; additional taxes are $100 annually. The system life is 25 years. The rate of return is most nearly

(A) 3.8%

(B) 5.0%

(C) 13%

(D) 16%

15. Heat losses through the walls in an existing building cost a company $1,300,000 per year. This amount is considered excessive, and two alternatives are being evaluated. Neither of the alternatives will increase the life of the existing building beyond the current expected life of six years, and neither of the alternatives will produce a salvage value. Improvements can be depreciated.

Alternative A: Do nothing, and continue with current losses.

Alternative B: Spend $2,000,000 immediately to upgrade the building and reduce the loss by 80%. Annual maintenance will cost $150,000.

Alternative C: Spend $1,200,000 immediately. Then, repeat the $1,200,000 expenditure 3 years from now. Heat loss the first year will be reduced 80%. Due to deterioration, the reduction will be 55% and 20% in the second and third years. (The pattern is repeated starting after the second expenditure.) There are no maintenance costs.

All energy and maintenance costs are considered expenses for tax purposes. The company's tax rate is 48%, and straight-line depreciation is used. 12% is regarded as the effective annual interest rate. Evaluate each alternative on an after-tax basis. Which alternative should be recommended?

(A) alternative A

(B) alternative B

(C) alternative C

(D) not enough information

16. A seven-year-old machine is being considered for replacement. The old machine is presumed to have a 10-year life. It has been depreciated on a straight-line basis from its original value of $1,250,000 to a current book value of $620,000. The present salvage value of the old machine is estimated at $400,000, and this is not expected to change over the next three years. The current operating costs are not expected to change from $200,000 per year. A new machine costs $800,000, with operating costs of $40,000 the first year, and increasing by $30,000 each year thereafter. The new machine is expected to have a salvage value of $100,000 after 10 years of useful life. Assume an annual interest rate of 12% before taxes. When should the old machine be replaced?

(A) Replace the old machine immediately.

(B) Replace the old machine after one year.

(C) Replace the old machine after two years.

(D) Replace the old machine after three years.

17. As production facilities move toward just-in-time manufacturing, it is important to minimize

(A) demand rate

(B) production rate

(C) inventory carrying cost

(D) setup cost

18. A company would like to purchase new equipment for $2,000,000 with a lifespan of 7 years. If the company sells the equipment in year 5 for $350,000 and the corporate tax rate is 48%, what is the after-tax salvage value? (Use the MACRS method to calculate depreciation.)

(A) $320,000

(B) $330,000

(C) $396,000

(D) $446,000

SOLUTIONS

1. Draw the cash flow diagram.

Using the factor tables, $(F/P, i, n) = 1.7908$ for $i = 6\%$ and $n = 10$ years. [Economic Factor Tables]

$$F = P(F/P, 6\%, 10) = (\$1000)(1.7908)$$
$$= \$1790.80 \quad (\$1800)$$

The answer is (D).

2. Draw the cash flow diagram.

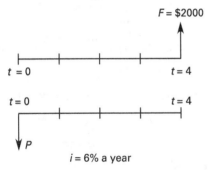

Using the factor tables, $(P/F, i, n) = 0.7921$ for $i = 6\%$ and $n = 4$ years. [Economic Factor Tables]

$$P = F(P/F, 6\%, 4)$$
$$P = (\$2000)(0.7921)$$
$$= \$1584.20 \quad (\$1600)$$

The answer is (C).

3. Draw the cash flow diagram.

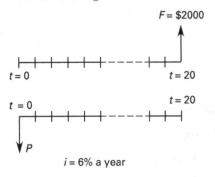

Using the factor tables, $(P/F, i, n) = 0.3118$ for $i = 6\%$ and $n = 20$ years. [Economic Factor Tables]

$$P = F(P/F, 6\%, 20)$$
$$P = (\$2000)(0.3118)$$
$$= \$623.60 \quad (\$620)$$

The answer is (A).

4. Draw the cash flow diagram.

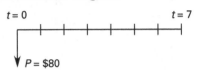

Calculate the F/P value. [Economic Factor Conversions]

$$(F/P, i\%, 7) = F/P = \frac{\$120}{\$80} = 1.5$$

Calculate the effective annual rate of return. [Economic Factor Conversions]

$$(F/P, i\%, n) = (1 + i)^n$$
$$(F/P, 5\%, 7) = (1 + 0.05)^7$$
$$= 1.4071$$
$$(F/P, 6\%, 7) = (1 + 0.06)^7$$
$$= 1.5036$$
$$(F/P, 7\%, 7) = (1 + 0.07)^7$$
$$= 1.6058$$

Therefore, $i \approx 6.0\%$.

The answer is (D).

5. The annual rent is

$$(\$75)\left(12 \, \frac{\text{mo}}{\text{yr}}\right) = \$900$$

The present worth of the land is

$$P = \$15,000$$

The annual costs are

$$A_{rent} = -\$900$$
$$A_{taxes,maint} = \$250 + \$150 = \$400$$

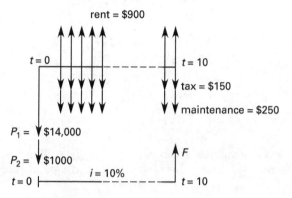

Find the future worth of the land.

$$F = (\$15,000)(F/P, 10\%, 10)$$
$$+ (\$400)(F/A, 10\%, 10)$$
$$- (\$900)(F/A, 10\%, 10)$$

Use the economic factor tables for $i = 10.00\%$ [Economic Factor Tables]

$$(F/P, 10\%, 10) = 2.5937$$
$$(F/A, 10\%, 10) = 15.9374$$

$$F = (\$15,000)(2.5937) + (\$400)(15.9374)$$
$$- (\$900)(15.9374)$$
$$= \$30,937 \quad (\$31,000)$$

The answer is (B).

6.

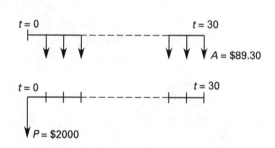

Find the present worth using economic factor conversions. [Economic Factor Conversions]

$$P = A\left(\frac{P}{A}, i, n\right)$$

Therefore,

$$\frac{P}{A}, i, 30 = \frac{\$2000}{\$89.30}$$
$$= 22.40$$

Using a trial-and-error method, find the value of P/A for different interest rates using Factor Tables. [Economic Factor Tables]

Try $i = 1.0\%$, from the Factor Tables, $(P/A, 1\%, 30) = 25.8077 \neq 22.40$

Try $i = 2.0\%$, from the Factor Tables, $(P/A, 2\%, 30) = 22.396 \neq 22.40$

Therefore, the interest rate is 2% per month. To find the effective annual interest rate, use the formula for non-annual compounding 2% per month, but because the rate is already per month, do not divide the rate by the number of compounding periods per year. [Economic Analysis: Nomenclature and Definitions]

$$i_e = (1 + r)^m - 1$$
$$= (1 + 0.02)^{12} - 1 = 0.2682 \quad (27\%)$$

The answer is (A).

7. Use the straight-line method to calculate the depreciation.

Depreciation: Straight Line

$$D = \frac{C - S_n}{n}$$

Each year depreciation will be the same.

$$D = \frac{\$500,000 - \$100,000}{25} = \$16,000$$

The answer is (B).

8. Find the depreciation using the sum-of-years digits method

$$\text{Depreciation: Sum-of-Years Digits Method}$$
$$D_j = \frac{2(C - S_n)(n - j + 1)}{(n)(n+1)}$$

$$D_1 = \frac{(2)(\$500{,}000 - \$100{,}000)(25 - 1 + 1)}{(25)(25 + 1)}$$
$$= \$30{,}769 \quad (\$31{,}000)$$

$$D_2 = \frac{(2)(\$500{,}000 - \$100{,}000)(25 - 2 + 1)}{(25)(25 + 1)}$$
$$= \$29{,}538 \quad (\$30{,}000)$$

$$D_3 = \frac{(2)(\$500{,}000 - \$100{,}000)(25 - 3 + 1)}{(25)(25 + 1)}$$
$$= \$28{,}308 \quad (\$28{,}000)$$

The answer is (C).

9. Find the annual operating costs of method A.

$$C = \left(24 \frac{\text{hr}}{\text{day}}\right)\left(365 \frac{\text{day}}{\text{yr}}\right)\left(\frac{\$10.50}{\text{hr}}\right)$$
$$= \$91{,}980 \text{ operational cost/yr}$$

Find the annual cost of method A. [Economic Factor Tables]

$$A = \$91{,}980 + (\$13{,}000 + \$10{,}000)(A/P, 6\%, 10)$$
$$\quad - (\$5000)(A/F, 6\%, 10)$$
$$= \$91{,}980 + (\$23{,}000)(0.1359)$$
$$\quad - (\$5000)(0.0759)$$
$$= \$94{,}726.20/\text{yr}$$

Find the annual operating costs of method B.

$$C = \left(24 \frac{\text{hr}}{\text{day}}\right)\left(365 \frac{\text{day}}{\text{yr}}\right)\left(\frac{\$8.00}{\text{hr}}\right)$$
$$= \$70{,}080 \text{ operational cost/yr}$$

Find the annual cost of method B. B is salvaged at both year 5 and year 10. [Economic Factor Tables]

$$A = \$70{,}080 + (\$6000 + \$2000)(A/P, 6\%, 5)$$
$$\quad + (\$7000 + \$2200 - \$2000)$$
$$\quad \times (P/F, 6\%, 5)(A/P, 6\%, 5)$$
$$\quad - (\$2000)(A/F, 6\%, 10)$$
$$= \$70{,}080 + (\$8000)(0.2374)$$
$$\quad + (\$7200)(0.7473)(0.2374)$$
$$\quad - (\$2000)(0.0759)$$
$$= \$73{,}104.74/\text{yr}$$

The following table can be used to determine which alternative is least expensive for each throughput range.

tons/yr	cost of using A	cost of using B	cheapest
0–20	\$94,726 (1×)	\$73,104.74 (1×)	B
20–40	\$94,726 (1×)	\$146,209.49 (2×)	A
40–50	\$94,726 (1×)	\$219,314.23 (3×)	A
50–60	\$189,452 (2×)	\$219,314.23 (3×)	A
60–80	\$189,452 (2×)	\$292,418.98 (4×)	A

Of the available answer options, only option D is accurate.

The answer is (D).

10. Calculate the breakeven fare per passenger for 80,000 passengers a year. [Economic Factor Tables][Economic Factor Conversions]

$$A_e = (\$60{,}000)(A/P, 6\%, 20) + A$$
$$\quad + G(P/G, 6\%, 20)(A/P, 6\%, 20)$$
$$\quad - (\$10{,}000)(A/F, 6\%, 20)$$
$$= (\$60{,}000)(0.0872) + \$37{,}440$$
$$\quad + (\$3744)(87.2304)(0.0872)$$
$$\quad - (\$10{,}000)(0.0272)$$
$$= \$70{,}878.70$$

$$\text{fare} = \frac{A_e}{80{,}000} = \frac{\$70{,}878.70}{80{,}000}$$
$$= \$0.885/\text{passenger} \quad (\$0.88/\text{passenger})$$

The passenger fare should go up each year by

$$\$0.885 = \$0.35 + G(A/G, 6\%, 20)$$
$$G = \frac{\$0.885 - \$0.35}{(A/G, 6\%, 20)}$$
$$= \frac{\$0.535}{7.6051}$$
$$= \$0.070 \quad (\$0.070 \text{ increase per year})$$

The answer is (D).

11. First, calculate the total Equivalent Uniform Annual Cost (EUAC) for the bus by considering the initial cost, salvage value, and the annual operating costs. [Economic Factor Tables]

$$\text{EUAC}_{\text{bus}} = (\$60,000)(A/P, 6\%, 20) + A$$
$$+ G(A/G, 6\%, 20)$$
$$- (\$10,000)(A/F, 6\%, 20)$$
$$\text{EUAC}_{\text{bus}} = (\$60,000)(0.0872) + \$37,440$$
$$+ (\$3744)(7.6051)$$
$$- (\$10,000)(0.0272)$$
$$= \$70,873$$

Calculate the break-even fare per passenger

$$\text{Break-even fare} = \frac{\text{EUAC}_{\text{bus}}}{80,000} = \frac{\$70,873}{80,000}$$
$$= \$0.886/\text{passenger} \quad (\$0.89/\text{passenger})$$

Find the needed subsidy. [Economic Factor Tables]

$$\text{subsidy} = \text{cost} - \text{revenue}$$
$$S = \$0.89 - (\$0.35 + (\$0.05)(A/G, 6\%, 20))$$
$$= \$0.89 - (\$0.35 + (\$0.05)(7.6051))$$
$$= \$0.1597 \quad (\$0.16)$$

The answer is (C).

12. Calculate the depreciation for the first two years.

Depreciation: Sum-of-Years Digits Method

$$D_j = \frac{2(C - S_n)(n - j + 1)}{(n)(n+1)}$$

$$D_1 = \frac{(2)(\$14,000 - \$1800)(8 - 1 + 1)}{(8)(8+1)}$$
$$= \$2711$$

$$D_2 = \frac{(2)(\$14,000 - \$1800)(8 - 2 + 1)}{(8)(8+1)}$$
$$= \$2372$$

Calculate the total depreciation from the first year and the second year. [Economic Factor Tables]

$$D = D_1 + D_2$$
$$= \$2711 + \$2372 = \$5083 \quad (\$5080)$$

The answer is (B).

13. Use the equivalent uniform annual cost (EUAC) method to find the best alternative. [Economic Factor Tables] [Economic Factor Conversions]

$$(A/P, 12\%, 15) = 0.1468$$

$$\text{EUAC}_{5\text{ft}} = (\$600,000)(0.1468)$$
$$+ \left(\frac{8}{15}\right)(\$600,000)$$
$$= \$408,080$$

$$\text{EUAC}_{10\text{ft}} = (\$710,000)(0.1468)$$
$$+ \left(\frac{3}{15}\right)(\$650,000)$$
$$= \$234,228$$

$$\text{EUAC}_{15\text{ft}} = (\$1,200,000)(0.1468)$$
$$+ \left(\frac{2}{15}\right)(\$900,000)$$
$$= \$296,160$$

$$\text{EUAC}_{20\text{ft}} = (\$1,500,000)(0.1468)$$
$$+ \left(\frac{1}{15}\right)(\$800,000)$$
$$= \$273,533$$

Build to 10 ft.

The answer is (B).

14. Find the annual insurance savings if fire sprinklers are installed.

$$\text{annual savings} = \left(\frac{0.69 - 0.47}{1000}\right)(\$3,500,000) = \$770$$

Calculate the P/A value.

$$P/A = (P/A, i\%, 25)$$
$$= \frac{\$7500}{\$770 - \$200 - \$100}$$
$$= 15.957$$

Find the rate of return that approximates the P/A value of 15.957 using economic factor conversions. [Economic Factor Tables] [Economic Factor Conversions]

$$(P/A, i\%, 25) = \frac{(1+i)^n - 1}{i(1+i)^n}$$

$$(P/A, 2\%, 25) = \frac{(1+0.02)^{25} - 1}{0.02(1+0.02)^{25}}$$
$$= 19.5235$$

$$(P/A, 4\%, 25) = \frac{(1+0.04)^{25} - 1}{0.04(1+0.04)^{25}}$$
$$= 15.6221$$

Interpolating from the above values, the rate of return is

$$i = 3.83\% \quad (3.8\%)$$

The answer is (A).

15. Calculate the present worth of each alternative using the economic factor tables. [Economic Factor Tables][Depreciation: Straight Line]

Calculate the present worth of alternative A.

$$P(A) = -(\$1.3)(1 - 0.48)(P/A, 12\%, 6)$$
$$= -(\$1.3)(0.52)(4.1114)$$
$$= -\$2.77 \quad \text{[millions]}$$

Calculate the present worth of alternative B. [Economic Factor Tables][Depreciation: Straight Line]

$$D_j = \frac{\$2}{6} = \$0.333$$
$$P(B) = -\$2 - (0.20)(\$1.3)(1 - 0.48)(P/A, 12\%, 6)$$
$$\quad - (\$0.15)(1 - 0.48)(P/A, 12\%, 6)$$
$$\quad + (\$0.333)(0.48)(P/A, 12\%, 6)$$
$$= -\$2 - (0.20)(\$1.3)(0.52)(4.1114)$$
$$\quad - (\$0.15)(0.52)(4.1114)$$
$$\quad + (\$0.333)(0.48)(4.1114)$$
$$= -\$2.22 \quad \text{[millions]}$$

Calculate the present worth of alternative C. [Economic Factor Tables][Depreciation: Straight Line]

$$D_j = \frac{1.2}{3} = 0.4$$
$$P(C) = -(\$1.2)(1 + (P/F, 12\%, 3))$$
$$\quad - (0.20)(\$1.3)(1 - 0.48)$$
$$\quad \times ((P/F, 12\%, 1) + (P/F, 12\%, 4))$$
$$\quad - (0.45)(\$1.3)(1 - 0.48)$$
$$\quad \times ((P/F, 12\%, 2) + (P/F, 12\%, 5))$$
$$\quad - (0.80)(\$1.3)(1 - 0.48)$$
$$\quad \times ((P/F, 12\%, 3) + (P/F, 12\%, 6))$$
$$\quad + (\$0.4)(0.48)(P/A, 12\%, 6)$$
$$= -(\$1.2)(1.7118)$$
$$\quad - (0.20)(\$1.3)(0.52)(0.8929 + 0.6355)$$
$$\quad - (0.45)(\$1.3)(0.52)(0.7972 + 0.5674)$$
$$\quad - (0.80)(\$1.3)(0.52)(0.7118 + 0.5066)$$
$$\quad + (\$0.4)(0.48)(4.1114)$$
$$= -\$2.54 \quad \text{[millions]}$$

Alternative B is superior.

The answer is (B).

16. This is a replacement study. Since this is a before-tax problem, depreciation is not a factor, nor is book value.

Find the equivalent uniform annual cost (EUAC) of keeping the old machine one more year. This includes the annual operating costs and the opportunity cost of not selling the machine, which is loss of interest for a year.

$$\text{EUAC(old)} = \$200,000 + (0.12)(\$400,000)$$
$$= \$248,000$$

Calculate the EUAC for the new machine. [Economic Factor Conversions][Economic Factor Tables]

$$\text{EUAC(new)} = (\$800,000)(A/P, 12\%, 10) + \$40,000$$
$$\quad + (\$30,000)(A/G, 12\%, 10)$$
$$\quad - (\$100,000)(A/F, 12\%, 10)$$

$$\text{EUAC(new)} = (\$800,000)(0.1770) + \$40,000$$
$$\quad + (\$30,000)(3.5847)$$
$$\quad - (\$100,000)(0.0570)$$
$$= \$283,441$$

Since the old machine is cheaper, keep it. The same analysis next year will give identical answers. Therefore, keep the old machine for the next three years, at which time the decision to replace it will be automatic.

The answer is (D).

17. With just-in-time manufacturing, production is one-at-a-time, according to demand. When one is needed, one is made. In order to make the economic order quantity (EOQ) approach zero, the **setup cost** must also approach zero.

The answer is (D).

18. Calculate the depreciation for the equipment using the MACRS method for years 1 through 5.

Depreciation: Modified Accelerated Cost Recovery System (MACRS)

$$D_j = (\text{factor})C$$
$$D_1 = (14.29\%)(\$2{,}000{,}000)$$
$$= \$285{,}800$$
$$D_2 = (24.49\%)(\$2{,}000{,}000)$$
$$= \$489{,}800$$
$$D_3 = (17.49\%)(\$2{,}000{,}000)$$
$$= \$349{,}800$$
$$D_4 = (12.49\%)(\$2{,}000{,}000)$$
$$= \$249{,}800$$
$$D_5 = (8.93\%)(\$2{,}000{,}000)$$
$$= \$178{,}600$$

Calculate the book value for years 1 through 5 using the respective depreciation value.

$$BV_j = C - \sum D_j$$
$$= \$2{,}000{,}000 - \$285{,}800$$
$$= \$1{,}714{,}200$$
$$BV_2 = \$1{,}714{,}200 - \$489{,}800$$
$$= \$1{,}224{,}400$$
$$BV_3 = \$1{,}224{,}400 - \$349{,}800$$
$$= \$874{,}600$$
$$BV_4 = \$874{,}600 - \$249{,}800$$
$$= \$624{,}800$$
$$BV_5 = \$624{,}800 - \$178{,}600$$
$$= \$446{,}200$$

Calculate the after-tax salvage value.

$$S - (S - BV)(\text{tax rate}) = \$350{,}000$$
$$- (\$350{,}000 - \$446{,}200)(0.48)$$
$$= \$396{,}176 \quad (\$396{,}000)$$

The answer is (C).

74 Professional Services, Contracts, and Engineering Law

Content in blue refers to the *NCEES Handbook*.

PRACTICE PROBLEMS

1. Different forms of company ownership include sole proprietorship, partnership and corporation.

Which of the following are true about the different forms of company ownership?

I. A sole proprietor personally assumes all debts and liabilities of the company.

II. A sole proprietorship can be transferred in the event of the death of the proprietor.

III. A partnership splits the debts and liabilities of the company between all partners.

IV. A corporation keeps the company and owner liability separate.

V. A corporation must dissolve at the death of a director.

(A) I, II, and III
(B) I, III, and IV
(C) I, III, IV, and V
(D) all of the above

2. The requirements needed for a contract to be legal and enforceable are

(A) offer, consideration, acceptance and voluntarily entered into
(B) offer, acceptance, escrow and execution of the contract
(C) public policy, acceptance, escrow and voluntarily entered into
(D) public policy, consideration, escrow and execution of the contract

3. The phrase "without expressed authority" means which of the following when used in regard to partnerships of design professionals?

(A) Each full member of a partnership is a general agent of the partnership and has complete authority to make binding commitments, enter into contracts, and otherwise act for the partners within the scope of the business.
(B) The partnership may act in a manner that it considers best for the client, even though the client has not been consulted.
(C) Only plans, specifications, and documents that have been signed and stamped (sealed) by the authority of the licensed engineer may be relied upon.
(D) Only officers to the partnership may obligate the partnership.

4. Which of the following statements is FALSE in regard to joint ventures?

(A) Members of a joint venture may be any combination of sole proprietorships, partnerships, and corporations.
(B) A joint venture is a business entity separate from its members.
(C) A joint venture spreads risk and rewards, and it pools expertise, experience, and resources. However, bonding capacity is not aggregated.
(D) A joint venture usually dissolves after the completion of a specific project.

5. Which of the following construction business types can have unlimited shareholders?

   I. S corporation
   II. LLC
   III. corporation
   IV. sole proprietorship

   (A) II and III only
   (B) I, II, and III
   (C) I, III, and IV
   (D) I, II, III, and IV

6. The phrase "or approved equal" allows a contractor to

   (A) substitute one connection design for another
   (B) substitute a more expensive feature for another
   (C) replace an open-shop subcontractor with a union subcontractor
   (D) install a product whose brand name and model number are not listed in the specifications

7. Cities, other municipalities, and departments of transportation often have standard specifications, in addition to the specifications issued as part of the construction document set, that cover such items as

   (A) safety requirements
   (B) environmental requirements
   (C) concrete, fire hydrant, manhole structures, and curb requirements
   (D) procurement and accounting requirements

8. What is intended to prevent a contractor from bidding on a project and subsequently backing out after being selected for the project?

   (A) publically recorded bid
   (B) property lien
   (C) surety bond
   (D) proposal bond

9. Which of the following is illegal, in addition to being unethical?

   (A) bid shopping
   (B) bid peddling
   (C) bid rigging
   (D) bid unbalancing

10. A contract has a value engineering clause that allows the parties to share in improvements that reduce cost. The contractor had originally planned to transport concrete on site for a small pour with motorized wheelbarrows. On the day of the pour, however, a concrete pump is available and is used, substantially reducing the contractor's labor cost for the day. This is an example of

   (A) value engineering whose benefit will be shared by both contractor and owner
   (B) efficient methodology whose benefit is to the contractor only
   (C) value engineering whose benefit is to the owner only
   (D) cost reduction whose benefit will be shared by both contractor and laborers

11. A material breach of contract occurs when the

   (A) contractor uses material not approved by the contract to use
   (B) contractor's material order arrives late
   (C) owner becomes insolvent
   (D) contractor installs a feature incorrectly

12. When an engineer stops work on a jobsite after noticing unsafe conditions, the engineer is acting as a(n)

   I. agent
   II. local official
   III. OSHA safety inspector
   IV. competent person

   (A) I only
   (B) I and IV
   (C) II and III
   (D) III and IV

13. A professional engineer is hired by a homeowner to design a septic tank and leach field. The septic tank fails after 18 years of operation. What sentence best describes what comes next?

(A) The homeowner could pursue a tort action claiming a septic tank should retain functionality longer than 18 years.

(B) The engineer will be protected by a statute of limitations law.

(C) The homeowner may file a claim with the engineer's original bonding company.

(D) The engineer is ethically bound to provide remediation services to the homeowner.

SOLUTIONS

1. I, III, and IV are true.

A *sole proprietor* is his or her own boss. This satisfies the proprietor's ego and facilitates quick decisions, but unless the proprietor is trained in business, the company will usually operate without the benefit of expert or mitigating advice. The sole proprietor also personally assumes all the debts and liabilities of the company. A sole proprietorship is terminated upon the death of the proprietor.

A *partnership* increases the capitalization and the knowledge base beyond that of a proprietorship, but offers little else in the way of improvement. In fact, the partnership creates an additional disadvantage of one partner's possible irresponsible actions creating debts and liabilities for the remaining partners.

A *corporation* has sizable capitalization (provided by the stockholders) and a vast knowledge base (provided by the board of directors). It keeps the company and owner liability separate. It also survives the death of any employee, officer, or director. Its major disadvantage is the administrative work required to establish and maintain the corporate structure.

The answer is (B).

2. To be legal, a contract must contain an *offer*, some form of *consideration* (which does not have to be equitable), and an *acceptance* by both parties. To be enforceable, the contract must be voluntarily entered into, both parties must be competent and of legal age, and the contract cannot be for illegal activities.

The answer is (A).

3. *Without expressed authority* means each member of a partnership has full authority to obligate the partnership (and the other partners).

The answer is (A).

4. One of the reasons for forming joint ventures is that the bonding capacity is aggregated. Even if each contractor cannot individually meet the minimum bond requirements, the total of the bonding capacities may be sufficient.

The answer is (C).

5. Normal corporations, S corporations, and limited liability companies (LLCs) can have unlimited shareholders. Sole proprietorships are for individuals.

The answer is (B).

6. When the specifications include a nonstructural, brand-named article and the accompanying phrase "or approved equal," the contractor can substitute something with the same functionality, even though it is not the brand-named article. "Or approved equal" would not be used with a structural detail such as a connection design.

The answer is (D).

7. Municipalities that experience frequent construction projects within their boundaries have standard specifications that are included by reference in every project's construction document set. This document set would cover items such as concrete, fire hydrants, manhole structures, and curb requirements.

The answer is (C).

8. *Proposal bonds*, also known as *bid bonds*, are insurance policies payable to the owner in the event that the contractor backs out after submitting a qualified bid.

The answer is (D).

9. *Bid rigging*, also known as *price fixing*, is an illegal arrangement between contractors to control the bid prices of a construction project or to divide up customers or market areas. *Bid shopping* before or after the bid letting is where the general contractor tries to secure better subcontract proposals by negotiating with the subcontractors. *Bid peddling* is done by the subcontractor to try to lower its proposal below the lowest proposal. *Bid unbalancing* is where a contractor pushes the payment for some expense items to prior construction phases in order to improve cash flow.

The answer is (C).

10. The problem gives an example of efficient methodology, where the benefit is to the contractor only. It is not an example of value engineering, as the change affects the contractor, not the owner. Performance, safety, appearance, and maintenance are unaffected.

The answer is (B).

11. *A material breach of the contract* is a significant event that is grounds for cancelling the contract entirely. Typical triggering events include failure of the owner to make payments, the owner causing delays, the owner declaring bankruptcy, the contractor abandoning the job, or the contractor getting substantially off schedule.

The answer is (C).

12. An engineer with the authority to stop work gets his/her authority from the agency clause of the contract for professional services with the owner/developer. It is unlikely that the local building department or OSHA would have authorized the engineer to act on their behalves. It is also possible that the engineer may have been designated as the jobsite's "competent person" (as required by OSHA) for one or more aspects of the job, with authority to implement corrective action.

The answer is (B).

13. Satisfactory operation for 18 years is proof that the design was partially, if not completely, adequate. There is no ethical obligation to remediate normal wear-and-tear, particularly when the engineer was not involved in how the septic system was used or maintained. While the homeowner can threaten legal action and even file a complaint, in the absence of a warranty to the contrary, the engineer is probably well-protected by a statute of limitations.

The answer is (B).

75 Engineering Ethics

Content in blue refers to the *NCEES Handbook*.

PRACTICE PROBLEMS

There are no problems in this book corresponding to Chap. 75 of the *Mechanical Engineering Reference Manual*.